Alkaloids From Medicinal Plants and Their Transformative Applications

Alkaloids are derived from plant, animal, and microbial sources and have extraordinary pharmacological qualities and therapeutic potentials. They are a focus of research in drug discovery and development due to their complex chemical structures and biosynthetic routes.

Alkaloids show potential as building blocks for developing new medications with increased efficacy and fewer side effects. *Alkaloids From Medicinal Plants and Their Transformative Applications* provides information on developments in the use of alkaloids in healthcare applications and discusses the extraction, properties, and characterization of alkaloids from medicinal herbs, shrubs, and trees around the world. It examines alkaloid-based drugs, showcasing their clinical efficacy and real-world applications, and explores emerging trends and prospects in alkaloid research, including sustainability and biotechnological advancements.

A volume in the *Exploring Medicinal Plants* series, this book is beneficial for healthcare professionals looking to study the medical uses of alkaloids and for evidence-based information on the use of medicinal plant-based alkaloids in the prevention and treatment of diseases around the world.

Exploring Medicinal Plants

Series Editor: Azamal Husen, Wolaita Sodo University, Ethiopia

Medicinal plants render a rich source of bioactive compounds used in drug formulation and development; they play a key role in traditional or indigenous health systems. As the demand for herbal medicines increases worldwide, supply is declining as most of the harvest is derived from naturally growing vegetation. Considering global interests and covering several important aspects associated with medicinal plants, the Exploring Medicinal Plants series comprises volumes valuable to academia, practitioners, and researchers interested in medicinal plants. Topics provide information on a range of subjects including diversity, conservation, propagation, cultivation, physiology, molecular biology, growth response under extreme environments, handling, storage, bioactive compounds, secondary metabolites, extraction, therapeutics, mode of action, and healthcare practices.

Led by Azamal Husen, PhD, this series is directed to a broad range of researchers and professionals consisting of topical books exploring information related to medicinal plants. It includes edited volumes, references, and textbooks available for individual print and electronic purchases.

Propagation to Pharmacopeia: Modern Approaches in Medicinal Plants
Edited by Abha Manohar K, Gopal Shukla, Saikat Gantait, A P Das, and Sumit Chakravarty

Exploring Medicinal Orchids
Edited by Azamal Husen and Manu Pant

Symbiotic Association of Microorganisms with Medicinal and Herbal Plants
Edited by Allah Ditta, Sajid Mehmood, and Azamal Husen

Tissue Culture Techniques and Medicinal Plants: Enhancing Propagation and Production
Edited by Azamal Husen and Manu Pant

Genetic Improvement and Conservation Practices of Medicinal Plants
Edited by Manu Pant, Sayed Sartaj Sohrab and Azamal Husen

Sustainable Use of Plants and Their Products in Neurodegenerative Diseases Management
Edited by Johnson O. Oladele, Ebenezer I. O. Ajayi, María L. Flores-López, and Oluwaseyi E. Okoro

Wild Edibles of Deserts: Food Security and Sustainability
Salman Majeed, Muhammad Zafar, Mushtaq Ahmad, Mohd Kafeel Ahmad Ansari, Shazia Sultana, and Andrea Pieroni

Impact of Climate Change on Medicinal and Herbal Plant micro-RNA
Kanchanlata Tungare, Parul Johri, Sachidanand Singh, and Surojeet Das

Exploring Medicinal Bamboos
Edited by Manu Pant and Azamal Husen

Exploring Herbal Synergies for Optimal Human Health
Edited by Art Devi, Nancy George, Zaved Ahmed Khan, and Vagish Dwibedi

Exploring Medicinal and Commercial Uses of Flowers
Edited by Rakesh Kumar Bachheti, Archana Bachheti, and Azamal Husen

Phenolic Compounds from Medicinal Plants: Pharmaceutical and Health Benefits
Edited by Rakesh Kumar Bachheti, Archana Bachheti, and Azamal Husen

Alkaloids From Medicinal Plants and Their Transformative Applications
Edited by Rakesh Kumar Bachheti, Archana Bachheti, and Azamal Husen

For more information about this series, please visit: www.routledge.com/Exploring-Medicinal-Plants/book-series/CRCEMP

Alkaloids From Medicinal Plants and Their Transformative Applications

Edited by
Rakesh Kumar Bachheti
Archana Bachheti
Azamal Husen

CRC Press
Taylor & Francis Group
Boca Raton London New York

CRC Press is an imprint of the
Taylor & Francis Group, an **informa** business

Designed cover image: Shutterstock

First edition published 2026
by CRC Press
2385 NW Executive Center Drive, Suite 320, Boca Raton FL 33431

and by CRC Press
4 Park Square, Milton Park, Abingdon, Oxon, OX14 4RN

CRC Press is an imprint of Taylor & Francis Group, LLC

ISBN: 978-1-032-86607-9 (hbk)
ISBN: 978-1-032-88697-8 (pbk)
ISBN: 978-1-003-53917-9 (ebk)

DOI: 10.1201/9781003539179

Typeset in Times
by SPi Technologies India Pvt Ltd (Straive)

Contents

Preface

From the earliest whispers of human civilization, plants have been companions, healers, and silent guides on the long journey of survival and discovery. Among the countless gifts they have offered, few are as fascinating and transformative as the alkaloids. These remarkable compounds, shaped in the mysterious laboratories of nature within the tissues of medicinal plants, have traveled across centuries, cultures, and continents, weaving themselves into the fabric of human health and medicine. The story of alkaloids is not merely one of chemistry or biology, but of resilience, innovation, and the constant dialogue between humankind and the natural world. Imagine ancient healers wandering through forests and fields, observing the behaviors of plants and animals, testing leaves, roots, and seeds with both courage and curiosity. Without the lens of modern science, they relied on experience, intuition, and tradition to uncover the secrets hidden in plants. Some brought relief from pain, others calmed the mind, and a few even possessed the mysterious ability to save lives in the face of deadly diseases. What they did not know was that in many cases, the invisible architects of these miraculous effects were alkaloids, the complex molecules that plants had evolved as part of their defense and survival strategies. Centuries later, as science blossomed in Europe and other parts of the world, the invisible began to be revealed. Chemists isolated the first alkaloids in their laboratories, giving names to substances that had been silently serving humanity for generations. Morphine from the opium poppy, quinine from the bark of the cinchona tree, and nicotine from tobacco plants became not just isolated compounds but symbols of how profoundly nature's chemical library could alter human destiny. Each discovery sparked new possibilities. Morphine reshaped pain management and surgery, quinine became a shield against the relentless scourge of malaria, and nicotine opened doors to understanding neurological pathways. What began as traditional knowledge slowly transformed into structured scientific inquiry, linking folklore to pharmacology.

The fascination with alkaloids only grew as scientists realized the sheer diversity of these compounds. No two structures seemed to tell the same story, yet all carried with them biological significance. Some were bitter to deter herbivores, some were potent enough to protect plants against pathogens, while others influenced growth and development in subtle ways. The same structural complexity that gave them roles in plants also provided countless opportunities for medicinal applications. In a way, plants had been perfecting a pharmacy long before humans recognized it. This book takes the reader on a journey through that pharmacy. It begins with the fundamental nature of alkaloids, unraveling their definition, origins, and the intricate roles they play in biology. It then guides the reader through their astonishing structural diversity and classification, showing how a simple nitrogen atom, embedded in different molecular frameworks, can give rise to endless possibilities. The story then deepens into the processes that create them, exploring the biosynthesis and regulation of alkaloids within medicinal plants. This window into the plant's inner life not only satisfies scientific curiosity but also offers clues for sustainable production and innovation. From there, the exploration turns practical. Extraction, isolation, purification, and characterization are not merely laboratory techniques; they are the bridges that connect the raw gifts of nature to usable medicines. These methods, refined over decades, remind us that every vial of purified alkaloid carries with it the effort of scientists, the wisdom of tradition, and the complexity of life itself. Yet, the true fascination of alkaloids lies in their pharmacological brilliance. Across the world of medicine, they have offered solutions to some of the most formidable health challenges. Cancer, a disease as ancient as humanity itself, has been met with the therapeutic potential of alkaloids that disrupt cell division and alter molecular pathways. Parasitic infections, which have plagued societies for centuries, have found remedies in plant-derived alkaloids that remain as relevant today as they were in traditional healing practices. Cardiovascular disorders, neurological diseases, and the endless quest for effective pain relief have all, in one way or another, been transformed by alkaloids. Their antioxidant properties extend this promise further, positioning them as guardians not only against disease but also as agents of prevention. The narrative does not stop with individual effects. Alkaloids tell a richer story when viewed through the lens of synergy and combination. Traditional medicine often did not use single compounds but whole plants, and in doing so, harnessed interactions that modern science is only beginning to fully appreciate. Combination therapies involving alkaloids open new doors to enhancing therapeutic potency, reducing side effects, and creating treatments that are more aligned with the body's natural systems. Drug discovery has celebrated numerous successes built on the foundation of alkaloid research, and their enduring relevance in medicinal chemistry proves that the library of nature is far from exhausted. What makes alkaloids even more captivating is how they stand at the intersection of tradition and innovation. Ethnobotanical studies continue to remind us that the knowledge of indigenous communities, passed through generations, carries within it untapped treasures. Every local remedy, every herbal preparation, might hold an alkaloid whose potential has yet to be realized. By listening to these traditions and combining them with cutting-edge science, we bridge the wisdom of the past with the possibilities of the future. At the same time, challenges loom on the horizon. The sustainability of alkaloid research and production requires careful consideration. Overharvesting of medicinal plants threatens biodiversity, while synthetic and biotechnological alternatives must rise to meet the demand without compromising ecological balance. Innovation in this field must go hand in hand with responsibility, ensuring that discoveries today do not become burdens tomorrow.

This book is not merely a compilation of facts, studies, and theories. It is a collaborative effort by experts from diverse fields, all united by their fascination with the hidden powers of alkaloids. Their collective voices bring together phytochemistry, pharmacology, biotechnology, and ethnobotany into a single tapestry. The narrative they weave is both educational and inspiring, intended for scholars, researchers, professionals, and students alike. For some, it may serve as a reference; for others, it may spark new questions, new projects, or even new breakthroughs. In presenting this work, we also extend gratitude to those who have supported and shaped it. The contributors, with their timely and up-to-date insights, have ensured that the chapters reflect both the depth of tradition and the cutting edge of science. Our appreciation also goes to Ms. Randy Brehm and the team at Taylor & Francis Group, LLC/CRC Press, whose cooperation has been steadfast. Ultimately, however, the true acknowledgment belongs to nature itself, whose quiet genius continues to inspire wonder and discovery. As readers embark on the journey through these chapters, they are invited to see alkaloids not just as molecules, but as storytellers. Each compound carries within it a history, an ecological purpose, and a potential to transform lives. From the forests where plants grow in quiet resilience, to the laboratories where researchers unravel molecular mysteries, to the clinics where patients find relief and hope, alkaloids remind us of the profound interconnectedness of life. Their story is not finished; it is evolving with every experiment, every discovery, and every application. In their complexity lies not only a scientific challenge but also the promise of healing, innovation, and sustainability. With great pleasure and anticipation, we present this volume to the readers. May it serve as both a guide and an inspiration, and may the story of alkaloids continue to unfold, shaping a healthier and more sustainable future for all.

Rakesh Kumar Bachheti
Archana Bachheti
Azamal Husen

About the Editors

Rakesh Kumar Bachheti graduated from Hemwati Nandan Bahuguna University, Garhwal, India, in 1996. He completed his M.Sc. in Organic Chemistry from the same university in 1998. Later, in 2001, he underwent a one-year Post Graduate Diploma in Pulp and Paper Technology from Forest Research Institute, Dehradun, India. Subsequently, he pursued his Ph.D. in Organic Chemistry from Kumaun University, Nainital, India, which he obtained in 2007. Bachheti has a total of 22 years of research and teaching experience. Before joining Addis Ababa Science and Technology University (AASTU) in Ethiopia, he worked as a Dean Project (Assistant) at Graphic Era University in Dehradun, India. It's worth noting that Graphic Era University is accredited with an "A+" grade by the National Assessment and Accreditation Council (NAAC). Dr. Bachheti is also a Research Fellow at INTI International University, Persiaran Perdana BBN, Putra Nilai, Nilai, Negeri Sembilan, Malaysia. Bachheti has an impressive research background and has presented papers at various international (Malaysia, Thailand, and India) and national conferences. His research interests primarily focus on natural products for industrial applications, biofuels and bioenergy, green synthesis of nanoparticles and their applications, and pulp and paper technology. His passion for natural products is evident in all of his research endeavors. As an academic mentor, Bachheti has successfully advised 40 M.Sc. and 9 Ph.D. students to completion. Additionally, numerous undergraduates have conducted research in his laboratory under his guidance. He actively contributes to curriculum development for B.Sc./M.Sc./Ph.D. programs. Bachheti is also recognized for his editorial work, having edited 10 books and authored 190 publications on various aspects of natural product chemistry, pulp and paper and nanotechnology. He has contributed 40 book chapters published by prestigious publishers like Springer, Elsevier, Taylor & Francis, and Nova Science Publishers. He is listed among the Top {2%} Scientists for the year 2025 and 2024 by Stanford University, USA, and Elsevier.

Archana Bachheti completed her B.Sc. in 1997 and M.Sc. in 1999 from Hemwati Nandan Bahuguna Garhwal University. She earned her Ph.D. from the Forest Research Institute, Dehradun, India, in 2006. Currently, she is a Professor at Graphic Era University in Dehradun, India. With over 20 years of extensive experience, she has served in various academic roles in India and has also provided expertise internationally. Her teaching portfolio includes subjects such as ecology and environment, environmental science, freshwater ecology, disaster management, and bryophytes and pteridophytes. Prof. Bachheti's research interests span a broad and interdisciplinary field of plant ecology, with a focus on ecorestoration, green chemistry—particularly the synthesis of nanomaterials—and exploring the medicinal properties of plants. Her research has involved ecological restoration of degraded lands, studying the physical and chemical properties of plant oils, and investigating plant-based nanomaterials. Throughout her career, she has been actively involved in mentoring students and researchers. Dr. Bachheti has published more than 130 research papers and edited over 8 books. She is listed among the Top 2% Scientists for the year 2024 by Stanford University, USA, and Elsevier.

Azamal Husen has served as Full Professor and Head of the Department of Biology at the University of Gondar, Ethiopia. He is currently affiliated as a Foreign Delegate with Wolaita Sodo University, Ethiopia, and holds an Adjunct Professorship at the Department of Biotechnology, Graphic Era (Deemed to be University) in Dehradun, Uttarakhand, India. His earlier academic roles include Full Professor at Sankalchand Patel University in Visnagar, India, and Visiting Faculty at the Forest Research Institute and Doon (P.G) College of Agriculture Science and Technology in Dehradun, India. With more than 25 years of experience in academia and research, he is highly regarded for his expertise in nanobiotechnology, plant physiology, environmental botany, medicinal plants, and plant propagation. His research has been supported by prominent national and international organizations, including the World Bank, the National Agricultural Technology Project, the Indian Council of Agricultural Research, the Indian Council of Forestry Research and Education, and the Japan Bank for International Cooperation. He is a prolific scholar, having published more than 350 academic works, including research articles, books, and book chapters. His editorial contributions are equally noteworthy; he serves on the editorial boards and peer review panels of numerous prestigious journals published by Elsevier, Frontiers Media, Springer Nature, Taylor & Francis, RSC, Oxford University Press, the Royal Society, CSIRO, MDPI, PLOS, Sciendo, John Wiley & Sons, and UPM Journals. He is also a member of the advisory board of Cambridge Scholars Publishing, United Kingdom. He is a fellow of the Plantae group under the American Society of Plant Biologists and holds membership in several scientific organizations, including the International Society of Root Research, the Asian Council of Science Editors, and the International Natural Product Sciences Taskforce. Throughout his career, he has delivered invited lectures, keynote addresses, and technical presentations at numerous international conferences, symposia, and workshops. In the field of academic development, he has made significant contributions through the design and delivery of orientation programs for university faculty, the development of educational modules, and the formulation of curricula at the undergraduate, postgraduate, and doctoral levels. In recognition of his contributions to teaching, research, and community service, he was honored with a special award from the University of Gondar. He currently serves as Editor-in-Chief of the *American Journal of Plant Physiology* and is Series Editor for several major academic collections, including *Exploring Medicinal Plants* (Taylor & Francis, USA), *Plant Biology, Sustainability, and Climate Change* (Elsevier, USA), and *Smart Nanomaterials Technology* (Springer Nature, Singapore). He also leads the *Handbook Series on Nanotechnology and Its Applications* (Springer Nature, Singapore) and *Plant Chemistry* (Taylor & Francis, USA). His outstanding contributions to science have earned him international acclaim, including the prestigious distinction of being named among the "World's Top 2% Scientists" for three consecutive years (2022 onward) by Stanford University, a recognition published in the Elsevier Data Repository. In addition to his regular academic responsibilities, he has served as Guest Editor for several high-impact special issues, including *Advances in Plant Metabolomics: Insights, Applications, and Integration* (*Plant Physiology and Biochemistry*), *Exploring Nanotechnology in Medicinal Plants* (*Journal of Herbal Medicine*), and *Integrating Nanotechnology with Plant Compounds: Enhancing Abiotic Stress Resilience and Crop Improvement via Signaling Molecules* (*Frontiers in Plant Science*), among others. His work continues to shape contemporary research at the intersection of plant science, biotechnology, and sustainable development. His academic leadership, research excellence, and international collaborations have positioned him as a prominent figure in global scientific circles.

Contributors

Asmaa S. Abd Elkarim
Chemistry of Tanning Materials and Leather Technology Department
National Research Centre
Giza, Egypt

Eman Y. Ahmed
Department of Natural Compounds Chemistry
National Research Centre
Dokki, Giza, Egypt

Rania F. Ahmed
Department of Natural Compounds Chemistry
National Research Centre
Dokki, Giza, Egypt

Anant
Rani Lakshmi Bai Central Agricultural University
Jhansi, India

Zeeshan Ansari
School of Pharmaceutical Sciences
Chhatrapati Shahu Ji Maharaj University
Kanpur, India

Piyush Bhalla
Chemistry and Bioprospecting Division
ICFRE-Forest Research Institute
Dehradun, India

Ruchi Bharti
Department of Chemistry, UIS
Chandigarh University
Mohali, Punjab, India

Surendra Singh Bisht
Chemistry and Bioprospecting Division
ICFRE-Forest Research Institute
Dehradun, India

Monica Butnariu
University of Life Sciences "King Mihai I" from Timisoara, Chemistry & Biochemistry Discipline
Timisoara, Romania

Noureddine Chaachouay
Agri-Food and Health Laboratory (AFHL)
ESEF of Berrechid, Hassan First University
Settat, Morocco

Kiran Chauhan
Chemistry and Bioprospecting Division
ICFRE-Forest Research Institute
Dehradun, India
and
Department of Chemistry, Government Degree College
Chamoli, India

Vikram Chauhan
Chemistry and Bioprospecting Division
ICFRE-Forest Research Institute
Dehradun, India

Omayma A. Eldahshan
Center for Drug Discovery Research and Development, Faculty of Pharmacy
and
Department of Pharmacognosy, Faculty of Pharmacy
Ain Shams University
Abbassia, Cairo, Egypt

Mai M. Elghonemy
Department of Natural Compounds Chemistry, National Research Centre
Dokki, Giza, Egypt

Esraa A. Elhawary
Department of Pharmacognosy, Faculty of Pharmacy, Ain Shams University
Abbassia, Cairo, Egypt

Heba A. S. El-Nashar
Department of Pharmacognosy, Faculty of Pharmacy, Ain Shams University
Abbassia, Cairo, Egypt

Abdelsamed I. Elshamy
Chemistry of Tanning Materials and Leather Technology Department
and
Department of Natural Compounds Chemistry
National Research Centre
Dokki, Giza, Egypt

Sherif E. Emam
Department of Pharmaceutics and Industrial Pharmacy, Faculty of Pharmacy
Zagazig University
Zagazig, Egypt

Ahmed F. Essa
Department of Natural Compounds Chemistry
National Research Centre
Dokki, Giza, Egypt

Abhishek Gupta
Plant Diversity Systematics and Herbarium Division
CSIR-National Botanical Research Institute
Lucknow, India

Bhanushree Gupta
Center for Basic Sciences
Pt. Ravishankar Shukla University
Raipur, India

Prakash Chandra Gupta
School of Pharmaceutical Sciences
Chhatrapati Shahu Ji Maharaj University
Kanpur, India

Howaida I. Abd-Alla
Department of Natural Compounds Chemistry
National Research Centre
Dokki, Giza, Egypt

Hiroshi Imagawa
Faculty of Pharmaceutical Sciences
Tokushima Bunri University
Yamashiro-cho, Tokushima, Japan

Yusuke Kasai
Faculty of Pharmaceutical Sciences
Tokushima Bunri University
Yamashiro-cho, Tokushima, Japan

Abdulsalam M. Kassem
Department of Pharmaceutics and Pharmaceutical Technology, Faculty of Pharmacy (Boys)
Al-Azhar University
Cairo, Egypt

Jay Kumar
School of Pharmaceutical Sciences
Chhatrapati Shahu Ji Maharaj University
Kanpur, India

Shankul Kumar
Department of Pharmacognosy, Narayan Institute of Pharmacy
Gopal Narayan Singh University
Bihar, India

Orchid A. Mahmoud
Department of Pharmacognosy, Faculty of Pharmacy
Ain Shams University
Abbassia, Cairo, Egypt

Shivani Naik
Department of Chemistry, UIS
Chandigarh University
Mohali, Punjab, India

Sumitra Nain
Department of Pharmaceutical Chemistry
Faculty of Pharmacy
Rajasthan, India

Sarvesh K. Pandey
Department of Chemistry
D. D. U. Gorakhpur University
Gorakhpur, India

Arvind Kumar Patel
Department of Pharmacology
Narayan Institute of Pharmacy
Gopal Narayan Singh University
Bihar, India

Vedant Kumar Prajapati
Department of Pharmaceutics
Narayan Institute of Pharmacy
Gopal Narayan Singh University
Bihar, India

Ahmed Sabt
Department of Natural Compounds Chemistry
National Research Centre
Dokki, Giza, Egypt

Baishakhi Sarkhel
Chemistry and Bioprospecting Division
ICFRE-Forest Research Institute
Dehradun, India

Preeti Saxena
Department of Chemistry
Doon (PG) College of Agriculture Science and Technology
Dehradun, India

Anshika Sharma
Guru Nanak College of Pharmaceutical Sciences, Jhajra
Dehradun, India

Nisha Sharma
School of Pharmaceutical Sciences
Chhatrapati Shahu Ji Maharaj University
Kanpur, India

Renu Sharma
Department of Chemistry, UIS
Chandigarh University
Mohali, Punjab, India

Abdel Nasser B. Singab
Department of Pharmacognosy, Faculty of Pharmacy
and
Center for Drug Discovery Research and Development, Faculty of Pharmacy
Ain Shams University
Abbassia, Cairo, Egypt

Shashi Ranjan Singh
Department of Pharmacology
Narayan Institute of Pharmacy
Gopal Narayan Singh University
Bihar, India

Swetza Singh
School of Pharmacy
Babu Banarasi Das University
Lucknow, India

Rehab F. Taher
Department of Natural Compounds Chemistry
National Research Centre
Dokki, Giza, Egypt

Shrejal Tiwari
Rani Lakshmi Bai Central Agricultural University, Jhansi
Jhansi, India

Sheetal Tyagi
Department of Chemistry, School of Basic & Applied Sciences
Shri Guru Ram Rai University
Dehradun, India

Deepti Singh Vashishth
Department of Environmental Science
Graphic Era (Deemed to be University)
Dehradun, India

Amit Verma
School of Pharmacy
Chandigarh University
Unnao Campus, India

Shivam Verma
School of Pharmaceutical Sciences
Chhatrapati Shahu Ji Maharaj University
Kanpur, India

Shweta Singh Verma
School of Pharmaceutical Sciences
Chhatrapati Shahu Ji Maharaj University
Kanpur, India

Nawal Zemmouli
Department of Cellular and Molecular Biology, Faculty of Natural and Life Sciences
El-Oued University
EL-Oued, Algeria

1 Alkaloids
Definition, Origin, and Biological Significance

Surendra Singh Bisht, Baishakhi Sarkhel, Vikram Chauhan, Sarvesh K. Pandey, and Bhanushree Gupta

1.1 INTRODUCTION

Alkaloids represent one of the most chemically diverse and biologically significant classes of secondary metabolites found in living organisms. Characterized by a wide array of structural frameworks, biosynthetic origins, and pharmacological activities, alkaloids have captivated scientific interest across multiple disciplines. Traditionally, these compounds have been isolated predominantly from plants; however, recent research has unveiled their presence in various other biological sources, including animals, insects, marine invertebrates, and microorganisms. Many alkaloids have a longstanding history of use in traditional medicine and continue to serve as critical components of modern pharmaceuticals.

The term *alkaloid* is etymologically rooted in the words "alkali" and the Greek *et8oσ* (meaning "type" or "similarity"), conveying the concept of a substance with alkali-like properties. The origin of "alkali" itself stems from the Arabic word *al-qali*, referring to plant ashes—a source of potash. In pre-industrial Europe, potash was commonly derived from wood ashes by leaching and evaporating the resulting lye, finding applications primarily in glassmaking and textile processing (Manfred Hesse, 1981).

The actual term "alkaloid" was first introduced by the German chemist Wilhelm Meissner in 1819 during his studies on the plant *Veratrum*. Meissner's initial definition was straightforward: alkaloids are plant-derived substances exhibiting basic (alkali-like) behavior (Clayden et al., 2001). At that time, only a limited number of such compounds, most notably morphine, had been identified, and those in impure forms. Interestingly, while the term "alkaloid" was coined to describe basic plant substances, no corresponding term like "acidoid" ever gained traction for plant-derived acids. Instead, plant acids and neutral compounds were eventually categorized based on structural criteria—for instance, flavonoids (C_6–C_3–C_6 framework), steroids, saccharides, peptides, and hydrocarbons, all of which represent chemically and structurally coherent classes (Manfred Hesse, 1981).

From a biological perspective, alkaloids are regarded as quintessential natural products. They are typically defined as nitrogen-containing, biologically active, heterocyclic compounds that often exhibit potent pharmacological properties and play ecological roles such as defense against herbivores or pathogens (Aniszewski, 1994). While this functional definition is broadly useful, it remains inherently imprecise due to the structural and biosynthetic heterogeneity of alkaloids. As a result, any attempt to define alkaloids strictly tends to be either overly inclusive or restrictive (Bentley, 1957; Cordell, 1981; Hegnauer, 1963, 1966, 1967).

DOI: 10.1201/9781003539179-1

To address this complexity, some scholars, such as Winterstein and Trier (Winterstein & Trier, 1910), have proposed defining alkaloids based on a set of common characteristics. These include the following:

1. A fundamental chemical nature.
2. The presence of nitrogen within a heterocyclic ring.
3. Biosynthesis is derived from amino acids or their immediate precursors.
4. Limited distribution across the plant and animal kingdoms.
5. Notable toxicity, particularly through action on the central nervous system.

Despite these defining traits, exceptions abound, making it difficult to articulate a concise, universally accepted definition. Nevertheless, the breadth of biological activity and the ongoing discovery of novel alkaloids across various life forms underscore the enduring importance of this compound class in both natural and applied sciences.

Over time, multiple definitions of alkaloids have emerged from different scientific disciplines, each offering valuable perspectives. Waller and Nowacki (Waller & Nowacki, 1978), for instance, highlighted several defining characteristics of alkaloids, particularly emphasizing the presence of nitrogen in the molecular structure. They noted that the nitrogen atom must be bonded to at least two carbon atoms, and that the molecule should contain at least one ring, although this ring is not necessarily heterocyclic (Figure 1.1). Importantly, they excluded compounds that are integral components of macromolecular cellular substances, vitamins, or hormones from the alkaloid category.

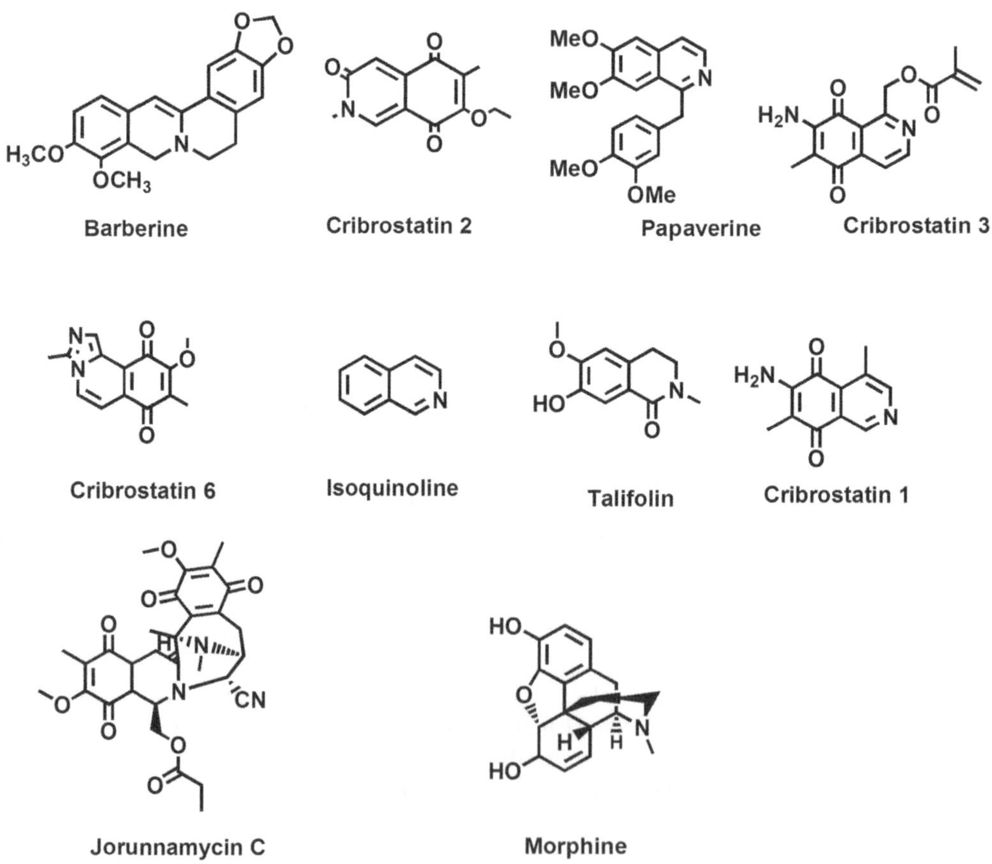

FIGURE 1.1 Representative isoquinoline-based alkaloid compounds.

Earlier, Sengbusch (Sengbusch, 2003) offered a simpler classification, describing alkaloids as nitrogen-containing bases, the majority of which exhibit pharmacological effects and are thus classified as drugs. From a chemical standpoint, alkaloids are defined as a group of structurally complex, nitrogenous heterocyclic compounds that possess significant physiological activity. Chemists have consistently underscored their basic (alkaline) character and toxicity, though they acknowledge a few exceptions to this rule (Hampel & Hawley, 1976). Another chemical definition simplifies alkaloids to nitrogen-containing compounds of natural origin, found in both plant and animal sources (Hesse, 1981). A more contemporary chemical definition refines this further, classifying alkaloids as biogenic nitrogen-containing molecules, predominantly heterocyclic, but explicitly excluding amino acids, peptides, nucleosides, amino sugars, and antibiotics from this category (Jakubke et al., 1994).

Despite subtle variations in terminology and emphasis across biology, chemistry, and medicine, these definitions converge on key elements. Alkaloids are broadly accepted as a large and diverse group of naturally occurring compounds that exhibit notable biological, pharmacological, physiological, and chemical activities. Their impact on human health is immense—without alkaloids, major advances in treating diseases such as malaria, leukemia, cancer, and Parkinson's disease would have been inconceivable. As such, alkaloids remain indispensable in drug discovery and therapeutic innovation, with continuing potential to contribute to human health and welfare.

To date, over 10,000 natural compounds and derivatives have been identified as alkaloids, though estimates vary between 8,000 and 14,000. New alkaloid molecules are being reported each year from various natural sources. These compounds typically occur as acid salts, but can also be found in glycosidic forms (bound to sugars), or as amides, esters, quaternary salts, or tertiary amine oxides. This structural and chemical diversity reflects the complexity of alkaloid classification and further reinforces their significance across multiple scientific domains.

1.1.1 Typology of Alkaloids

The classification of bioactive chemical compounds, including alkaloids, is essential for organizing structurally and functionally similar molecules. Effective categorization facilitates their morphological recognition, elucidation of biological activity, and utility in both scientific research and applied fields such as pharmacology, agriculture, and biotechnology. Alkaloids, due to their structural diversity and functional complexity, can be classified according to four principal criteria: (a) biological and ecological activity, (b) relevance to chemical and technological innovation, (c) chemical structure, and (d) biosynthetic origin.

1.1.1.1 Bioecological Classification of Alkaloids

From a biological and ecological perspective, alkaloids can be classified into several distinct categories based on their origin and functional role. These include neutral or weakly basic alkaloids, such as lactams like ricinine and certain N-oxides like indicine; animal-derived alkaloids, which are produced by organisms such as amphibians, mammals, and arthropods; and marine alkaloids, found in marine invertebrates and algae. Additional groups include moss alkaloids, which exhibit unique structural features and ecological functions; fungal and bacterial alkaloids, commonly involved in symbiotic or pathogenic interactions; and nonnatural alkaloids, which comprise structurally modified compounds and synthetic analogues created through chemical innovation. The category of nonnatural alkaloids is expanding rapidly due to advances in bio-organic chemistry and stereochemistry. These molecules are central to modern drug discovery, especially in response to the growing resistance of pathogens to conventional antibiotics. As pathogenic microorganisms continue to evolve, the development of novel alkaloid-based therapeutics remains a critical area of pharmaceutical innovation (Aniszewski, 2015).

1.1.1.2 Chemotechnological Classification of Alkaloids

Alkaloids may also be classified based on their origin and the degree of technological intervention involved in their synthesis (Pelletier, 1983). This classification includes the following:

- **Natural alkaloids** are biosynthesized by living organisms and have evolved naturally over time. These may still be undiscovered or poorly characterized.
- **Biomimetic alkaloids**, which are exact synthetic copies of natural alkaloids created in the laboratory.
- **Bionic alkaloids** are chemically engineered molecules inspired by natural alkaloids but modified through advanced technological processes. These compounds are structurally distinct from their natural counterparts.
- **Synthetic alkaloids** are entirely designed and synthesized through artificial means. These do not occur in nature and are engineered for specific functions or properties.

This chemotechnological framework supports both fundamental research and practical applications, especially in pharmaceutical and materials science.

1.1.1.3 Chemo-Molecular Classification of Alkaloids

Chemo-molecular classification organizes alkaloids based on their fundamental chemical structure, typically characterized by the primary nitrogen-containing nucleus or scaffold. This method of classification encompasses a wide array of structural classes, each with distinct chemical and biological properties. Major categories include acridones, aromatics, carbolines, ephedras, ergots, and imidazoles. The indole group comprises several subtypes such as bisindoles, indolizidines, oxindoles, and pyrroloindoles. Additional classes include manzamines; quinoline-related compounds such as quinolines, quinazolines, and quinolizidines; and various phenyl-based structures, including phenylisoquinolines and phenylethylamines. The classification also includes nitrogenous ring systems like piperidines, pyrrolidines, pyrrolizidines, and pyridines, along with purines and simple tetrahydroisoquinolines. Other important categories are steroids, tropanes, and a broad range of terpenoid-based alkaloids, including diterpenes, triterpenes, and sesquiterpenes. This chemo-molecular framework not only facilitates structural elucidation and classification but also enhances our understanding of biosynthetic pathways, chemical reactivity, and pharmacological potential across diverse biological systems (Aniszewski, 2015).

1.1.1.4 Biosynthetic Classification of Alkaloids

Biosynthetically, alkaloids are grouped according to their precursor molecules and metabolic pathways. This method recognizes three principal categories:

1. **True Alkaloids**

   True alkaloids are biosynthesized from amino acids and contain nitrogen within a heterocyclic ring. These compounds typically exhibit strong biological activity, often even at low concentrations. They usually possess a bitter taste and are solid at room temperature, although exceptions like nicotine (a brown liquid) exist. True alkaloids commonly form water-soluble salts and often crystallize as well-defined substances.

   These alkaloids occur in plants in various forms, including as free bases, as acid salts, and as N-oxides. Each form influences the alkaloid's solubility, stability, and bioavailability, and may play a role in the compound's storage, transport, and biological function within the plant.

   True alkaloids are typically found in a restricted number of plant species and families. They arise from amino acid precursors, including L-ornithine, L-lysine, L-phenylalanine/tyrosine, L-tryptophan, and L-histidine (Pelletier, 1983). During biosynthesis, decarboxylated amino acids are condensed with non-nitrogenous structures to yield these compounds.

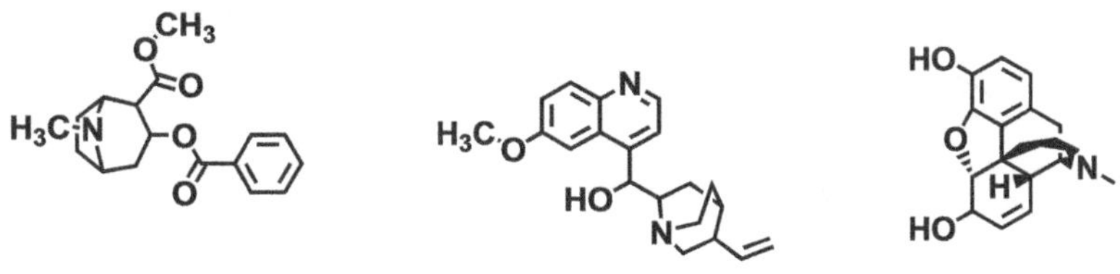

Cocaine Quinine Morphine

FIGURE 1.2 Chemical structures of representative alkaloids: cocaine, quinine, and morphine.

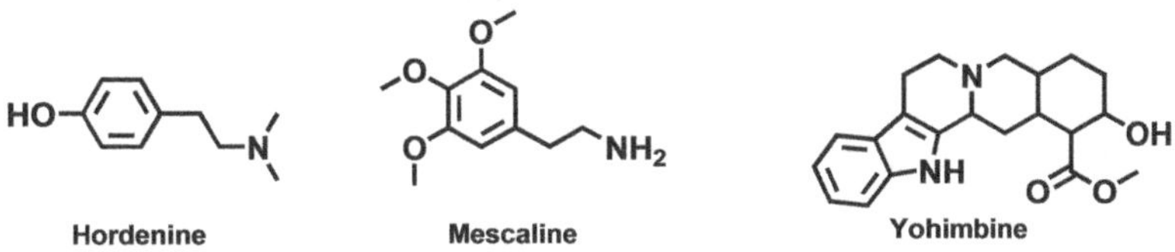

Hordenine Mescaline Yohimbine

FIGURE 1.3 Representative chemical structures of protoalkaloids.

Examples include: Cocaine, Quinine, Morphine (Figure 1.2). True alkaloids may be found in their natural state or may be synthesized as biomimetic, bionic, or synthetic derivatives for medicinal and industrial applications.

2. **Protoalkaloids**
   Protoalkaloids are a subclass of alkaloids in which the nitrogen atom, although derived from an amino acid, is not incorporated into a heterocyclic ring. These compounds are typically derived from amino acids such as L-tyrosine and L-tryptophan. Structurally, protoalkaloids possess a closed-ring system and are considered relatively simple compared to true alkaloids. Despite their simplicity, they are fully functional alkaloids and exhibit notable biological activity. However, they represent only a small fraction of all known alkaloids. Common examples of protoalkaloids include hordenine, mescaline, and yohimbine (Figure 1.3) (Chini et al., 1992). Like other alkaloid types, protoalkaloids may occur naturally or be synthesized as biomimetic, bionic, or fully synthetic compounds.
3. **Pseudoalkaloids**
   Pseudoalkaloids are a distinct group of alkaloids whose core carbon skeletons are not directly derived from amino acids. Although they are not synthesized from amino acids in the classical sense, pseudoalkaloids are often linked to amino acid metabolic pathways (Jakubke et al., 1994). They may originate from precursors or degradation products (postcursors) of amino acids, and can also form through amination or transamination reactions involving intermediates connected to these pathways. In some cases, pseudoalkaloids arise from non-amino-acid precursors, with the nitrogen atom being introduced at a relatively late stage in the biosynthesis, such as in compounds with steroidal or terpenoid backbones. The nitrogen source in these reactions may still be an amino acid, contributing via transamination if an appropriate aldehyde or ketone is available. Pseudoalkaloids can be classified into several structural types, including those derived from acetate, phenylalanine, terpenoids, and steroids. Representative examples include coniine, capsaicin, ephedrine, solanidine, caffeine, theobromine, and pinidine (Figure 1.4). Like other alkaloid classes, pseudoalkaloids may occur naturally or be produced as biomimetic, bionic, or fully synthetic compounds.

This chapter serves as a comprehensive and multidisciplinary resource on alkaloids, highlighting their structural diversity, biosynthetic origins, ecological functions, and pharmacological relevance. It integrates classical knowledge with modern advancements

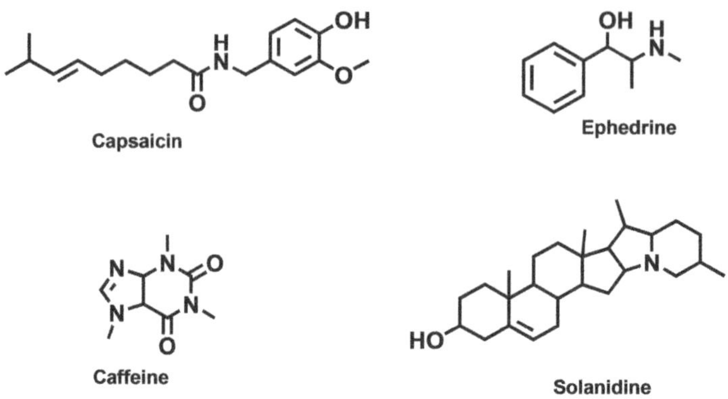

FIGURE 1.4 Representative structures of pseudoalkaloids.

in extraction technologies, analytical methods, and synthetic biology. By presenting a unified framework for the classification and application of alkaloids, the chapter aims to support researchers, academicians, and industry professionals engaged in drug discovery, phytochemistry, and natural product research. Its relevance is further underscored by the growing interest in sustainable bioactive compounds and the urgent need for novel therapeutic agents to combat emerging health challenges.

1.2 HISTORICAL IMPORTANCE OF ALKALOIDS

Human recognition of alkaloids is as old as civilization, since these substances have been used as drugs, in potions, medicines, teas, poultices, and poisons for 4,000 years. It is likely that, in the hunt for food and in dealings with enemies, particular use was made of plants containing alkaloids for arrow poisons, and this use probably preceded their medicinal use. Even today, these poisons are still in use in Africa and South America. Arrow poisons have provided men with ouabain and k-strophanthin for acute cardiac insufficiency, physostigmine for the treatment of glaucoma and myasthenia gravis, D-tubocurarine as a muscle relaxant in anesthesia, reserpine as an antihypertensive and psychotropic drug, and ajmaline in cases of cardiac rhythm disturbances. Sooner or later, arrow poisons will disappear, and there is a real need to continue to evaluate these poisons for active constituents.

As far as the use of alkaloids in medicine is concerned, it is only relatively recently (early 19th century) that therapeutically active substances have been isolated. The first crude drug to be investigated chemically was opium, the dried latex of the opium poppy, *Papaver somniferum*, a drug that had been used for centuries for both its analgesic and narcotic properties. In 1803, Derosne isolated a semipure alkaloid from opium, and in 1805, Sertürner isolated and characterized this constituent as morphine and recognized its basic nature. It was another 12 years before this lead was pursued. Between the years 1817 and 1820, the laboratory of Pelletier and Caventou at the Faculty of Pharmacy in Paris isolated so many active principles from crude drugs that even today no other laboratory has isolated as many active principles of pharmaceutical importance. Among the alkaloids obtained in this brief period were strychnine, emetine, brucine, piperine, caffeine, quinine, cinchonine, and colchicine. These alkaloids are the cornerstone of all that has occurred in alkaloid chemistry to the present day (Cordell, 1981). In 1939, the number of alkaloids that had been isolated and structurally identified was on the order of 200 (Manske, 1950). By 1989, the *Dictionary of Alkaloids* (Southon & Buckingham, 1989) listed details of 10,000 alkaloids, and new structures continue to be found.

1.3 EXTRACTION TECHNIQUES FOR ALKALOIDS

Alkaloids are important secondary metabolites found in plants, widely recognized for their diverse biological and pharmacological activities. Accurate quantification of alkaloids in plant tissues or biological samples is crucial for applications such as drug development, pharmacological evaluation, plant standardization, and chemotaxonomic studies. Since the biological effects of alkaloids are often dose-dependent, reliable extraction and quantification methods are essential to ensure consistent and reproducible results.

1.3.1 Extraction from Plant Materials

The extraction of bioactive alkaloids from plants traditionally relies on methods such as maceration, percolation, decoction, Soxhlet extraction, and hydro-distillation. Maceration is a simple and energy-efficient technique requiring minimal equipment, though it is time-consuming and solvent-intensive (Hurkul et al., 2024a, 2024b; Grujić et al., 2024). Infusion, a related method using hot or cold solvents, is commonly applied for caffeine extraction (Custodio-Mendoza et al., 2024). Percolation offers faster extraction using water–alcohol mixtures but demands skilled handling and precise control of plant material size (Azwanida, 2015; Bitwell et al., 2023). Decoction, involving boiling plant material in water, is suitable for heat-stable compounds but not for heat-sensitive ones (Azwanida, 2015). Soxhlet extraction allows for efficient, large-scale extraction with solvent recycling, although extended heating may degrade sensitive compounds (Mohammed, 2019). Hydro-distillation, widely used for essential oils, avoids organic solvents and yields high-quality oils, though it may lead to hydrolysis with prolonged boiling (Rasul, 2018).

Typical solvents include ethanol and methanol, often used in combination with sonication, vortexing, or refluxing (Alam et al., 2023; Devaliya & Shirsat, 2017; Pan et al., 2016; Matos et al., 2024). Plant samples are typically freeze-dried before extraction, and post-extraction acid–base procedures can enrich alkaloid content. Despite their wide use, conventional methods are often labor-intensive, time-consuming, and require large solvent volumes. Additionally, prolonged heating can degrade heat-sensitive phytochemicals, reducing overall efficiency and yield.

1.3.2 Modern Green Extraction Techniques for Alkaloids

Recent advancements in extraction technologies have introduced green and efficient methods for isolating alkaloids from plant matrices. Techniques such as supercritical fluid extraction (SFE), microwave-assisted extraction (MAE), ultrasound-assisted extraction (UAE), accelerated solvent extraction (ASE), enzyme-assisted extraction, magnetic solid phase extraction (MSPE), and pressurized solvent extraction have significantly reduced extraction time, solvent usage, and environmental impact while enhancing efficiency and yield. For instance, ASE utilizes high pressure and temperature to improve diffusion and solvent penetration (Bitwell et al., 2023), while MAE employs microwave energy to rapidly heat samples, making it ideal for extracting polar compounds like quercetin (Trusheva et al., 2007). SFE, especially with supercritical CO_2, enables precise extraction under controlled conditions (Farooq et al., 2023), and UAE improves cell wall permeability using ultrasonic waves, resulting in faster and higher-yielding extractions (Chemat et al., 2017; Koetz et al., 2021).

Innovative techniques such as MSPE, using magnetic nanoparticles, have shown superior performance over traditional liquid–liquid extraction, especially in extracting compounds like atropine and berberine (Mahmoudi et al., 2024; Wang et al., 2021). Modified QuEChERS methods and advanced sorbents like graphene–boron nitride hybrids further enhance extraction from complex matrices (González-Gómez et al., 2022). Additionally, supercritical fluid chromatography (SFC) offers rapid, eco-friendly separation of alkaloids, making it an excellent tool for green analytical chemistry (Murauer & Ganzera, 2018). Together, these modern methods, often integrated with analytical platforms like LC-MS and UPLC, are revolutionizing phytochemical research by offering sustainable, precise, and high-throughput solutions for alkaloid extraction and analysis.

1.4 CONTEMPORARY RELEVANCE AND APPLICATIONS OF ALKALOIDS IN MEDICINE

Alkaloids continue to play a significant role in modern pharmacotherapy. While some earlier compounds have been discontinued, they have often served as valuable templates for designing synthetic analogues with improved safety profiles or more targeted effects. For instance, local anesthetics such as mepivacaine (Meaverine™), commonly used in dental and ocular procedures, were developed by mimicking structural components of cocaine.

Several alkaloids are still extensively employed in treating life-threatening conditions. Atropine is used as an antidote for poisoning by nerve agents or insecticides. Paclitaxel and its derivatives are frontline treatments for ovarian and breast cancers. The dimeric alkaloids vincristine and vinblastine, derived from the Madagascar periwinkle (*Catharanthusroseus*, family Apocynaceae), along with their semisynthetic variants, are widely used in the treatment of leukemia and lymphoma.

Morphine, obtained from the opium poppy (*Papaver somniferum*, family Papaveraceae), remains indispensable for managing severe pain (Murauer & Ganzera, 2018). Recently, another alkaloid from the same plant—noscapine, a phthalideisoquinoline—has attracted significant scientific interest due to its antimitotic activity. Current research is focused on evaluating noscapine and its semisynthetic derivatives as potential treatments for various forms of cancer (DeBono et al., 2015).

However, the cultivation of alkaloid-producing plants poses challenges, as they exhibit significant variability in growth patterns and alkaloid content. Consequently, industrial-scale production of alkaloids requires diverse strategies and technologies, tailored to the specific plant species and desired compounds, to ensure consistent and efficient yield (Dewick, 2002).

1.4.1 Mechanism of Action

Alkaloids exert diverse effects across animal metabolic systems, and their toxic mechanisms vary substantially. Toxicity can result from enzymatic disruptions that interfere with physiological processes, inhibition of DNA synthesis and repair through nucleic acid intercalation, or direct impacts on the nervous system. Many alkaloids exert multifaceted toxic effects (Mithöfer & Boland, 2012).

Taxines, a class of alkaloids, act as calcium channel antagonists, causing an increase in cytoplasmic calcium levels (Wilson et al., 2001). Pyrrolizidine alkaloids, on the other hand, undergo hepatic biotransformation via oxidases into highly reactive pyrrolic intermediates. These reactive pyrroles alkylate nucleic acids and proteins, leading to pronounced toxic effects (Cushnie et al., 2014).

The antibacterial mechanisms of alkaloids vary among structural classes. Synthetic quinolone alkaloids have been shown to inhibit respiratory processes. Isoquinoline derivatives—such as berberine, sanguinarine, protoberberine, and benzophenanthridine—interfere with cell division by disrupting the Z-ring. The phenanthridine isoquinoline alkaloid ungeremine inhibits nucleic acid synthesis, while the indolizidine alkaloids pergularinine and tylophorinidine target dihydrofolate reductase, also impairing DNA synthesis.

1.5 CONCLUSION AND FUTURE DIRECTION

Alkaloids comprise a vast and structurally diverse class of secondary metabolites with significant biological roles across plants, animals, and humans. Their pharmacological relevance continues to drive research, particularly in drug discovery and therapeutic innovation. A deeper understanding of their biosynthetic pathways—including key enzymes, transcription factors, and regulatory genes—enhances both our capacity to optimize production and our insight into their ecological functions.

To address challenges in sustainable supply, synthetic biology is now being employed to engineer microbial "factories" capable of producing complex plant-derived alkaloids. By reconstructing metabolic pathways in microorganisms such as *Escherichia coli* and *Saccharomyces cerevisiae*, high-value compounds like benzylisoquinoline alkaloids and noscapine have been successfully synthesized. These microbial systems offer scalable and efficient alternatives to traditional plant extraction. Interestingly, the inherent toxicity of many alkaloids—once considered a limitation—is now viewed as a therapeutic opportunity. Toxic compounds previously overlooked are being revisited as potential drug candidates. For example, indole alkaloids from *Psychotria* species demonstrate potent antioxidant activity and appear to play key roles in plant stress responses. Alkaloid production is also influenced by developmental stages, environmental stressors, and ecological interactions, reflecting a finely tuned genetic and metabolic regulation. This dynamic variability supports both plant defense and ecological fitness while offering new avenues for pharmacological exploitation.

In sum, the convergence of metabolic engineering, synthetic biology, and ecological insights is reshaping how alkaloids are studied, produced, and applied. Embracing this complexity will be essential to unlocking their full potential as structurally versatile and therapeutically valuable compounds.

COMPETING INTERESTS

The authors declare that they have no competing interest.

DECLARATION OF GENERATIVE AI AND AI-ASSISTED TECHNOLOGIES IN THE WRITING PROCESS

During the preparation of this work, the book chapter author(s) have not taken any assistance from AI, and the book chapter author(s) reviewed and edited the content as needed and take(s) full responsibility for the content of the publication.

REFERENCES

Alam, S.D., Beg, M.A., Bagadi, M., Locatelli, M., Alothman, Z.A., Mustaqeem, M., & Ali, I. (2023). Facile extraction of berberine from different plants, separation, and identification by thin-layer chromatography, high-performance liquid chromatography, and biological evaluation against leishmaniosis. *Journal of Separation Science*, 46, 2300582.

Alkaloid. (n.d.).*Nature Curse or Blessing? Manfred Hesse*. VHCA, Verlag Helvetica ChimicaActa. WILEY-VCH, Weinheim (Federal Republic of Germany), ISBN:3-906390-24-1

Aniszewski, T. (1994). The biological basis of quinolizidine alkaloids. *Science Legumes*, 21, 1–24.

Aniszewski, T. (2015). *Alkaloids: Chemistry, Biology, Ecology, and Applications*. Elsevier Science.

Azwanida, N.N.(2015). A review on the extraction methods use in medicinal plants, principle, strength and limitation. *Med Aromat Plants (Los Angel)*, 4, 2167–2412.

Bentley, K.W. (1957). *The Alkaloids*. Wiley Interscience.

Bitwell, C., SenIndra, S., Luke, C., & Kakoma, M.K. (2023). A review of modern and conventional extraction techniques and their applications for extracting phytochemicals from plants. *Scientific African*, 19, e01585.

Chemat, F., Rombaut, N., Sicaire, A.G., Meullemiestre, A., Fabiano-Tixier, A.S., & Abert-Vian, M. (2017). Ultrasound assisted extraction of food and natural products. Mechanisms, techniques, combinations, protocols and applications. A review. *Ultrasonics Sonochemistry*, 34, 540–560.

Chini, C., Bilia, A.R., Keita, A., & Morelli, I. (1992). Protoalkaloids from *Boscia angustifolia. PlantaMedica*, 58, 476.

Clayden, J., Greeves, N, Warren, S, & Wothers, P. (2001). *Organic Chemistry*. Oxford University Press.

Cordell, G.A. (1981). *Introduction to Alkaloids: A Biogenetic Approach*. Wiley.

Cushnie, T.P.T., Cushnie, B., & Lamb, A.J. (2014). Alkaloids: An overview of their antibacterial, antibiotic-enhancing and antivirulence activities. *International Journal of Antimicrobial*, 44, 377–386.

Custodio-Mendoza, J.A., Pokorski, P., Aktas, H., Carro, A.M., & Kurek, M.A. (2024). Simultaneous determination of six catechins and caffeine in tea and wine using salting-out assisted liquid–liquid extraction and high-performance liquid chromatography with ultraviolet detection. *Journal of Separation Science*, 47. https://doi.org/10.1002/jssc.202400142

DeBono, A., Capuano, B., & Scammells, P.J. (2015). Progress toward the development of noscapine and derivatives as anticancer agents. *Journal of Medicinal Chemistry*, 58, 5699–5727.

Devaliya, R., & Shirsat, M. (2017). Development and validation of RP-HPLC method for quantification of berberine in ethanol fraction of methanol extract and developed formulation of Tinospora cordifolia. *Oriental Journal of Chemistry*, 33, 989–994. https://doi.org/10.13005/ojc/330253

Dewick, P.M. (2002). *Medicinal Natural Products* (2nd eds., p. 362). John Wiley & Sons.

Farooq, S., Farooq, S., Rather, S.A., and Ganaie, T.A. (2023). *Supercritical CO2 extraction of natural products, in: Extraction of Natural Products from Agro-Industrial Wastes: A Green and Sustainable Approach* (pp. 79–90). Elsevier. https://doi.org/10.1016/B978-0-12-823349-8.00011-3

González-Gómez, L., Morante-Zarcero, S., Pereira, J.A., Câmara, J.S., & Sierra, I. (2022). Improved analytical approach for determination of tropane alkaloids in leafy vegetables based on μ-QuEChERS combined with HPLC-MS/MS. *Toxins (Basel)*, 14.

Grujičić, M., Yayla, S., Cetinkaya, A., Hurkul, M.M., Jovanović, S.,& Ozkan, S.A. (2024). A novel co/Zn-ferrite molecularly imprinted polymer-based electrochemical assay for sensing of gallic acid in plant extracts, wine, and herbal supplement. *Analytical and Bioanalytical Chemistry*, 417, 1327–1339.

Hampel, C.A., & Hawley, G.G. (1976). *Glossary of Chemical Terms*. Nostrand Reinhold Company.

Hegnauer, R.(1963). The taxonomic significance of alkaloids. In T. Swain (Ed.), *Chemical Plant Taxonomy* (pp. 389–399). Academic Press.

Hegnauer, R. (1966). Comparative phytochemistry of alkaloids. In T. Swain (Ed.), *Comparative Phytochemistry*. Academic Press.

Hegnauer, R. (1967). Chemical characters in plant taxonomy: Some possibilities and limitations. *Pure and Applied Chemistry*, 14, 173–187.

Hesse, M. (1981). *Alkaloid Chemistry*. John Wiley & Sons.

Hurkul, M.M., Cetinkaya, A., Yayla, S., Kaya, S.I., Budak, F., Tok, K.C., Gumustas, M., Uzun, L., & Ozkan, S.A. (2024a). Highly selective and sensitive molecularly imprinted sensors for the electrochemical assay of quercetin in methanol extracts of Rubus sanctus and Fragaria vesca. *Talanta*, 273, 125883.

Hurkul, M.M., Yayla, S., Cetinkaya, A., Kaya, S.I., Uzun, L., & Ozkan, S.A.(2024b). A novel electrochemical sensor based on a molecularly imprinted polymer for highly selective and sensitive determination of rutin from herbal supplements and plant extracts. *Analytical Methods*, 16(10), 1480–1488.

Jakubke, H.-D., Jeschkeit, H., & Eagelson, M. (1994). *Concise Encyclopedia Chemistry*. Walter de Gruyter.

Koetz, M., Klein-Junior, L.C., Santos, M.C., da Silva, T.A., Toson, N.S.B., & Henriques, A.T. (2021). An ultrasound assisted extraction–solid-phase extraction–ultra-performance liquid chromatography combined strategy for atropine determination in Atropa belladonna leaves. *Biomedical Chromatography*, 35, e5053.

Mahmoudi, S., Chaichi, M.J., Shamsipur, M., Nazari, O.L., & SamadiMayodi, A.R. (2024) Determination of atropine by HPLC in plant of datura by liquid-liquid extraction and magnet solid-phase extraction. *Journal of Chromatographic Science*, 62, 182–190.

Manske, R.H.F. (1950). *The Alkaloids* (Vols. 1–5).Academic Press.

Matos, R.A., de Souza, T.L., Barreto, L.H.S., de Souza, L.B., dos Santos, D.C.M.B., Brandão, H.N., de Santana, D.A., & de Freitas Santos Júnior, A. (2024). Sustainable analytical method for determination of bioactive compounds in solid oral formulations (capsules and tablets) containing green tea (Camellia sinensis) marketed in Brazil. *Sustainable Chemistry and Pharmacy*, 40, 101655.

Mithöfer, A., & Boland, W. (2012). Plant defense against herbivores: Chemical aspects. *Annual Review of Plant Biology*, 63, 431–450.

Mohammed, Z.H. (2019). Determination of nicotine extracted from eggplant and green pepper by HPLC. *Baghdad Science Journal*, 16, 61–67.

Murauer, A., & Ganzera, M. (2018). Quantitative determination of major alkaloids in Cinchona bark by supercritical fluid chromatography. *Journal of Chromatography. A*, 1554, 117–122.

Pan, Q., Saiman, M.Z., Mustafa, N.R., Verpoorte, R., & Tang, K. (2016). A simple and rapid HPLC-DAD method for simultaneously monitoring the accumulation of alkaloids and precursors in different parts and different developmental stages of Catharanthusroseus plants. *Journal of Chromatography, B: Analytical Technologies in the Biomedical and Life Sciences*, 1014, 10–16.

Pelletier, S.W. (1983). The nature and definition of an alkaloid. In S.W. Pelletier (Ed.), *Alkaloids. Chemical and Biological Perspectives* (Vol. 1, pp. 1–31. John Wiley & Sons.

Rasul, M.G. (2018). Conventional extraction methods use in medicinal plants, their advantages and disadvantages. *International Journal of Basic Sciences and Applied Computing*, 2(6), 10–14.

Sengbusch, P. (2003). *Alkaloids. Botany Online*. WEB: University of Hamburg.

Southon, I.W., & Buckingham, J. (eds.) (1989). *Dictionary of Alkaloids*. Chapman & Hall.

Trusheva, B., Trunkova, D., & Bankova, V.(2007). Different extraction methods of biologically active components from propolis; a preliminary study. *Chemistry Central Journal*, 1, 13.

Waller, G.R., & Nowacki, E.K. (1978). *Alkaloid Biology and Metabolism in Plants*. Plenum Press.

Wang, Y., Yin, S.J., Zhao, C.P., Chen, G.Y., & Yang, F.Q. (2021). Preparation of a zirconium terephthalate metal-organic framework coated magnetic nanoparticle for the extraction of berberine prior to high-performance liquid chromatography analysis. *Journal of Separation Science*44, 1220–1230.https://doi.org/10.1002/jssc.202001026

Wilson, C.R., Sauer, J., & Hooser, S.B. (2001). Taxines: A review of the mechanism and toxicity of yew (Taxus spp.) alkaloids. *Toxicon*, 39, 175–185.

Winterstein, E., & Trier, G. (1910). *Die Alkaloide. EineMonographie der natürlichenBasen*. Borntraeger.

2 Classification and Structural Diversity of Alkaloids

Ahmed Sabt, Ahmed F. Essa, Abdulsalam M. Kassem, Sherif E. Emam, Rania F. Ahmed, Eman Y. Ahmed, Asmaa S. Abd Elkarim, Mai M. Elghonemy, Rehab F. Taher, Yusuke Kasai, Hiroshi Imagawa, Howaida I. Abd-Alla, and Abdelsamed I. Elshamy

2.1 INTRODUCTION

Herbal and medicinal plants are a crucial source of various metabolites that perform different physiological functions. Among the metabolites that have been extensively studied are phytochemicals, primarily secondary metabolites synthesized by plants (Abate et al., 2022; Asfaw et al., 2022). These phytochemicals exhibit considerable structure, concentration, distribution, and biological activity variability, even within the same cultivar. They are generally classified into three main categories: phenolic compounds, terpenes, and nitrogen- and sulfur-containing compounds (Mazid et al., 2011; Ncube et al., 2015). Alkaloids constitute a diverse group of natural compounds originating from secondary metabolic processes. They are characterized by a basic nitrogen atom in their molecular structure, excluding nitrogen that is part of an amide bond or peptide. Although alkaloids are predominantly derived from plant sources, they have also been discovered in animals, insects, marine invertebrates, and certain microorganisms (Bribi, 2018; Lu et al., 2012; Roberts & Wink, 1998). These compounds can be found in various cellular organelles, including mitochondria, vesicles, chloroplasts, and vacuoles. Their precursors, primarily amino acids, are products of metabolic pathways such as glycolysis.

Alkaloids are an extensive category of chemical compounds that exist naturally and have a minimum of one nitrogen atom. Alkaloids can refer to several artificial substances with an analogous structure (Vrabec et al., 2023). Numerous creatures, such as bacteria, fungi, plants, and animals, generate alkaloids. Acid–base extraction or solvent extraction, followed by silica-gel column chromatography, are two methods used to purify these organisms' crude extracts (Pereira et al., 2023; Zhang et al., 2018; Pereira et al., 2023; Zhang et al., 2018). Alkaloids possess a multitude of pharmacological properties, such as antimalarial, antiasthma, anticancer, cholinomimetic, vasodilatory, antiarrhythmic, analgesic, antibacterial, and antihyperglycemic properties (Pereira et al., 2023; Uzor, 2020). In conventional and modern medicine, many alkaloids have been used as beginning points for developing new drugs. Many alkaloids have been employed in entheogenic ceremonies and for recreational purposes because they contain psychotropic and stimulant properties (Pereira et al., 2023). Additionally, alkaloids can be poisonous (Muñoz et al., 2020; Pereira et al., 2023). Despite their varied effects on human and animal metabolism systems, alkaloids virtually always produce a bitter taste (Muñoz et al., 2020).

The exploration of this class of metabolites commenced in 1806 when Sertürner successfully isolated morphine from opium. Since then, numerous plant extracts have been utilized for their toxic and medicinal properties, attributed to their alkaloid content. Chemically, alkaloids are characterized as crystalline, colorless substances with a bitter taste, capable of forming salts upon

 DOI: 10.1201/9781003539179-2

reaction with acids. Plants may exist in various forms, including free states, salts, or N-oxides (Amirkia & Heinrich, 2014; Bribi, 2018; Kutchan, 1995).

Alkaloids are found with various chemical structures and connected functioning components in all kingdoms, and they exhibit an extensive spectrum of biological features. Alkaloids like pinnatoxins, pinnamine, and halochlorine, which are present in marine life like shellfish and sponges, have been proven to be effective in alleviating inflammatory conditions of the heart and noncardiac (Kuramoto et al., 2004). Alkaloids serve as protective compounds in several ladybird beetles, releasing unpleasant alkaloids that contain hemolymph when molested. In its toxic venom, a small number of ant species contain alkaloids such as cis- and trans-2-methyl-6-alkylpiperidines in addition to protein-based compounds that can be employed both offensively and defensively (Braekman et al., 1998).

Alkaloids are categorized according to several factors, chief among them being their chemical characteristics, taxonomy, and botanical and medicinal uses, all of which will be covered in detail in this chapter. The most widely used classification scheme divides alkaloids into heterocyclic (typical) and non-heterocyclic (atypical) alkaloids based on where nitrogen is located in their structure (Evans & Ferguson, 2018). Furthermore, according to their biosynthesis processes, alkaloids, terpenoid indole alkaloids, benzylisoquinoline alkaloids, tropane alkaloids, and purine alkaloids can be divided into groups (Facchini & Griffiths, 1981).

Throughout history, different societies have used plants containing alkaloids in their roots, leaves, stems, fruits, and seeds for both toxic and medicinal purposes, including making teas, poultices, and potions to treat specific illnesses (Goyal, 2013; Jing et al., 2014; Roberts & Wink, 1998). The beneficial effects of alkaloids for healthcare are intimately related to their biological significance. Alkaloids are now known in medicine for many different applications as stimulants, anesthetics, antibacterials, antimalarials, analgesics, antihypertensive agents, spasmolytics, anticancer agents, antiasthmatic therapies, vasodilators, and antiarrhythmic agents. There is still much to learn about these characteristics and the toxicities that go along with them (Kuete, 2014).

2.2 ALKALOIDS CLASSIFICATION

The existence of a carbon–nitrogen (CN) skeleton is the main factor influencing this structural classification (Bribi, 2018; Cushnie et al., 2014). According to their chemical structure and biosynthesis processes, alkaloids can be systematically divided into three primary categories: (a) true alkaloids, which are heterocyclic compounds; (b) protoalkaloids, which are non-heterocyclic compounds; and (c) pseudoalkaloids (Ranjitha & Sudha, 2015).

2.2.1 True Alkaloids

Derived from cyclic amino acids, true alkaloids are heterocyclic molecules distinguished by their pharmacological action and intricate chemical structures. These substances are frequently found in nature as salts with different organic acids, such as oxalic, lactic, malic, tartaric, acetic, and citric acids, and they include an intracyclic nitrogen atom (Henning et al., 2013; Talapatra & Talapatra, 2015). According to Kukula-Koch and Widelski (2017) (Kukula-Koch & Widelski, 2017), the alkaloids in this group are derivatives of particular amino acids, including glycine/aspartic acid, L-ornithine, L-tyrosine, L-phenylalanine, L-lysine, L-histidine, L-tryptophan, and L-arginine. Tryptophan, for example, is the precursor of indole, quinoline, and pyrroloindole alkaloids; tyrosine is the basis for isoquinoline alkaloids; ornithine is the precursor of tropane, pyrrolizidine, and pyrrolidine alkaloids; lysine is linked to quinolizidine and piperidine alkaloids; aspartate is the precursor of pyridine alkaloids; anthranilic acid produces quinazoline, quinoline, and acridone alkaloids; and histidine derivatives result in imidazole alkaloids as well (Aniszewski, 2015; Böttger et al., 2018; Kaur et al., 2019; Kaur & Arora, 2015). There are other classes of heterocyclic alkaloids, such as pyrrole, pyrrolidine, pyrrolizidine, pyridine, piperidine, tropane, aporphine, quinolone, isoquinoline, quinolizidine, indole, indolizidine, and imidazole (Figure 2.1). These categories contain

FIGURE 2.1 Classes of the true alkaloids.

notable alkaloids such as morphine, cocaine, quinine, dopamine, lobeline, nantenine, berberine, salsoline, geissospermine, piperine, and nicotine (Hussain et al., 2018).

2.2.1.1 Pyrrolidine Alkaloids

The amino acid L-ornithine is the primary source of pyrrolidine alkaloids. At the same time, certain molecules may also come from arginine and lysine by incorporating acetate or malonate units, which create ring structures. Plant groups such as Fabaceae, Erythroxylaceae, and Moraceae encompass pyrrolidine alkaloids, such as putrescine, hygrine, and cuscohygrine (Figure 2.2) (Kaur & Arora, 2015; Kukula-Koch & Widelski, 2017).

2.2.1.2 Pyrrole Alkaloids

A substantial subgroup of heterocyclic alkaloids, pyrrole alkaloids, is present in a wide range of pharmacologically relevant natural and synthesized substances. Heme and chlorophyll derivatives, which have four pyrrole groups joined by methine bridges, are the most well-known pyrrole alkaloids (Estévez et al., 2014). According to Bauer and Knölker (2012), the pyrrole ring is the main structural component of carbazole alkaloids, which are primarily found in the Rutaceae family, namely in the genera *Murraya, Glycosmis*, and *Micromelum* (Bauer & Knölker, 2012).

2.2.1.3 Pyridine Alkaloids

These heterocyclic substances are analogs of L-ornithine and have an unsaturated nitrogen radical in their nucleus. According to Kaur and Arora (2015), alkaloids include nicotine, lobeline, cytisine, guvacine, coniine, trigonelline, arecoline, arecaidine, anabasine, sparteine, and pelletierine (Figure 2.3) (Kaur & Arora, 2015). The plant families Aizoaceae, Annonaceae, Apocynaceae, Araceae,

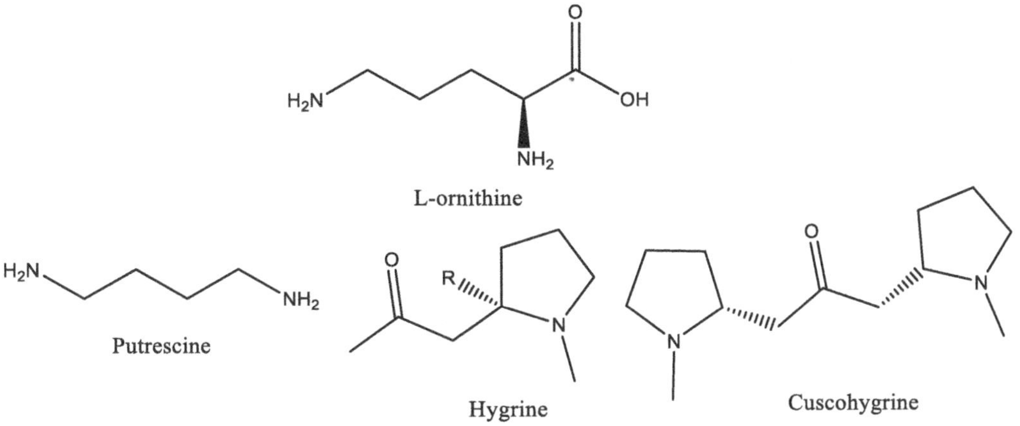

FIGURE 2.2 Primary sources of plant pyrrolidine alkaloids.

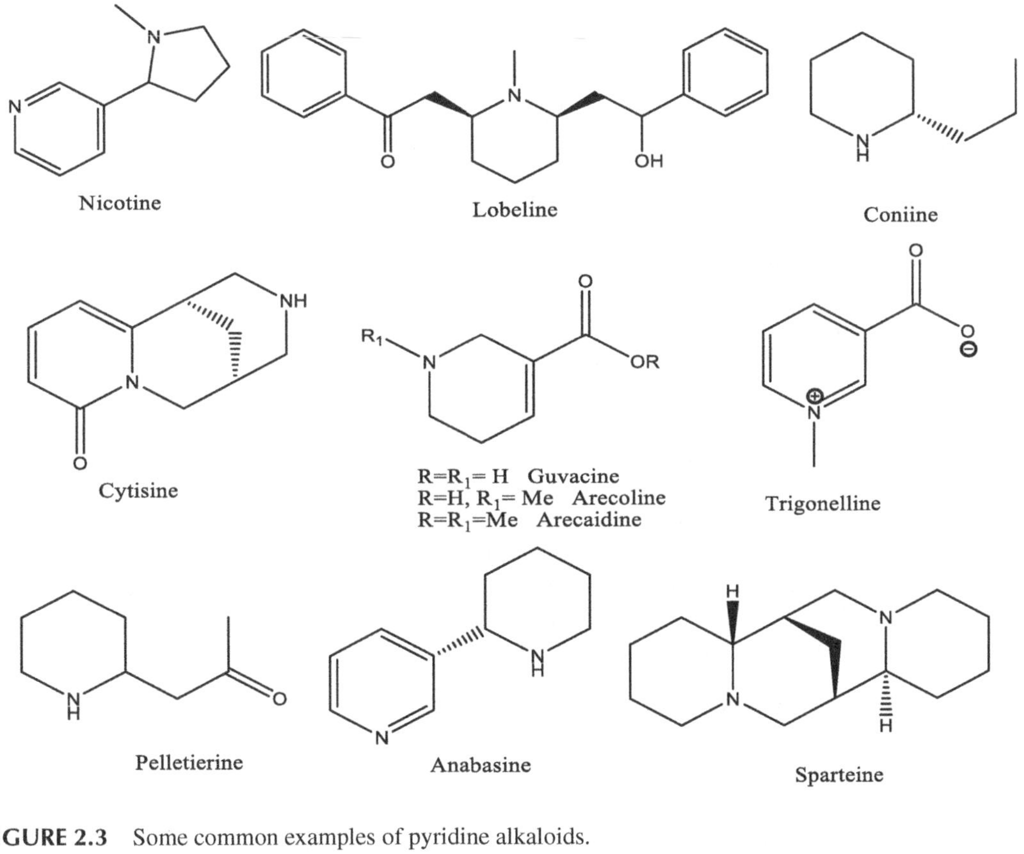

FIGURE 2.3 Some common examples of pyridine alkaloids.

Bignoniaceae, Dipsacaceae, Gramineae, Palmae, and Umbelliferae are among those that contain these alkaloids (Silva et al., 2013).

2.2.1.4 Pyrrolizidine Alkaloids

The amino acid ornithine is the parent of over 500 compounds categorized as pyrrolizidine alkaloids. As esters of 1-hydroxymethyl-1,2-dehydropyrrolizidine (necines), pyrrolizidine alkaloids often have a hydroxyl group at location 7, but they can also have hydroxylated groups at positions C-2 and C-6 (Koleva et al., 2012; Schramm et al., 2019). Two rings of five members with a nitrogen

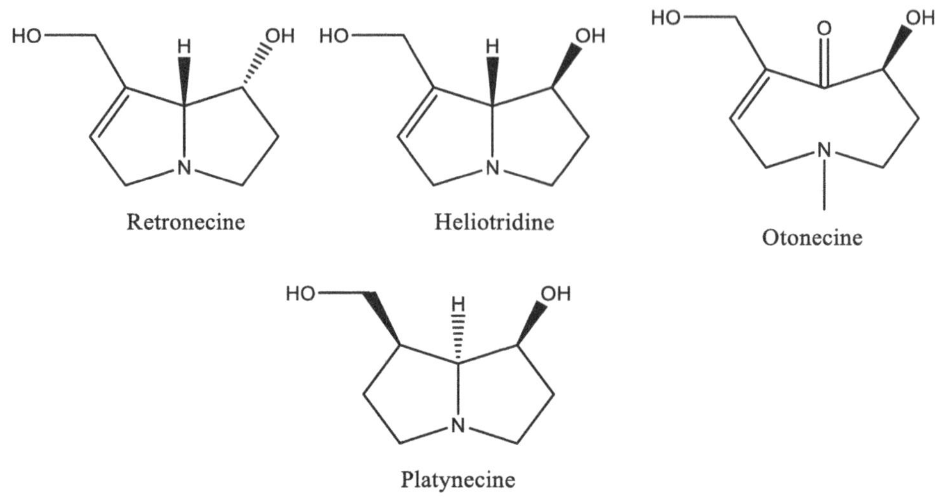

FIGURE 2.4 The four categories of necines.

atom in between define their structure; the toxicity of this alkaloid is determined by a double ligation in C-1 and C-2. The four categories of necines are retronecine, heliotridine, otonecine, and platynecine (Figure 2.4), based on their bases (Moreira et al., 2017). The Asteraceae and Boraginaceae families contain the majority of pyrrolizidine alkaloids (Koleva et al., 2012).

2.2.1.5 Tropane Alkaloids

Ornithine is the precursor from which tropane alkaloids are derived. *N*-methyl-8-azabicyclo[3,2,1]octane is the skeleton that characterizes tropane alkaloids, which have a complicated structure (Mao et al., 2014). Tropane alkaloids, which include more than 200 different types of alkaloids, are mostly found in angiosperms belonging to several families, including the Proteaceae, Solanaceae, Erythroxylaceae, Convolvulaceae, Brassicaceae, and Euphorbiaceae (Jirschitzka et al., 2012). The Solanaceae family contains the most researched, primarily in the genera *Atropa* (atropine alkaloid), *Hyoscyamus* (scopolamine), and *Datura* (hyoscyamine and scopolamine) (Figure 2.5) (Ajungla et al., 2009; Guirimand et al., 2010; Wink, 2010). Cocaine (Figure 2.5) is another well-known tropane alkaloid from the *Erythroxylum coca* (Erythroxylaceae) plant. Interestingly, *Erythroxylum coca* alone contains 186 alkaloids (Jirschitzka et al., 2012; Oliveira et al., 2010). Another prevalent tropane alkaloid on the World Health Organization's basic medication list is atropine (Ranjitha & Sudha, 2015).

2.2.1.6 Quinoline Alkaloids

This class of alkaloids originates from L-tryptophan and consists of heterocyclic aromatic compounds created when a pyridine ring and a benzene ring fuse. One of this group's most significant compounds, quinine, is found in *Cinchona ledgeriana*, whereas *C. officinalis* yields the quinidine alkaloid. Additionally, bulaquine, tafenoquine, pamaquine, and chloroquine are quinoline alkaloids (Figure 2.6) (Debnath et al., 2018; Kaur & Arora, 2015; Marella et al., 2013). The Rutaceae family and plants in the genus *Melodinus* are significant sources of these alkaloids (Byler et al., 2009; Cai et al., 2011).

2.2.1.7 Piperidine Alkaloids

Piperidine alkaloids are a class of over 700 compounds that have been extensively investigated. The structure of these derivatives of L-lysine includes one amine, five methylene groups, and a ring of six radicals. This group comprises alkaloids such as coniine, lobeline, cynapine, and solenopsin.

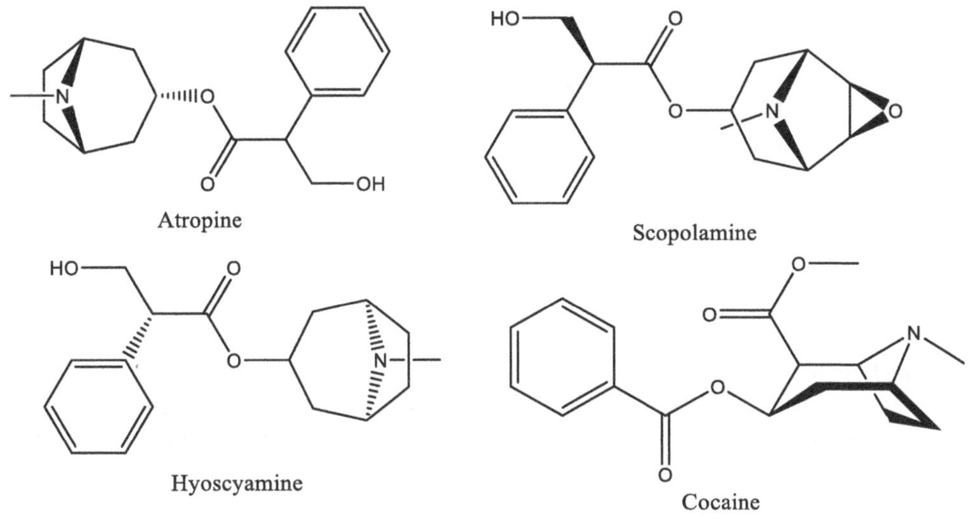

FIGURE 2.5 Some examples of tropane alkaloids.

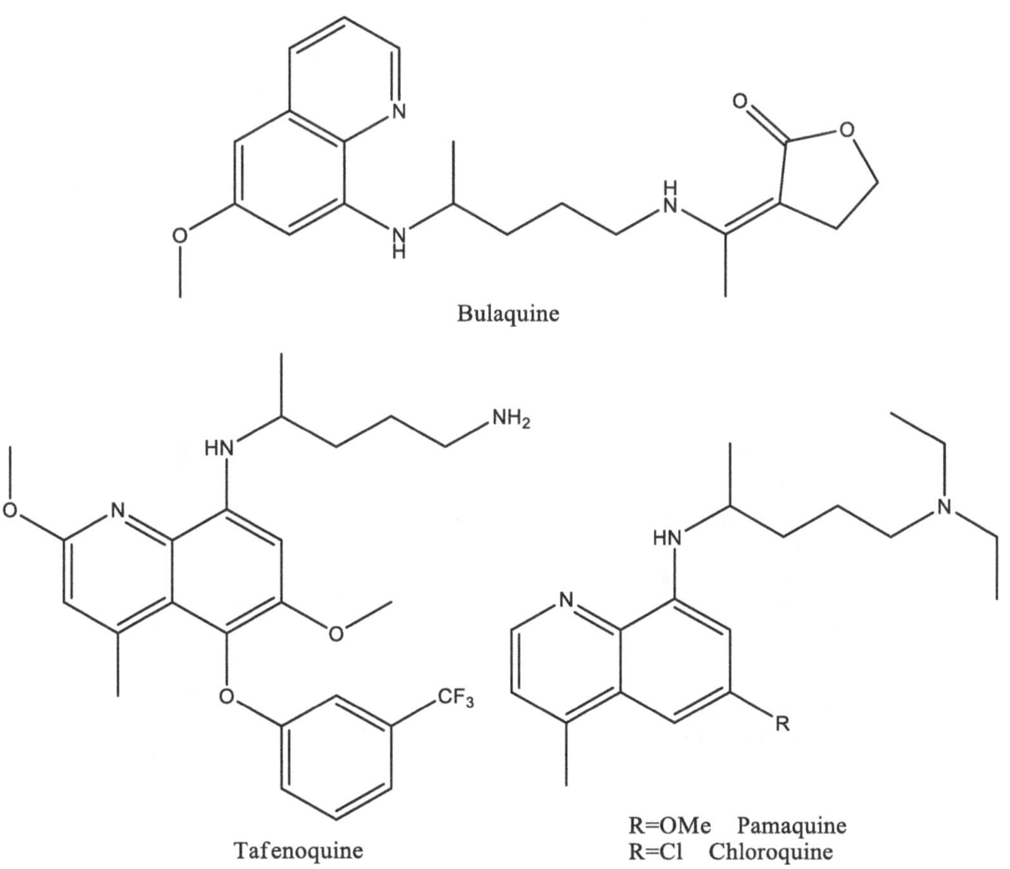

FIGURE 2.6 Some examples of quinidine alkaloids.

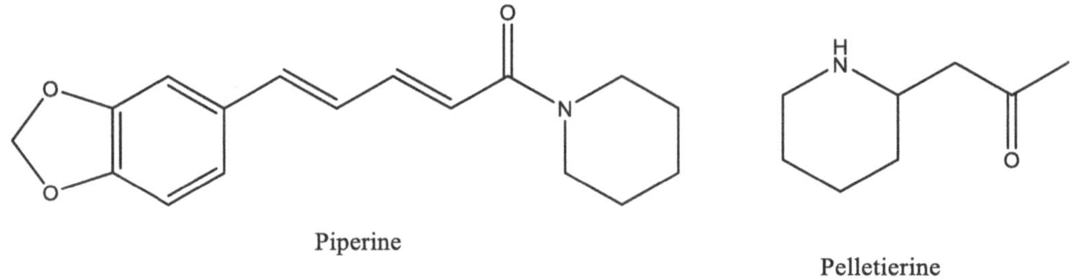

FIGURE 2.7 Some examples of piperidine alkaloids.

Plants in the Lobeliaceae family, especially *Lobelia inflata*, which includes the alkaloid lobeline, are known to contain piperidine alkaloids (Kaur & Arora, 2015; Liu et al., 2010). They are also present in *Punica granatum* as pelletierine and black pepper (*Piper nigrum*) as piperine (Figure 2.7). Similarly, piperidine residues can be found in the skeleton of many natural products (Böttger et al., 2018; Goel et al., 2018).

2.2.1.8 Isoquinoline Alkaloids

Among the most prevalent in the kingdom of plants, they belong to the following families: Annonaceae, Rutaceae, Amarryllidaceae, Fumariaceae, Ranunculaceae, and Papaveraceae (Kukula-Koch & Widelski, 2017). The heterocyclic aromatic compounds known as isoquinoline alkaloids are produced from 3,4-dihydroxytyramine (dopamine) and the amino acids tyrosine and phenylalanine. Isoquinoline alkaloids can be classified into two groups based on their structural makeup: benzylisoquinolines, which have a second aromatic ring, and simple isoquinolines, which have a benzene ring joined to a pyridine ring (Hussain et al., 2018; Kukula-Koch & Widelski, 2017). Similarly, these can be classified into subgroups of benzylquinoline, aporphine, protoberberine, benzo[c]phenanthridine, protopine, phthalide isoquinoline, morphine, and emetine alkaloids due to their structural non-homogeneity and dependence on the degree of oxygenation, internal rearrangement, distribution, and the existence of extra rings that are connected to the main system (Bhadra & Kumar, 2011; Hussain et al., 2018; Koleva et al., 2012; Kukula-Koch & Widelski, 2017). Morphine (Figure 2.8),

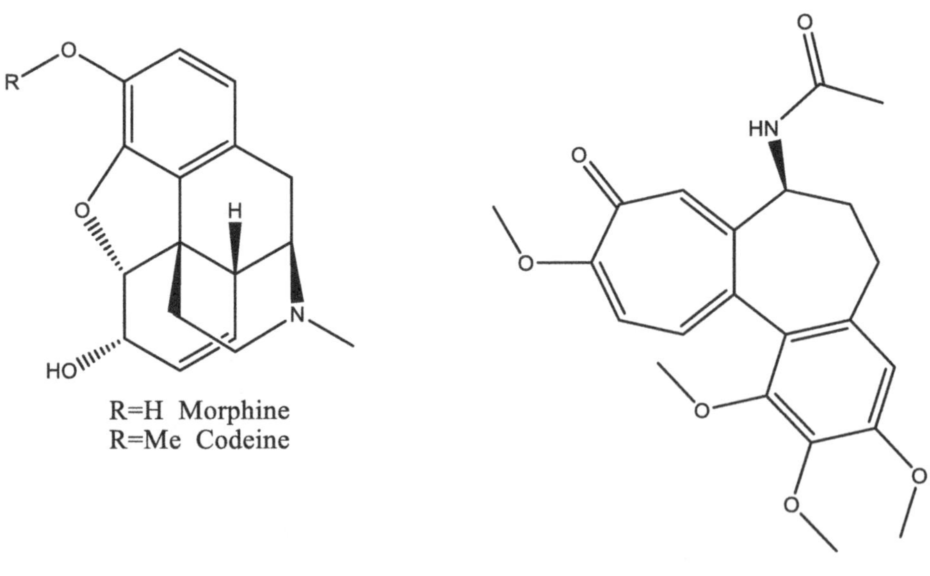

FIGURE 2.8 Some examples of isoquinoline alkaloids.

an isoquinoline alkaloid derived from opium (*Papaver somniferum*), is one of the most well-known alkaloids. According to Baros et al. (2012), Diamond and Desgagné-Penix (2016), and Debnath et al. (2018), this species also contains the benzylisoquinoline alkaloid codeine (Figure 2.8), berberine in *Berberis aristata, B. lyceum*, and *B. tinctoria*, and colchicine (Figure 2.8) in *Colchicum autumnale* (Baros et al., 2012; Debnath et al., 2018; Diamond & Desgagné-Penix, 2016).

2.2.1.9 Indole Alkaloids

They are bicyclic, with a six-membered benzene ring linked to a five-membered pyrrole ring that contains nitrogen. The latter allows them to be basic, and tryptophan is their precursor (Hamid et al., 2017). Approximately 2000 compounds known as indole alkaloids have been identified and are found in various families, including the Apocynaceae, Loganiaceae, Rubiaceae, and Nyssaceae. Additionally, there are plant species, including *Rauvolfia serpentina* (ajmalicine, reserpine, serpentinine, ajmaline, corynanthine, and yohimbine) (Figure 2.9), *Passiflora incarnata* (harman, harmol, harmine, and harmaline) (Figure 2.9), *Rauwolfia speciosa* (mitragynine) (Figure 2.9), and *Catharanthus roseus* (vinblastine and vincristine alkaloids) (Guirimand et al., 2010; Hamid et al., 2017; Sagi et al., 2016)

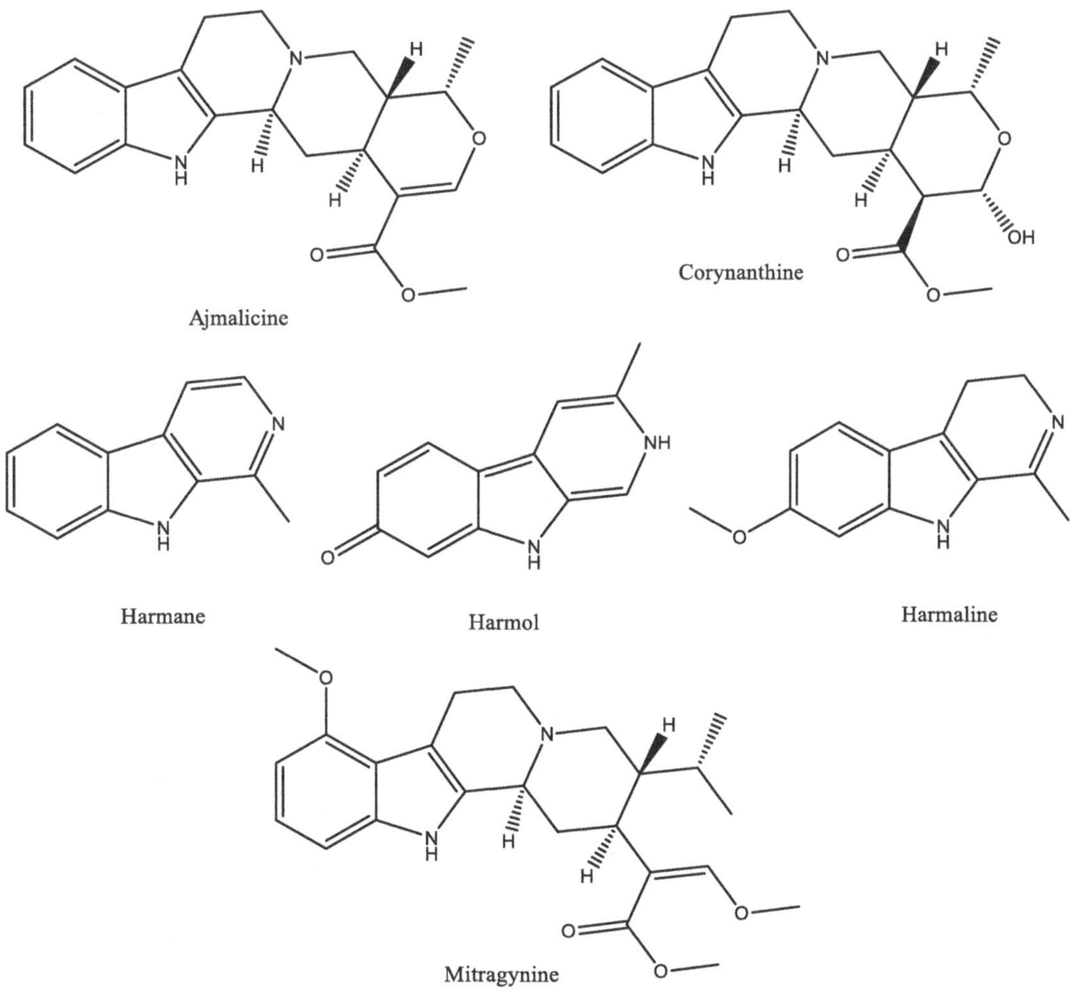

FIGURE 2.9 Some examples of indole alkaloids.

2.2.1.10 Aporphine Alkaloids

Isoquinoline alkaloids are the source of aporphine alkaloids, which are heterocyclic alkaloids. Annonaceae, Menispermaceae, Papaveraceae, Ranunculaceae, Lauraceae, Monimiaceae, Magnoliaceae, and Berberidaceae are merely a few of the varied families (around 20) from which these compounds have been identified. Lirinidine, asimilobine, *N*-methyl-asimilobine, nornuciferine, nuciferine, anonaine, magnoflorine, dicentrine, boldine, galucine, and neolitsine are a few examples of common aporphine alkaloids (Figure 2.10). The substituents in each of the aforementioned vary depending on where the nitrogen atom is located, including -H, $-CH_3$, $-COOH_3$, etc. (Teles et al., 2019; Muthna et al., 2013).

2.2.1.11 Quinolizidine Alkaloids

Combining six members with a nitrogen atom results in these molecules, which are derivatives of the amino acid L-lysine and have structural variations ranging from simple to complicated. Although they are also found in the genera *Baptisia, Thermopsis, Genista, Cystus*, and *Sophora*, they are primarily found in the Leguminosae family, particularly in the genus *Lupinus*. Lupanine (*Lupinus luteus*), cytisine (*Laburnum* species), and sparteine alkaloids (*Sarothamnus scoparius*) are the three most well-known quinolizidine alkaloids (Figure 2.11) (Bunsupa et al., 2012; Kaur & Arora, 2015; Szőke et al., 2013).

2.2.1.12 Imidazole Alkaloids

L-histidine is the source of these alkaloids, primarily sourced from the Rutaceae and Cactaceae families, including histamine, histidine, pilocarpine, and pilosine (Aniszewski, 2015). According to Silva et al. (2013) and Debnath et al. (2018), the primary imidazole alkaloid, pilocarpine, was extracted from the leaves of the Rutaceae plant *Pilocarpus microphyllus* (Debnath et al., 2018; Silva

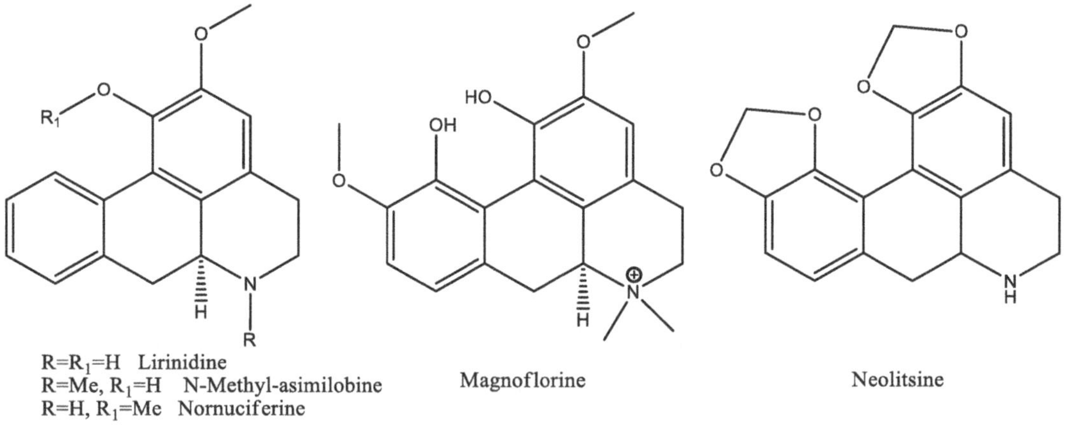

FIGURE 2.10 Some examples of aporphine alkaloids.

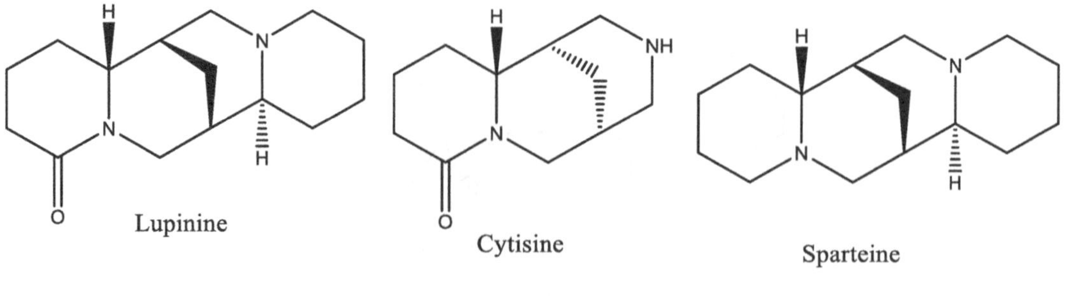

FIGURE 2.11 Some examples of quinolizidine alkaloids.

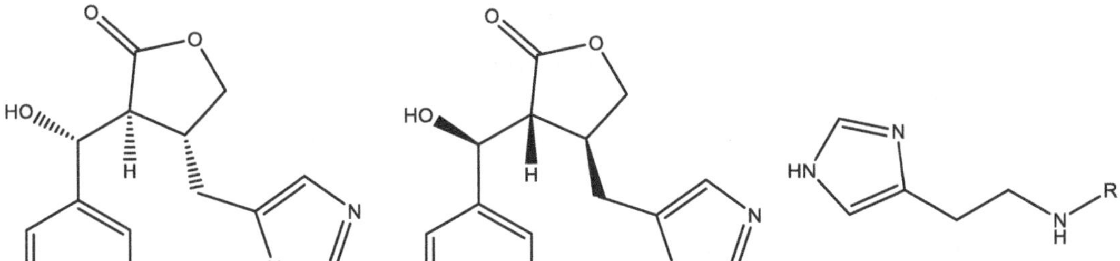

R=H Histamine
R=COOH Histidine

Iopilosine Epiisopilosine

FIGURE 2.12 Some examples of imidazole alkaloids.

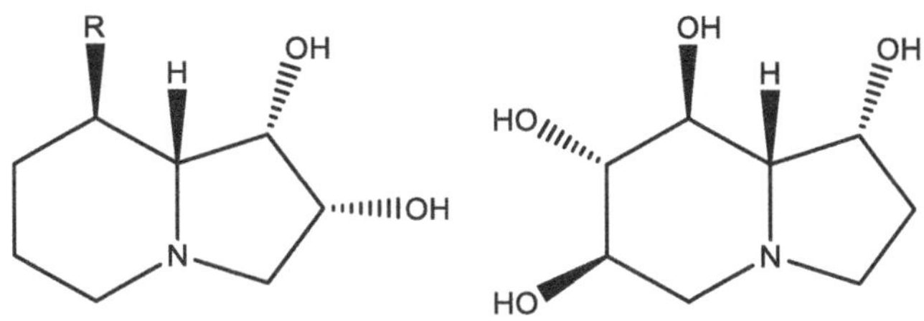

R=OH Swainsonine
R=H Lentiginosine

Castanospermine

FIGURE 2.13 Some examples of indolizidine alkaloids.

et al., 2013). Other alkaloids of the same group, such as isopilosine, epiisopilosine, and epiisopiloturine, were also extracted from this plant (Figure 2.12).

2.2.1.13 Indolizidine Alkaloids

L-Lysine is an amino acid that is the precursor of indolizidine alkaloids. Heterocyclic indolizidine alkaloids are distinguished by the fusion of rings with six and five members and the presence of a nitrogen atom between ring fusions. They have been discovered in many *Ipomoea* plants. In particular, *Ipomoea alba* (seeds) contains alkaloids such as ipalbine, ipalbidine, and isoipomine. Among other things, *Ipomoea carnea* hosts the alkaloid swainsonine, which is present in the majority of this genus's plants but was first identified in *Swainsona* plants. Swainsonine, castanospermine, and lentiginosine (Figure 2.13) are the three primary indolizidine alkaloids (Meira et al., 2012).

2.3 PROTOALKALOIDS

Protoalkaloids, which can be produced from amino acids or biogenic amines, are chemically defined by the presence of a nitrogen atom outside the ring that stays as a side chain rather than as a component of the heterocyclic system. Although they are not very prevalent in nature, some protoalkaloids include mes caline, ephedrine, colchicine, cathinone, etc. (Jayakumar & Murugan, 2016; Kukula-Koch & Widelski, 2017; Talapatra & Talapatra, 2015; Wansi et al., 2013). According to Alves de Almeida et al. (2017), protoalkaloids can be generated from L-tyrosine and L-tryptophan, which respectively provide terpenoid indole and phenylethylamine (de Almeida et al., 2017). *Lophophora williamsii*, also called peyote, is the source of mescaline, one of the most prevalent phenylethylamine alkaloids (Beyer et al., 2009). In contrast, the Apocynaceae, Loganiaceae, and Rubiaceae families contain about 3000 monoterpenoid indole alkaloids, making them a reasonably large category (Pan et al., 2016).

2.4 PSEUDOALKALOIDS

Although heterocyclic and include nitrogen, pseudoalkaloids are not produced from amino acids; they are typically acetate, pyruvic acid, adenine/guanine, or geraniol derivatives (Aniszewski, 2015). The diterpenoid alkaloids (of 18, 19, and 20 carbons) derived from a range of sources, including the genus *Aconitum, Consolida*, and *Delphinium*, serve as an illustration of this group (Wang et al., 2010).

2.5 PLANT-DERIVED ALKALOIDS

More than 40,000 compounds are included in the Dictionary of Alkaloids as of October 2023 (Buckingham et al., 2010). Many of these chemicals' designation depends on the plant species and botanical family from whence they originated (Table 2.1). Furthermore, alkaloids can be categorized based on their origin (Talapatra & Talapatra, 2015). Alternative classification techniques are used because of plants' wide range of alkaloids and the lack of a standardized taxonomy framework for their classification. These techniques can be based on their chemical structures, the most common method, or their natural or biochemical origins.

As highlighted in the document, several important alkaloids and their plant sources demonstrate this significance. Quinine, obtained from *Cinchona pubescens* (Rubiaceae) (Table 2.1), is well known for its antimalarial activity and has played a historic role in combating malaria, with additional analgesic and anti-inflammatory benefits (Noriega et al., 2015). Berberine, from *Berberis asiatica* (Berberidaceae) (Table 2.1), is another prominent alkaloid with diverse pharmacological effects, including antimicrobial, cardioprotective, and anti-diabetic properties (Mazumder et al., 2011). Anonaine, isolated from *Annona squamosa* (Annonaceae) (Table 2.1), has been reported to possess anticancer and antimicrobial activities, showing potential for drug development in modern therapeutics (Porwal & Kumar, 2015). Similarly, morphine from *P. somniferum* (Papaveraceae) (Table 2.1) remains the most widely used analgesic in clinical medicine. However, its addictive nature and associated health concerns necessitate cautious use and further research into safer alternatives (Baros et al., 2012).

As highlighted in the document, solanine from *Solanum melongena* (Solanaceae) (Table 2.1) has been reported to possess both toxic and therapeutic properties, particularly showing potential in antimicrobial and anticancer studies (Boulaaba et al., 2019). Cocaine, isolated from *Erythroxylum coca* (Erythroxylaceae) (Table 2.1), is one of the most well-known alkaloids, valued historically for its local anesthetic effects but also recognized for its strong psychoactive and addictive properties, which limit its clinical use today (Jirschitzka et al., 2012). Piperine, the active alkaloid of *P. nigrum* (Piperaceae) (Table 2.1), is widely studied for its bioenhancing ability, improving the absorption and efficacy of various drugs, in addition to its antioxidant and anti-inflammatory effects (Li et al., 2011). Echimidine, derived from *Echium hypertropicum* (Boraginaceae) (Table 2.1), has been noted for its biological activities. However, its toxicity as a pyrrolizidine alkaloid raises safety concerns that require careful pharmacological evaluation (Carvalho et al., 2013).

Lycorine, obtained from *Amaryllis belladonna* (Amaryllidaceae) (Table 2.1), has been noted for its potential anticancer, antiviral, and anti-inflammatory effects, making it a promising candidate for therapeutic applications (Koetz et al., 2017). Atropine, from *Atropa belladonna* (Amaryllidaceae) (Table 2.1), is one of the most historically significant alkaloids, widely used as an anticholinergic agent in treating conditions such as bradycardia and in ophthalmology for pupil dilation (Koetz et al., 2017). Lupanine, derived from *Lupinus angustifolius* (Fabaceae) (Table 2.1), belongs to the quinolizidine alkaloids and has shown both toxic and beneficial pharmacological effects, including potential roles in modulating glucose metabolism (Jansen et al., 2012). Vincristine, isolated from *Vinca rosea* (Apocynaceae) (Table 2.1), is one of the most important alkaloids in cancer chemotherapy, used extensively in the treatment of leukemia, lymphoma, and other malignancies (Adewusi & Afolayan, 2010).

TABLE 2.1
Some Plant-Derived Alkaloids Based on Their Occurrences and Sources

Alkaloid	Plant Source (Family)
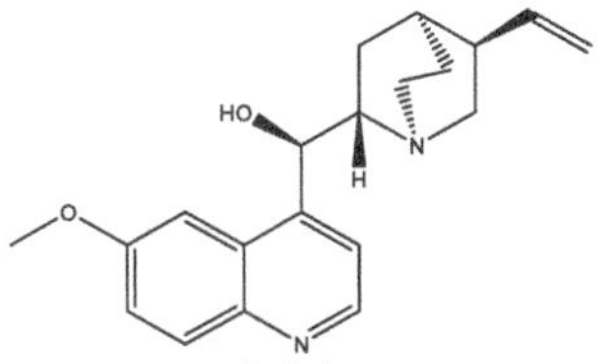 Quinine	 *Cinchona pubescens* (Rubiaceae)
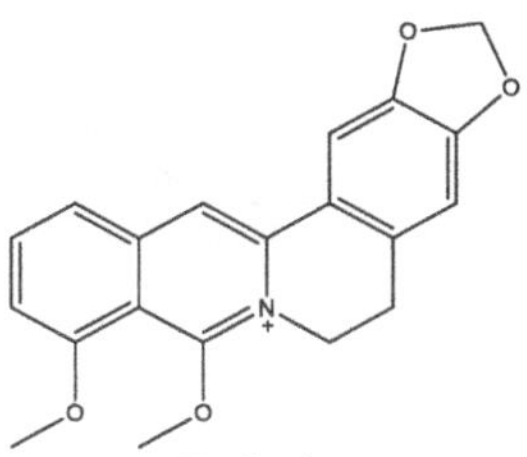 Berberine	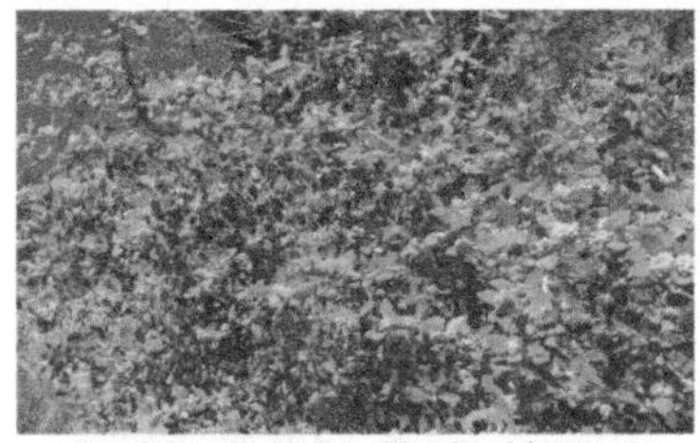 *Berberis asiatica* (Berberidaceae)
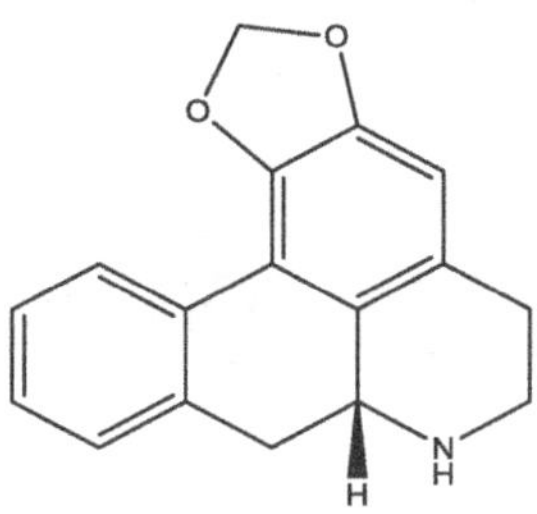 Anonaine	 *Annona squamosa* (Annonaceae)
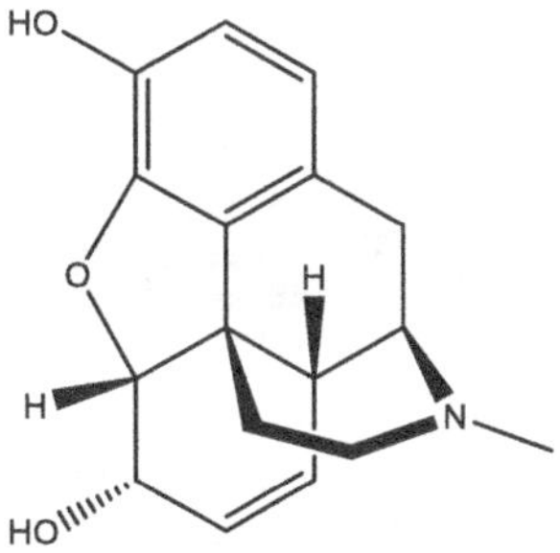 Morphine	 *Papaver somniferum* (Papaveraceae)
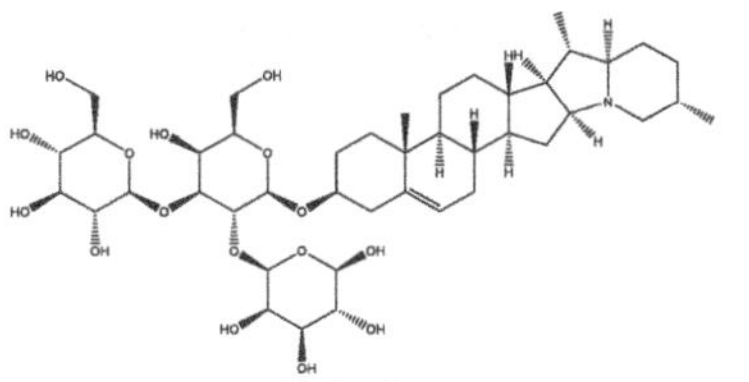 Solanine	 *Solanum melongena* (Family: Solanaceae)

(Continued)

TABLE 2.1 (Continued)
Some Plant-Derived Alkaloids Based on Their Occurrences and Sources

Alkaloid	Plant Source (Family)
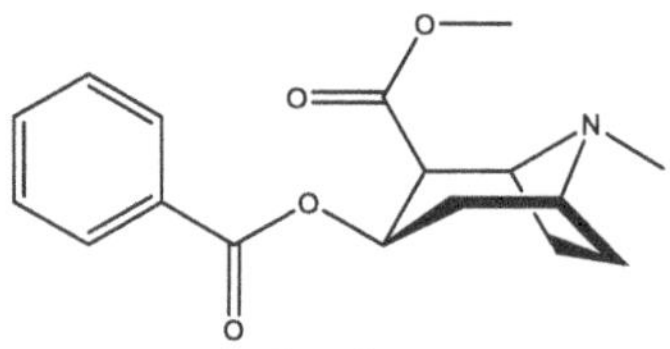 Cocaine	 *Erythroxylum coca* (Family: Erythroxylaceae)
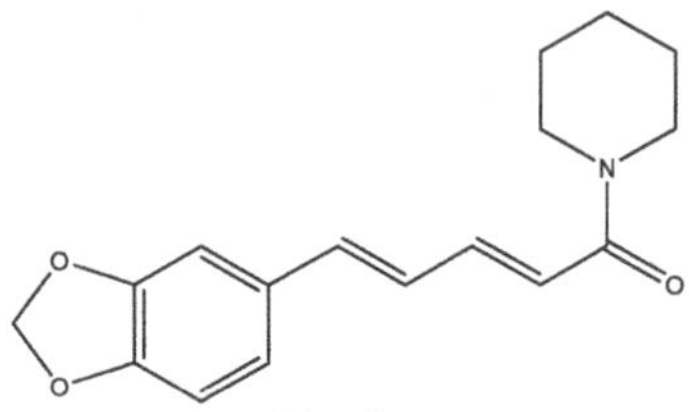 Piperine	 *Piper nigrum* (Family: Piperaceae)
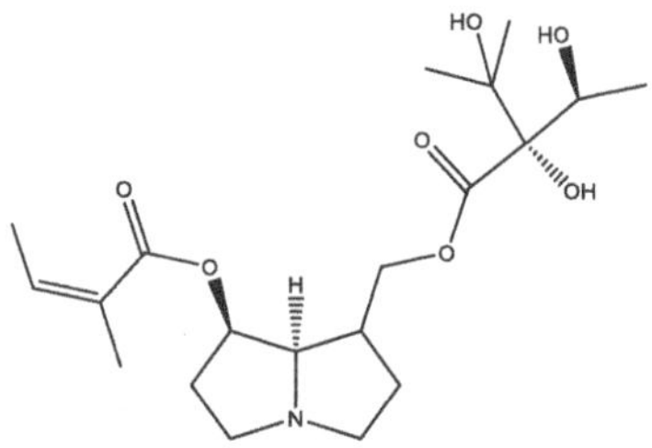 Echimidine	 *Echium hypertropicum* (Family: Boraginaceae)
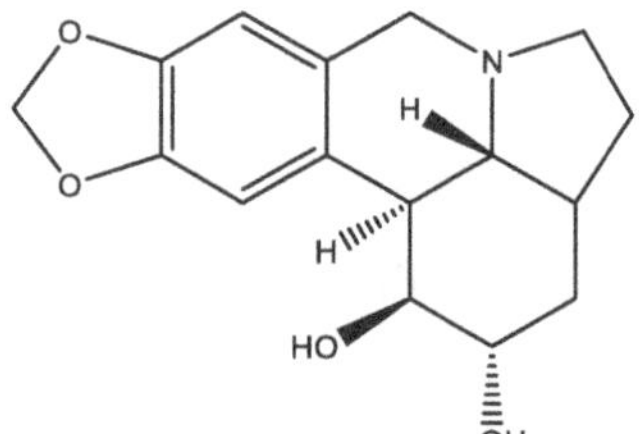 Lycorine	 *Amaryllis belladonna* (Family: Amaryllidaceae)
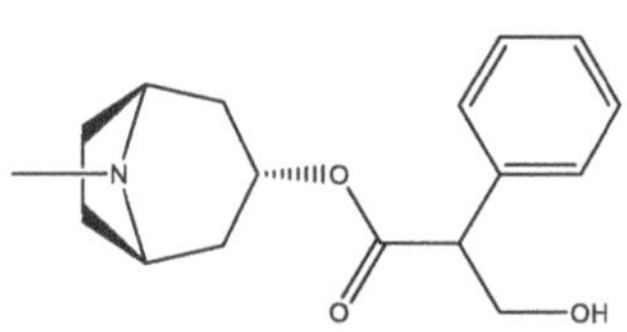 Atropine	 *Atropa belladonna* (Family: Amaryllidaceae)

(Continued)

TABLE 2.1 (Continued)
Some Plant-Derived Alkaloids Based on Their Occurrences and Sources

Alkaloid	Plant Source (Family)
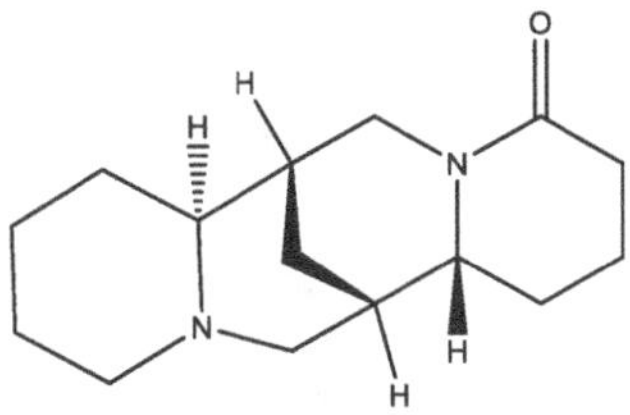 Lupanine	 *Lupinus angustifolius* (Family: Fabaceae)
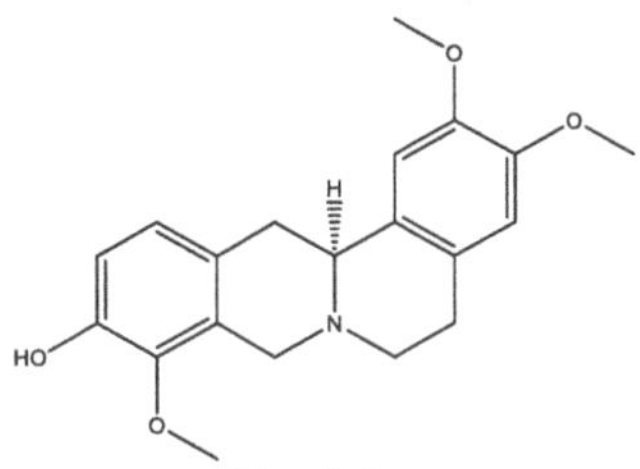 Vincristine	 *Vinca rosea* (Family: Apocynaceae)

2.6 MARINE-DERIVED ALKALOIDS

It has been discovered that marine species are abundant in special alkaloids. There have been reports of several interesting classes of marine alkaloids, such as manzamine and bromotyrosine-derived alkaloids (Fan et al., 2008). In the past few years, the lamellarins and related pyrrole-derived alkaloids (lukianols, polycitrins, polycitones, storniamides, didemnimides, ningalins, and purpurone) have been considered a group of marine alkaloids with numerous biological functions (Fan et al., 2008). The number of reported marine alkaloids is still growing at an accelerating rate. Consequently, the chemistry and pharmacology of these interesting compounds have been studied in several earlier studies (Bailly, 2004; Cironi et al., 2005; Yang et al., 2005). Additionally, much interest has been in synthesizing lamellarins and similar pyrrole-derived alkaloids.

In 1985, Andersen and his team identified the first lamellarin, a class of pyrrole alkaloids generated by DOPA-(2-amino-3-(3′,4′-dihydroxyphenyl)propionic acid) from the prosobranch mollusk *Lamella* sp (Andersen et al., 1985). Since then, there have been reports of over 70 distinct lamellarins and associated naturally occurring alkaloids originating from pyrrole. The structural subunit shared by the majority of lamellarins, lukianols, polycitrins, polycitones, storniamides, pinnatoxins, and ningalins is a 3,4-diarylated pyrrole 2-carboxylic acid ester or amide moiety. They have been isolated from many marine creatures and places, including sponges, ascidians (tunicates), and marine prosobranch mollusks (Fan et al., 2008). The variety of the generating species suggests a possible microbial connection to their production. Multiple biological functions, including cytotoxicity and antitumor activity, the reverse of multidrug resistance (MDR), suppression of HIV-1 integrase, antibiotic activity, inhibition of human aldose reductase (h-ALR2), resistance to cell division, immunomodulation, antioxidant activity, and feeding deterrence, have been demonstrated by lamellarins and related pyrrole-derived alkaloids (Fan et al., 2008).

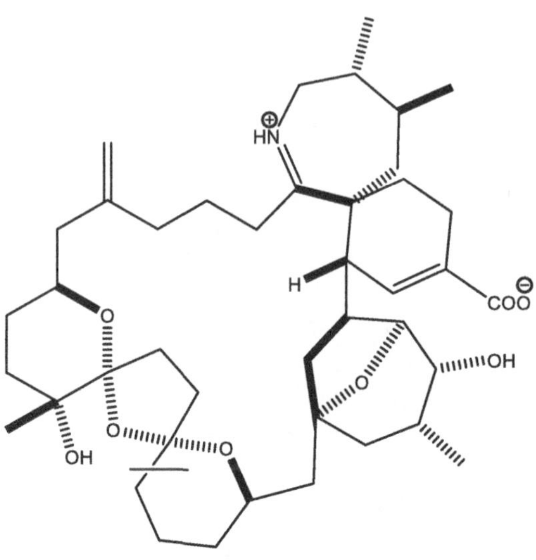

Pinnatoxin A

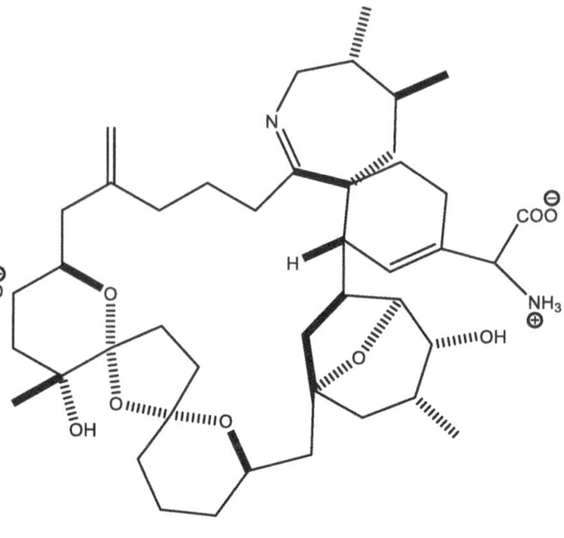

34*S* isomer Pinnatoxin B
34*R* isomer Pinnatoxin C

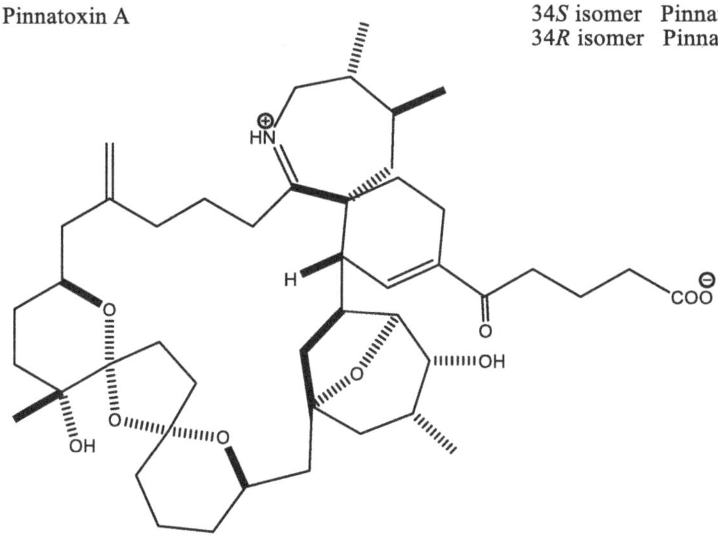

Pinnatoxin D

FIGURE 2.14 Chemical structures of pinnatoxins A, B, C, and D.

The genus *Pinna* of shellfish is primarily found in shallow seas in the Indian and Pacific Oceans' subtropical and tropical areas (Kuramoto et al., 2004). A common cause of food-related illness is consuming this bivalve's adductor muscle in China and Japan (Zheng et al., 1990). A poisonous extract from *P. attenuata*, pinnatoxin, is a Ca^{2+} channel-related stimulator (Zheng et al., 1990). Pinnatoxin A and a combination mixture including pinnatoxin B, pinnatoxin C, and pinnatoxin D (Figure 2.14) were derived from *P. muricata*. The pinnatoxin alkaloids comprise a 20-membered ring, which includes moieties of the 6,7-azaspiro, 5,6-bicyclo, and 6,5,6-triketal rings. A carboxylate anion and either an ammonium or iminium cation are specifically found in pinnaic acids (Kuramoto et al., 2004).

A further instance of marine alkaloids is halichlorine, originating from the sea sponge *Halichondria okadai* Kadota (Kuramoto et al., 1996). Halichlorine comprises a [5.6]-spiro ring

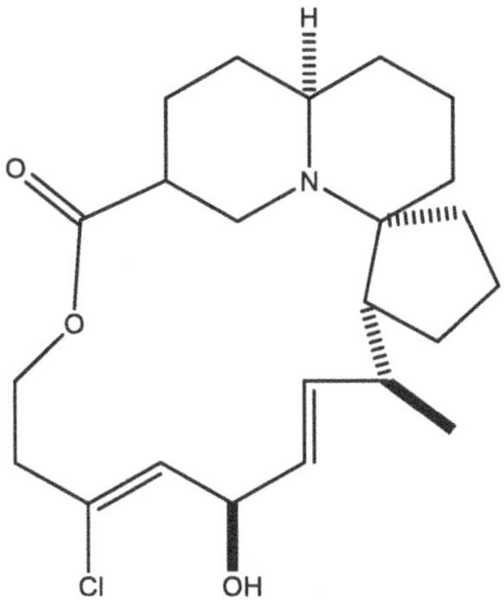

Halichlorine

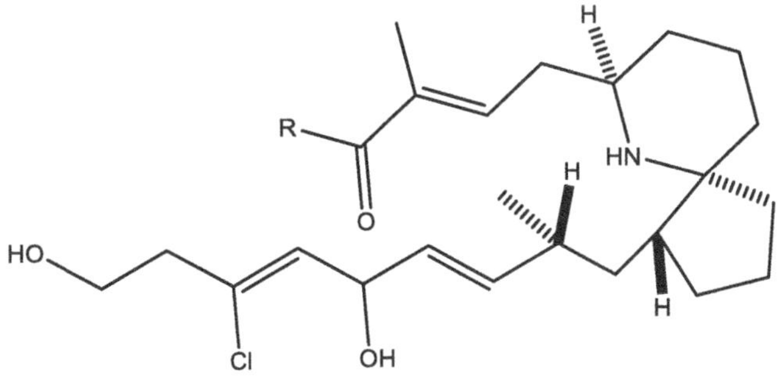

R = OH Pinnaic acid
R = $NHCH_2CH_2SO_3H$ Tauropinnaic acid

FIGURE 2.15 Chemical structures of halichlorine and pinnaic acids.

moiety, an azabicyclo [4.4.0] ring, and a chemically hindered 15-membered lactone. Pinnaic acids (Figure 2.15) were also obtained from *Pinna muricata* viscera (Kim et al., 1990). Taurine and a 6-azaspiro [4.5]decane molecule are the two parts that make up tauropinnaic acid (Figure 2.15).

2.7 FUNGAL-DERIVED ALKALOIDS

Many recent research efforts on identifying alkaloids from fungi have concentrated on their potential as cytotoxic agents for creating novel disease treatments. As of the present time, the following structural categories appear through the comparatively few studies that reveal various biological functions for fungal-derived alkaloids: Purine alkaloids, pyrrolidine, pyrrolizidine, indole, quinazoline, quinoline, and diketopiperazine. Out of the different fungal extracts, approximately 35 distinct substances that are fungal-derived alkaloids were extracted and identified (Willems et al., 2020). Additionally, numerous alkaloids with superior biological characteristics, including antibacterial, insecticidal, cytotoxic, and anticancer activity, have been identified recently from endophytic fungi (Zhang et al., 2012).

2.8 CONCLUSION

Alkaloids represent one of the most significant classes of naturally occurring compounds, distinguished by their nitrogen content, structural diversity, and wide distribution across living organisms. Their complex biosynthetic origins, involving multiple pathways such as the shikimate, ornithine, lysine, and terpenoid routes, reflect the biochemical versatility that underpins their wide range of biological activities. Historically, plant-derived alkaloids formed the foundation of traditional medicine, valued for their therapeutic properties in treating ailments from coughs to digestive and nervous system disorders. In modern pharmacology, these compounds continue to demonstrate immense value, serving not only as direct therapeutic agents but also as structural templates for drug development. Examples such as codeine, morphine, quinine, brucine, and ephedrine underscore the enduring relevance of alkaloids in addressing critical healthcare needs. As research advances, the exploration of both known and novel alkaloids holds great promise for the discovery of innovative treatments, ensuring that these natural compounds remain central to the evolution of medicine.

COMPETING INTERESTS

The authors declare that they have no competing interest.

DECLARATION OF GENERATIVE AI AND AI-ASSISTED TECHNOLOGIES IN THE WRITING PROCESS

During the preparation of this work, the book chapter author(s) have not taken any assistance from AI, and the book chapter author(s) reviewed and edited the content as needed and take(s) full responsibility for the content of the publication.

REFERENCES

Abate, L., Tadesse, M. G., Bachheti, A., & Bachheti, R. K. (2022). Traditional and phytochemical bases of herbs, shrubs, climbers, and trees from Ethiopia for their anticancer response. *BioMed Research International, 2022*. https://doi.org/10.1155/2022/1589877

Adewusi, E. A., & Afolayan, A. J. (2010). A review of natural products with hepatoprotective activity. *Journal of Medicinal Plants Research, 4*(13), 1318–1334.

Ajungla, L., Patil, P. P., Barmukh, R. B., & Nikam, T. D. (2009). Influence of biotic and abiotic elicitors on accumulation of hyoscyamine and scopolamine in root cultures of Datura metel L. *Indian Journal of Biotechnology, 8*(3), 317–322.

Amirkia, V., & Heinrich, M. (2014). Alkaloids as drug leads--A predictive structural and biodiversity-based analysis. *Phytochemistry Letters, 10*, xlviii–liii.

Andersen, R. J., Faulkner, D. J., He, C. H., Van Duyne, G. D., & Clardy, J. (1985). Metabolites of the marine prosobranch mollusk Lamellaria sp. *Journal of the American Chemical Society, 107*(19), 5492–5495.

Aniszewski, T. (2015). *Alkaloids: Chemistry, biology, ecology, and applications*. Elsevier.

Asfaw, T. B., Esho, T. B., Bachheti, A., Bachheti, R. K., Pandey, D. P., & Husen, A. (2022). Exploring important herbs, shrubs, and trees for their traditional knowledge, chemical derivatives, and potential benefits. In Tilahun Belayneh Asfaw, Tarekegn Berhanu Esho, Archana Bachheti, Rakesh Kumar Bachheti, D. P. Pandey, & Azamal Husen (eds) *Herbs, shrubs, and trees of potential medicinal benefits* (pp. 1–26). CRC Press.

Bailly, C. (2004). Lamellarins, from A to Z: A family of anticancer marine pyrrole alkaloids. *Current Medicinal Chemistry. Anti-Cancer Agents, 4*(4), 363–378.

Baros, S., Karsayová, M., Jomová, K., Gáspár, A., & Valko, M. (2012). Free radical scavenging capacity of Papaver somniferum L. and determination of pharmacologically active alkaloids using capillary electrophoresis. *The Journal of Microbiology, Biotechnology and Food Sciences, 1*, 725.

Bauer, I., & Knölker, H.-J. (2012). Synthesis of pyrrole and carbazole alkaloids. In H. J. Knölker (ed) *Alkaloid Synthesis. Topics in Current Chemistry* (vol. 309). Springer. https://doi.org/10.1007/128_2011_192

Beyer, J., Drummer, O. H., & Maurer, H. H. (2009). Analysis of toxic alkaloids in body samples. *Forensic Science International, 185*(1–3), 1–9.

Bhadra, K., & Kumar, G. S. (2011). Therapeutic potential of nucleic acid-binding isoquinoline alkaloids: Binding aspects and implications for drug design. *Medicinal Research Reviews, 31*(6), 821–862.

Böttger, A., Vothknecht, U., Bolle, C., & Wolf, A. (2018). *Lessons on caffeine, cannabis & Co* (pp. 171–178). Springer.

Boulaaba, M., Medini, F., Hajlaoui, H., Mkadmini, K., Falleh, H., Ksouri, R., Isoda, H., Smaoui, A., & Abdelly, C. (2019). Biological activities and phytochemical analysis of phenolic extracts from Salsola kali L. Role of endogenous factors in the selection of the best plant extracts. *South African Journal of Botany, 123*, 193–199.

Braekman, J. C., Daloze, D., & Pasteels, J. M. (1998). Alkaloids in animals. In Margaret F. Roberts & Michael Wink (eds) *Alkaloids: Biochemistry, ecology, and medicinal applications* (pp. 349–378). Springer.

Bribi, N. (2018). Pharmacological activity of alkaloids: A review. *Asian Journal of Botany, 1*(1), 1–6.

Buckingham, J., Baggaley, K. H., Roberts, A. D., & Szabo, L. F. (2010). *Dictionary of Alkaloids with CD-ROM*. CRC Press.

Bunsupa, S., Yamazaki, M., & Saito, K. (2012). Quinolizidine alkaloid biosynthesis: Recent advances and future prospects. *Frontiers in Plant Science, 3*, 239.

Byler, K. G., Wang, C., & Setzer, W. N. (2009). Quinoline alkaloids as intercalative topoisomerase inhibitors. *Journal of Molecular Modeling, 15*, 1417–1426.

Cai, X.-H., Li, Y., Su, J., Liu, Y.-P., Li, X.-N., & Luo, X.-D. (2011). Novel indole and quinoline alkaloids from Melodinus yunnanensis. *Natural Products and Bioprospecting*, *1*, 25–28.

Carvalho, J. C. B., dos Santos Almeida, H., Lobo, J. F. R., Ferreira, J. L. P., Oliveira, A. P., & Rocha, L. (2013). Pyrrolizidine alkaloids in two endemic capeverdian Echium species. *Biochemical Systematics and Ecology*, *50*, 1–6.

Cironi, P., Albericio, F., & Alvarez, M. (2005). Lamellarins: Isolation, activity and synthesis. In W. Gribble Gordon, & A. Joule John (eds) *Progress in heterocyclic chemistry* (vol. 16, pp. 1–26). Elsevier.

Cushnie, T. P. T., Cushnie, B., & Lamb, A. J. (2014). Alkaloids: An overview of their antibacterial, antibiotic-enhancing and antivirulence activities. *International Journal of Antimicrobial Agents*, *44*(5), 377–386.

de Almeida, A. C., de Faria, F. M., Dunder, R. J., Manzo, L. P. B., Souza-Brito, A. R. M., & Luiz-Ferreira, A. (2017). Recent trends in pharmacological activity of alkaloids in animal colitis: Potential use for inflammatory bowel disease. *Evidence-based Complementary and Alternative Medicine*, *2017*(1), 8528210.

Debnath, B., Singh, W. S., Das, M., Goswami, S., Singh, M. K., Maiti, D., & Manna, K. (2018). Role of plant alkaloids on human health: A review of biological activities. *Materials Today Chemistry*, *9*, 56–72.

Diamond, A., & Desgagné-Penix, I. (2016). Metabolic engineering for the production of plant isoquinoline alkaloids. *Plant Biotechnology Journal*, *14*(6), 1319–1328.

Estévez, V., Villacampa, M., & Menéndez, J. C. (2014). Recent advances in the synthesis of pyrroles by multi-component reactions. *Chemical Society Reviews*, *43*(13), 4633–4657.

Evans, L. W., & Ferguson, B. S. (2018). Food bioactive HDAC inhibitors in the epigenetic regulation of heart failure. *Nutrients*, *10*(8), 1120.

Facchini, V., & Griffiths, L. A. (1981). The involvement of the gastro-intestinal microflora in nitro-compound-induced methaemoglobinaemia in rats and its relationship to nitrogroup reduction. *Biochemical Pharmacology*, *30*(9), 931–935.

Fan, H., Peng, J., Hamann, M. T., & Hu, J.-F. (2008). Lamellarins and related pyrrole-derived alkaloids from marine organisms. *Chemical Reviews*, *108*(1), 264–287.

Goel, P., Alam, O., Naim, M. J., Nawaz, F., Iqbal, M., & Alam, M. I. (2018). Recent advancement of piperidine moiety in treatment of cancer-A review. *European Journal of Medicinal Chemistry*, *157*, 480–502.

Goyal, S. (2013). Ecological role of alkaloids. In Kishan Gopal Ramawat, & Jean-Michel Mérillon (eds) *Natural products: Phytochemistry, botany and metabolism of alkaloids, phenolics and terpenes* (pp. 149–171). Springer.

Guirimand, G., Courdavault, V., St-Pierre, B., & Burlat, V. (2010). Biosynthesis and regulation of alkaloids. *Plant Developmental Biology-Biotechnological Perspectives*, *2*, 139–160.

Hamid, H. A., Ramli, A. N. M., & Yusoff, M. M. (2017). Indole alkaloids from plants as potential leads for antidepressant drugs: A mini review. *Frontiers in Pharmacology*, *8*, 96.

Henning, C. P., Ringuelet, J. A., & Viña, S. Z. (2013). *Compuestos secundarios nitrogenados: Alcaloides*. Libros de Cátedra.

Hussain, G., Rasul, A., Anwar, H., Aziz, N., Razzaq, A., Wei, W., Ali, M., Li, J., & Li, X. (2018). Role of plant derived alkaloids and their mechanism in neurodegenerative disorders. *International Journal of Biological Sciences*, *14*(3), 341.

Jansen, G., Jürgens, H.-U., Schliephake, E., & Ordon, F. (2012). Effect of the soil pH on the alkaloid content of Lupinus angustifolius. *International Journal of Agronomy*, *2012*(1), 269878.

Jayakumar, K., & Murugan, K. (2016). Solanum alkaloids and their pharmaceutical roles: A review. *Journal of Analytical & Pharmaceutical Research*, *3*(6), 75.

Jing, H., Liu, J., Liu, H., & Xin, H. (2014). Histochemical investigation and kinds of alkaloids in leaves of different developmental stages in Thymus quinquecostatus. *The Scientific World Journal*, *2014*(1), 839548.

Jirschitzka, J., Schmidt, G. W., Reichelt, M., Schneider, B., Gershenzon, J., & D'Auria, J. C. (2012). Plant tropane alkaloid biosynthesis evolved independently in the Solanaceae and Erythroxylaceae. *Proceedings of the National Academy of Sciences*, *109*(26), 10304–10309.

Kaur, R., & Arora, S. (2015). Alkaloids-important therapeutic secondary metabolites of plant origin. *Journal of Critical Reviews*, *2*(3), 1–8.

Kaur, R., Matta, T., & Kaur, H. (2019). Plant derived alkaloids. *Saudi Journal Life Science*, *2*(5), 158–189.

Kim, D. K., Kudo, I., Fujimori, Y., Mizushima, H., Masuda, M., Kikuchi, R., Ikizawa, K., & Inoue, K. (1990). Detection and subcellular localization of rabbit platelet phospholipase A2 which preferentially hydrolyzes an arachidonoyl residue. *The Journal of Biochemistry*, *108*(6), 903–906.

Koetz, M., Santos, T. G., Rayane, M., & Henriques, A. T. (2017). Quantification of atropine in leaves of atropa belladonna: Development and validation of method by high-perfomance liquid chromatography (HPLC). *Drug Analytical Research*, *1*(1), 44–49.

Koleva, I. I., van Beek, T. A., Soffers, A. E. M. F., Dusemund, B., & Rietjens, I. M. C. M. (2012). Alkaloids in the human food chain--natural occurrence and possible adverse effects. *Molecular Nutrition & Food Research*, *56*(1), 30–52.

Kuete, V. (2014). Health effects of alkaloids from African medicinal plants. In *Toxicological survey of African medicinal plants* (pp. 611–633). Elsevier.

Kukula-Koch, W. A., & Widelski, J. (2017). Alkaloids. In *Pharmacognosy* (pp. 163–198). Elsevier.

Kuramoto, M., Arimoto, H., & Uemura, D. (2004). Bioactive alkaloids from the sea: A review. *Marine Drugs*, *2*(1), 39–54.

Kuramoto, M., Tong, C., Yamada, K., Chiba, T., Hayashi, Y., & Uemura, D. (1996). Halichlorine, an inhibitor of VCAM-1 induction from the marine sponge Halichondria okadai Kadata. *Tetrahedron Letters*, *37*(22), 3867–3870.

Kutchan, T. M. (1995). Alkaloid biosynthesis [mdash] the basis for metabolic engineering of medicinal plants. *The Plant Cell*, *7*(7), 1059.

Li, S., Lei, Y., Jia, Y., Li, N., Wink, M., & Ma, Y. (2011). Piperine, a piperidine alkaloid from Piper nigrum re-sensitizes P-gp, MRP1 and BCRP dependent multidrug resistant cancer cells. *Phytomedicine*, *19*(1), 83–87.

Liu, H.-L., Huang, X.-Y., Dong, M.-L., Xin, G.-R., & Guo, Y.-W. (2010). Piperidine alkaloids from Chinese mangrove Sonneratia hainanensis. *Planta Medica*, *76*(09), 920–922.

Lu, J.-J., Bao, J.-L., Chen, X.-P., Huang, M., & Wang, Y.-T. (2012). Alkaloids isolated from natural herbs as the anticancer agents. *Evidence-Based Complementary and Alternative Medicine*, *2012*(1), 485042.

Mao, Z., Huang, S., Gao, L., Wang, A., & Huang, P. (2014). A novel and versatile method for the enantioselective syntheses of tropane alkaloids. *Science China Chemistry*, *57*, 252–264.

Marella, A., Tanwar, O. P., Saha, R., Ali, M. R., Srivastava, S., Akhter, M., Shaquiquzzaman, M., & Alam, M. M. (2013). Quinoline: A versatile heterocyclic. *Saudi Pharmaceutical Journal*, *21*(1), 1–12.

Mazid, M., Khan, T. A., & Mohammad, F. (2011). Role of secondary metabolites in defense mechanisms of plants. *Biology and Medicine*, 3(2), 232–249.

Mazumder, P. M., Das, S., Das, M. K., et al. (2011). Phyto-pharmacology of Berberis aristata DC: A review. *Journal of Drug Delivery and Therapeutics*, *1*(2), 46–50.

Meira, M., da Silva, E. P., David, J. M., & David, J. P. (2012). Review of the genus Ipomoea: Traditional uses, chemistry and biological activities. *Revista Brasileira de Farmacognosia*, *22*, 682–713.

Moreira, H., Slezak, A., Szyjka, A., Oszmianski, J., & Gasiorowski, K. (2017). Antioxidant and cancer chemopreventive activities of cistus and pomegranate polyphenols. *Acta Poloniae Pharmaceutica*, *74*(2), 688–698.

Muñoz, I. J., Schilman, P. E., & Barrozo, R. B. (2020). Impact of alkaloids in food consumption, metabolism and survival in a blood-sucking insect. *Scientific Reports*, *10*(1), 9443.

Muthna, D., Cmielova, J., Tomsik, P., & Rezacova, M. (2013). Boldine and related aporphines: From antioxidant to antiproliferative properties. *Natural Product Communications*, *8*(12), 1934578X1300801235.

Ncube, B., Nair, J. J., Rárová, L., Strnad, M., Finnie, J. F., & Van Staden, J. (2015). Seasonal pharmacological properties and alkaloid content in Cyrtanthus contractus NE Br. *South African Journal of Botany*, *97*, 69–76.

Noriega, P., Sola, M., Barukcic, A., Garcia, K., & Osorio, E. (2015). Cosmetic antioxidant potential of extracts from species of the Cinchona pubescens (Vahl). *International Journal of Phytocosmetics and Natural Ingredients*, *2*(1), 14.

Oliveira, S. L., da Silva, M. S., Tavares, J. F., Sena-Filho, J. G., Lucena, H. F. S., Romero, M. A. V., & Barbosa-Filho, J. M. (2010). Tropane alkaloids from Erythroxylum genus: Distribution and compilation of 13C-NMR spectral data. *Chemistry & Biodiversity*, *7*(2), 302–326.

Pan, Q., Mustafa, N. R., Tang, K., Choi, Y. H., & Verpoorte, R. (2016). Monoterpenoid indole alkaloids biosynthesis and its regulation in Catharanthus roseus: A literature review from genes to metabolites. *Phytochemistry Reviews*, *15*, 221–250.

Pereira, A. G., Cassani, L., Garcia-Oliveira, P., Otero, P., Mansoor, S., Echave, J., Xiao, J., Simal-Gándara, J., & Prieto, M. A. (2023). Plant alkaloids: Production, extraction, and potential therapeutic properties. In *Natural Secondary Metabolites: From Nature, Through Science, to Industry* (pp. 157–200). Springer.

Porwal, M., & Kumar, A. (2015). Neuroprotective effect of Annona squamosa & (-) Anonaine in decreased GABA receptor of epileptic rats. *Journal of Applied Pharmaceutical Science*, *5*(1), 18–23.

Ranjitha, D., & Sudha, K. (2015). Alkaloids in foods. *International Journal of Pharmaceutical, Chemical & Biological Sciences*, 5(4), 896–906.

Roberts, M. F., & Wink, M. (1998). *Alkaloids: Biochemistry, ecology, and medicinal applications*. Plenum Press.

Sagi, S., Avula, B., Wang, Y.-H., & Khan, I. A. (2016). Quantification and characterization of alkaloids from roots of Rauwolfia serpentina using ultra-high performance liquid chromatography-photo diode array-mass spectrometry. *Analytical and Bioanalytical Chemistry*, *408*, 177–190.

Schramm, S., Köhler, N., & Rozhon, W. (2019). Pyrrolizidine alkaloids: Biosynthesis, biological activities and occurrence in crop plants. *Molecules*, *24*(3), 498.

Silva, V. G., Silva, R. O., Damasceno, S. R. B., Carvalho, N. S., Prudêncio, R. S., Aragão, K. S., Guimarães, M. A., Campos, S. A., Véras, L. M. C., Godejohann, M., et al. (2013). Anti-inflammatory and antinociceptive activity of epiisopiloturine, an imidazole alkaloid isolated from Pilocarpus microphyllus. *Journal of Natural Products*, *76*(6), 1071–1077.

Szőke, É., Lemberkovics, É., & Kursinszki, L. (2013). *Alkaloids derived from lysine: Piperidine alkaloids*. Springer.

Talapatra, S. K., & Talapatra, B. (2015). *Chemistry of plant natural products*. Springer.

Teles, M. M. R. S., Pinheiro, A. A. V., Dias, C. D. S., Tavares, J. F., Barbosa Filho, J. M., & Da Cunha, E. V. L. (2019). Alkaloids of the lauraceae. *The Alkaloids. Chemistry and Biology*, *82*, 147–304.

Uzor, P. F. (2020). Alkaloids from plants with antimalarial activity: A review of recent studies. *Evidence-based Complementary and Alternative Medicine*, *2020*(1), 8749083.

Vrabec, R., Blunden, G., & Cahlíková, L. (2023). Natural alkaloids as multi-target compounds towards factors implicated in Alzheimer's disease. *International Journal of Molecular Sciences*, *24*(5), 4399.

Wang, F.-P., Chen, Q.-H., & Liu, X.-Y. (2010). Diterpenoid alkaloids. *Natural Product Reports*, *27*(4), 529–570.

Wansi, J. D., Devkota, K. P., Tshikalange, E., & Kuete, V. (2013). Alkaloids from the medicinal plants of Africa. In *Medicinal plant research in Africa* (pp. 557–605). Elsevier.

Willems, T., De Mol, M. L., De Bruycker, A., De Maeseneire, S. L., & Soetaert, W. K. (2020). Alkaloids from marine fungi: Promising antimicrobials. *Antibiotics*, *9*(6), 340.

Wink, M. (2010). *Annual plant reviews, functions and biotechnology of plant secondary metabolites*. John Wiley & Sons.

Yang, G., Wang, A.-L., Chen, H.-L., & You, Y.-C. (2005). Recent research progresses on lamellarins and analogous alkaloids of bioactive marine mollusc. *Chinese Journal of Organic Chemistry*, *25*(6), 641.

Zhang, Q.-W., Lin, L.-G., & Ye, W.-C. (2018). Techniques for extraction and isolation of natural products: A comprehensive review. *Chinese Medicine*, *13*, 1–26.

Zhang, Y., Han, T., Ming, Q., Wu, L., Rahman, K., & Qin, L. (2012). Alkaloids produced by endophytic fungi: A review. *Natural Product Communications*, *7*(7), 1934578X1200700742.

Zheng, S., Huang, F., Chen, S., Tan, X., Zuo, J., Peng, J., & Xie, R. (1990). The isolation and bioactivities of pinnatoxin. *Chinese Journal of Marine Drugs/Zhongguo Haiyang Yaowu. Qingdao*, *9*(1), 33–35.

3 Biosynthesis and Regulation of Alkaloids in Medicinal Plants

Monica Butnariu

3.1 INTRODUCTION

The plants that, through their content in various substances, cause transient or fatal illnesses to humans and animals are called toxic or poisonous plants. It seems interesting to point out that some of the harmful species are food or fodder plants. They are rarely found in mushrooms [*Claviceps purpurea* (Fr.) Tul., *Psilocybe mexicana* Heim.], in *Pteridophytes*, they are very rare (only in some species of the families *Lycopodiaceae* and *Equisetaceae*), and in *Gymnospermae*, they are widespread in species of *Taxaceae* and *Ephedraceae*. Between 10% and 20% of *Angiosperms* synthesize alkaloids (ALKs), with these being frequently found in dicotyledons, especially from the *Apocynaceae, Berberidaceae, Erytroxylaceae, Lauraceae, Loganiaceae, Papaveraceae, Rubiaceae, Rutaceae, Ranunculaceae*, and *Solanaceae* families; among monocots, only the *Liliaceae* and *Amarylidaceae* families produce ALKs. A number of ALKs are characteristic of a single species (morphine [$C_{17}H_{19}NO_3$]), a small number of genera in a family (hyoscyamine [$C_{17}H_{23}NO_3$]), or a number of species belonging to a certain genus (thebaine [$C_{19}H_{21}NO_3$]) (Huo et al., 2021). Other ALKs are found in genera or species belonging to a range of taxonomically distant families (caffeine [$C_8H_{10}N_4O_2$], berberine [$C_{20}H_{18}NO_4^+$], nicotine [$C_{10}H_{14}N_2$]) (Andriamaharavo et al., 2015). In general, in producing plants, one or more ALKs predominate quantitatively and are referred to as main ALKs. As a rule, all ALKs in a plant have the same biogenetic precursor, whether they are primary or secondary, and only apparently, their structures differ. In *Papaver* species, morphinan (phenanthrene [$C_{14}H_{10}$]) ALKs are found alongside benzylisoquinoline ones, in *Cinchona*, quinoline, and indole ALKs.

3.2 HISTORY, DISTRIBUTION, AND NOMENCLATURE

The term ALK was introduced by the pharmacist Meisner in 1819, on the occasion of the extraction of veratrine [$C_{36}H_{51}NO_{11}$] from *Veratrum album*, and shows the special interest given to the basic (alkaline) character of these substances (Gaillard & Pepin, 2001). The name ALK is historical and not rational, but it is maintained because it is helpful from a practical point of view. The first ALK, later known as narcotine, was isolated by Derasne (1803), as quoted by Buchanan et al. (2024). This was followed by morphine (1806), isolated by Sertürner, also as quoted by Buchanan et al. (2024). The isolation of other ALKs, especially by Pelletier and Caventon, soon followed, for example (according to Figure 3.1).

Coniine was the first ALK whose chemical structure was established by Schiff, in 1870, as quoted by O'Callaghan (2001). The ALKs have a relatively narrow distribution.

 DOI: 10.1201/9781003539179-3

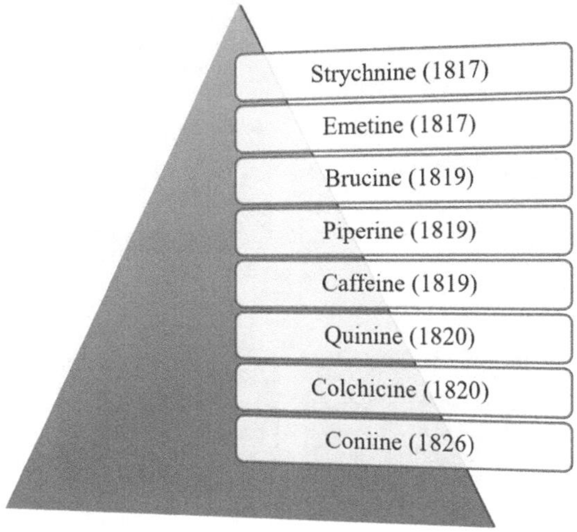

FIGURE 3.1 Alkaloids and the year of isolation.

Bacteria, algae, fungi, lichens, and mosses are absent, with the exception of ALKs from the *C. purpurea* mushroom. In pteridophytes and gymnosperms, they appear sporadically in the families *Equisetaceae, Lycopodiaceae, Ephedraceae, and Taxaceae*. Among the monocots, the families *Liliaceae* and *Amaryllidaceae* are rich in ALKs. They are more widespread in dicotyledons, with the richest families including *Raminculaceae, Berberidaceae, Solanaceae, Rubiaceae, Leguminosae, Asclepiadaceae, Apocynaceae, Loganiaceae*, etc. The families *Labiatae, Rosaceae*, and *Compositae*, the richest in species, are practically devoid of ALKs (O'Callaghan, 2001). Regarding the spread of ALKs taken separately, some, especially those with a complex structure, have been found so far only in one family. An example is strychnine [$C_{21}H_{22}N_2O_2$], which has been isolated to date only from species of the genus *Strychnos* (Philippe et al., 2004). Instead, other ALKs with a simpler structure are found in different families with no phylogenetic relationship. An example is nicotine [$C_{10}H_{14}N_2$], which, apart from the *Solanaceae* family, has been found in various species of *Sedum, Asclepios, Equisetum, and Lycopodium*. The location of ALKs in the plant varies: sometimes they are spread in all the organs of the plant, as in *Solanaceae*, sometimes they are located in certain organs, as in the seeds of *Strychnos nux vomica*, in fruits of *Papaver somniferum*, in leaves of *Erytroxilon coca*, in tubers of *Aconitum tauricum*, in rhizomes and roots of *V. album*, in bark of *Cinchona officinalis*. Usually, ALKs are found in plants grouped in substances with N with a similar structure. It is rare for a plant to contain only one ALK (ricin from *Ricinus communis*) (Chowański et al. 2016). Usually, from the groups of ALK substances, an ALK called "main alkaloid" predominates in the plant, and it is accompanied by ALKs called "secondary alkaloids" which differ by the number or position of the substituents (OH or CH_3), the number or position of double bonds, etc. Such a family of ALKs has the same biogenetic origin. Sometimes, two or more types of ALKs, with different structures and different biogenetic origins, are found in the same plant. Due to the numerous secondary ALKs that accompany the main ALK, the number known so far is over 2000, despite their relatively limited distribution. ALKs are not rationally named after their chemical constitution because, in the past and even today, their chemical constitution was determined later, after they were isolated (Marciniak et al., 2019). That is why ALKs are named with names that are formed like this (according to Figure 3.2).

The names of secondary ALKs are usually formed after the name of the main ALK to which a prefix or suffix is added, such as quinine, quinidine, and hydroquinine. The isomeric ALKs are named with prefixes such as pseudo-, iso-, neo-, alo-, and nor- (Liu et al., 2021).

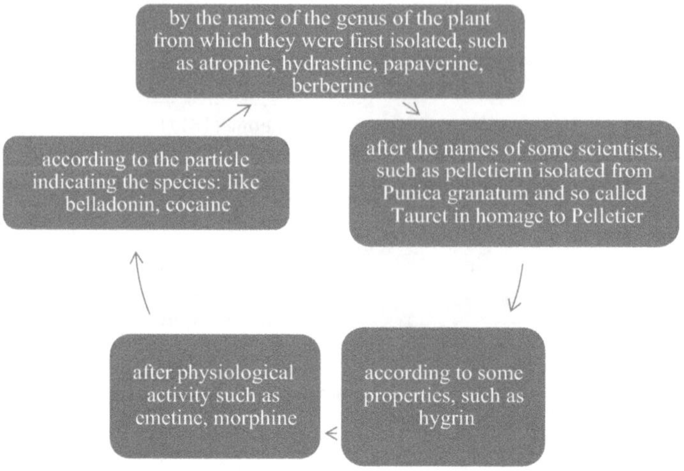

FIGURE 3.2 Methods of naming alkaloids.

3.3 PHYSICAL AND CHEMICAL PROPERTIES OF ALKALOIDS

Most ALKs are crystalline substances, a small number are amorphous, and very few are liquid (nicotine, cocaine, scopolamine, sparteine, hygrine, arecoline). The ALKs are colorless substances, except for berberine, which is yellow, and sanquinarine salts, which are red.

Many ALKs are optically active and are found in plants as one of the antipodes or as a racemate. In general, ALKs are insoluble or sparingly soluble in water. They are soluble in nonpolar organic solvents, but they form salts with acids that are sparingly soluble in water and sparingly soluble in nonpolar solvents. This difference in solubility between the base ALKs and their salts is used for their isolation and purification (Konan et al., 2019).

The most important chemical property of ALKs is the basic character of the molecule. This basicity varies depending on the nitrogen atom that is incorporated in the molecule as a primary, secondary, tertiary, or quaternary amine. Due to their basicity, ALKs form salts with acids. They are mainly found in the plant in the form of salts of oxalic, malic, tannic, succinic, meconic, aconitic, chelidonic acids, etc. Although they have different structures, most ALKs behave similarly towards some reagents called "general reagents of alkaloids", with which they form colored complexes or precipitates. The more commonly used precipitation reagents are Dragendorff reagent (aqueous solution of potassium iodobismuthate), Bouchardat reagent (aqueous solution of potassium iodoiodurate), Mayer reagent (aqueous solution of potassium tetraiodomercurate), phosphomolybdenic acid, phosphotungstic acid, silicotungstic acid, as well as the salts of some heavy metals (mercuric chloride). The most commonly used color reagents are Fröhde reagent (ammonium molybdate in concentrated sulfuric acid), Mandelin reagent (ammonium vanadate in sulfuric acid), and Wenzel reagent (potassium permanganate solution) (Chen et al., 2024). Neither precipitation nor color reactions is specific enough to identify an ALK, but, used in conjunction with other methods, they are helpful.

3.4 EXTRACTION AND DOSAGE OF ALKALOIDS

The extraction of ALKs is carried out by two classic processes:

- the plant material is treated with alkaline aqueous solution, such as $Ca(OH)_2$, Na_2CO_3, NH_4OH, and then with an organic solvent, most often benzene or toluene; the extract is then stirred with acidic aqueous solution (organic or inorganic acid) when the ALKs in the form of salts pass into the water, and then they are precipitated by changing the pH to 9–10, separated and dried;

- plant material is extracted with diluted acid solution in water or hydroalcoholic solution; pigments or other impurities are removed by stirring the aqueous solution with chloroform or another non-polar solvent, after which the ALKs are precipitated by adding alkali to the aqueous solution and separated by filtration (Jones & Kinghorn, 2012).

The crude ALKs thus separated are purified by fractional precipitation, crystallization as salts at different pH levels, liquid–liquid partition, or column chromatography with various adsorbents. The methods of quantitative determination of ALKs are the following:

- gravimetric methods, which provide for their extraction and separation in the form of base, salt, or complex, which is dried or weighed;
- titrimetric methods, in aqueous or aprotic medium, based on the basic nature of ALKs;
- colorimetric methods, which are based on the color reaction they give with a reagent, the color absorption being determined spectrally;
- biological methods, based on the specific physiological properties of ALKs;
- polarographic, fluorimetric, potentiometric, chromatographic, etc. methods are less used (Dymek et al., 2021).

The science that studies the nature and properties of toxic substances, their action on living organisms, their identification and dosing methods, and the means to combat the effects produced on organisms is called toxicology. The notion of "poison", synonymous with "toxic", has been known since antiquity. Thus, according to Dioscorioles, the word "toxic", which was known to the ancient Greeks, derives from "toxon", which means bow and is related to the practice of using poisoned arrows. Plinius Secundus believed that the term "toxicus" comes from "*Taxus*" – the herbaceous plant whose seeds were used to poison arrows.

Other authors believe that the origin of the term "toxic" was the Egyptian word "tako", which means destruction or death. In the Romanian language, the synonym for the word "toxic" is the word "otrava", which derives from the Slavonic verb "otraviti" and means "to embitter" or "to cause grief". It is known that many substances, in small doses, have a toxic and even deadly action. Phytotoxicology is of particular importance, especially for herbivores and, in part, for omnivores and carnivores. The frequency of plant poisoning is still relatively high. Many plants can produce pathological conditions, but the designation as a toxic plant is relative, as toxicity depends on pedoclimatic conditions, season, meteorological conditions, soil fertilization practices, atmospheric pollution levels, and the stage of development (Zhang et al., 2023a).

Several criteria were considered in defining the plants that cause poisoning. Thus, among the toxic plants were included those that cause diseases when sensitive animals consume the fruits, seeds, aerial parts, or roots; plants that contain factors that limit or inhibit the optimal morphoproductive performance of animals; and plants that contain substances that interfere with reproductive function and, in certain circumstances, produce infertility. The toxicity of plants is also influenced by factors dependent on the body. Apart from the difference in receptivity, some animals do not have gustatory and olfactory sensitivity that would allow them to discern toxic plants. This category includes greedy animals, those who suffer from the perversion of taste. Toxic plants are also consumed if they are mixed with other plants whose taste and smell mask the organoleptic properties. The most frequent cases of plant poisoning occur in young animals, especially at the beginning of the grazing period (Takla et al., 2018).

In order to reduce the incidence of plant poisoning, it is necessary to avoid pastures where poisonings occurred in previous years, botanical analysis of pastures and remove toxic plants, taking out for grazing in the case of using new pastures for a few days of reform animals, avoiding taking hungry animals out to graze, knowing the phytosanitary actions that have been carried out on pastures and taking animals out to graze or fodder harvesting after the remanence period has passed, the lands near the enterprises that pollute the atmosphere should not be used for the cultivation of fodder plants.

3.5 THE SPREAD, BIOSYNTHESIS, AND BIOGENESIS OF ALKALOIDS

Plants producing ALKs are much more widespread; most of them belong to the dicotyledonous and, less often, monocotyledonous and cryptogamous plants. Over 1000 ALKs have been identified in plants, and it is estimated that only 2% of all known plant species have been researched regarding their ALK content. An ALK occurs in a single plant or a few species belonging to the same botanical genus. Only a few ALKs with a simple structure can be found in plants belonging to very different species. A plant rarely contains a single ALK. Usually, it contains several ALKs with a similar structure, as is the case with coca leaves, tobacco, or quinine bark. The ALK content of a plant varies a lot with the region, the climate, and the season: many tropical plants. However, they develop normally, lose the ability to synthesize ALKs when they are moved to regions with a temperate or cold climate (Rodés-Bachs & Van der Fels-Klerx, 2023). For example, seven ALKs have been found in ephedra, and more than 60 ALKs have been isolated from the anticancer drug periwinnia. Many plants do not contain ALKs, and from this fact it follows that they are not a vital factor, as are, for example, carbohydrates, amino acids, fats, and purines. Opinions regarding the physiological role of ALKs in plants are still not clear. The ALKs could be a means of defense for the plant against insects or higher animals. However, at the same time, they can be considered products of protein dissimilation that the plant, due to the lack of excretion possibilities, deposits in its organs.

There are several objections to the second option: it is unlikely that the plant would build huge and complicated molecules to fix a single nitrogen atom, or why, for an important physiological function of the excretion process, would it be carried out in such diverse ways in related botanical species, when other vitally important processes (carbon dioxide assimilation, glycolysis) are carried out in all plants through an identical mechanism (Jank & Rath, 2021).

To synthesize ALKs, plants use natural amino acids or their biochemical degradation products as raw materials. In the biosynthesis of ALKs, the Mannich reaction plays a big role, in which an amine ($R\text{-}NH_2$), an aldehyde ($R'\text{-}CH{=}O$), and a compound capable of taking part in an aldol condensation in the form of a methylene component participate. The ALKs are mostly located in certain organs of the plant, such as leaves, seeds, roots, or bark. In plants, ALKs are found as salts of common plant acids such as citric acid, malic acid, tartaric acid, and oxalic acid (Chirkin et al., 2015). According to the biosynthetic precursor, ALKs can be grouped according to Table 3.1.

To isolate the ALKs, the crushed plant or its aqueous extract is treated with a base (alkaline hydroxide or ammonia) to release the ALK from its salts. Then the solution is extracted with a solvent (ether, chloroform). In the case of volatile ALKs, such as nicotine, steam distillation is also used. The study of the composition of ALKs, their chemical structure and activity, as well as their

TABLE 3.1
Classification of Alkaloids According to the Biosynthetic Precursor

No.	Precursor	Representatives – Type of Alkaloid
1	Ornithine	pyrrolidine, pyrrolizidine, and tropane alkaloids
2	Lysine	piperidine, pyridine, quinolizidine alkaloids
3	Phenylalanine and Tyrosine	isoquinoline, galantamine alkaloids; some protoalkaloids also come from these precursors (ephedrine, capsaicin)
4	Tryptophan	indole alkaloids of various types: beta carbolinic, yohimbanic, strychnic, lysergic, but also the protoalkaloids psilocin and psilocybin, anthranilic acid, acridine, and quinazolone alkaloids
5	Histidine	imidazole alkaloids
6	Mevalonic acid or AcetylCoA	terpene and sterol alkaloids
7	Purine bases purine alkaloids	purine alkaloids

biogenesis, has raised a continuous interest in organic chemistry and biochemistry. The isolation of ALKs was based on theoretical and applied considerations, taking into account the stimulating action on material and energy metabolism, the pharmacological effects, and the toxicological implications of these compounds.

Currently, the notion of ALKs defines only heterocyclic nitrogenous substances distinguished by their basic character, specific physicochemical properties (colorless, optically active, bitter taste, precipitates with certain salts, for example, KI, $PtCl_2$, etc., and with acids, for example, phosphotungstic, phosphomolybdenic, silico-tungstic, picric, etc.), biologically active properties on animal organisms and the human body (Mason & Weinreb, 2019).

Details on ALKs can be found in specialized treatises in the field of organic chemistry, plant biochemistry, pharmacology, toxicology, etc. The production and location of ALKs in plants are explained by the fact that plant organisms do not have specific ways of eliminating nitrogenous products, similar to animals (e.g., renal elimination of urea, uric acid, and other non-protein nitrogenous derivatives). Taking into account the definition, the classification of natural nitrogenous substances as ALKs is arbitrary, among them not including biogenic amines, pyrazines, pterins, nitrogenous vitamins and their derivatives, aminoglycosides and antibiotics with nitrogen in the molecule, even if from the point of view of origin, all are natural substances with nitrogen in the molecule, pharmacologically active and some even toxic in overdose. Their biosynthesis takes place in specialized cells (chloroplasts, growth roots, laticifers), from where they are later transported to the storage tissues (Schmidt et al., 2012). The pharmaceutical forms in which ALKs are found are tablets, dragees, suppositories, and ampoules for injections (e.g., Scobutil, which is based on scopolamine extracted from *Solanaceae* plants).

In the classification of ALKs, various authors take into account: the biogenesis of the substance, starting from the amino acid or other precursor molecules, when we distinguish:

- The ALKs derived from glycocol (caffeine, theophylline, theobromine);
- The ALKs derived from tryptophan (quinine, quinidine, ergometrine, ergotamine, reserpine, ajmalicine, psilocybin, psilocin); ALKs derived from phenylalanine (morphine, codeine, emetine, ephedrine, capsaicin, colchicine); ALKs derived from lysine (sparteine); ALKs derived from ornithine (atropine, scopolamine, hyoscyamine, cocaine); ALKs derived from acetyl-CoA (terpene ALKs, sterols); ALKs derived from anthranilic acid (acronicine, evoxanthin, vasicine); ALKs derived from histidine (pilocarpine) (Wang & Chen, 2014).
- Their chemical structure, concerning the basic skeleton that can be distinguished in their molecule, when talking about pyridine ALKs (nicotine);
- indolic ALKs (strychnine, reserpine, lysergic ALKs, vincamine, vincristine, vinblastine, ajmaline);
- purine ALKs (caffeine, theophylline, theobromine);
- tropane ALKs (atropine, scopolamine, hyoscyamine);
- quinoline ALKs (quinine, quinidine);
- isoquinoline ALKs (benzylisoquinoline, bisbenzylisoquinoline, morphinanic, proaporphinic, aporphinic, aporphindimer, pavinic, erythrinic, protoberberinic, protopinic, phthalidetetrahydroisoquinoline, spirobenzyltetrahydroisoquinoline, indenobenzazepine, readinic, benzophenanthridinic, galantamine, emetine) (Yuan et al., 2019);
- quinolizidine ALKs (sparteine, cytisine);
- imidazole ALKs (pilocarpine);
- pyrrolizidine ALKs (monocrotaline, senecionine);
- indolizidine ALKs (securinine);
- quinazoline ALKs (vasicine);
- acridine ALKs (acronicine);

- sterol ALKs (solanine, protoverine);
- terpene ALKs (monoterpenes – skitanthin, sesquiterpenes – nufaridine, diterpenes – aconitine) (Yu et al., 2013).

Their presence in certain genera or families (ALKs from species of *Papaveraceae, Solanaceae, Strychnos*, rye horn, *Rauwolfia*, or *Catharanthus*); their pharmacological characteristics (e.g., sympathomimetic ALKs, parasympatholytic, cytostatic/cytotoxic, antihypertensive, oxytocic, etc.).

Depending on the biogenetic origin and the integration of the nitrogen atom in the heterocycle, Hegnauer R. classified ALKs into protoALKs, pseudoALKs, and proper ALKs, meaning that:

- ProtoALKs are simple bases, amines, or amides that are formed from amino acids by decarboxylation, N-methylation, oxidation, or ring closure and are characterized by intense pharmacological activity. The nitrogen atom is not integrated into the heterocycle, being placed in the side chain of the heterocycle. This category includes: calamine, capsaicin, cathinone, colchicine, choline, cyanogenic compounds, ephedrine, physostigmine, histamine, mescaline, psilocybin, spermidine, spermine, and tyramine. Also in this group, some authors include betaines, which are obtained by quaternization of a nitrogen atom belonging to an amino acid;
- PseudoALKs include bases whose fundamental structures belong to non-ALK derivatives and whose nitrogen content represents a random characteristic; most of these compounds derive from terpenoid metabolism, the most well-known representatives being sterol ALKs (*Veratrum* sp.) and terpenic ALKs (*Aconitum* sp.);
- The actual ALKs are always present with at least one nitrogen atom included in the cycle; it can be tertiary (atropine) or quaternary (berberine). The skeleton of the ALKs themselves derives biogenetically from an amino acid that couples with a second non-nitrogenous reaction partner, leading, through successive passage through several biosynthetic steps, to the final ALK. The ALKs can contain one, two, or even more nitrogen atoms in the molecule. Their chemical structures are very diverse, so their pharmacological properties are also different and very varied (Hegnauer, 1981).

Seen in this context, ALKs can be estimated as "storage substances" of nitrogen from heterocyclic compounds. At the origin of the N-heterocyclic structure of various ALKs are, as precursors, various amino acids. The biogenesis of ALKs has been and remains a vast field of scientific research. The nitrogen atoms derive mostly from the metabolism of one/some amino acids. At the same time, the rest of the skeleton is provided by non-nitrogen precursors, such as acetate, propionate, or isoprene residues. It was discovered that, in general, biosynthesis processes occur in young plants at the tissue level with more intense metabolic activity – amino acids. Thus, for example, pyrrolidine and piperidine heterocycles are formed from L-ornithine; L-aspartic acid also yields pyridine; isoquinoline is synthesized from L-tyrosine; indole is synthesized from L-tryptophan, etc. (Clarke & Mann, 1959). However, there are situations when heterocyclic compounds are synthesized from other precursors. Moreover, in this case, it is exemplified: the purine heterocycle results from xanthine; quinazoline and quinoline are formed from anthranilic acid. Next, ALKs are formed from these intermediate metabolites, represented by various heterocyclic compounds. An overview of the various ALKs, considering the types of compounds, the molecular species of the ALKs, and the main plant sources of extraction, is given in Table 3.2.

It is noted that amino acids or other metabolic intermediates are involved in the biosynthesis of ALKs in many cases. The action of ALKs on the human body as biologically active substances (with physiological, pharmacological, or toxicological action) is exemplified by the effects induced by them. In this sense, it is mentioned that: morphine – acts on the central nervous system (CNS) producing drowsiness; papaverine – has a similar action; cocaine – acts on the peripheral nervous system and lends itself to use as an anesthetic; coniin and conhydrin (present in hemlock) – are

TABLE 3.2
The Source of Some Alkaloids

The Original Compound	The Main Derived Alkaloids	References
Tryptophan	6,6′-dibromoindigo, auxin, C-curarine, camptothecin, evodiamine, harmine, ibogaine, indigo, LSD, nigaquinone; physostigmine, psilocin, psilocybin, pyrrolnitrine, quinine, reserpine, rhyncophylline; rutecarpine, serotonin, taurosporine; streptonigrin, strychnine, trichotomin, VCR, VLB; vohimbine	(Almeida et al., 2021)
Lysine	Anabasine, lobeline, matrine, peletierine, piperine, sparteine	(Bunsupa et al., 2017)
Proline	Prodigiosin, pyrrole-2-carboxylic acid, stahydrin	(Angle et al., 2007)
Glutamic acid	Acromelic acid, GABA, ibotenic acid, kainic acid	(Pérez et al., 2016)
Histidine	Cimetidine, histamine, pilocarpine	(Cordell & Lamahewage, 2022)
Phenylalanine and tyrosine	Aritolochic acid, Berberine, chelidonine, coclaurine, colchicine, D-tubocurarine, dopamine, emetine, galantamine, L-DOPA, lycorine, morphine, phenylethylamine, thyroxine	(Cordell, 2010)
Ornithine and arginine	Agrobatin, atropine, cocaine, ephedradine, kukoamine, nicotine, putrescine, spermine, tetrodotoxin	(Tiburcio & Galston, 1986)
2,3-diaminopropionic acid	Albizin, chiscalic acid	(DiMaio et al., 1982)
Anthranilic acid	Acridone, acronicin, echinopsin, febrifugin, iodinine, kokusagin, meloquinone, orixin, phenazinomycin, pyocyanin, rutacridone, vasicin	(Chen et al., 2021)
Nicotinic acid	Arecoline, nicotinamide, nicotinic acid, ricinine, vitamin B6	(Lehrfeld, et al., 1964)
Compounds with purine nucleus	5-fluorouracil, caffeine, folic acid, pteridine, vitamin B1	(Yao et al., 2019)
Compounds with a porphinic skeleton	Bilirubin, chlorophyll, heme, porphin, vitamin B12	(Montrucchio et al., 2013)
Terpenoid derivatives	1-deoxyxylulose, aconitin, actinidin, atisin, batracotoxin, caconin, conesin, delphinin, gentianin, homobatracotoxin, ignavine, jesaconitin, loganin, mesaconitin, mevalonic acid, napelin, nufaridine, peiminoside, secologanin, solanine, sweroside, swertiamarin, taxol, verticin, gypsy	(Wang et al., 2024)
Compounds of polyketides	Coniin, nigrifactin, piericidin A1, pinidine	(Matsumoto et al., 2024)

poisons that act on the CNS with direct effects on the respiratory system; strychnine – is also poison; quinine – used as medicine in case of paudisism; codeine – is used in antitussive medicinal therapy, etc. Additionally, in the research on ALKs, the synthesis of compounds with properties and effects similar to those of natural products was pursued through semi-synthesis in the laboratory (Zhang et al., 2023b).

In the case of opium ALKs, for example, it is considered that they can be

- natural compounds: morphine, codeine, thebaine, papaverine, narcotine, etc.;
- semi-synthetic compounds: hydromorphone, hydrocodone, heroin;
- synthetic compounds: fentanyl, pethidine, methadone, propoxyphene.

Most ALKs are, from a biosynthetic point of view, substances derived from amino acids or compounds with amino groups resulting from transamination, which couple with a second non-nitrogen component whose number of C atoms can differ greatly. A characteristic of the formation of these ALKs is the formation of C-N bonds with the help of intra- and intermolecular reactions. If initially it was considered that ALKs derive only from the amino acids tryptophan, phenylalanine, ornithine, lysine, and glycine, today anthranilic acid, cysteine, methionine, glutamic acid, aspartic acid, proline, histidine, as well as derivatives of tryptamine and phenylalanine, tyrosine, or DOPA (Xing et al., 2008).

The biosynthesis of these amino acids in the plant organism takes place starting from an organic acid (citric acid), which in turn originates in photosynthesis; this is in the presence of ammonia or by attaching an amino group transferred from glutamic or aspartic acids with the help of a transaminase, giving birth to one of the above-mentioned amino acids. The non-nitrogen structures that participate in the formation of ALKs are – [C_1] (methyl) units, provided by methionine or formate, – [C_2] (acetate) units, – [C_6-C_3] units, which can be derived from phenylpropane or acid cinnamic, – [C_6-C_1] units, represented by benzoic acid and derivatives, – [C_5] units, hemiterpenes and their dimers [C_{10}].

The ALKs – biogenic substances of therapeutic interest. In general, the amino acids lysine and ornithine preferentially couple with [C_2] units, phenylalanine with phenylpropane derivatives, and tryptophan with monoterpenes [C_{10}] (Chaudhary et al., 2021). The chemical reactions that take place in the synthesis of ALKs are mainly of five types, namely, the formation of Schiff bases, as in the case of nicotine biosynthesis. The reaction takes place between a primary amine function and a ketone group, according to the model:

$$R-CH_2-CO-R_1+H_2N-R_2 \rightarrow R-CH_2-C\left(R_1\right)=N-R_2$$

In the formation of an ALK, a single amino acid molecule (hygrine), two molecules of the same amino acid (quinolizidine, benzylisoquinoline ALKs) or, more rarely, two or even more molecules of different amino acids (sparteine) may be involved. The process of biosynthesis of ALKs also involves other types of reactions, such as oxidative couplings, oxidations of aromatic nuclei, esterifications, etherifications, and allylic oxidations, all contributing to the achievement of great structural diversity. In general, the name of ALKs comes from the name of the genus or species from which they were isolated for the first time, to which the suffix -ina is added:

- atropine → *Atropa belladonna* L.
- belladonna → *Atropa belladonna*
- papaverine → *Papaver somniferum* L.
- cocaine → *Erythroxylum coca* Lam.

Some ALKs are named according to their pharmacological action: emetine, morphine, and others owe their name to a researcher (peletierine, hörhamerine) or promoter (nicotine) (Shen et al., 2020). Since, in addition to a so-called main ALK, a series of secondary ALKs belonging to the same chemical structure or close to it also appear in the same plant, the name given to the compounds considered secondary is often derived from the common name of the ALK to which it is associated either a prefix (ephedrine/pseudoephedrine), either a suffix (quinine/quinidine) or syllables are transposed (anagram-like), as in the case of narcotine/cotarnine/tarconine. Prefixes such as iso-, pseudo-, neo-, allo-, and epi- are used for isomers. The prefix nor- is used in the case of the demethylated ALK (ephedrine/norephedrine).

The common names of all ALKs end in "in". Considering the compounds that are at the origin of ALKs, we can distinguish ALKs whose biogenesis starts from various amino acids or, sometimes, from compounds resulting from various amination reactions. It is known that N-heterocyclic compounds of the pyridinic, piperidic, and mixed quinolinic type (benzene-pyridinic) are formed from lysine and then ALKs.

The N-heterocyclic precursors result from ornithine: pyrrolidine, tropane, and pyrrolizine. Next, specific ALKs are formed. In the case of tyrosine [$C_9H_{11}NO_3$], the ALKs have as intermediate precursors the N-heterocycles tetrahydroisoquinoline, phenylethylisoquinoline (isoquinoline being an isomer of quinoline) (Wittmann et al., 2015). Tryptophan [$C_{11}H_{12}N_2O_2$], in turn, can generate complex indolic N-heterocycles with ergoline, yohimbine, reserpine, ajmaline, and quinoline structure – from which, in the end, typical ALKs derive. Histidine [$C_6H_9N_3O_2$] generates an imidazole nucleus, then specific ALKs.

Purine [$C_5H_4N_4$] with a specific nucleus. Purine with a mixed specific nucleus (pyrimidinic and imidazolic) generates xanthine as a precursor from which caffeine, theobromine, and theophylline result. Compounds resulting from amination reactions can generate intermediates, which evolve into ALKs with a terpenoid core or with a steroid core. Anthranilic acid, in turn, can generate intermediates with a quinoline N-heterocycle or acridine N-heterocycle structure.

Finally, they evolve into specific ALKs. The ALKs have found use in phytotherapy (Qiu et al., 2014). Gradually, with the advances made in pharmaceutical biochemistry and pharmacognosy, various extractive compounds of pharmaceutical interest were obtained.

3.6 CLASSIFICATION AND CHEMICAL STRUCTURE OF ALKALOIDS

The chemical structure of ALKs is very different and sometimes very complicated. ALKs contain one or more nitrogen atoms in their structure, which can exist as a primary, secondary, tertiary, or quaternary amine. Almost all heterocycles with nitrogen are found in the structure of ALKs, such as pyrrole, pyrroline, pyrrolidine, pyrrolizidine, pyridine, piperidine, quinoline, isoquinoline, quinolizidine, indole, acridine, pyrimidine, imidazole, and purine. These heterocyclic cores can be combined with each other to form complex structures (Bhambhani et al., 2021). ALKs can be divided into two groups:

3.6.1 NON-HETEROCYCLIC ALKALOIDS (ATYPICAL)

These ALKs are sometimes called protoALKs. The most important representatives of this group are:

- *Hordein* is found in the embryo of *Hordeum vulgarae*.

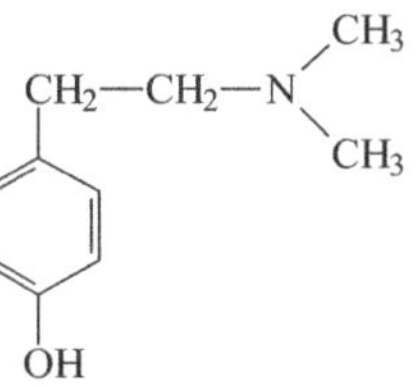

Hordein

- *Muscaline*, which was isolated from cacti of the genus *Anhalonium*.

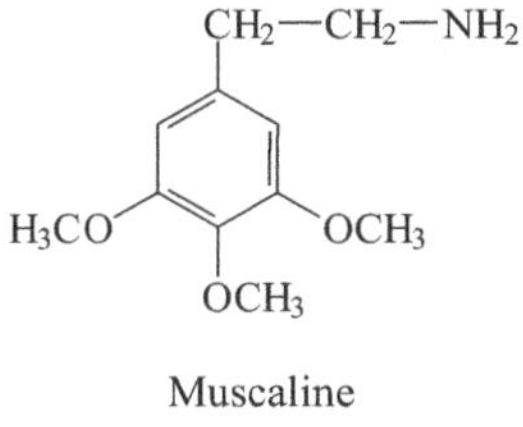

Muscaline

- *Ephedrine* found, along with other related ALKs, in *Ephedra vulgaris* and other ephedra species.

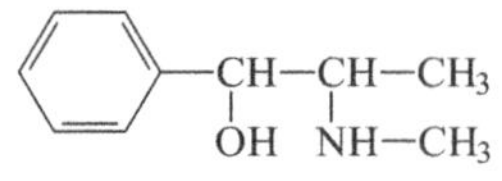

Ephedrine

The ephedrine molecule contains two asymmetric carbon atoms with a non-identical structure. Consequently, there are two pairs of enantiomers, forming two racemes: DL-ephedrine (erythro) and DL-pseudoephedrine (threo form):

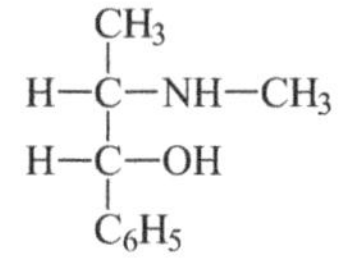

D(–)Ephedrine

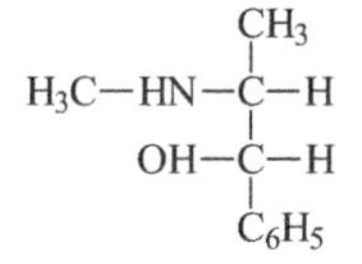

L(+)Ephedrine

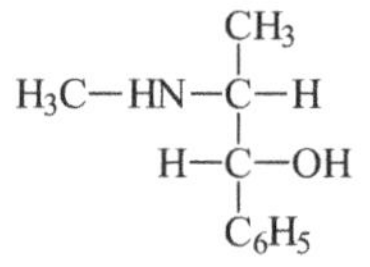

D(–)Pseudoephedrine

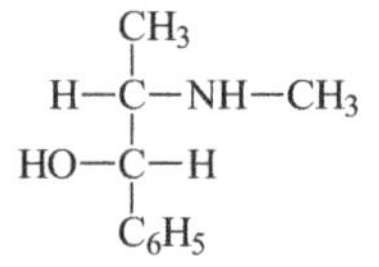

L(+)Pseudoephedrine

The main ALK in *Ephedra vulgaris*, the one that possesses physiological activity, is D(–) ephedrine. The plant also contains, in small quantities, L(+)pseudoephedrine and four other related ALKs, which differ from ephedrine by the number of methyl groups linked to nitrogen; two of them, D (–)norephedrine and L (+)norpseudoephedrine, are not methylated at nitrogen; the other two, D(–)methylephedrine and L(+)methylpseudoephedrine, each has a methyl group at nitrogen (Kudo et al., 2023).

- *Colchicine* is the ALK of *Colchicum autumnale*. It contains two rings of seven atoms, one of which is tropolonic.

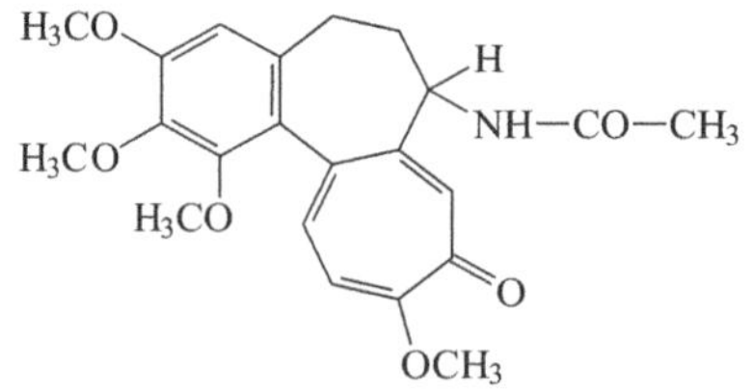

Colchicine

Other non-heterocyclic ALKs, in which the nitrogen atom is placed in a side chain, are erythromycin, jurubin, and pachysandrin.

3.6.2 Heterocyclic Alkaloids (Typical)

These ALKs are divided into 12 groups corresponding to the basic structures.

3.6.2.1 Pyrrole and Pyrrolidine

pyrrole pyrrolidine

The most representative ALKs in this group are:

- *Higrine*, which is contained, along with other ALKs, in the leaves of *Erythroxylum coca*.

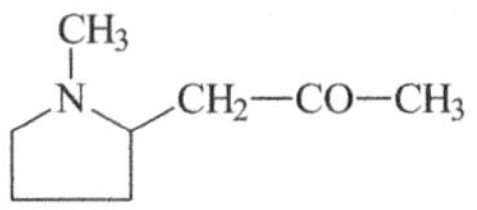

Higrine

- *Stachygrina* has been isolated from many plants, including *Betonica officinalis, Iris germanica*, and *Stachys tuberifera*.

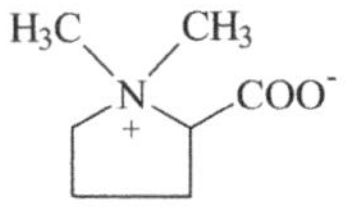

Stachygrina

- *Trigonelline*, also known as N-methylbetaine of nicotinic acid, is widely distributed in plants, including *Pisum sativum, Cannabis sativa, Trigonella foenum graecum*, and some species of coffee.

3.6.2.2 Pyridine and Piperidine

Pyridine Piperidine

ALKs in this group are found in various plant species.

- *Coniine* has been found in many plants, especially in *Conium maculatum*, in *Aethusa cynapium, Arum maculatum*, as well as in *Arisarum vulgarae* and *Caladium bulbosum*:

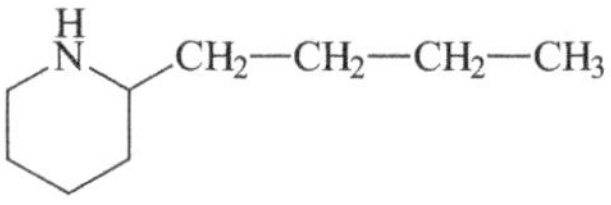

Coniine

- *Arecoline* is a very basic oil that was extracted from the fruits of the *Areca catechu* species:

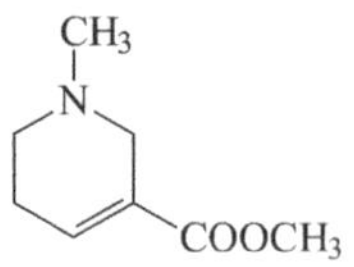

Arecoline

- *Lobelin* is the main ALK in *Lobelia inflata* and has the structure of a hydroxy-ketone deriving from N-methylpiperidine.

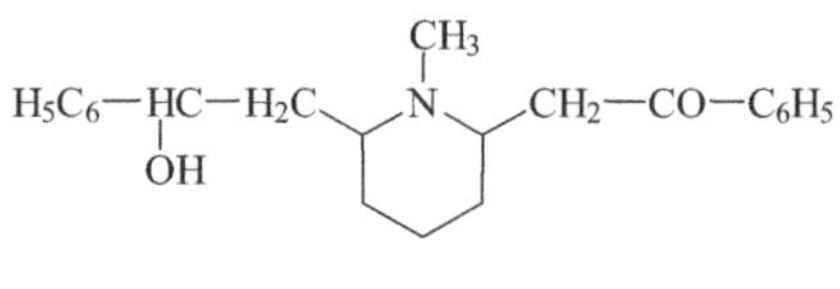

Lobelin

- *Peletierin* was isolated from the root bark of *Punica granatum* and contains an aldehydic function.

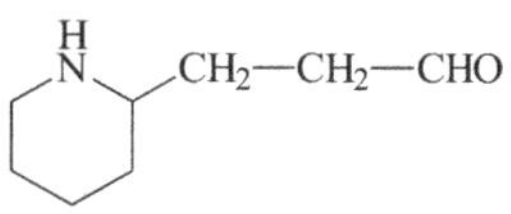

Peletierin

- *Nicotine* is found in different species, such as *Nicotiana tabacum* and *Nicotiana rustica*, as well as in several species of *Lycopodium*, such as *Lycopodium clavatum*, but always in small quantities.

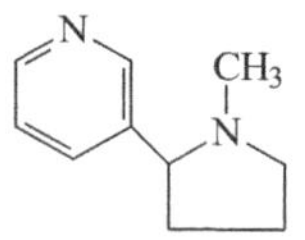

Nicotine

This ALK contains a pyridine ring linked to a pyrrolidine ring.

- *Anabasine* is found in tobacco in a tiny proportion, but it is the main ALK in *Anabasis aphylla*. This ALK is a derivative of nicotine.

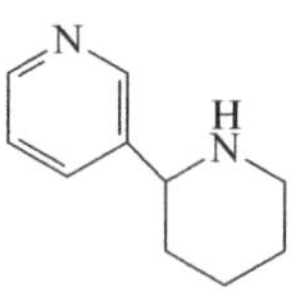

Anabasine

- *Piperine* is one of the burning-tasting ALKs in black pepper, a creeper from the tropics that clings to tree trunks in the forests where it grows.

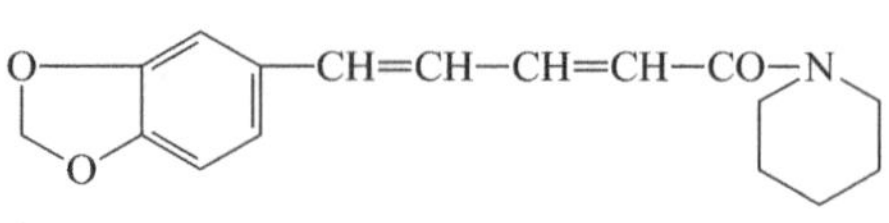

Piperine

Pepper is one of the oldest seasoning plants (spices), sought after for its hot-tasting substances and the essential oils from the fruits, which give it its aroma.

- *Ricinin* was isolated from the seeds and young plants of *R. communis* (Wachira et al., 2014).

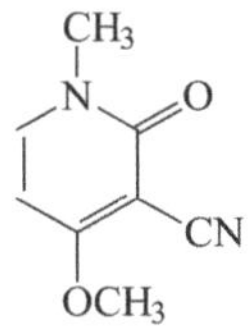

Ricinin

3.6.2.3 Tropane

About 25 ALKs have been isolated in different plant species from the *Solanaceae, Convolvulaceae, Dioscoreaceae*, and *Erythroxylaceae* families, which all have the composition of esters and are derivatives of the bicyclic tertiary amine tropane.

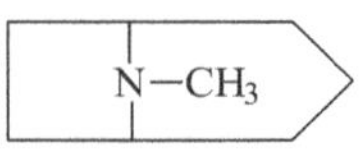

Tropane

- *Hyoscyamine* is the ester of tropine with tropic levogyric acid and is the form in which the ALK occurs in plants.

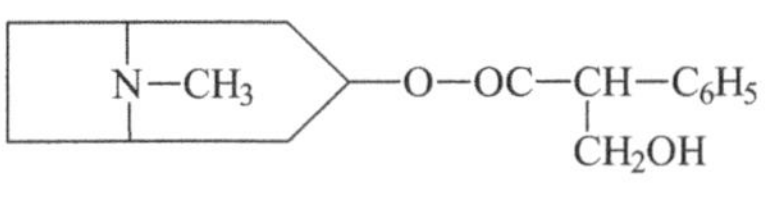

Hyoscyamine

- *Atropine* is the ester of tropine with tropic racemic acid; it is found in small quantities in plants, but it is formed from hyoscyamine by racemization during the isolation operations.
- Hyoscine is also known as scopolamine or atroscine; from a chemical point of view, hyoscine is the ester of tropic levogyric acid with scopine.

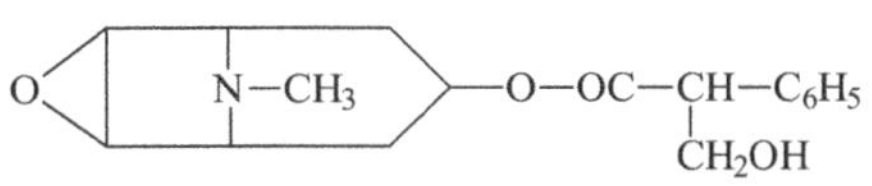

Hyoscine

Scopolamine is racemized, like hyoscyamine, very easily (Srinivasan & Smolke, 2020).

- *Cocaine* is the main ALK in the leaves of *Erythroxylon coca*, a shrub native to the mountainous region of South America. This ALK is a double ester of a hydroxyamino acid, levogyr ecgonine, with methyl alcohol and with benzoic acid.

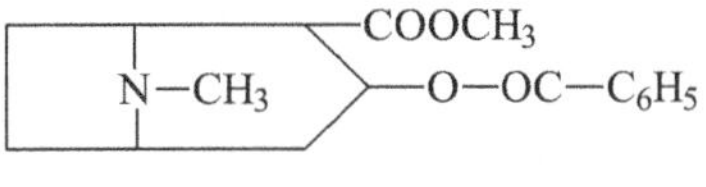

Cocaine

- *Pseudopeletierine* or N-methyl granatonin is the main ALK from the root bark of *Punica granatum*.

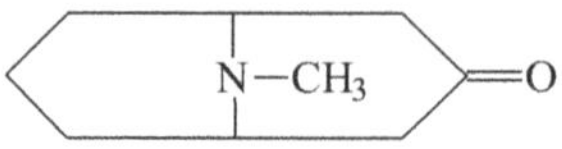

Pseudopeletierine

In pseudopelletierin, there is an unbranched chain of eight carbon atoms that is part of a cyclooctane ring.

3.6.2.4 Quinoline

The bark of some trees belonging to the *Cincona* and *Remijia* species, originating from the high regions of the Andes, contains over 20 ALKs with a quinoline nucleus.

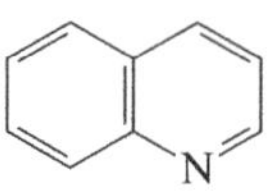

Quinoline

The most important representatives of this group are:

- *Quinine* has two parts in its molecule: one quinoline and another composed of a non-aromatic bicyclic system. In an alcoholic solution, it is strongly levorotatory.

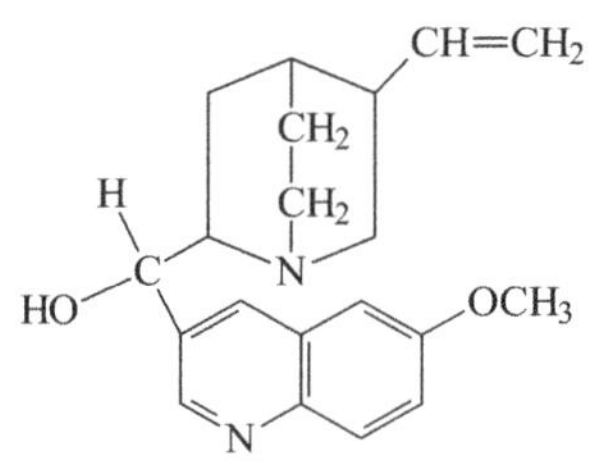

Quinine

- *Quinidine* is an isomer of quinine; it is also known as conquinine.

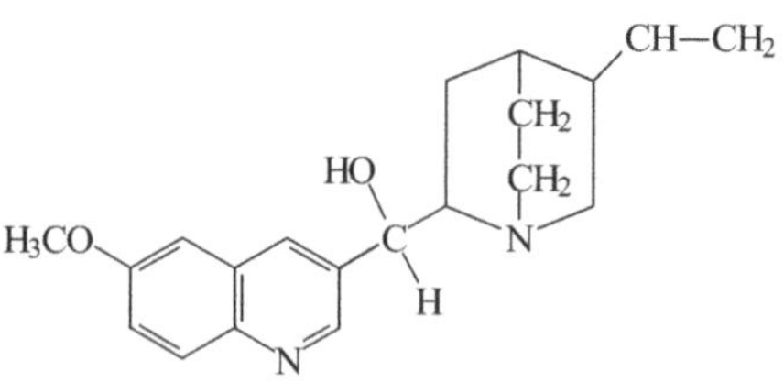

Quinidine

- *Cinchonine* is the basic substance from which most of the ALKs in quinine bark are derived. It differs from quinine by the absence of the methoxyl group in position 6 of the quinoline nucleus:

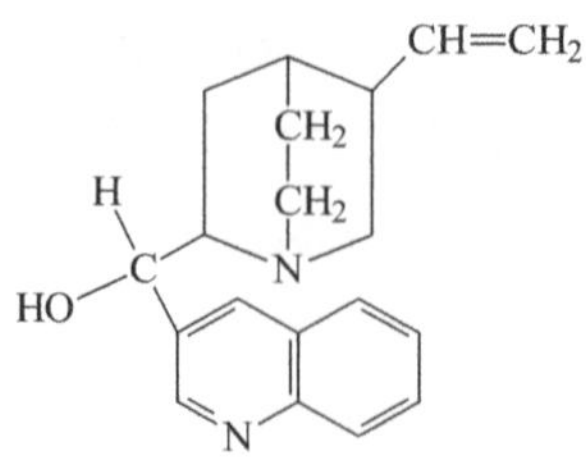

Cinchonine

- *Cinconidine* is an isomer of cinchonene, which differs from quinidine by the absence of the methoxyl group in position 6 of the quinoline nucleus.

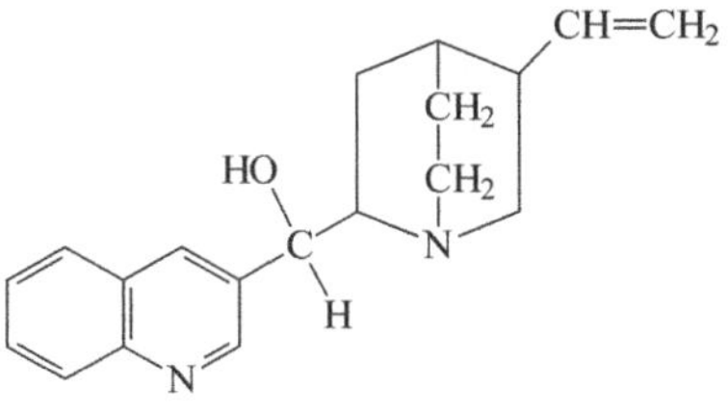

Cinchonidine

Another representative ALK from this group is cusparin (Shang et al., 2022).

3.6.2.5 Isoquinoline

The group of ALKs containing an isoquinoline nucleus is characterized by a large number of its representatives. Several types of ALKs can be distinguished in this group: some are simple derivatives of isoquinoline, and others contain complicated cyclic systems such as those of ALKs from the morphine group.

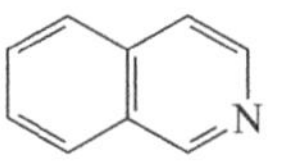

Isoquinoline

- *Papaverine* was discovered in 1848 by Merck; it is found in *P. somniferum*, but in a small proportion; it is an optically inactive substance.

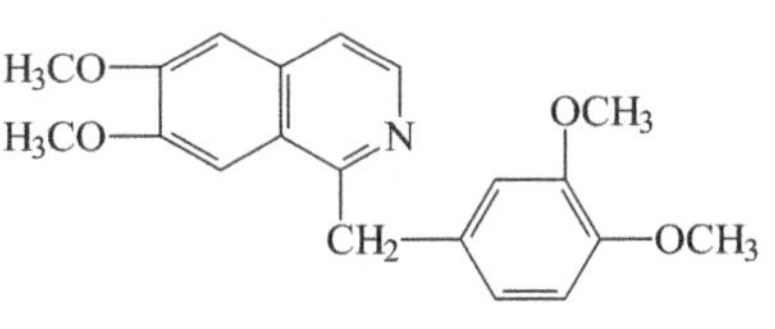

Papaverine

- *Narcotine* is one of the main ALKs in *P. somniferum*; this ALK is strongly levorotatory.

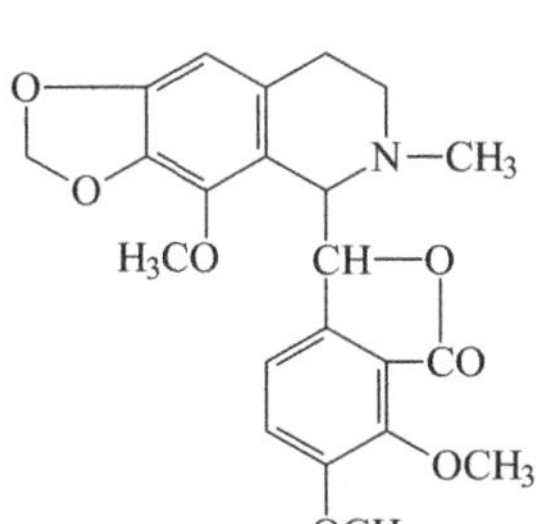

Narcotine

- *Hydrastine* is the ALK from *Hydrastis canadensis*; this substance is related to narcotine, namely, it lacks the CH_3O group in the isoquinoline nucleus.

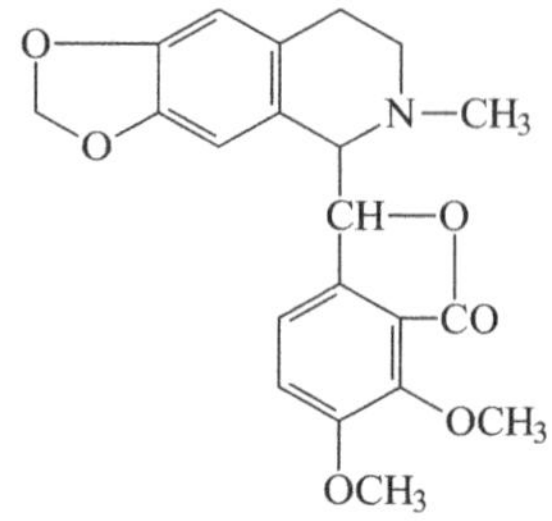

Hydrastin

- *Berberine* was discovered in the shrub *Berberis vulgaris* and then in other plants; berberine is a weak, optically inactive base.

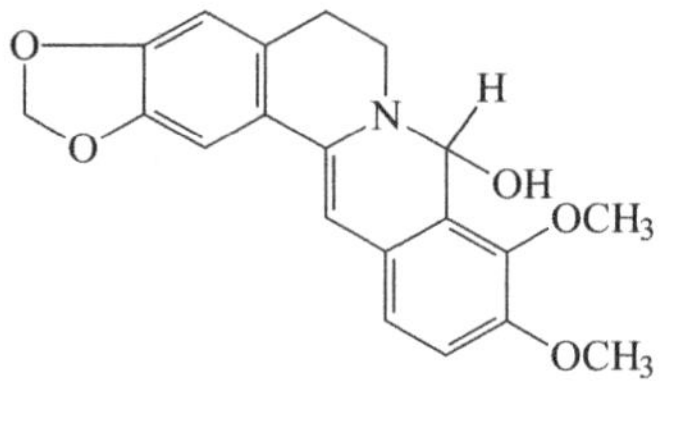

Berberine

- *Coridalina* was isolated from *Corydalis bulbosa* and other species of *Corydalis*; it is a dextrorotatory compound that has a structure similar to that of berberine, from which it differs by the presence of a CH_3 group at one of the saturated rings and possesses four methoxyl groups.

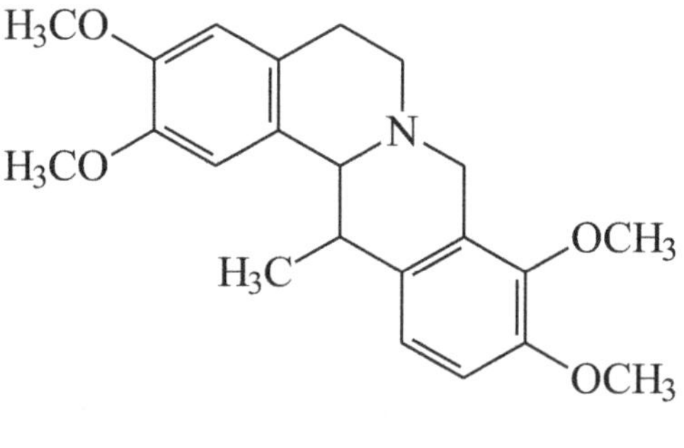

Corydaline

- *Emetine* is found in a South American climbing plant; it is a levorotatory substance that has the following structure:

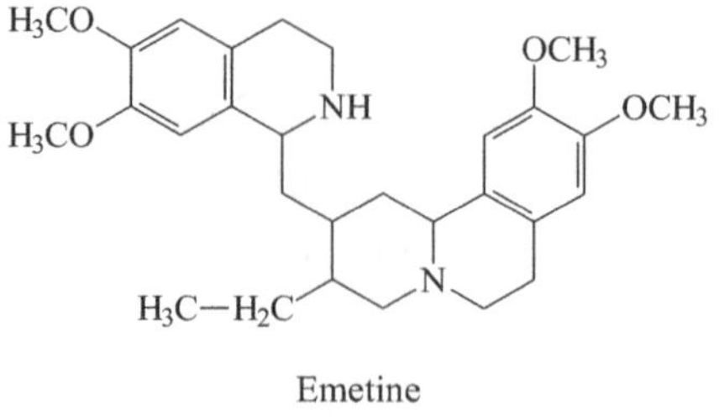

Emetine

- *Morphine* was discovered by Sertürner in 1817; it is a tertiary amine with a methyl group attached to the nitrogen atom; one of the three oxygen atoms belongs to a phenolic OH group, the second oxygen atom belongs to a secondary alcohol, and the third is chemically indifferent and belongs to an ether; the molecule also contains an aliphatic double bond.

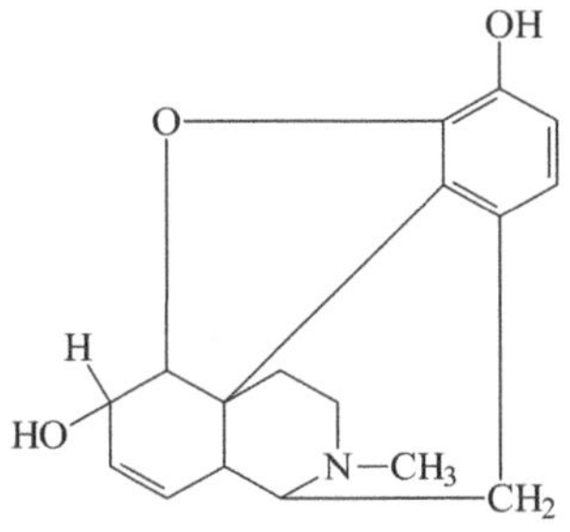

Morphine

The atoms of the morphine skeleton are distributed in two approximately perpendicular planes, one comprising the octahydroisoquinolinic system, and the other the tetralinic and dihydrofuranic ring. Morphine is a levorotatory substance.

- *Codeine* is the monomethyl ether of morphine in the phenolic OH group.
  Codeine is also a levorotatory compound.

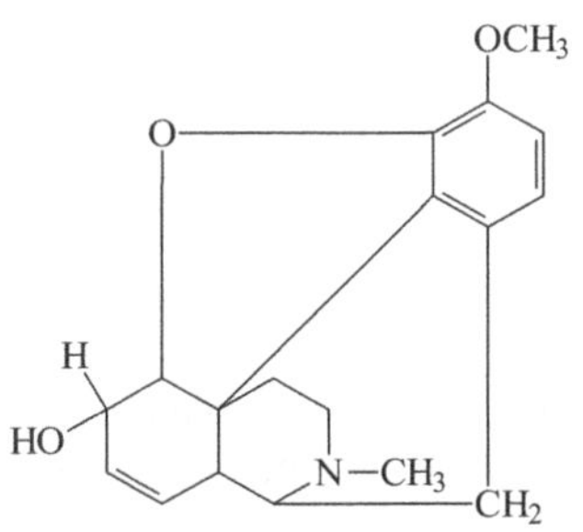

Codeine

- *Thebaine* is found in opium only in small amounts and is also levorotatory.

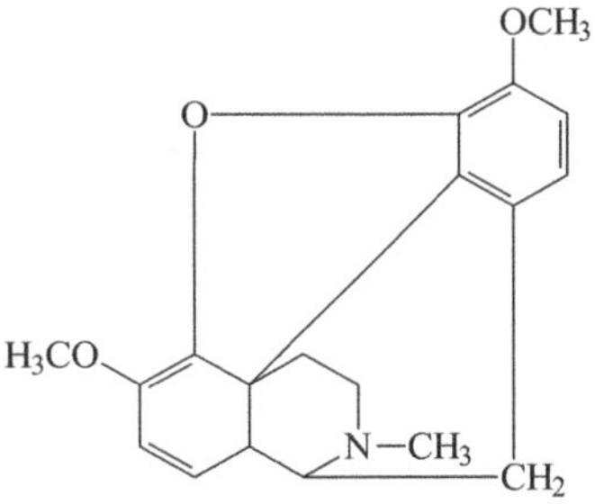

Thebaine

Structurally, thebaine is the methyl ether of codeinone, a ketone obtained by the oxidation of codeine (Shang et al., 2020).

3.6.2.6 Norlupinan

The so-called lupine ALKs are found in several genera of the legume family, such as *Lupinus luteus, Laburnum vulgare*, and *Spartium scoparium*. However, they have also been found in plants from the Chenopodiaceae, *Berberidaceae*, and *Papaveraceae* families.

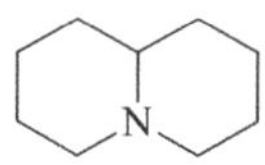

Norlupian

The stereochemistry of these compounds is remarkable because the nitrogen atom in the basic ring (norlupinane or quinolizidine) changes its configuration very easily:

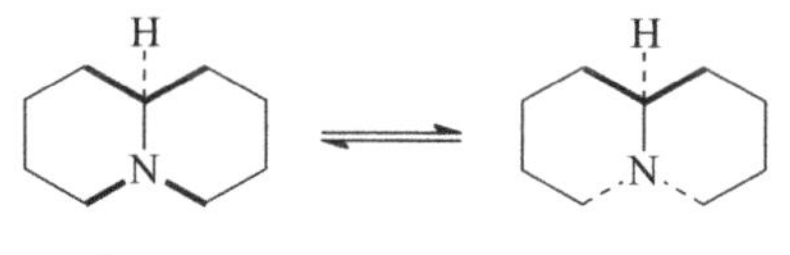

To date, about 50 lupine ALKs have been isolated, among which the most widespread are lupinine, cytisine, sparteine, and laburnine.

- *Lupinine* is a levorotatory compound that was isolated in 1835. This ALK has one of the following two structures.

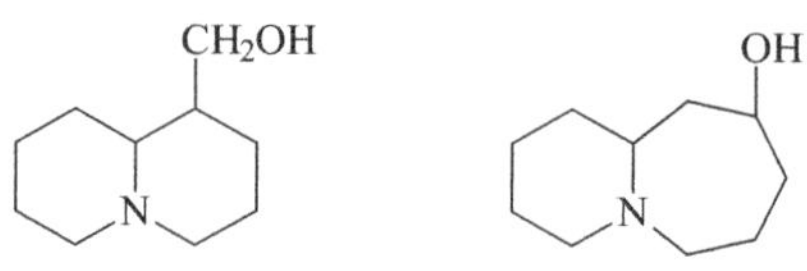

Lupine

The lupinin molecule (formula I) contains two asymmetric carbon atoms. Four optically active or two racemic isomers are therefore possible, corresponding to the formulas:

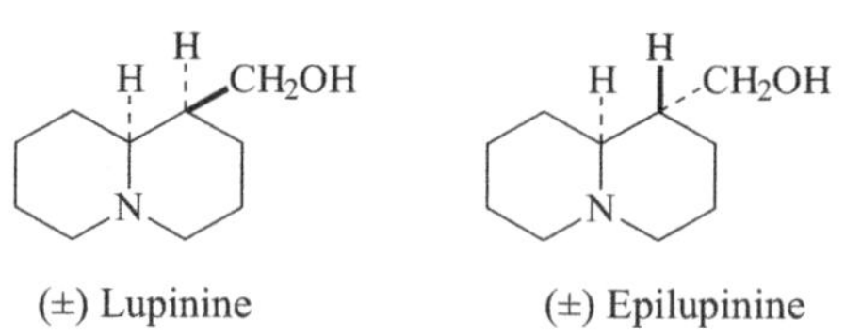

- *Cystisine* contains two nitrogen atoms, one of which functions as a secondary amine. The second nitrogen atom is tertiary, and the oxygen is inert. The compound has an aromatic character.

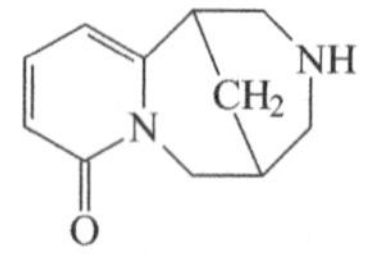

Cystisine

- *Sparteine* has been found in nature in both dextrorotatory and levorotatory forms.

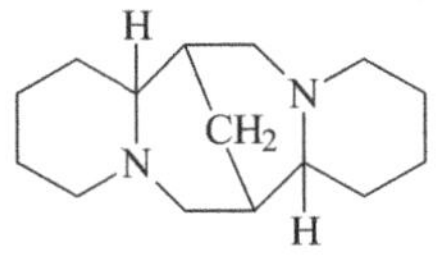

Spartein

The correct formula of sparteine was proposed by Clemo in 1933 (Michael, 2016).

3.6.2.7 *Indol* Sau *Benzopirol*

A large number of ALKs contain an indole nucleus, usually fused with other rings, one of which is a second nitrogenated heterocyclic ring.

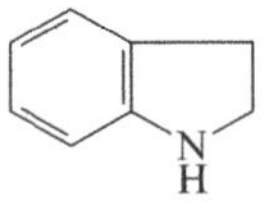

Indole

- *Ergometrine*, also called ergobasine or ergonovine, was discovered simultaneously in several laboratories. It is a levorotatory substance found in *Secale cornutum*, the dry sclerotium (mycelium) of a parasitic fungus that grows on various grasses, but especially on rye. The structure of this ALK contains (+)2-amino-1-propanol linked as an amide to lysergic acid.

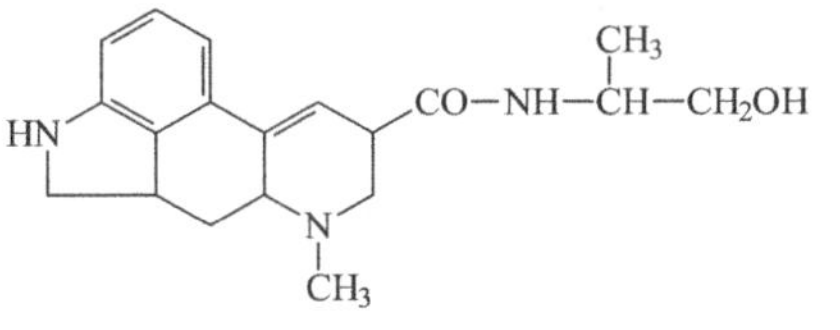

Ergometrine

In ergot, along with ergometrine, there is also the dextrorotatory isomer called ergometrinine, ergobasinine, or ergonovinine.

In ergometrinin, (+)2-amino-1-propanol is amide-linked to isolisergic acid. The two acids, lysergic acid and isolysergic acid, are stereoisomers, distinguished by the position of the carboxyl group on one side or the other of the ring to which it is attached.

- *Ergotamine* is another ALK contained in *Secale cornutum*. This ALK yields lysergic acid, ammonia, pyruvic acid, L-proline, and L-phenylalanine upon hydrolysis.

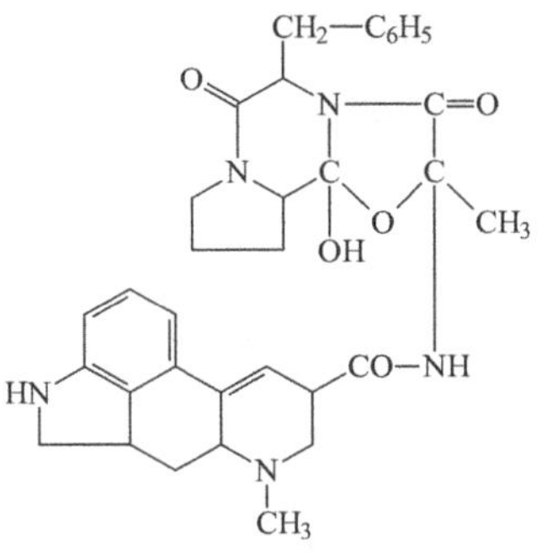

Ergotamine

In this case, along with levorotamine ergotamine, in the corn of rye, there is dextrorotatory ergotamine, which, instead of lysergic acid, contains isolisergic acid (Passie et al., 2008).

- *Ergotoxin* is a group of ALKs from *Secale cornutanu*, which includes the following pairs of levorotatory and dextrorotatory ALKs.
- Ergocristine – Ergocristinine
- Ergocryptine – Ergocriptinine
- Ergocornine – Ergocorninine

These ALKs give, by hydrolysis, lysergic acid (and, respectively, isolisergic acid), ammonia, dimethylpyruvic acid, L-proline, and a second amino acid, namely, L-phenylalanine, L-leucine, and L-valine.

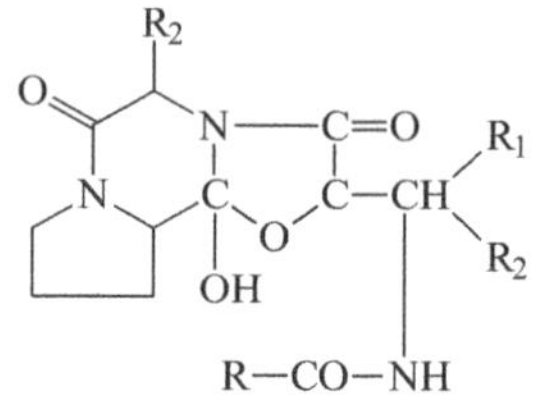

In this formula, R represents the lysergic acid residue, and R1 and R2 differ for the three pairs of amino acids as

- ergocristine: $R_1 = –CH_3$; $R_2 = –CH_2–C_6H_5$;
- ergocriptine: $R_1 = –CH_3$; $R_2 = –CH_2–CH(CH_3)_2$;
- ergocornine: $R_1 = –CH_3$; $R_2 = –CH(CH_3)_2$.

α-ketonic acids are not contained in ALK molecules as such but are part, together with the two amino acids, of a more complicated and unusual structure.

- *Physostigmine* or *eserine* is the most important ALK from the fruits of *Physostigma venenosum*. This ALK contains a hydrogenated and methylated indolic nucleus, condensed with a pyrrolidine nucleus, also N-methylated:

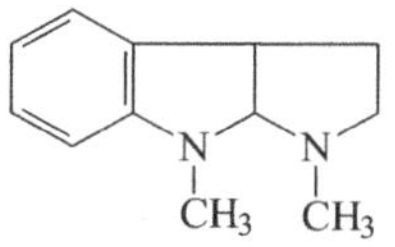

Physostigmine

- *Yohimbine* is the main ALK in the bark of the *Pansinystalia yohimba* tree, synonymous with *Corynanthe yohimbe*. Other ALKs that accompany the main ALK in yohimbelia bark, and which are mostly stereoisomers with it, are: isohimbine, aloyohimbine, corinantidin (α-yohimbine), β, γ, and δ-yohimbine, and pseudoyohimbine.

From another tree, *Pseudocinchona africane*, besides corinantidine, corinanthine, corinantein, and coinanteidin were isolated (Jabir et al., 2022). The structure of yohimbine is as follows:

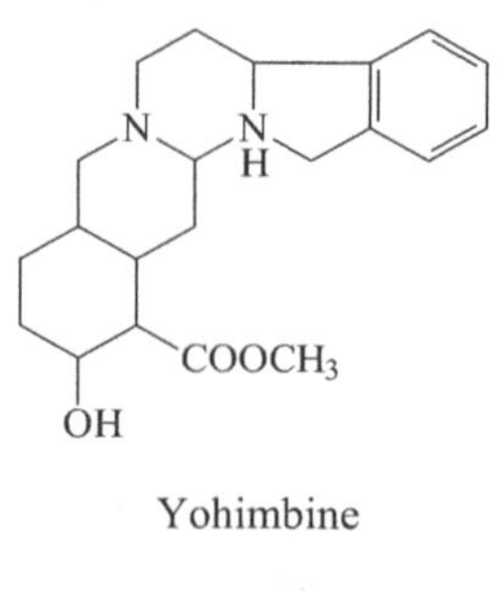

Yohimbine

- *Reserpine* is one of over 30 ALKs isolated from different species of *Rauwolfia*, especially from *Rauwolfia serpentina*, which grows in India, Ceylon, South Africa, and other tropical regions.

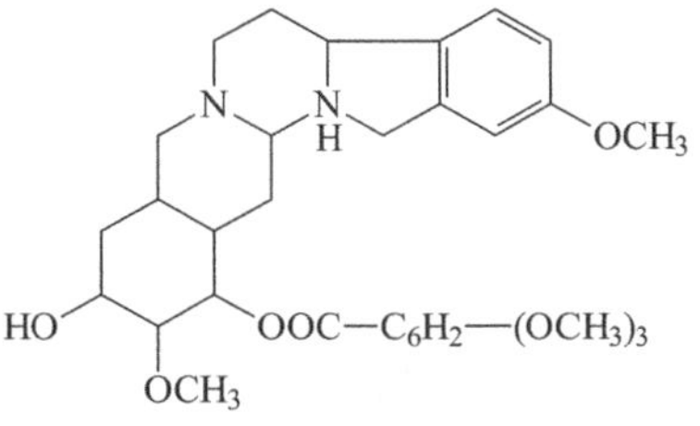

Reserpine

Other ALKs found in *Rauwolfia serpentina* ranwolfia are ajmaline, ajmaline, serpentine, deserpentine, and rescinamine.

- *Alstonine* is the most important ALK found in the bark of various alstonia species, especially *Alstonia constricta, Alstonia scholaris*, and *Alstonia spectabilis*.
- *Strychnine* is a strong, monoacid base containing two nitrogen atoms. The second nitrogen atom, non-basic, is contained in the molecule in the form of an amide group –NH–CO–. The second oxygen atom, also inert, belongs to an ether.

Strychnine is an indole ALK with a heptacyclic structure found in the seeds and leaves of *Strychnos nux vomica* (Guo et al., 2018).

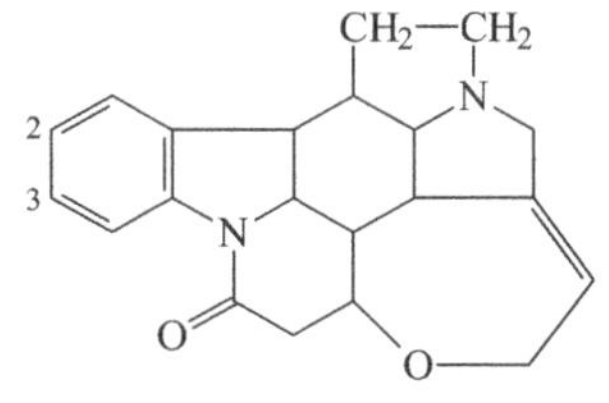

Strychnine

- *Brucine* [$C_{23}H_{26}N_2O_4$] is a dimethoxystrychnine, with -OCH_3 groups in positions 2 and 3.

3.6.2.8 *Imidazole* and *Glyoxaline*

This group of ALKs has numerous representatives.

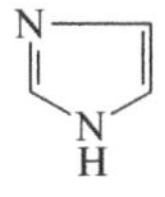

Imidazole

The most important representatives of this group are:

- *Pilocarpine* and *isopilocarpine*, which are found in the leaves of the *South American plants Pilocarpus jaborandi* and *Pilocarpus microphyllus*. Pilocarpine is dextrorotatory; the two ALKs are stereoisomers. The isomerism of the two ALKs is a cis-trans isomerism, due to the two asymmetric carbon atoms in the lactonic ring. Pilocarpine has the cis configuration, and isopilocarpine has the trans configuration.

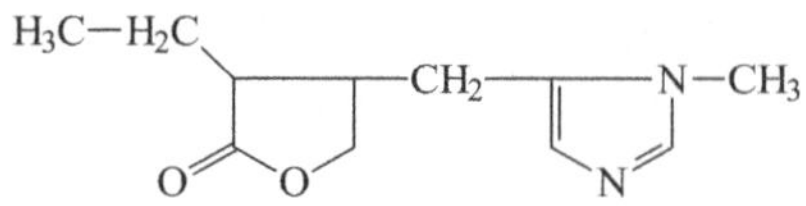

Pilocarpine

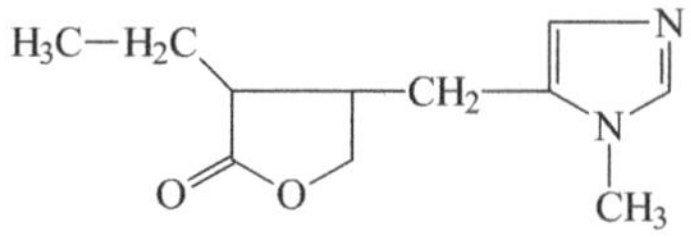

Isopilocarpine

3.6.2.9 Purine

It is the basic skeleton of the nitrogenous bases adenine and guanine in nucleic acids.

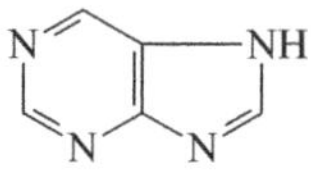

Purine

The most important representatives of this group of ALKs are:

- *Caffeine*, or 1,3,7-trimethylxanthine, is found in the seeds of various coffee species, especially *Arabica coffee* (about 1%), and in other tropical plants. It is also known as theine, guaranine, or methyltheobromine.

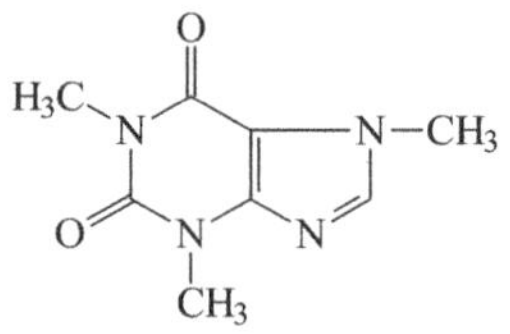

Caffeine

- *Theobromine* or 3,7-dimethylxanthine is contained in the seeds of the *Theobroma cacao* plant (about 1.8%). It is presented as a crystalline powder with a bitter taste.

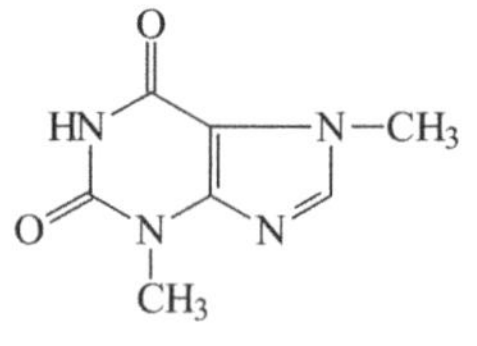

Theobromine

- *Theophylline* or 1,3-dimethylxanthine is found, along with caffeine, in the leaves of *Thea sinensis*. It is also known as teocin.

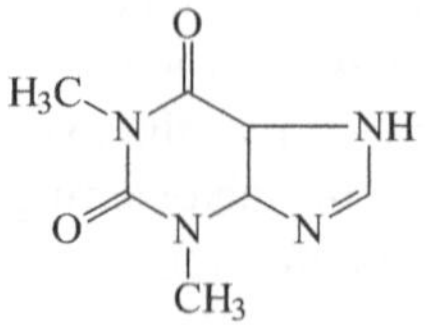

Theophylline

3.6.2.10 Steroids

From nightshades, such as *Solanum tuberosum, Solanum lycopersicum*, and other related species, several glycosides have been isolated whose aglycones contain molecules of 27 carbon atoms and one nitrogen atom. All possess the cholesterol skeleton and an OH group in position 3 thereof (Delbrouck et al., 2023).

- *Solanine* is found in large quantities in young plants of *Solanum Toberosum*. It is the glycoside of a trisaccharide (composed of glucose, galactose, and rhamnose) with an aglycone, solanidine, which has the following structure:

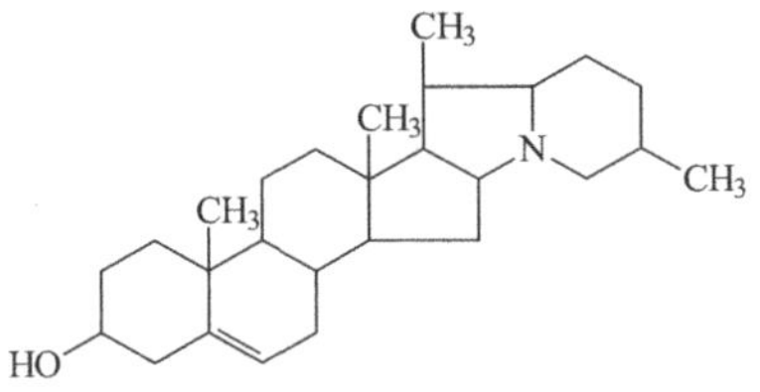

Solanidine

- *Tomatine* was isolated from the leaves of *Solanum lycopersicum* and other related plants. Through hydrolysis, this glycoside gives two molecules of glucose and one each of galactose and xylose, together with the tomatidine aglycone.

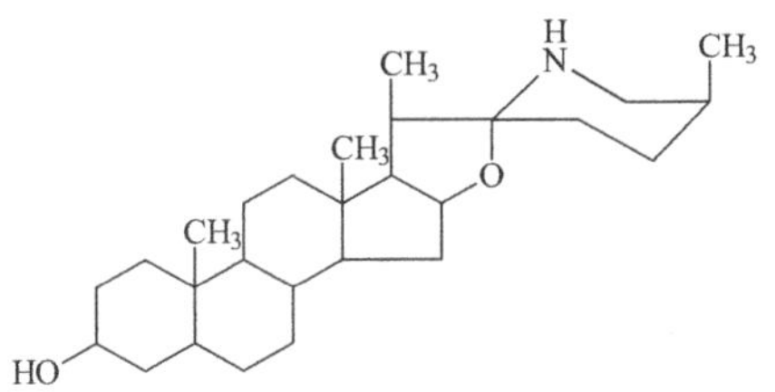

Tomatidine

- *Conesine* is an ALK with a steroid nucleus, isolated from the bark and seeds of an *Indian shrub, Holarrhena antidysenterica*, and from other plants:

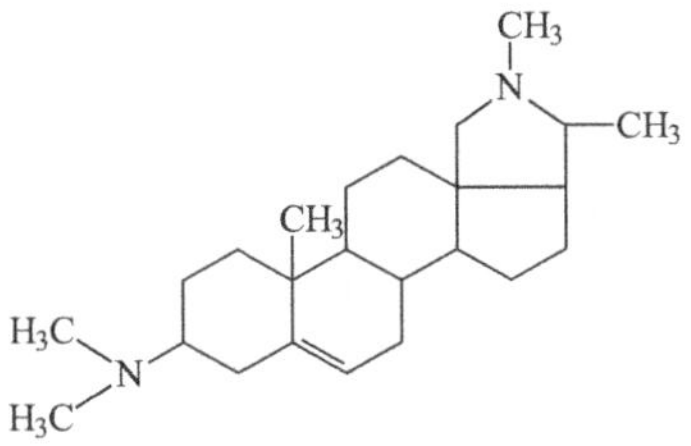

Conesina

- *Rubijervina* was obtained by extraction from *V. album*, a plant that grows in countries with a temperate climate.

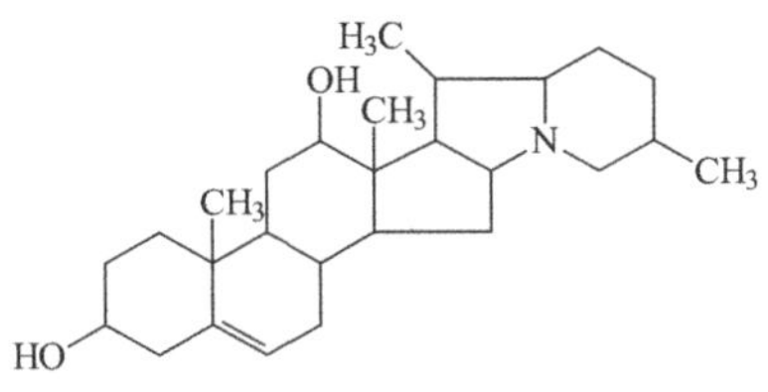

Rubijervina

This group also includes other ALKs, among which funtumine, alkamines, and their glycosides can be mentioned (Grobosch et al., 2008).

3.6.2.11 Terpenoids

From the group of ALKs with terpenic structure, aconitine ($C_{34}H_{47}O_{11}$), atisine ($C_{22}H_{33}O_2N$), and lictonine can be mentioned.

3.6.2.12 Aporphine

Aporphinic ALKs represent a group of natural substances made up of over 100 compounds spread over 18 families. Among them, the richest ALKs with such a structure are the *Papaveraceae, Annonaceae, Lauraceae*, and *Monimiaceae* families.

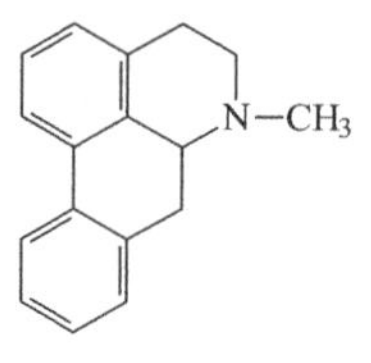

Aporphina

The nitrogen in the structure of these compounds is, in most cases, methylated, but a series of compounds is also known in which the methyl group is replaced by hydrogen.

The ALKs in this group have the following general structure:

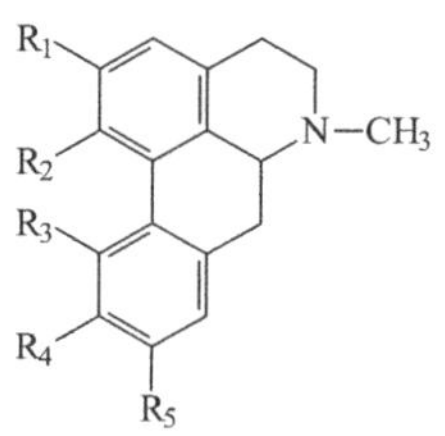

The most important representatives of this structure are the following:

Alkaloid Name	R_1	R_2	R_3	R_4	R_5
Glaucine	OCH_3	OCH_3	H	OCH_3	OCH_3
Isotebaine	OCH_3	OH	OCH_3	H	H
Isoboldine	OCH_3	OH	H	OCH_3	OH
Corytuberin	OCH_3	OH	OH	OCH_3	H
Coridine	OCH_3	OH	OCH_3	OCH_3	H
Boldin	OH	OCH_3	H	OCH_3	OH
Laurothetanin	OCH_3	OCH_3	H	OCH_3	OH

Another structure of these acids is:

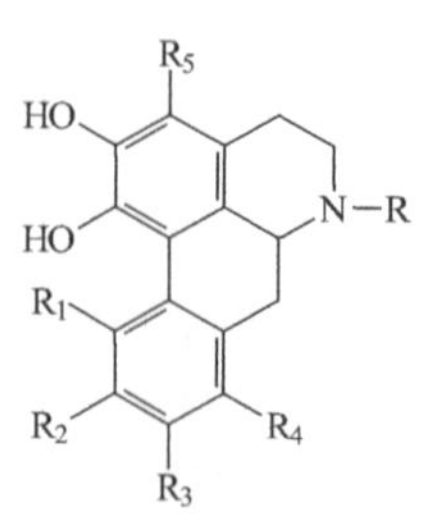

The most important representatives in this structure are

Alkaloid Name	R	R_1	R_2	R_3	R_4	R_5
Bulbocapnin	CH_3	OH	OCH_3	H	H	H
Stefanini	CH_3	H	H	H	OCH_3	H
Xylopin	H	H	H	H	OCH_3	H
Ochoteina	CH_3	H	H	OCH_3	OCH_3	OCH_3

3.6.3 Genalkaloids

They are natural ALKs in which the trivalent nitrogen becomes pentavalent by fixing an oxygen atom:

$$R{\equiv}N \xrightarrow{O} R{\equiv}N{=}O$$

Genalkaloids are amino acid combinations which can be found as such in some plants.

The first genalkaloid, geneserine, was isolated from the seeds of *Physostigma venenosum*, along with the ALK eserine (physostigmine).

Genhyoscyamine from *Atropa belladonna*, genmorphine from *P. somniferum*, and genstrychnine from the seeds of *Strychnos nux vomica* were also isolated.

Genalkaloids are weak bases that do not give salts with acids. They are soluble in water and insoluble in non-polar solvents (Kwakye et al., 2018).

Genalkaloids corresponding to ALKs with higher water solubility (atropine, scopolamine) are hygroscopic and can be obtained in the hydrated form:

$$R{\equiv}N\begin{matrix} \diagup OH \\ \diagdown OH \end{matrix}$$

These amino acid combinations show all the physiological properties of the ALKs from which they come, but have a much lower toxicity.

Administered orally or parenterally, genalkaloids are reduced to the respective ALKs at the level of specific cells that are very rich in reducing diastases.

$$R{\equiv}N{=}O \xrightarrow{H_2} R{\equiv}N + H_2O$$

Unlike some ALKs (morphine, cocaine), the corresponding genalkaloids are not habit-forming.

ALKs can be classified according to their chemical structure, according to the systematic units to which plants with ALKs belong, or according to pharmacodynamic criteria (Amssayef & Eddouks, 2023).

3.6.4 Classification of Alkaloids According to Chemical Structure

The ALKs can be heterocyclic or non-heterocyclic. According to the nature of the main heterocyclic nucleus that makes up the heterocyclic ALKs, they can be grouped into 12 large classes. According to the nature of the amides, the non-heterocyclic ALKs are grouped into two classes. Table 3.3 lists only the main toxic ALKs found in plants.

TABLE 3.3
The Main Types of Alkaloids Found in Toxic Plants

Name of Alkaloids	Molecular Chemical Formula	Name of the Species in Which They Are Found
I. Heterocyclic alkaloids		
1. Alkaloids with a piperidine and pyridine nucleus		
Coniine	$C_8H_{17}N$	*Conium maculatum, Aethusa cynapium*
Conicine	$C_8H_{15}N$	*Conium maculatum*
Conhydrin	$C_8H_{17}ON$	*Conium maculatum*
Nicotine	$C_{10}H_{14}N_2$	*Nicotiana tabacum, Equisetum arvense, Atropa belladonna*
Ricinin	$C_8H_8O_2N_2$	*Ricinus communis*
lobeline	$C_{22}H_{27}O_2N$	*Lobelia inflata*
Lobelanin	$C_{22}H_{25}O_2N$	*Lobelia inflata*
Lobelanidine	$C_{22}H_{20}O_2N$	*Lobelia inflata*
2. Alkaloids with a tropane nucleus		
Atropine	$C_{17}H_{23}O_3N$	*Atropa belladonna*
Belladonna	$C_{34}H_{42}O_4N_2$	*Atropa belladonna*
Apoatropin	$C_{17}H_{21}O_2N$	*Atropa belladonna, Datura stramonium, Hyoscyamus niger*
Scopina	$C_8H_{13}O_2N$	*Scopolia carniolica*
Tropin	$C_8H_{15}ON$	*Scopolia carniolica*
Hyoscyamine	$C_{17}H_{23}O_3N$	*Hyoscyamus niger, Atropa belladonna, Datura stramonium, Scopolia carniolica*
Scopolamine	$C_{17}H_{21}O_4N$	*Atropa belladonna, Hyoscyamus niger, Datura stramonium, Scopolia carniolica*
3. Alkaloids with a quinoline nucleus		
Dictamnine	$C_{12}H_9O_2N$	*Dictamnus albus*
4. Alkaloids with an isoquinoline nucleus		
Papaverine	$C_{20}H_{21}O_4N$	*Papaver somniferum*
Codeine	$C_{18}H_{21}O_3N$	*Papaver somniferum*
Morphine	$C_{17}H_{19}O_3N$	*Papaver somniferum*
thebaine	$C_{19}H_{21}O_3N$	*Papaver somniferum*
Laudanine	$C_{20}H_{25}O_4N$	*Papaver somniferum*
Narcotine	$C_{22}H_{23}O_7N$	*Papaver somniferum*
Narcotolin	$C_{21}H_{21}O_7N$	*Papaver somniferum*
Berbamine	$C_{37}H_{40}O_6N_2$	*Berberis vulgaris*
Oxycanthin	$C_{37}H_{40}O_6N_2$	*Berberis vulgaris*
Magnoflorina	$C_{20}H_{24}O_4N$	*Berberis vulgaris*
Chelidonine	$C_{20}H_{19}O_5N$	*Chelidonium majus*
chelerythrine	$C_{21}H_{18}O_4N$	*Chelidonium majus*
Hemochelidonin	$C_{21}H_{23}O_5N$	*Chelidonium majus*
Oxychelidonine	$C_{20}H_{17}O_6N$	*Chelidonium majus*
Sanguinarine	$C_{20}H_{14}O_4N$	*Chelidonium majus*
5. Alkaloids with a quinolizidine core		
Berberubin	$C_{19}H_{15}O_4N$	*Berberis vulgaris*
Cytisine	$C_{11}H_{14}ON_2$	*Cytisus sp., Laburnum anagyroides*

(Continued)

TABLE 3.3 (Continued)
The Main Types of Alkaloids Found in Toxic Plants

Name of Alkaloids	Molecular Chemical Formula	Name of the Species in Which They Are Found
Octalupine	$C_{14}H_{24}O_2N_2$	*Lupinus polyphyllus*
Lupanina	$C_{15}H_{24}ON_2$	*Lupinus angustifolius, Lupinus albus*
Spartein	$C_{15}H_{26}N_2$	*Spartium junceum, Sarothamnus scoparius, Lupinus albus*
6. Protopine-type alkaloids		
Cryptopine	$C_{21}H_{23}O_5N$	*Papaver somniferum*
Protopine	$C_{26}H_{19}O_5N$	*Papaver somniferum*
7. Alkaloids with an indolic nucleus		
Bufotenin	$C_{12}H_{16}ON$	*Amanita muscaria*
Clivonine	$C_{17}H_{19}O_5N$	*Clivia miniata*
Hipeastrine	$C_{17}H_{17}O_5N$	*Galanthus nivalis, Leucojum aestivum*
Homolycorine	$C_{18}H_{21}O_4N$	*Leucojum aestivum*
Lycorenine	$C_{18}H_{23}O_4N$	*Narcissus poeticus, Amaryllis belladonna, Leucojum aestivum*
Masonine	$C_{17}H_{17}O_4N$	*Galanthus nivalis*
Nivalina	$C_{18}H_{19}O_5N$	*Galanthus nivalis*
8. Alkaloids with a pyrrolizidine nucleus		
Jacobin	$C_{18}H_{25}O_6N$	*Senecoi jacobea*
Jacoline	$C_{18}H_{27}O_5N$	
Jaconina	$C_{18}H_{26}O_8NCl$	
Jacodina	$C_{18}H_{23}O_5N$	*Senecio vulgaris*
Senecionine	$C_{18}H_{25}O_5N$	
9. Alkaloids with an indolizidine nucleus		
Amaryllidin	$C_{17}H_{19}O_5N$	*Amaryllis belladonna*
Belamarin	$C_{18}H_{19}O_4N$	*Amaryllis belladonna*
Caranina	$C_{16}H_{17}O_3N$	
Galantine	$C_{18}H_{23}O_4N$	*Amaryllis belladonna, Narcissus poeticus*
Licorice	$C_{16}H_{17}O_4N$	*Clivia miniata, Galanthusnivalis, Narcissus poeticus*
Narcisidine	$C_{18}H_{23}O_5N$	*Narcissus poeticus*
10. Galantamine-type alkaloids		
Galantamine	$C_{17}H_{21}O_3N$	*Amaryllis belladonna, Leucojum aestivum*
Narcisamine	$C_{16}H_{19}O_3N$	*Narcissus pseudonarcissus*
Narvedina	$C_{17}H_{19}O_3N$	*Galanthus nivalis*
11. Alkaloids with a steroid core		
Germinin	$C_{37}H_{59}O_{11}N$	*Veratrum album, Veratrum nigrum*
Germitetrin	$C_{41}H_{63}O_{14}N$	
Neogermbudin	$C_{37}H_{59}O_{12}N$	
Protoveratridine	$C_{32}H_{51}O_9N$	*Veratrum album*
Gerbutrin	$C_{39}H_{61}O_{14}N$	
Protoveratrin A and B	$C_{41}H_{63}O_{14}N$	
Protroveratrin B	$C_{41}H_{63}O_{15}N$	
Veratridine	$C_{36}H_{51}O_{11}N$	

(Continued)

TABLE 3.3 (Continued)
The Main Types of Alkaloids Found in Toxic Plants

Name of Alkaloids	Molecular Chemical Formula	Name of the Species in Which They Are Found
12. Terpenoid alkaloids		
Ajaconin	$C_{22}H_{33}O_3N$	*Delphinium ajacis*
Napelina	$C_{22}H_{33}O_3N$	
Zongorina	$C_{21}H_{29}O_3N$	*Aconitum napellus*
Aconine	$C_{35}H_{41}O_9N$	
Aconitine	$C_{34}H_{47}O_{11}N$	
Delcosin	$C_{24}H_{39}O_7N$	*Delphinium consolida, Delphinium ajacis*
Ajacina	$C_{34}H_{48}O_9N_2$	*Delphinium ajacis*
Neoline	$C_{24}H_{39}O_6N$	*Aconitum napellus*
II. Non-heterocyclic alkaloids		
1. Acyclic amides		
Galician	$C_6H_{13}N_3$	*Galega officinalis*
Guanidine	CH_5N_3	
2. Cyclic amides		
Colchicine	$C_{22}H_{25}O_6N$	*Colchicum autumnale*
Colchicine	$C_{21}H_{23}O_6N$	*Colchicum autumnale*
Pulustrina	$C_{17}H_{31}O_2N_3$	*Equisetum arvense*

Currently, depending on the origin of the nitrogen atom in the structure of the molecule, among the ALKs are

1. True ALKs are compounds that are made of amino acids and contain a nitrogen atom in the heterocycle (hyoscyamine, caffeine, platyphylline).
2. ProtoALKs are compounds that are made of amino acids and contain an aliphatic nitrogen atom in the side chain (ephedrine, capsaicin).
3. PseudoALKs are nitrogen compounds of terpene and steroid nature (solasodine).

In classifying ALKs, there are two approaches. Chemical classification is based on the structure of the carbon–nitrogen skeleton:

1. Pyridine and piperidine derivatives (anabasine, nicotine).
2. With condensed pyrrolidine and piperidine rings (tropane derivatives) – atropine, cocaine, hyoscyamine, scopolamine.
3. Quinoline derivatives (quinine).
4. Isoquinoline derivatives (morphine, codeine, papaverine).
5. Indole derivatives (strychnine, brucine, reserpine).
6. Purine derivatives (caffeine, theobromine, theophylline).
7. Imidazole derivatives (pilocarpine).
8. Steroid ALKs (solasonine).
9. Acyclic ALKs and ALKs with an exocyclic nitrogen atom (ephedrine, spherophysin, colchamine).

Another type of classification of ALKs is based on the botanical sign, according to which ALKs are united by plant sources. Most ALKs have basic properties related to their name. In plants, ALKs are present in the form of organic acid salts (lemon, apple, wine, oxalic).

3.6.5 Isolation from Plant Raw Materials

The first method (extraction in the form of salts)

The second method (extraction as bases): The basic (alkaline) properties of ALKs are expressed to varying degrees. In nature, ALKs are more common, which refer to tertiary bases, less often secondary or quaternary ammonium (Sakai & Hirose, 2003). Due to their basic nature, ALKs form salts with acids of varying degrees of strength. The ALK salts are easily decomposed under the action of caustic alkalis and ammonia. At the same time, free bases are allocated. Due to their basic nature, ALKs, when they interact with acids, form salts. This property is used in the isolation and purification of ALKs, their quantitative determination, and preparation. Salt ALKs are soluble in water and ethanol (especially in dilute solutions) when heated, insufficiently or completely insoluble in organic solvents (chloroform, ethyl ether, etc.). As an exception, scopolamine hydrobromide, cocaine hydrochloride, and other opium ALKs may be mentioned. The ALK bases usually do not dissolve in water but dissolve easily in organic solvents. Exceptions are nicotine, ephedrine, anabasine, and caffeine, which are well soluble in both water and organic solvents. Quinine – an ALK isolated from cinchona bark (*Cinchona oficinalis*) – is a colorless crystal with a very bitter taste. Quinine and its derivatives have antipyretic and antimalarial action. Nicotine is the main ALK of tobacco and machorka (Parveen et al., 2024). Nicotine is very toxic; the lethal dose for humans is 40 mg/kg, with natural nicotine 2–3 times more toxic than synthetic nicotine. Atropine is a racemic form of hyoscyamine; it has anticholinergic (spasmolytic and mydriatic) action.

3.7 CONCLUSION AND FUTURE REMARKS

The ALKs present the most incredible variety of structures and the most significant number of individual representatives of natural products. These natural products of plant origin have as their only common characteristic the fact that they possess a basic nitrogen atom, which, in most cases, is found in a heterocycle. However, some substances in which the nitrogen atom is placed in a side chain are considered ALKs because they have the chemical and physiological properties of this class of substances. Although the ALK content of plants is small, it is closely related to human beings. Many ALKs are good medicines for curing diseases. For example, Coptidis rhizome root is the main component of berberine, which has antibacterial and anti-inflammatory effects. Reserpine in raufu can lower blood pressure; garantamine from lycoris sativa has a curative effect for the sequelae of poliomyelitis, the morphine base contained in the poppy husk is the famous analgesic; quinine is a valuable antipyretic; tricusperzine and vinblastine are good for cancer; colchicine (alkali) can produce polyploids by artificial mutation. Some ALKs can be used to make agricultural pesticides. ALKs have also been isolated in vertebrates and invertebrates, some of which are related to the plants they eat, and those found in flu, salamanders, and some fish are true animal metabolites.

REFERENCES

Almeida, M. C., Resende, D. I. S. P., da Costa, P. M., Pinto, M. M. M., & Sousa, E. (2021). Tryptophan derived natural marine alkaloids and synthetic derivatives as promising antimicrobial agents. *European Journal of Medicinal Chemistry*, *209*, 112945. https://doi.org/10.1016/j.ejmech.2020.112945

Amssayef, A., & Eddouks, M. (2023). Alkaloids as promising agents for the management of insulin resistance: A review. *Current Pharmaceutical Design*, 29(39), 3123–3136. https://doi.org/10.2174/0113816128270340231121043038

Andriamaharavo, N. R., Garraffo, H. M., Spande, T. F., Giddings, L. A., Vieites, D. R., Vences, M., & Saporito, R. A. (2015). Individual and geographic variation of skin alkaloids in three swamp-forest species of Madagascan poison frogs (Mantella). *Journal of Chemical Ecology*, *41*(9), 837–847. https://doi.org/10.1007/s10886-015-0616-4

Angle, S. R., Bensa, D., & Belanger, D. S. (2007). New access to indolizidine and pyrrolizidine alkaloids from an enantiopure proline: total syntheses of (-)-lentiginosine and (1R,2R,7aR)-dihydroxypyrrolizidine. *The Journal of Organic Chemistry*, *72*(15), 5592–5597. https://doi.org/10.1021/jo070462w

Bhambhani, S., Kondhare, K. R., & Giri, A. P. (2021). Diversity in chemical structures and biological properties of plant alkaloids. *Molecules (Basel, Switzerland)*, *26*(11), 3374. https://doi.org/10.3390/molecules26113374

Buchanan, W. W., Rainsford, K. D., Kean, C. A., & Kean, W. F. (2024). Narcotic analgesics. *Inflammopharmacology*, *32*(1), 23–28. https://doi.org/10.1007/s10787-023-01304-y

Bunsupa, S., Yamazaki, M., & Saito, K. (2017). Lysine-derived alkaloids: Overview and update on biosynthesis and medicinal applications with emphasis on quinolizidine alkaloids. *Mini Reviews in Medicinal Chemistry*, *17*(12), 1002–1012. https://doi.org/10.2174/1389557516666160506151213

Chaudhary, N. K., Taylor, W. C., Mander, L. N., & Karuso, P. (2021). Isolation and structure elucidation of additional alkaloids from the tropical rainforest tree *Galbulimima baccata*. *Journal of Natural Products*, *84*(9), 2525–2535. https://doi.org/10.1021/acs.jnatprod.1c00537

Chen, P. N., Hao, M. J., Li, H. J., Xu, J., Mahmud, T., & Lan, W. J. (2021). Biotransformations of anthranilic acid and phthalimide to potent antihyperlipidemic alkaloids by the marine-derived fungus Scedosporium apiospermum F41-1. *Bioorganic Chemistry*, *116*, 105375. https://doi.org/10.1016/j.bioorg.2021.105375

Chen, X., Qin, Y., Wang, L., Zhu, Y., Zhang, H., Liu, W., Zeng, M., & Dai, Q. (2024). Co-amorphous systems of sulfasalazine with matrine-type alkaloids: Enhanced solubility behaviors and synergistic therapeutic potential. *European Journal of Pharmaceutics and Biopharmaceutics: Official Journal of Arbeitsgemeinschaft fur Pharmazeutische Verfahrenstechnik e.V*, *203*, 114475. https://doi.org/10.1016/j.ejpb.2024.114475

Chirkin, E., Atkatlian, W., & Porée, F. H. (2015). The Securinega alkaloids. *The Alkaloids. Chemistry and Biology*, *74*, 1–120. https://doi.org/10.1016/bs.alkal.2014.11.001

Chowański, S., Adamski, Z., Marciniak, P., Rosiński, G., Büyükgüzel, E., Büyükgüzel, K., Falabella, P., Scrano, L., Ventrella, E., Lelario, F., & Bufo, S. A. (2016). A review of bioinsecticidal activity of solanaceae alkaloids. *Toxins*, *8*(3), 60. https://doi.org/10.3390/toxins8030060

Clarke, A. J., & Mann, P. J. (1959). Plant enzyme reactions leading to the formation of heterocyclic compounds. 3. Plant amine oxidase and the formation of pyrrolidine and piperidine alkaloids. *The Biochemical Journal*, *71*(4), 596–609. https://doi.org/10.1042/bj0710596

Cordell G. A. (2010). Important groups of alkaloids derived biosynthetically from phenylalanine and tyrosine. Preface. *The Alkaloids. Chemistry and Biology*, *68*, ix. https://doi.org/10.1016/s1099-4831(10)06812-4

Cordell, G. A., & Lamahewage, S. N. S. (2022). Ergothioneine, ovothiol A, and selenoneine-histidine-derived, biologically significant, trace global alkaloids. *Molecules (Basel, Switzerland)*, *27*(9), 2673. https://doi.org/10.3390/molecules27092673

Delbrouck, J. A., Desgagné, M., Comeau, C., Bouarab, K., Malouin, F., & Boudreault, P. L. (2023). The therapeutic value of solanum steroidal (Glyco)Alkaloids: A 10-year comprehensive review. *Molecules (Basel, Switzerland)*, *28*(13), 4957. https://doi.org/10.3390/molecules28134957

DiMaio, J., Nguyen, T. M., Lemieux, C., & Schiller, P. W. (1982). Synthesis and pharmacological characterization in vitro of cyclic enkephalin analogues: effect of conformational constraints on opiate receptor selectivity. *Journal of Medicinal Chemistry*, *25*(12), 1432–1438. https://doi.org/10.1021/jm00354a008

Dymek, A., Widelski, J., Wojtanowski, K. K., Płoszaj, P., Zhuravchak, R., & Mroczek, T. (2021). Optimization of pressurized liquid extraction of Lycopodiaceae alkaloids obtained from two *Lycopodium* species. *Molecules (Basel, Switzerland)*, *26*(6), 1626. https://doi.org/10.3390/molecules26061626

Gaillard, Y., & Pepin, G. (2001). LC-EI-MS determination of veratridine and cevadine in two fatal cases of Veratrum album poisoning. *Journal of Analytical Toxicology*, *25*(6), 481–485. https://doi.org/10.1093/jat/25.6.481

Grobosch, T., Binscheck, T., Martens, F., & Lampe, D. (2008). Accidental intoxication with Veratrum album. *Journal of Analytical Toxicology*, *32*(9), 768–773. https://doi.org/10.1093/jat/32.9.768

Guo, R., Wang, T., Zhou, G., Xu, M., Yu, X., Zhang, X., Sui, F., Li, C., Tang, L., & Wang, Z. (2018). Botany, phytochemistry, pharmacology and toxicity of strychnos nux-vomica L.: A review. *The American Journal of Chinese Medicine*, 46(1), 1–23. https://doi.org/10.1142/S0192415X18500015

Hegnauer, R. (1981). Chemotaxonomy of Erythroxylaceae (including some ethnobotanical notes on Old World species). *Journal of Ethnopharmacology*, *3*(2-3), 279–292. https://doi.org/10.1016/0378-8741(81)90058-1

Huo, Z. Q., Zhao, Q., Zhu, W. T., Hao, X. J., & Zhang, Y. (2021). Bousmekines A-E, new alkaloids from two bousigonia Species: B.angustifolia and B. mekongensis. *Natural products and bioprospecting*, *11*(2), 207–213. https://doi.org/10.1007/s13659-020-00278-6

Jabir, N. R., Firoz, C. K., Zughaibi, T. A., Alsaadi, M. A., Abuzenadah, A. M., Al-Asmari, A. I., Alsaieedi, A., Ahmed, B. A., Ramu, A. K., & Tabrez, S. (2022). A literature perspective on the pharmacological applications of yohimbine. *Annals of Medicine*, *54*(1), 2861–2875. https://doi.org/10.1080/07853890.2022.2131330

Jank, B., & Rath, J. (2021). Emerging tropane alkaloid contaminations under climate change. *Trends in Plant Science*, *26*(11), 1101–1103. https://doi.org/10.1016/j.tplants.2021.08.001

Jones, W. P., & Kinghorn, A. D. (2012). Extraction of plant secondary metabolites. *Methods in Molecular Biology (Clifton, N.J.)*, *864*, 341–366. https://doi.org/10.1007/978-1-61779-624-1_13

Konan, K. V., Le, T. C., & Mateescu, M. A. (2019). Enhanced solubility of alkaloids by complexation with polycarboxylic materials for controlled release formulations: Case of Peschiera fuchsiaefolia. *AAPS PharmSciTech*, *20*(3), 108. https://doi.org/10.1208/s12249-019-1315-1

Kudo, Y., Umemoto, K., Obata, T., Kaneda, A., Ni, S. R., Mikage, M., Sasaki, Y., & Ando, H. (2023). Seasonal variation of alkaloids and polyphenol in Ephedra sinica cultivated in Japan and controlling factors. *Journal of Natural Medicines*, *77*(1), 137–151. https://doi.org/10.1007/s11418-022-01656-9

Kwakye, G. F., Jiménez, J., Jiménez, J. A., & Aschner, M. (2018). Atropa belladonna neurotoxicity: Implications to neurological disorders. *Food and Chemical Toxicology: An International Journal Published for the British Industrial Biological Research Association*, *116*(Pt B), 346–353. https://doi.org/10.1016/j.fct.2018.04.022

Lehrfeld, J., Burkman, A. M., & Gearien, J. E. (1964). Synthesis of 6-substituted nicotinic acid derivatives as analogs of ergot alkaloids. *Journal of Medicinal Chemistry*, *7*, 150–154. https://doi.org/10.1021/jm00332a007

Liu, J., Flegel, J., Otte, F., Pahl, A., Sievers, S., Strohmann, C., & Waldmann, H. (2021). Combination of pseudo-natural product design and formal natural product ring distortion yields stereochemically and biologically diverse pseudo-sesquiterpenoid alkaloids. *Angewandte Chemie (International ed. in English)*, *60*(39), 21384–21395. https://doi.org/10.1002/anie.202106654

Marciniak, P., Kolińska, A., Spochacz, M., Chowański, S., Adamski, Z., Scrano, L., Falabella, P., Bufo, S. A., & Rosiński, G. (2019). Differentiated effects of secondary metabolites from *Solanaceae* and *Brassicaceae* plant families on the heartbeat of *Tenebrio molitor* pupae. *Toxins*, *11*(5), 287. https://doi.org/10.3390/toxins11050287

Mason, J. D., & Weinreb, S. M. (2019). The alstoscholarisine alkaloids: Isolation, structure determination, biogenesis, biological evaluation, and synthesis. *The Alkaloids. Chemistry and Biology*, *81*, 115–150. https://doi.org/10.1016/bs.alkal.2018.09.001

Matsumoto, T., Hamai, M., Inui, Y., Ohta, T., Yoshida, T., Kojima, N., Yoshikawa, H., Kutose, R., Ando, S., & Watanabe, T. (2024). Chemical structures of polyketides and alkaloids isolated from a culture of fungus Curvularia moringae. *Fitoterapia*, *179*, 106263. https://doi.org/10.1016/j.fitote.2024.106263

Michael J. P. (2016). Simple indolizidine and quinolizidine alkaloids. *The Alkaloids. Chemistry and Biology*, *75*, 1–498. https://doi.org/10.1016/bs.alkal.2014.12.001

Montrucchio, D. P., Córdova, M. M., & Santos, A. R. (2013). Plant derived aporphinic alkaloid S-(+)-dicentrine induces antinociceptive effect in both acute and chronic inflammatory pain models: evidence for a role of TRPA1 channels. *PloS One*, *8*(7), e67730. https://doi.org/10.1371/journal.pone.0067730

O'Callaghan J. P. (2001). Evolution of a rational use of opioids in chronic pain. *European Journal of Pain (London, England)*, *5*(Suppl A), 21–26. https://doi.org/10.1053/eujp.2001.0275

Parveen, S., Maurya, N., Meena, A., & Luqman, S. (2024). Cinchonine: A versatile pharmacological agent derived from natural cinchona alkaloids. *Current Topics in Medicinal Chemistry*, 24(4), 343–363. https://doi.org/10.2174/0115680266270796231109171808

Passie, T., Halpern, J. H., Stichtenoth, D. O., Emrich, H. M., & Hintzen, A. (2008). The pharmacology of lysergic acid diethylamide: a review. *CNS Neuroscience & Therapeutics*, *14*(4), 295–314. https://doi.org/10.1111/j.1755-5949.2008.00059.x

Pérez, S. J., Purino, M. A., Cruz, D. A., López-Soria, J. M., Carballo, R. M., Ramírez, M. A., Fernández, I., Martín, V. S., & Padrón, J. I. (2016). Enantiodivergent synthesis of (+)- and (-)-pyrrolidine 197B: Synthesis of trans-2,5-disubstituted pyrrolidines by intramolecular hydroamination. *Chemistry (Weinheim an der Bergstrasse, Germany)*, *22*(43), 15529–15535. https://doi.org/10.1002/chem.201602708

Philippe, G., Angenot, L., Tits, M., & Frédérich, M. (2004). About the toxicity of some Strychnos species and their alkaloids. *Toxicon: Official Journal of the International Society on Toxinology*, *44*(4), 405–416. https://doi.org/10.1016/j.toxicon.2004.05.006

Qiu, S., Sun, H., Zhang, A. H., Xu, H. Y., Yan, G. L., Han, Y., & Wang, X. J. (2014). Natural alkaloids: Basic aspects, biological roles, and future perspectives. *Chinese Journal of Natural Medicines*, *12*(6), 401–406. https://doi.org/10.1016/S1875-5364(14)60063-7

Rodés-Bachs, C., & Van der Fels-Klerx, H. J. (2023). Impact of environmental factors on the presence of quinolizidine alkaloids in lupins: A review. *Food Additives & Contaminants. Part A, Chemistry, Analysis, Control, Exposure & Risk Assessment*, *40*(6), 757–769. https://doi.org/10.1080/19440049.2023.2217273

Sakai, T., & Hirose, A. (2003). Mixed ternary ion associate formation between xanthene dye, cinchona-alkaloid and quaternary ammonium ion and its application to the determination of trace amounts of quaternary ammonium salts in pharmaceuticals. *Talanta*, *59*(1), 167–175. https://doi.org/10.1016/s0039-9140(02)00474-5

Schmidt, A. W., Reddy, K. R., & Knölker, H. J. (2012). Occurrence, biogenesis, and synthesis of biologically active carbazole alkaloids. *Chemical Reviews*, *112*(6), 3193–3328. https://doi.org/10.1021/cr200447s

Shang, X. F., Yang, C. J., Morris-Natschke, S. L., Li, J. C., Yin, X. D., Liu, Y. Q., Guo, X., Peng, J. W., Goto, M., Zhang, J. Y., & Lee, K. H. (2020). Biologically active isoquinoline alkaloids covering 2014–2018. *Medicinal Research Reviews*, *40*(6), 2212–2289. https://doi.org/10.1002/med.21703

Shang, X. F., Morris-Natschke, S. L., Liu, Y. Q., Li, X. H., Zhang, J. Y., & Lee, K. H. (2022). Biology of quinoline and quinazoline alkaloids. *The Alkaloids. Chemistry and Biology*, *88*, 1–47. https://doi.org/10.1016/bs.alkal.2021.08.002

Shen, Y., Liang, W. J., Shi, Y. N., Kennelly, E. J., & Zhao, D. K. (2020). Structural diversity, bioactivities, and biosynthesis of natural diterpenoid alkaloids. *Natural Product Reports*, *37*(6), 763–796. https://doi.org/10.1039/D0NP00002G

Srinivasan, P., & Smolke, C. D. (2020). Biosynthesis of medicinal tropane alkaloids in yeast. *Nature*, *585*(7826), 614–619. https://doi.org/10.1038/s41586-020-2650-9

Takla, S. S., Shawky, E., Hammoda, H. M., & Darwish, F. A. (2018). Green techniques in comparison to conventional ones in the extraction of Amaryllidaceae alkaloids: Best solvents selection and parameters optimization. *Journal of Chromatography. A*, *1567*, 99–110. https://doi.org/10.1016/j.chroma.2018.07.009

Tiburcio, A. F., & Galston, A. W. (1986). Arginine decarboxylase as the source of putrescine for tobacco alkaloids. *Phytochemistry*, *25*(1), 107–110. https://doi.org/10.1016/s0031-9422(00)94511-4

Wachira, S. W., Omar, S., Jacob, J. W., Wahome, M., Alborn, H. T., Spring, D. R., Masiga, D. K., & Torto, B. (2014). Toxicity of six plant extracts and two pyridone alkaloids from Ricinus communis against the malaria vector Anopheles gambiae. *Parasites & Vectors*, *7*, 312. https://doi.org/10.1186/1756-3305-7-312

Wang, F. P., & Chen, Q. H. (2014). Stemona alkaloids: Biosynthesis, classification, and biogenetic relationships. *Natural Product Communications*, *9*(12), 1809–1822.

Wang, X., Xin, J., Sun, L., Sun, Y., Xu, Y., Zhao, F., Niu, C., & Liu, S. (2024). Exploring the biomedical potential of terpenoid alkaloids: Sources, structures, and activities. *Molecules (Basel, Switzerland)*, *29*(9), 1968. https://doi.org/10.3390/molecules29091968

Wittmann, I., Schierling, A., Dettner, K., Göhl, M., Schmidt, J., & Seifert, K. (2015). Detection of a new piperideine alkaloid in the pygidial glands of some Stenus beetles. *Chemistry & Biodiversity*, *12*(9), 1422–1434. https://doi.org/10.1002/cbdv.201400391

Xing, J., Yang, Z., Lv, B., & Xiang, L. (2008). Rapid screening for cyclo-dopa and diketopiperazine alkaloids in crude extracts of Portulaca oleracea L. using liquid chromatography/tandem mass spectrometry. *Rapid Communications in Mass Spectrometry: RCM*, *22*(9), 1415–1422. https://doi.org/10.1002/rcm.3526

Yao, C. P., Zou, Z. X., Zhang, Y., Li, J., Cheng, F., Xu, P. S., Zhou, G., Li, X. M., Xu, K. P., & Tan, G. S. (2019). New adenine analogues and a pyrrole alkaloid from *Selaginella delicatula*. *Natural Product Research*, *33*(14), 1985–1991. https://doi.org/10.1080/14786419.2018.1482892

Yu, Y., Zhu, C., Wang, S., Song, W., Yang, Y., & Shi, J. (2013). Homosecoiridoid alkaloids with amino acid units from the flower buds of Lonicera japonica. *Journal of Natural Products*, *76*(12), 2226–2233. https://doi.org/10.1021/np4005773

Yuan, M. X., Qiu, Y., Ran, Y. Q., Feng, G. K., Deng, R., Zhu, X. F., Lan, W. J., & Li, H. J. (2019). Exploration of indole alkaloids from marine fungus *Pseudallescheria boydii* F44-1 using an amino acid-directed strategy. *Marine Drugs*, *17*(2), 77. https://doi.org/10.3390/md17020077

Zhang, J., Chen, J., Zhao, W., Wang, J., Pan, X., Lai, M., Nguemdjo Chimeze, V. W., & Pameté, E. (2023a). Optimizing the production of antioxidant and antibacterial alkaloids from Macleaya cordata by subcritical water extraction technology. *Chemistry & Biodiversity*, *20*(8), e202300048. https://doi.org/10.1002/cbdv.202300048

Zhang, Q., Jiang, Q., Sa, K., Liang, J., Sun, D., Li, H., & Chen, L. (2023b). Research progress of plant-derived natural alkaloids in central nervous system diseases. *Phytotherapy Research: PTR*, *37*(11), 4885–4907. https://doi.org/10.1002/ptr.7955

4 Extraction, Isolation, Purification, and Characterization of Alkaloids from Medicinal Plants

Noureddine Chaachouay

4.1 INTRODUCTION

W. Meissner, who was the first to define the term alkaloids, described them as all organic bases of plant origin in 1818 (Kukula-Koch & Widelski, 2017). Subsequent chemical analyses demonstrated the presence of nitrogen in a heterocyclic ring within the structure of alkaloids. Based on this definition, they were categorized as primary, secondary, tertiary, or quaternary amines. Alkaloids are a complex class of naturally occurring organic compounds, almost exclusively of plant origin. They are characterized by at least one heterocyclic ring containing nitrogen (Zenk & Juenger, 2007). Secondary metabolites are synthesized by plants primarily as a defense against herbivores and pathogens, and they influence plant physiology and ecology. The structural diversity and extensive biological activities of alkaloids have drawn the attention of investigators working on medicinal plants and the pharmaceutical industry (Yang & Stöckigt, 2010).

Alkaloid-containing plants have played a crucial role in the production of traditional medicinal systems around the world for many years. As examples of this long history, it is believed that Ayurvedic medicine employs the use of *Rauvolfia serpentina* to treat hypertension and or mental disorders. At the same time, Indigenous South American communities utilized Cinchona bark to reduce fevers (Kiran et al., 2018). The isolation of pure alkaloids (e.g., morphine from the opium poppy and quinine from Cinchona bark) and other compounds during the 19th century was a significant milestone in pharmacology, serving as an impetus for the development of modern pharmaceuticals (Patil, 2012).

The growing awareness that most of these remedies had strong biological activity triggered a growth in demand for systematic and scientific testing of those remedies, therefore, creating the discipline of molecular ethnopharmacology. Such a shift toward laboratory science has successfully provided medicinal plausibility to the plants rich in alkaloids and even led to the search for different biologically valid channels (Amirkia, 2016). This can be attributed to the existing information on thousands of alkaloids extracted from various plants under different ecological conditions, as well as the discovery of hundreds of alkaloids with remarkable medicinal properties, metabolic synthesis, and potential future clinical applications and syntheses (Crossley & Shenvi, 2015).

Because of the potent pharmacological properties of the alkaloids, they are still of utmost importance in modern medicine (Bhambhani et al., 2021; Chaachouay et al., 2023a). They are the precursors for many therapeutic agents, such as analgesics (morphine), antimalarials (quinine), anticancer agents (vincristine and vinblastine), and antihypertensives (reserpine). The continued exploration and characterization of these compounds serve to deepen biological understanding and uncover novel therapeutic agents. This chapter focuses on compiling information about the various methods

DOI: 10.1201/9781003539179-4

used for extracting, isolating, purifying, and identifying alkaloids from medicinal plants. Furthermore, we aim to emphasize the significance of alkaloids as integral bioactive compounds in drug discovery and their ongoing influence on both traditional and modern medicine.

4.2 CLASSIFICATION AND BIOSYNTHESIS OF ALKALOIDS

Alkaloids can be classified into four major categories: biological activity, chemical structure, biosynthetic pathway, and natural occurrence, which distinguishes them as non-heterocyclic or heterocyclic alkaloids, biological amines, or protoalkaloids (Gutiérrez-Grijalva et al., 2020). The framework of alkaloids leads to their division into two substance categories, namely non-heterocyclic and heterocyclic structures (Figure 4.1). Heterocyclic alkaloids contain one or more nitrogen atoms within ring structures, representing higher-compartmentalized molecules (Kukula-Koch & Widelski, 2017). The biosynthesis of alkaloids emerges from different amino acid precursors, which include the alkaloid morphine from various isoquinolines together with vincristine from indoles, as well as the tropanes producing atropine, and the alkaloid quinine results from quinolines. The ring structures containing nitrogen within these compounds exhibit various pharmacological properties, leading to analgesic and antimalarial actions (Chaachouay et al., 2021; El-Sakka, 2010). Due to their complex structure, heterocyclic alkaloids can act on specific biological targets and possess medicinal value. Non-heterocyclic alkaloids do not contain a nitrogen ring, yet still contain nitrogen within their bioactive structures. This family covers phenylethylamines (ephedrine), capsaicinoids (capsaicin), and colchicine-type alkaloids (Chaudhary et al., 2024). These compounds are less structurally complex than their heterocyclic counterparts, but they are known to exhibit various physiological activities, including serving as neurotransmitters, stimulants, and defensive agents in plants (Waller, 2012). Their biological effects are commonly achieved through receptor or enzymatic interactions, rather than multi-component molecular binding. Both types of alkaloids illustrate how minor differences in nitrogen incorporation generate radically different chemical and biological profiles, emphasizing nature's adaptability in producing biologically active molecules (Cushnie et al., 2014).

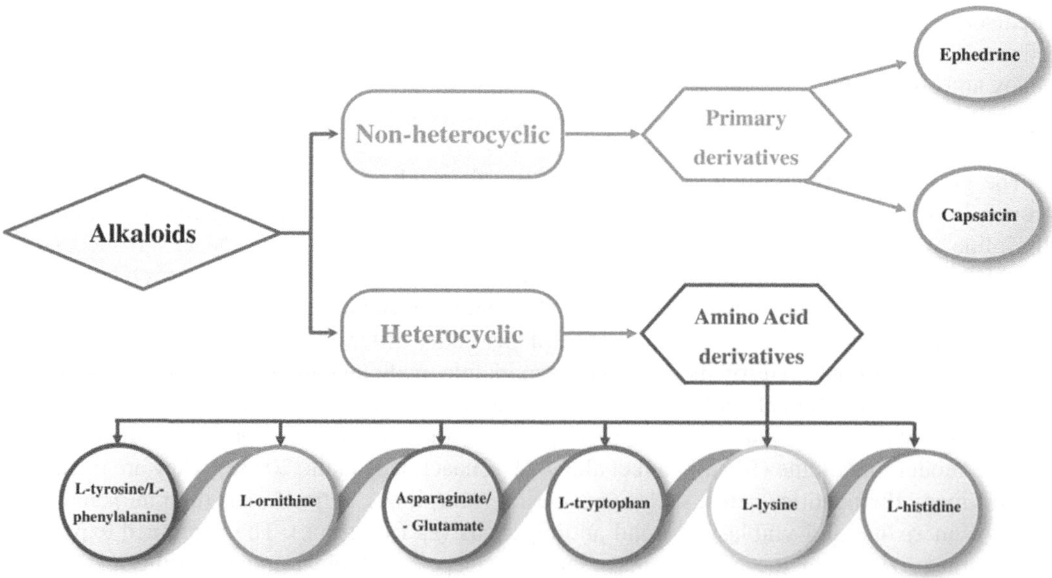

FIGURE 4.1 Classification of alkaloids.

The biosynthesis of plant alkaloids often initiates with primary metabolites, mainly amino acids (e.g., tryptophan, tyrosine, lysine, and ornithine), which are converted into intricate secondary metabolites via specialized metabolic pathways (Lichman, 2021). Thus, some secondary metabolites are formed by condensing two precursor (primary) metabolites, such as indole alkaloids, which are produced by coupling tryptamine (derived from tryptophan) with secologanin to produce ajmalicine and serpentine. Isoquinoline alkaloids are derived from the condensation of two tyrosine-like molecules, whereas tropane alkaloids are derived from ornithine via a putrescine intermediate (Waller, 2012). These biosynthetic pathways are often multistep processes, involving hydroxylation, methylation, oxidation, and glycosylation in specialized tissues or compartments. Deeper insights into plant metabolic engineering are now emerging through advances in genomics and metabolomics.

Alkaloids are present in plant tissues as water-soluble salts of organic acids (acetic, tartaric, malic, lactic acids, oxalic, and citric), esters (e.g., atropine, cocaine, aconitine, scopolamine), or combined with tannins (*Cinchona* bark) or sugars (e.g., the glycoalkaloids of *Solanum* species) rather than as free bases (Chaachouay et al., 2023b; Kukula-Koch & Widelski, 2017). Alkaloids are not equally distributed among plant families; some plant families are especially notable for their high alkaloid content. The Papaveraceae family is most noted for isoquinoline alkaloids such as morphine and sanguinarine (Bisset, 1985; Li et al., 2020). The Solanaceae represent a group of plants that contain both tropane alkaloids and steroidal alkaloids, which also encompasses the genera *Datura* and *Solanum*. Rubiaceae: alkaloids indoliques. Amaryllidaceae and Apocynaceae are other families that produce a significant amount of natural alkaloids (Bhambhani et al., 2021; Li et al., 2020). This distribution is not arbitrary; it reflects evolutionary pressures and ecological functions, including deterring herbivores and defending against microbes. Investigation of them is complex because of the effects of environmental factors, developmental stages, and genetic regulation on their biosynthesis and accumulation. Nevertheless, it is also crucial for the development of pharmacognosy and drug discovery.

4.3 PREPARATION OF PLANT MATERIAL

Phytomedicinal science has become a well-established field that has evolved through extensive efforts and the accurate collection and authentication of phytomedicinal specimens. The identification must be correct, as misidentification may yield false results and pose safety implications (Williamson et al., 1996). Conventional research practices require voucher specimens (i.e., preserved source samples deposited in herbaria for future reference). These physical examples serve as a verifiable signature of the plant material used in such studies and are critical for reproducibility (Zygmunt & Namieśnik, 2003). Morphological characters and anatomical studies can facilitate authentication, while the molecular approach of DNA barcoding enables the accurate identification of plant species. When threats to biodiversity are filtered through good authentication and identification practices, the botanical materials used in scientific research are authentic and traceable (Mishra et al., 2016).

After authentication, the plant material is dried, powdered, and stored for stability and subsequent extraction (Figure 4.2). Drying is a necessary process that eliminates water, which prevents pathogenic growth and oxidation of bioactive compounds. Different drying techniques are employed, including air-drying, oven-drying, and freeze-drying; each method has its own merits, depending on the plant's nature and the sensitivity of its constituents (Nurhaslina et al., 2022). Freeze-drying is an example of a low-temperature process that is highly effective in maintaining the integrity of thermolabile substances (Hussain et al., 2020). The plant material is then ground when dry to convert it into a fine powder, increasing the surface area and enabling better solvent extraction (Inyang et al., 2017). Appropriate storage in sealed containers and protection from light and moisture are essential to preserve the stability and integrity of the phytochemicals until analysis.

The initial treatment method removes unwanted constituents that could disrupt the extraction of the compound. Defatting is one of several pretreatment procedures, particularly in seed processing

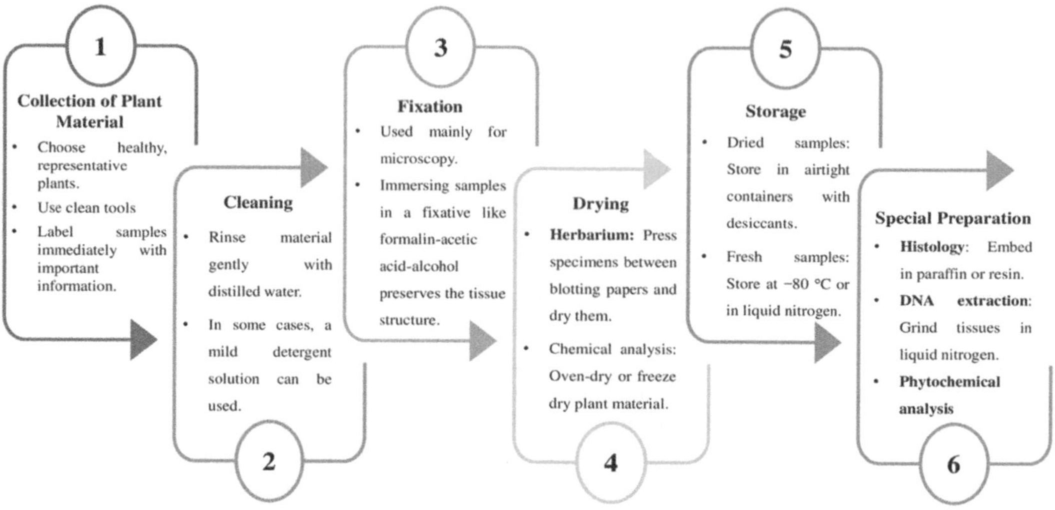

FIGURE 4.2 Different stages of preparation of plant material.

and the extraction of lipid-rich plant matter (Azwanida, 2015). Lipids and oils must be stripped out because extracting them with other compounds complicates the analysis, while reducing the efficiency of extracting polar compounds, particularly for polysaccharides. Appropriate pre-extraction treatments enhance the purity and yield of phytochemical responses, leading to more reliable and efficient downstream analysis methods (Sipoloni et al., 2025).

4.4 EXTRACTION OF ALKALOIDS

4.4.1 Traditional Extraction Methods

Traditional methods for alkaloid extraction from medicinal plants, such as maceration, percolation, and Soxhlet extraction, are fundamentally based on the principle of using a suitable solvent to dissolve and separate these bioactive compounds from the plant matrix (Kalaskar et al., 2025). Maceration, the most straightforward technique, involves soaking chopped plant matter in a solvent for an extended period with frequent mixing. This extraction is accomplished using passive diffusion, where the solvent can enter the cells to solubilize the alkaloids (Kalaskar et al., 2025). Although it is simple and relatively easy, requiring little equipment, the main disadvantages include a long extraction time and limited yield, as well as the use of a higher amount of solvents (Dagdag et al., 2024). An approach to resolving some of the drawbacks of maceration is percolation, which uses a flow of fresh solvent through the packed plant material in a percolator (Hidayat & Wulandari, 2021).

A key advance in continuous solid–liquid extraction is the Soxhlet extraction technique, developed by Franz von Soxhlet in 1879 and still considered the benchmark for comparison with more modern methods (Taylor, 2020). Under this design, the sample is put in a thimble, and a continuous flow of fresh solvent is passed through the sample via distillation in a boiling flask. The refluxing phase ensures that the sample is constantly exposed to the correct solvent, thereby ensuring that very high extraction yields are obtained. The Soxhlet apparatus is available, particularly for the extraction of non-polar alkaloids using organic solvents such as chloroform or dichloromethane (Thorat, 2018). The most significant advantages of this approach are that it can be applied with a high sample size and does not require filtration. The first drawbacks are extremely long extraction times, the use of large amounts of possibly toxic solvents, and high temperature stresses, which cause the loss of thermolabile compounds. These have been highlighted as the main drawbacks in the available literature. These shortcomings have driven the development of more effective and environmentally benign methodologies.

4.4.2 Modern Extraction Techniques

Modern extraction techniques have been developed to overcome the limitations of traditional methods, with a focus on enhancing efficiency, yield, selectivity, and sustainability. Ultrasound-assisted extraction (UAE) utilizes cavitation bubbles generated by high-frequency sound waves to disrupt plant cell walls, enhancing solvent penetration and mass transfer (M. Siddique et al., 2025). This results in significantly shorter extraction times and reduced solvent consumption at low temperatures, preserving thermolabile alkaloids. Similarly, MAE employs electromagnetic radiation to cause instantaneous heating of the moisture within plant cells, creating high internal pressure that ruptures the cells and rapidly releases bioactive compounds into the solvent (Destandau & Michel, 2022). MAE is known to be extremely fast and efficient; however, it requires solvents with high dielectric constants.

Further advanced techniques offer even greater selectivity and green credentials. Supercritical fluid extraction (SFE), predominantly using carbon dioxide (CO_2) as the supercritical fluid, provides a tunable and solvent-free alternative (Herzyk et al., 2024). The solvating power of supercritical CO_2 can be finely adjusted by varying temperature and pressure, allowing for the highly selective extraction of target alkaloids, often without the need for extensive post-processing (Laurintino, 2024). Pressurized liquid extraction (PLE), also known as accelerated solvent extraction (ASE), operates by using solvents at elevated temperatures and pressures well above their normal boiling points (Alarcon et al., 2023). This maintains the solvents in a liquid state, dramatically enhancing their solubilizing power and kinetics, which leads to fast, efficient extractions with minimal solvent volume compared to Soxhlet.

The efficiency of all these techniques, whether modern or traditional, is governed by a complex interplay of several key factors. The choice of solvent is paramount, as its polarity must match that of the target alkaloids (e.g., pure or acidified polar solvents for hydrophilic alkaloids, and less polar solvents for lipophilic ones). Temperature is a double-edged sword; while it generally increases solubility and diffusion rates, it can also lead to the degradation of sensitive compounds. Extraction time must be optimized, as modern methods like UAE and MAE achieve equilibrium much faster than traditional ones (Pereira et al., 2023). Furthermore, parameters such as pH are critical for alkaloid extraction, as adjusting them can convert alkaloids into their salt forms (acidic pH) for better solubility in polar solvents or into their free base forms (basic pH) for extraction into organic solvents (Yubin et al., 2014). Other plant-specific factors, including particle size and the plant-to-solvent ratio, also significantly influence the extraction yield and kinetics.

4.5 ISOLATION AND PURIFICATION OF ALKALOIDS

The isolation and purification of alkaloids from plants involves a series of systematic steps, each designed to further utilize the unique chemical characteristics and activities of these compounds (Phukan et al., 2023). Acid–base extraction is first used, based on the amphoteric nature of alkaloids. In this extraction process, plant material is subjected to acidic treatment, which converts the alkaloids into their water-soluble salts. The salts are then extracted from the plant material, removing non-alkaloid, non-aqueous constituents (Zahari et al., 2016).

4.5.1 Solvent Partitioning and Liquid–Liquid Extraction

Solvent partitioning, the largest part of the process, is a critical procedure in the isolation and purification of alkaloids from medicinal plants, as it involves liquid–liquid extraction (Yubin et al., 2014). This approach has been developed based on the selectivity of alkaloids in distinguishing between the relative behaviors of an aqueous-soluble salt form under acidic conditions and the organic-soluble free base form under basic conditions. Typically, the separation occurs between an aqueous phase and an organic, immiscible solvent, such as dichloromethane or chloroform, under various pH conditions. An extract is then acidified to protonate the alkaloids, causing them to precipitate into the aqueous layer, thereby separating them from neutral and acidic contaminations (Yubin et al., 2014). The basification of the aqueous phase then deprotonates the alkaloids, allowing them to revert to their lipophilic form and be extracted into a fresh organic solvent (Yubin et al., 2014). This yields a significantly enriched total alkaloid fraction, with its efficiency hinging on the solvent choice, number of extraction stages, and the strength of the acids and bases used.

4.5.2 Precipitation Techniques

Precipitation techniques serve as a classical and effective method for the initial isolation and purification of alkaloids from crude plant extracts (Yubin et al., 2014). This approach leverages the ability of alkaloids to form insoluble complexes with specific reagents, enabling their selective separation from other extract components. Common precipitants include Mayer's reagent (potassium mercuric iodide), Dragendorff's reagent (potassium bismuth iodide), and Wagner's reagent (iodine–potassium iodide), which react with alkaloids to form distinctive crystals or amorphous solids that can be collected via filtration or centrifugation (Alamgir, 2018). The isolated precipitate is then treated to decompose the complex, often using hydrogen sulfide to break down metal-based complexes or appropriate solvents to liberate and recover the pure alkaloid. While highly selective for certain alkaloid classes, a key limitation is that these reagents are not universally applicable to all alkaloid types, and the process may require subsequent steps to remove reagent-derived impurities. Despite the advent of chromatography, precipitation remains a valuable tool for rapid initial concentration and purification, particularly in the early stages of alkaloid investigation or for the targeted isolation of specific alkaloid groups (Leitão et al., 2021).

4.5.3 Chromatographic Methods

Chromatographic methods are the cornerstone of modern alkaloid isolation and purification, enabling the separation of complex mixtures based on differential partitioning between a stationary and mobile phase (Subash et al., 2025). The process typically begins with column chromatography, a versatile and scalable preparative technique that utilizes a silica gel or alumina-packed column with solvents of increasing polarity to fractionate the crude alkaloid extract into broader groups based on polarity. Thin-layer chromatography (TLC) is an indispensable parallel analytical tool used to rapidly monitor the progress of column chromatography, determine the number of constituents in a mixture, and optimize solvent systems by visualizing separated bands under UV light or with specific alkaloid-detecting sprays, such as Dragendorff's reagent (Bele & Khale, 2011). For more efficient and rapid separations, flash chromatography, a medium-pressure variant of column chromatography, is employed, utilizing smaller particle sizes and pressurized gas to achieve higher resolution and faster run times than conventional gravity-fed columns. This serves as a crucial intermediate purification step.

For the final stages of purification, high-performance liquid chromatography (HPLC) is the gold standard due to its superior resolving power (Aguilar & Hearn, 1996). Operating at high pressure with precisely controlled pumps, HPLC forces the sample through a tightly packed column containing very fine stationary phase particles, achieving exceptional separation of structurally similar alkaloids that other methods cannot resolve. Analytical HPLC is used to assess purity and identify compounds, often coupled with mass spectrometry (HPLC-MS) or diode array detection (HPLC-DAD) for characterization (Wolfender, 2009). Preparative HPLC is used when a large quantity of pure compound, measured in milligrams or grams, is required. Larger columns and higher flow rates are employed to isolate specific alkaloids to an extremely high level of purity, which is necessary for definitive structure elucidation through NMR and biological testing. The choice of phase used in the mobile phase, column chemistry, and gradient profile are essential variables that determine the success of HPLC separation of any particular alkaloid mixture (Zhao et al., 2020).

4.5.4 Advanced Techniques

The most recent methods for the ultimate isolation and purification of alkaloids from medicinal plants are largely controlled by Counter-Current Chromatography (CCC) and preparative High-Performance Liquid Chromatography (prep-HPLC), which offer better resolution and scalability (Sotenjwa, 2020). CCC is a support-free liquid–liquid partitioning method that does not depend on adsorption or denaturation of solid stationary phases, instead relying on the dependence of the solubility of compounds between two immiscible liquid layers; thus, it is unusually effective with large and crude samples and in the extraction of soluble and fragile alkaloids with high productivity (Hu & Pan, 2012). At the same time, pre-purification HPLC is considered the ultimate high-resolution technology for obtaining final purity, as it utilizes strong columns packed with small particles of the stationary phase to isolate the most structurally related analogs of alkaloids through sharp gradient elution under high pressure (Stead, 1998). The two methods are very complementary, and both are very likely to be chosen jointly. CCC also offers an easy way to effectively fractionate mixed complexes, initially, and finally purify them by prep-HPLC to procure analytically pure compounds for structural characterization and bioactivity tests (Sutherland & Fisher, 2009).

4.6 CHARACTERIZATION AND STRUCTURAL ELUCIDATION

4.6.1 Spectroscopic Techniques

On the grounds of its arrangement, the characterization and structure elucidation of purified alkaloids relies on a compilation of the current literature conducted by a search in major scientific databases, and on a multifaceted approach based on the application of spectroscopic techniques, which are complementary and critical in nature of the information (Verpoorte & Schripsema, 1994). The process typically begins with Ultraviolet-Visible (UV-Vis) spectroscopy, which provides initial indications of chromophores and conjugated systems within the molecule (Molyneux et al., 2019). Potential classes of an aeroplane alumina, e.g., indoles, isoquinolines, or pyrrols, can be indicated by their absorption maxima, depending on their profiles against known reference sunlight. They can be used as an excellent preliminary step towards determining the properties of the core compound.

Fourier transform infrared spectroscopy (FTIR) is employed to give a more detailed analysis of the functional groups (Khan et al., 2018; Siddique, 2024). This method values the typical frequencies of vibration of the covalent bonds that recently provided a functional group fingerprint of the molecule. The FTIR spectra may be used to determine the identity of amine groups and carbonyl groups (Berthomieu & Hienerwadel, 2009). The molecular boots can be reconstructed, and the different sub-classes of alkaloids can be distinguished by researchers depending on their unique infrared absorption patterns.

The final tool for complete structural elucidation, however, is Nuclear Magnetic Resonance (NMR) spectroscopy (Elyashberg, 2015). One-dimensional NMR experiments, primarily performed on ^{1}H and ^{13}C NMR, provide valuable information about the number and type of hydrogen and carbon atoms in the molecule, their chemical environments, and their connectivity (Huang et al., 2007). This basic information is then enormously elaborated by two-dimensional (2D) NMR methods. Correlation spectroscopy provides proton–proton couplings, Heteronuclear Single Quantum Coherence offers direct carbon–hydrogen bonds, Heteronuclear Multiple Bond Correlation detects long-range couplings, and the construction of the carbon skeleton can be achieved. All atoms can be unequivocally assigned in an often very complex and unknown alkaloid molecule, and its identity can thus be demonstrated (Byrne, 2007; Tarighat et al., 2024).

4.6.2 Spectrometric Techniques

Mass spectrometry (MS) is a valuable method for the structural elucidation of alkaloids, providing crucial data on molecular weight and compound fragmentation (Budzikiewicz, 2015). Techniques are frequently paired with separation techniques to test complex mixtures (Zaikin & Halket, 2006).

The Gas Chromatography-Mass Spectrometry (GC-MS) technique is used for volatile and thermally stable compounds. At the same time, the method of Liquid Chromatography-Mass Spectrometry (LC-MS) is found to be helpful for the analysis of polar, non-volatile, and heat-labile alkaloids (Frolova et al., 2025). The fragmentation pattern generated by methods such as tandem MS (MS/MS) generates a diagnostic fingerprint, which carries with it structural information either by the loss of characteristic neutral fragments or by the formation of specific product ions, which can either be compared to spectral libraries or may be used to suggest a new structure (T. Li et al., 2023).

High-Resolution Mass Spectrometry (HR-MS) is the preferred method for determining molecular formula. HR-MS enables researchers to differentiate ions that have the same mass but differ in terms of the elements composing the molecule by determining the mass of the entire molecule with unprecedented accuracy (Wu et al., 2012). A shortlist of possible elemental formulas is obtained using the exact mass data. It is often the most important and first step in characterizing a newly discovered or unknown alkaloid. Combined with the fragmentation data from MS/MS and the structural information from NMR spectroscopy, HR-MS provides clear information on the molecular formula, completing the entire dataset needed to fully elucidate the structure (Kumar et al., 2023).

4.7 APPLICATIONS AND FUTURE PERSPECTIVES

One of the pharmacological effects is represented by alkaloids, a heterogeneous group of naturally occurring compounds (Chaachouay et al., 2020; Manske & Holmes, 2014). Anti-inflammatory and neuroprotective, many plant alkaloids are considered first-choice agents in the management of neurodegenerative diseases, such as Alzheimer's and Parkinson's (Utpal et al., 2025). They are known to act by inhibiting major enzymes, such as acetylcholinesterase (AChE), and through the modulation of nuclear factor-kappa B and cyclooxygenase-2 (COX-2)-mediated inflammatory pathways (Maphetu et al., 2022). For example, berberine and harmine have demonstrated potent neuroprotective activity by modulating these pathways, resulting in the inhibition of inflammation and oxidative stress in brain tissue (Mohi-Ud-Din et al., 2022). Furthermore, alkaloids have been shown to have neuroprotective effects and have demonstrated promise in treating other diseases. Compounds, such as tetrandrine and sinomenine, have anti-inflammatory actions, and galantamine has already been recognized for clinical use in symptomatic management treatments for Alzheimer's disease (Plazas et al., 2022). The multipurpose therapeutic potential of alkaloids has highlighted their use in drug discovery and development. New technologies are expected to play a crucial role in alkaloid research and application. Nanotechnology, for instance, has been utilized to enhance the bioavailability and efficacy of alkaloid treatment. Nanoparticulation of Berberine leads to improved antimicrobial activity compared to the bulk drug treatment (Behl et al., 2022; Mohi-Ud-Din et al., 2022). Furthermore, producing alkaloids based on plant cells is a sustainable and controllable way of synthesizing these compounds, as it adapts to seasonal variations and environmental conditions. Such progress, combined with the application of artificial intelligence and machine learning in drug discovery, should lead to the accelerated development of alkaloid-derived therapeutics in the future, thereby overcoming current limitations and facilitating their widespread use in the clinic.

4.8 CONCLUSION

Based on the detailed phytochemical processes outlined, it is concluded that the successful isolation of pure, characterized alkaloids from medicinal plants is a complex yet achievable multidisciplinary endeavor. The efficacy of initial extraction is paramount and is highly dependent on the careful selection of solvents and methods (e.g., maceration, Soxhlet) based on the target compound's polarity and stability. Subsequent isolation and purification, primarily through chromatographic techniques such as TLC, column chromatography, and increasingly HPLC, are crucial for separating the complex mixture of plant metabolites and obtaining individual alkaloids in high purity.

Final structural elucidation and characterization rely heavily on advanced spectroscopic and spectrometric analyses. Techniques such as Mass Spectrometry (MS) provide molecular mass and fragmentation patterns, while NMR spectroscopy offers detailed information on carbon and proton frameworks. Additionally, FTIR confirms the presence of functional groups. The rigorous application of this sequential protocol—extraction, isolation, purification, and characterization—is fundamental to authenticating the chemical identity of the isolated alkaloid. This confirmed that purity and structure are the essential prerequisites for any subsequent pharmacological, toxicological, or clinical research aimed at validating the traditional use of the plant and developing these natural compounds into potential therapeutic agents.

COMPETING INTERESTS

The authors declare that they have no competing interest.

DECLARATION OF GENERATIVE AI AND AI-ASSISTED TECHNOLOGIES IN THE WRITING PROCESS

During the preparation of this work, the book chapter author(s) have not taken any assistance from AI, and the book chapter author(s) reviewed and edited the content as needed and take(s) full responsibility for the content of the publication.

LIST OF ABBREVIATIONS

2D	Two-Dimensional
AChE	Acetylcholinesterase
ASE	Accelerated Solvent Extraction
CCC	Counter-Current Chromatography
COX-2	Cyclooxygenase-2
DAD	Diode Array Detection
ESI	Electrospray Ionization
FTIR	Fourier Transform Infrared Spectroscopy
HPLC	High-Performance Liquid Chromatography
HR-MS	High-Resolution Mass Spectrometry
MAE	Microwave-Assisted Extraction
MS	Mass Spectrometry
NMR	Nuclear Magnetic Resonance
NMR Spectroscopy	Nuclear Magnetic Resonance Spectroscopy
PLE	Pressurized Liquid Extraction
Prep-HPLC	Preparative High-Performance Liquid Chromatography
SFE	Supercritical Fluid Extraction
TLC	Thin-Layer Chromatography
UAE	Ultrasound-Assisted Extraction
UV-Vis	Ultraviolet-Visible

REFERENCES

Aguilar, M.-I., & Hearn, M. T. W. (1996). High-resolution reversed-phase high-performance liquid chromatography of peptides and proteins. In *Methods in Enzymology* (Vol. 270, pp. 3–26). Academic Press. https://doi.org/10.1016/S0076-6879(96)70003-4

Alamgir, A. N. M. (2018). Methods of qualitative and quantitative analysis of plant constituents. In A. N. M. Alamgir (Ed.), *Therapeutic Use of Medicinal Plants and Their Extracts: Volume 2* (Vol. 74, pp. 721–804). Springer International Publishing. https://doi.org/10.1007/978-3-319-92387-1_9

Alarcon, N. Á., Ramírez, J. S. O., Ortega-Barbosa, J. P., Ballesteros-Vivas, D., Rodríguez-Varela, L. I., & Alfonso, F. P. (2023). Pressurized liquid extraction (PLE). In *Green Extraction Techniques in Food Analysis* (pp. 357–395). Bentham Science Publishers. https://www.benthamdirect.com/content/books/9789815049459.chap9

Amirkia, V. (2016). *Plant extracts and natural products-Predictive structural and biodiversity-based analyses of uses, bioactivity, and'research and development'potential* [PhD Thesis, UCL (University College London)]. https://discovery.ucl.ac.uk/id/eprint/1527357/

Azwanida, N. N. (2015). A review on the extraction methods use in medicinal plants, principle, strength and limitation. *Med Aromat Plants*, *4*(196), 2167–2412.

Behl, T., Singh, S., Sharma, N., Zahoor, I., Albarrati, A., Albratty, M., Meraya, A. M., Najmi, A., & Bungau, S. (2022). Expatiating the pharmacological and nanotechnological aspects of the alkaloidal drug berberine: Current and future trends. *Molecules*, *27*(12), 12. https://doi.org/10.3390/molecules27123705

Bele, A., & Khale, A. (2011). An overview on thin layer chromatography. *International Journal of Pharmacy and Pharmaceutical Sciences*, *2*, 256.

Berthomieu, C., & Hienerwadel, R. (2009). Fourier transform infrared (FTIR) spectroscopy. *Photosynthesis Research*, *101*(2–3), 157–170. https://doi.org/10.1007/s11120-009-9439-x

Bhambhani, S., Kondhare, K. R., & Giri, A. P. (2021). Diversity in chemical structures and biological properties of plant alkaloids. *Molecules*, *26*(11), 3374.

Bisset, N. G. (1985). Plants as a source of isoquinoline alkaloids. In *The Chemistry and Biology of Isoquinoline Alkaloids* (pp. 1–22). Springer. https://link.springer.com/content/pdf/10.1007/978-3-642-70128-3_1?pdf=chapter%20toc

Budzikiewicz, H. (2015). Mass spectrometry in natural product structure elucidation. In A. D. Kinghorn, H. Falk, & J. Kobayashi (Éds.), *Progress in the Chemistry of Organic Natural Products* (Vol. 100, pp. 77–221). Springer International Publishing. https://doi.org/10.1007/978-3-319-05275-5_2

Byrne, L. T. (2007). Nuclear magnetic resonance spectroscopy: Strategies for structural determination. In *Bioactive Natural Products* (pp. 91–126). CRC Press. https://www.taylorfrancis.com/chapters/edit/10.1201/9781420006889-7/nuclear-magnetic-resonance-spectroscopy-strategies-structural-determination-lindsay-byrne

Chaachouay, N., Azeroual, A., Bencharki, B., Douira, A., & Zidane, L. (2023a). Various Metabolites and or Bioactive Compounds from Vegetables, and Their Use Nanoparticles Synthesis, and Applications. In: Husen, A. (eds) *Nanomaterials from Agricultural and Horticultural Products. Smart Nanomaterials Technology* (pp. 187–209). Springer Nature. https://doi.org/10.1007/978-981-99-3435-5_10

Chaachouay, N., Azeroual, A., Benkhnigue, O., & Zidane, L. (2023b). Alkaloids: A suitable precursor for nanomaterials synthesis, and their various applications. In A. Husen (Éd.), *Secondary Metabolites Based Green Synthesis of Nanomaterials and Their Applications* (pp. 23–48). Springer Nature. https://doi.org/10.1007/978-981-99-0927-8_2

Chaachouay, N., Douira, A., Hassikou, R., Brhadda, N., Dahmani, J., Belahbib, N., Ziri, R., & Zidane, L. (2020). *Etude floristique et ethnomédicinale des plantes aromatiques et médicinales dans le Rif (Nord du Maroc)* [PhD Thesis, Département de Biologie-Université Ibn Tofail-Kénitra]. https://scholar.google.com/scholar?cluster=12967211105625020518&hl=en&oi=scholarr

Chaachouay, N., Douira, A., & Zidane, L. (2021). COVID-19, prevention and treatment with herbal medicine in the herbal markets of Salé Prefecture, North-Western Morocco. *European Journal of Integrative Medicine*, *42*, 101285.

Chaudhary, B., Patial, B., Sharma, R., & Chawla, A. (2024). An emerging class of antimicrobial heterocycles derived from natural sources. *Letters in Drug Design & Discovery*, *21*(13), 2521–2536.

Crossley, S. W., & Shenvi, R. A. (2015). A longitudinal study of alkaloid synthesis reveals functional group interconversions as bad actors. *Chemical Reviews*, *115*(17), 9465–9531.

Cushnie, T. T., Cushnie, B., & Lamb, A. J. (2014). Alkaloids: An overview of their antibacterial, antibiotic-enhancing and antivirulence activities. *International Journal of Antimicrobial Agents*, *44*(5), 377–386.

Dagdag, O., Kim, H., Haldhar, R., Daoudi, W., & Berdimurodov, E. (2024). Phytochemicals/plant extracts: Preparation, characterization, and properties. In Chandrabhan Verma, Ashish Kumar, & Abhinay Thakur (eds) *Phytochemistry in Corrosion Science*. CRC Press.

Destandau, E., & Michel, T. (2022). *Microwave-Assisted Extraction*. https://books.rsc.org/books/edited-volume/1999/chapter/4578767

El-Sakka, M. A. (2010). *Phytochemistry (3) Alkaloids*. Al Azhar University, Faculty of Pharmacy, Department of Pharmacognosy. https://alazharpharmacy.com/documents/level-4/semester-2/phytochemistry-3/phytochemistry-3-book.pdf

Elyashberg, M. (2015). Identification and structure elucidation by NMR spectroscopy. *TrAC Trends in Analytical Chemistry*, *69*, 88–97. https://doi.org/10.1016/j.trac.2015.02.014

Frolova, N., Orlova, A., Popova, V., Bilova, T., & Frolov, A. (2025). *GC-MS-Based Metabolomics–Part 1 Gas Chromatography-Mass Spectrometry (GC-MS) and Its Place in the Plant Metabolomics Toolbox.* https://chemrxiv.org/engage/chemrxiv/article-details/6787437981d2151a0231c5f7

Gutiérrez-Grijalva, E. P., López-Martínez, L. X., Contreras-Angulo, L. A., Elizalde-Romero, C. A., & Heredia, J. B. (2020). Plant alkaloids: Structures and bioactive properties. In *Plant-Derived Bioactives: Chemistry and Mode of Action* (pp. 85–117). Springer Singapore.

Herzyk, F., Piłakowska-Pietras, D., & Korzeniowska, M. (2024). Supercritical extraction techniques for obtaining biologically active substances from a variety of plant byproducts. *Food*, *13*(11), 1713.

Hidayat, R., & Wulandari, P. (2021). Methods of extraction: Maceration, percolation and decoction. *Eureka Herba Indonesia*, *2*(1), 68–74.

Hu, R., & Pan, Y. (2012). Recent trends in counter-current chromatography. *TrAC Trends in Analytical Chemistry*, *40*, 15–27. https://doi.org/10.1016/j.trac.2012.07.018

Huang, X., Powers, R., Tymiak, A., Espina, R., & Roongta, V. (2007). Introduction to NMR and its application in metabolite structure determination. In D. Zhang, & M. Zhu (Eds.), *Drug Metabolism in Drug Design and Development: Basic Concepts and Practice* (pp. 369–405). John Wiley & Sons, Inc.

Hussain, Q., Shi, J., Scheben, A., Zhan, J., Wang, X., Liu, G., Yan, G., King, G. J., Edwards, D., & Wang, H. (2020). Genetic and signalling pathways of dry fruit size: Targets for genome editing-based crop improvement. *Plant Biotechnology Journal*, *18*(5), 1124–1140.

Inyang, U., Oboh, I., & Etuk, B. (2017). Drying and the different techniques. *International Journal of Food Nutrition and Safety*, *8*(1), 45–72.

Kalaskar, M., Yele, S. U., Ayyanar, M., Gurav, N., Beldar, V., & Surana, S. J. (2025). Methods of Extraction. In Uchenna E. Odoh, Shailendra S. Gurav, & Michael O. Chukwuma (eds) *Pharmacognosy and Phytochemistry: Principles, Techniques, and Clinical Applications* (pp. 121–142). John Wiley & Sons, Inc.

Khan, S. A., Khan, S. B., Khan, L. U., Farooq, A., Akhtar, K., & Asiri, A. M. (2018). Fourier transform infrared spectroscopy: Fundamentals and application in functional groups and nanomaterials characterization. In S. K. Sharma (Éd.), *Handbook of Materials Characterization* (pp. 317–344). Springer International Publishing. https://doi.org/10.1007/978-3-319-92955-2_9

Kiran, K., Priya, A. J., & Devi, R. G. (2018). Medicinal and therapeutic uses of Rauwolfia serpentina-Indian snakeroot. *Drug Invention Today*, *10*. https://search.ebscohost.com/login.aspx?direct=true&profile=ehost&scope=site&authtype=crawler&jrnl=09757619&AN=133549239&h=%2BZTP5OA3IGcRvdoqmuXBnfQzuT2lPJK66WLr3SIEypA1KKwgDLVlV%2FPutl6lnaJfeCUitrhgzYC4qJJnRlTNAw%3D%3D&crl=c

Kukula-Koch, W. A., & Widelski, J. (2017). Alkaloids. In *Pharmacognosy* (pp. 163–198). Elsevier. https://www.sciencedirect.com/science/article/pii/B9780128021040000093

Kumar, S., Gajjela, R., Kumar, H., Arulraj, R. A., Subramaniam, S., Kothandaramachandran, T., Chauthe, S. K., Gupta, A., Mathur, A., & Roy, A. (2023). A strategy for evaluation of isotopic enrichment and structural integrity of deuterium labelled compounds by using HR-MS and NMR. *Analytical Methods*, *15*(11), 1470–1477.

Laurintino, T. K. S. (2024). *Supercritical Extraction of Costus Spicatus Leafs Combined with Ultrasound Pretreatment: Experimentation and Modeling.* https://repositorio.ufsc.br/handle/123456789/263633

Leitão, G. G., Leal, C. M., Mendonça, S. C., & Pereda-Miranda, R. (2021). Purification of alkaloids by countercurrent chromatography. *Revista Brasileira de Farmacognosia*, *31*(5), 625–647. https://doi.org/10.1007/s43450-021-00163-4

Li, T., Zhang, K., Niu, X., Chen, W., Yang, X., Gong, X., Tu, P., Wang, Y., Liu, W., & Song, Y. (2023). MS/MS fingerprint comparison between adjacent generations enables substructure identification: Flavonoid glycosides as cases. *Journal of Pharmaceutical and Biomedical Analysis*, *234*, 115559.

Li, Y., Winzer, T., He, Z., & Graham, I. A. (2020). Over 100 million years of enzyme evolution underpinning the production of morphine in the Papaveraceae family of flowering plants. *Plant Communications*, *1*(2). https://www.cell.com/plant-communications/fulltext/S2590-3462(20)30010-9

Lichman, B. R. (2021). The scaffold-forming steps of plant alkaloid biosynthesis. *Natural Product Reports*, *38*(1), 103–129.

Manske, R. H. F., & Holmes, H. L. (2014). *The Alkaloids: Chemistry and Physiology*. Elsevier.

Maphetu, N., Unuofin, J. O., Masuku, N. P., Olisah, C., & Lebelo, S. L. (2022). Medicinal uses, pharmacological activities, phytochemistry, and the molecular mechanisms of *Punica granatum* L. (pomegranate) plant extracts: A review. *Biomedicine & Pharmacotherapy*, *153*, 113256. https://doi.org/10.1016/j.biopha.2022.113256

Mishra, P., Kumar, A., Nagireddy, A., Mani, D. N., Shukla, A. K., Tiwari, R., & Sundaresan, V. (2016). DNA barcoding: An efficient tool to overcome authentication challenges in the herbal market. *Plant Biotechnology Journal*, *14*(1), 8–21.

Mohi-Ud-Din, R., Mir, R. H., Wani, T. U., Shah, A. J., Banday, N., & Pottoo, F. H. (2022). Berberine in the treatment of neurodegenerative diseases and nanotechnology enabled targeted delivery. *Combinatorial Chemistry & High Throughput Screening*, *25*(4), 616–633. https://doi.org/10.2174/1386207324666210804122539

Molyneux, R. J., Beck, J. J., Colegate, S. M., Edgar, J. A., Gaffield, W., Gilbert, J., Hofmann, T., McConnell, L. L., & Schieberle, P. (2019). Guidelines for unequivocal structural identification of compounds with biological activity of significance in food chemistry (IUPAC Technical Report). *Pure and Applied Chemistry*, *91*(8), 1417–1437. https://doi.org/10.1515/pac-2017-1204

Nurhaslina, C. R., Bacho, S. A., & Mustapa, A. N. (2022). Review on drying methods for herbal plants. *Materials Today: Proceedings*, *63*, S122–S139.

Patil, P. N. (2012). *Discoveries in Pharmacological Sciences*. World Scientific. https://books.google.fr/books?hl=fr&lr=&id=PVG6CgAAQBAJ&oi=fnd&pg=PR5&dq=The+isolation+of+pure+alkaloids+(e.g.,+morphine+from+opium+poppy+and+quinine+from+Cinchona+bark)+and+other+compounds+during+the+19th+century+were+significant+milestones+in+pharmacology+and+was+an%E2%80%82impetus+for+the+invention+of+modern+pharmaceuticals&ots=hp6Zd3VZ7A&sig=2uL1sUbnPBcL7VMEWWbquoAmi6g

Pereira, A. G., Cruz, L., Cassani, L., Chamorro, F., Lourenço-Lopes, C., Freitas, V., Otero, P., Fraga-Corral, M., Prieto, M. A., & Simal-Gandara, J. (2023). Comparative study of microwave-assisted extraction and ultrasound-assisted extraction techniques (MAE vs. UAE) for the optimized production of enriched extracts in phenolic compounds of Camellia japonica var Eugenia de Montijo. *Engineering Proceedings*, *37*(1), 124.

Phukan, M. M., Sangma, S. R., Kalita, D., Bora, P., Das, P. P., Manoj, K., Pankaj, P. P., Jamir, I., Verma, D. K., Bupesh, G., & Meenakshi Sundaram, K. (2023). Chapter 8 - Alkaloids and terpenoids: Synthesis, classification, isolation and purification, reactions, and applications. In C. Verma & D. K. Verma (Éds.), *Handbook of Biomolecules* (pp. 177–213). Elsevier. https://doi.org/10.1016/B978-0-323-91684-4.00017-7

Plazas, E., Avila M. M. C., Muñoz, D. R., & Cuca S. L. E. (2022). Natural isoquinoline alkaloids: Pharmacological features and multi-target potential for complex diseases. *Pharmacological Research*, *177*, 106126. https://doi.org/10.1016/j.phrs.2022.106126

Siddique, I. (2024). Exploring functional groups and molecular structures: A comprehensive analysis using FTIR spectroscopy. *Chemistry Research Journal*, *9*(2), 70–76.

Siddique, M., Rashid, R., & Ali, A. (2025). Fundamentals of acoustic cavitation, ultrasound-assisted processes, and sonochemistry. In *Modeling and Simulation of Sono-Processes* (pp. 3–17). Elsevier. https://www.sciencedirect.com/science/article/pii/B9780443236518000012

Sipoloni, V. M., Bassicheto, M. C., de Oliveira, M. V., de Souza, A. B. D. S., Alves, G. M., & Veiga, T. (2025). Non-conventional techniques in plant alkaloid extraction: A decade of progress (2014–2023). *Chemistry & Biodiversity*, e202403225.

Sotenjwa, V. Z. (2020). *Phytochemical Variation and Quality Control of Non-Volatile Constituents of Artemisia Afra*. https://www.researchgate.net/profile/Zimkhitha-Sotenjwa/publication/365862363_Artemisia_afra/links/649e7f75b9ed6874a5e9a20c/Artemisia-afra.pdf

Stead, P. (1998). Isolation by preparative HPLC. In R. J. P. Cannell (Éd.), *Natural Products Isolation* (Vol. 4, pp. 165–208). Humana Press. https://doi.org/10.1007/978-1-59259-256-2_6

Subash, P., Kumar, K. S., Rao, K. S., & Khute, S. (2025). *Advanced Chromatography Analytical Methods for the Isolation and Identification of Natural Drug Molecules*. https://www.intechopen.com/online-first/1206559

Sutherland, I. A., & Fisher, D. (2009). Role of counter-current chromatography in the modernisation of Chinese herbal medicines. *Journal of Chromatography A*, *1216*(4), 740–753. https://doi.org/10.1016/j.chroma.2008.11.095

Tarighat, M. A., Upadhyay, D., Andhare, P., Kumar, H., Tariq, M., Jain, M., & Abdi, G. (2024). NMR spectroscopy for characterization of metabolite. In V. Singh (Éd.), *Advances in Metabolomics* (pp. 179–203). Springer Nature. https://doi.org/10.1007/978-981-97-7459-3_8

Taylor, B. M. (2020). *Novel Strategies using Micro Pressurized Liquid Extraction to Rapidly Prepare Solid Samples for Analysis*. https://ucalgary.scholaris.ca/items/0a12f666-084e-4be6-b5be-db05f5903a76

Thorat, B. (2018). Chemical extraction and biomedical importance of secondary organic metabolites from plants: A review. *Journal of Biomedical and Therapeutic Sciences*, *5*(1), 9–42.

Utpal, B. K., Al Amin, M., Zehravi, M., Sweilam, S. H., Arjun, U. V. N. V., Madhuri, Y. B., Gupta, J. K., Yaidikar, L., Tummala, T., Suseela, R., Durairaj, A., Reddy, K. T. K., Al Fahaid, A. A. F., Rab, S. O., Almahjari, M. S., & Emran, T. B. (2025). Alkaloids as neuroprotectors: Targeting signaling pathways in neurodegenerative diseases. *Molecular and Cellular Biochemistry*. https://doi.org/10.1007/s11010-025-05258-3

Verpoorte, R., & Schripsema, J. (1994). Isolation, identification, and structure elucidation of alkaloids a general overview. In H. F. Linskens & J. F. Jackson (Éds.), *Alkaloids* (Vol. 15, pp. 1–24). Springer, Berlin, Heidelberg. https://doi.org/10.1007/978-3-642-84226-9_1

Waller, G. R. (2012). *Alkaloid Biology and Metabolism in Plants*. Springer Science & Business Media. https://books.google.fr/books?hl=fr&lr=&id=BcvkBwAAQBAJ&oi=fnd&pg=PR13&dq=The+biosynthesis+of+plant+alkaloids+often+initiates+with+primary%E2%80%82metabolites,+mainly+amino+acids+(e.g.,+tryptophan,+tyrosine,+lysine,+ornithine)+that+are+converted+to+intricate+secondary+metabolites+via+specialized+metabolic+pathways&ots=QMxPxlQpa7&sig=HQyXvtJJgIfIA9TtTxIruGVK8cs

Williamson, E. M., Okpako, D. T., & Evans, F. J. (1996). *Selection, Preparation and Pharmacological Evaluation of Plant Material, Volume 1* (Vol. 1). John Wiley & Sons. https://books.google.fr/books?hl=fr&lr=&id=_tTELyhJxOoC&oi=fnd&pg=PP9&dq=Preparation+of+Plant+Material&ots=Kev7NXufF0&sig=kSZwG0g5JoJTH2TECMx4GPLNe1I

Wolfender, J.-L. (2009). HPLC in natural product analysis: The detection issue. *Planta Medica*, *75*(07), 719–734. https://doi.org/10.1055/s-0028-1088393

Wu, A. H., Gerona, R., Armenian, P., French, D., Petrie, M., & Lynch, K. L. (2012). Role of liquid chromatography–high-resolution mass spectrometry (LC-HR/MS) in clinical toxicology. *Clinical Toxicology*, *50*(8), 733–742. https://doi.org/10.3109/15563650.2012.713108

Yang, L., & Stöckigt, J. (2010). Trends for diverse production strategies of plant medicinal alkaloids. *Natural Product Reports*, *27*(10), 1469–1479.

Yubin, J. I., Miao, Y., Bing, W., & Yao, Z. (2014). The extraction, separation and purification of alkaloids in the natural medicine. *Journal of Chemical and Pharmaceutical Research*, *6*(1), 338–345.

Zahari, A., Ablat, A., Omer, N., Nafiah, M. A., Sivasothy, Y., Mohamad, J., Khan, M. N., & Awang, K. (2016). Ultraviolet-visible study on acid-base equilibria of aporphine alkaloids with antiplasmodial and antioxidant activities from Alseodaphne corneri and Dehaasia longipedicellata. *Scientific Reports*, *6*(1), 21517. https://doi.org/10.1038/srep21517

Zaikin, V. G., & Halket, J. M. (2006). Derivatization in mass spectrometry—8. Soft ionization mass spectrometry of small molecules. *European Journal of Mass Spectrometry*, *12*(2), 79–115. https://doi.org/10.1255/ejms.798

Zenk, M. H., & Juenger, M. (2007). Evolution and current status of the phytochemistry of nitrogenous compounds. *Phytochemistry*, *68*(22–24), 2757–2772.

Zhao, W.-J., Chen, X.-Y., Liu, Y.-Q., Li, P., & Li, H.-J. (2020). Liquid chromatographic separation of alkaloids in herbal medicines: Current status and perspectives. *Journal of Separation Science*, *43*(9–10), 1755–1772. https://doi.org/10.1002/jssc.202000081

Zygmunt, B., & Namieśnik, J. (2003). Preparation of samples of plant material for chromatographic analysis. *Journal of Chromatographic Science*, *41*(3), 109–116.

5 Pharmacological Activities of Alkaloids
An Overview

Shivani Naik, Ruchi Bharti, and Renu Sharma

5.1 INTRODUCTION

Alkaloids are primarily found in nature as organic compounds containing nitrogen within plants, fungi, bacteria, and other organisms (Dang et al., 2012). Over 50,000 are recognized as natural products, and more than 12,000 are known as alkaloids (Wink, 2008). A larger number of alkaloids are found in plants of higher groups, particularly dicotyledons, with a smaller number occurring in lower plants (Shi et al., 2014). Alkaloids represent the largest category of secondary metabolites found in living organisms. They rarely exist in isolation, typically as mixtures of a few major alkaloids and several minor ones. Alkaloids exhibit a broad range of structural types and pharmacological activity (Zhang et al., 2014). To be classified as an alkaloid, a molecule must contain a simple nitrogen atom that is not part of a peptide bond. This definition encompasses a category of structurally versatile and biologically unrelated compounds, many of which have significant pharmacological effects (Wink, 2010). Most alkaloids are characterized by a bitter taste, weak basicity, colorlessness, poor solubility in water, and high solubility in organic solvents like chloroform (Bruneton, 1999). The defining characteristic of alkaloids is the presence of nitrogen, which is obtained from amino acids embedded in a heterocyclic ring. Pelletier et al. in 1983 defined alkaloids as cyclic structures containing nitrogen that have a negative oxidation state and are limited in availability across all living organisms. Many alkaloids are properly specified crystalline substances that produce salts when reacted with acids and can exist as salts in plants. In addition to hydrogen, carbon, and nitrogen, many alkaloids also consist of oxygen (Weina et al., 2017). Due to their basic nature, alkaloids can produce salts with both mineral acids (like sulfates and nitrates) and organic acids (like sulfamates and maleates). Typically, these alkaloid salts are water-soluble, while being generally insoluble in organic solvents, except in rare instances (Badri et al., 2019). Understanding the solubility of alkaloids and their salts is vital for their medicinal applications (Dias et al., 2012). Morphine, the first alkaloid extracted from *Papaver somniferum*, was followed by the discovery of various other alkaloids, including benzylisoquinoline and papaverine, which established this plant as a key model for studying alkaloid biosynthesis (Beaudoin & Facchini, 2014).

Recent natural product chemistry and biotechnology advancements have expanded the scope of alkaloid research. Various modern techniques in extraction, structural analysis, and isolation have accelerated the discovery of some new alkaloids with enhanced pharmacological properties. From ancient civilizations to modern pharmaceutical industries, alkaloids were crucial in producing therapeutic agents (Lin et al., 2024). Additionally, sustainable methods for alkaloid production, including microbial fermentation and plant cell culture, are gaining attention and offering eco-friendly and scalable alternatives to traditional extraction methods. This chapter aims to give an in-depth review of the pharmacological uses of alkaloids, focusing on their therapeutic applications. Most alkaloids are biosynthetically derived from tryptophan, nicotinic acid, phenylalanine, tyrosine, and anthranilic acid, as shown in Figure 5.1.

 DOI: 10.1201/9781003539179-5

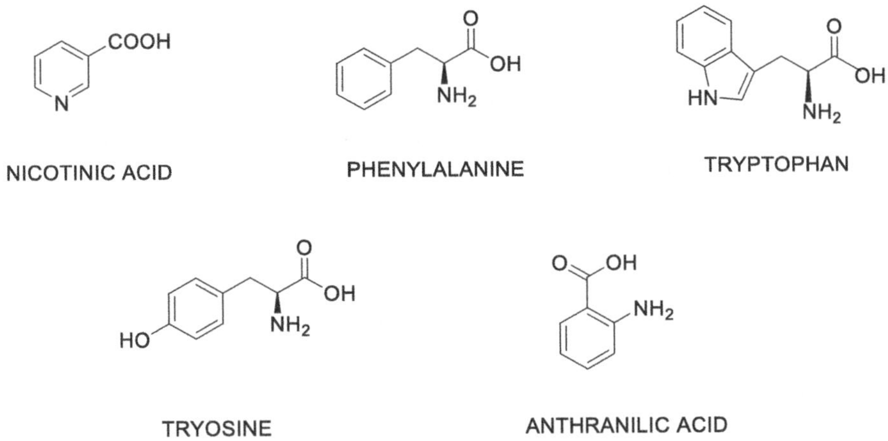

FIGURE 5.1 Some basic amino acids present in alkaloids (Seigler & Seigler, 1998).

5.1.1 History of Alkaloids

The history of alkaloids initiates beginning with the early use of plant-based substances for medicinal purposes. The name "alkaloid" was first discovered in the early nineteenth century, with Friedrich Sertürner's extraction of morphine from *opium poppy* marking the origin of modern alkaloid chemistry. Subsequent discoveries, such as quinine from *Cinchona* bark, advanced the field significantly (Sertürner, 1804). The 19th and early 20th centuries saw further progress with the elucidation of alkaloid structures by researchers like Adolf von Baeyer and Richard Willstätter (Bauer & Loew, 1937). The 20th century brought advancements in synthetic methodologies, enabling the laboratory synthesis of alkaloids, including quinine and morphine. Modern research has expanded to explore the therapeutic applications of alkaloids in areas such as cancer treatment, pain management, and neurodegenerative diseases (Cragg & Newman, 2005), Gordon and Rath's discussion of synthetic alkaloids (1994), and Schmid and Keller's review of modern medical applications (Gordon & Rath, 1994). Most of the alkaloids are produced in plants as a part of their secondary metabolite. Animals eliminate metabolic waste through urine, bile, and many other means, whereas plants retain their metabolic by-products within their cellular structures. This retention process allows plants to accumulate secondary metabolites, including alkaloids. Due to their typically high toxicity, alkaloids are sequestered in metabolically inactive parts of the cell to prevent damage to the plant itself. The synthesis of alkaloids may be random; many plant alkaloids act as a defense mechanism, deterring herbivores and protecting against harmful micro-organisms (Schmid & Keller, 2013).

5.1.2 Properties of Alkaloids

1. Alkaloids generally have colorless, crystalline solids, though some are oily or liquid at room temperature.
2. They exhibit bitter taste and are optically active due to chiral centers.
3. Alkaloids may exhibit toxicity at high concentrations, which serves as a defense mechanism in plants against pathogens.
4. They form complexes with metals, which influences their bioavailability and reactivity.
5. Many alkaloids have high melting points and have different UV absorbance patterns because of the conjugated systems.
6. These compounds are known for their strong physiological activities.

All these properties make alkaloids valuable for medical potential (William, 2009).

5.1.3 Overview of Pharmacological Activities

Plant alkaloids serve a diverse range of pharmacological functions, as illustrated in Figure 5.2, primarily noted for their protective roles against herbivores. Many alkaloids possess distinctive characteristics such as a sour taste, the potential to disrupt protein function upon ingestion and metabolism, and effects on the CNS. These properties underscore their varied mechanisms of action across different biological activities (Harborne, 1993). One notable aspect of alkaloids is their muscle-relaxing effect. For instance, D-tubocurarine, an alkaloid, functions as an anti-paralytic agent by blocking acetylcholine receptors, which facilitates muscle relaxation at neuromuscular junctions. Similarly, aporphine alkaloids isolated from *Mahonia aquifolium*, which include magnoflorine, isothebaine, and isocorydine, demonstrate the ability to ease contractions induced by noradrenaline more effectively (Zuo et al., 2006).

Apart from their muscle-relaxing uses, alkaloids are also known for their antioxidant potential. They work by scavenging free radicals and donating electrons. For example, a quinoline alkaloid derived from the aleurone base of *Oryza sativa* has significant antioxidant activity, as evidenced by its potential using the radical substrate DPPH (Yang et al., 2006).

Alkaloids also have specific therapeutic applications. Vinblastine and Vincristine, obtained from *Catharanthus roseus*, are useful in the treatment of leukemia and Hodgkin's disease. This group of alkaloids gives a chemopreventive effect by disrupting the protein microtubules that are used during cell division for the formation of the mitotic spindle, leading to their depolymerization and inhibition of tumor cell division and differentiation, thus reducing cancer incidence (Pradeep & Kuttan, 2004). Furthermore, quinoline alkaloids such as skimmianine and kokisaginine are isolated from *Raulinoa echinata*, exhibit antifungal properties against the symbiotic fungus of leaf-cutting ants (*Leucoagaricus gongylophorus*), and have in vitro activity against *Trypanosoma cruzi*. In addition, research by Wirasathien et al. highlights that aporphine-type alkaloids isolated from the aerial parts of *Pseuduvaria setosa* show anti-tuberculosis potential against *Mycobacterium tuberculosis*, antimalarial against *Plasmodium falciparum*, and cytotoxic against various cancer cells, including breast cancer and lung cancer cells (Zhao et al., 2015).

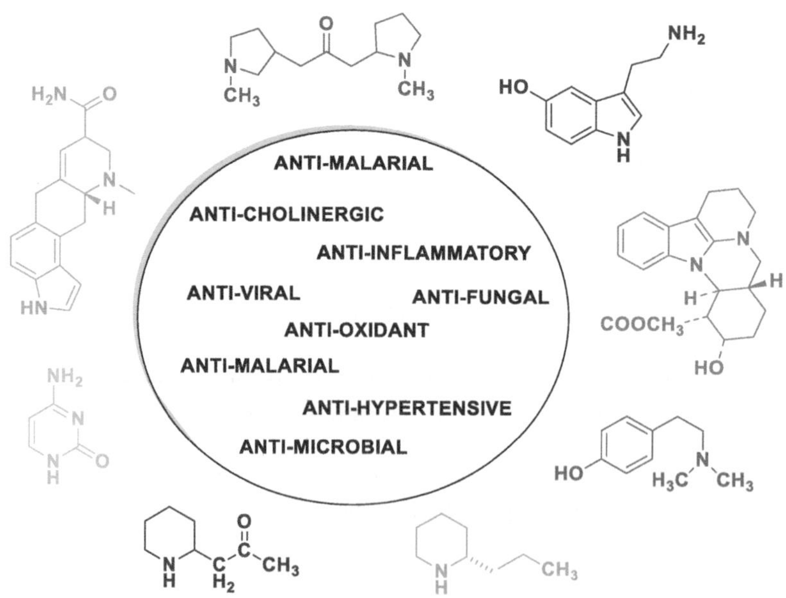

FIGURE 5.2 Pharmacological importance of plant alkaloids.

5.2 CLASSIFICATION OF ALKALOIDS

Alkaloids are categorized based on several criteria, such as their natural sources and structure. The widely accepted and accurate category is grouped based on their core structure, specifically the carbon–nitrogen skeleton. This classification categorizes alkaloids into several major categories: pyridine, pyrrolidine, isoquinoline, quinoline, indole, quinazoline, steroidal, and diterpenoid. Each one of them is further subdivided into smaller categories based on the unique structural properties of the members (Shakil, 1998).

Different Types of Alkaloids

True Alkaloids

This group contains a heterocyclic ring infused with nitrogen and sourced from amino acids (Henning, 2013).

Examples: Atropine, Nicotine, Morphine.

Protoalkaloids

This group has no heterocyclic ring but is still sourced from amino acids.

Examples: Ephedrine, Mescaline, Adrenaline (Talapatra & Talapatra, 2015)

Pseudoalkaloids

This group contains a heterocyclic ring fused with nitrogen, but is not sourced from amino acids.

Examples: Theobromine, Theophylline

Alkaloids are categorized into two main groups: Heterocyclic and Non-Heterocyclic alkaloids. Heterocyclic alkaloids have one or more rings with elements other than carbon, such as nitrogen, and are obtained from amino acids. Examples of this category are morphine, nicotine, and atropine, whereas non-heterocyclic alkaloids lack these ring structures and are derived from simple amines or non-cyclic groups. Examples of non-heterocyclic alkaloids are ephedrine, adrenaline, and mescaline. This classification helps in understanding their various biological activities and therapeutic potential (Van Rensburg & Van Zyl, 2013).

5.2.1 Heterocyclic Alkaloids

Alkaloids are chemically intricate, biologically active compounds obtained from cyclic amino acids. These compounds contain a nitrogen atom in the ring and are mainly established in nature as salts, specifically with organic acids like oxalic acid, lactic acid, malic acid, and citric acid. Their complex structure and physiological potential contribute to their versatile pharmacological effects. Heterocyclic alkaloids are classified into several categories based on their structural features, which include pyrrole, pyrrolidine, pyrrolizidine, pyridine, piperidine, tropane, quinolone, isoquinoline, aporphine, quinolizidine, and imidazole. Each of the above groups is characterized by a particular heterocyclic ring structure, where nitrogen is infused into the cycle, contributing to the pharmacological properties of these alkaloids (Majid, 2021). We will now discuss several of these alkaloid classes in more detail.

5.2.1.1 Pyrrole Alkaloids

This group of heterocyclic alkaloids is significant due to their latency in a wide range of natural and synthetic combinations with broad pharmacological properties (Estevez et al., 2014). Their structural properties, involving a conjugated system, are well-identified for their antimicrobial, anti-inflammatory, and antitumor activities. Moreover, pyrrole heterocyclic alkaloids, like those derived from heme and chlorophyll, are among the well-known. These compounds have four pyrrole rings connected by a methine bridge (Bauer & Knölker, 2012).

5.2.1.2 Pyridine Alkaloids

Pyridine alkaloids are a group of nitrogen-containing heterocyclic compounds with a pyridine ring as a structural nucleus. They are derived from nicotinic acid or related compounds, widely found in plants, microorganisms, and marine organisms (Kaur & Arora, 2015). Ordinary pyridine alkaloids include piperine, trigonelline, guvacine, cytisine, lobeline, anabasine, pelletierine, and nicotine (Silva Teles et al., 2019). Pyridine alkaloids have diverse biological activities, such as antimicrobial and insecticidal, and also in the medicine and agriculture fields.

5.2.1.3 Quinoline Alkaloids

This category of alkaloids, which are found in L-tryptophan, contains heterocyclic aromatic compounds created by the merger of a benzene ring with a pyridine ring. Quinine, an important member of this group, is found in *Cinchona ledgeriana*, while quinidine is derived from *Cinchona officinalis*. Quinoline alkaloids also show antimicrobial, anti-inflammatory, and anticancer activities. Spare quinoline alkaloids include chloroquine, tafenoquine, and bulaquine. Due to their biological significance, they remain a focus of extensive research in drug development and therapeutic potential (Debnath et al., 2018).

5.2.1.4 Piperidine Alkaloids

Piperidine alkaloids are usually studied with approx. 700 known compounds in this group. They are sourced from L-lysine and feature a structure with a six-membered ring, five methylene groups, and one amine group (Goel et al., 2018). These alkaloids are broadly distributed in nature, specifically in plants like black pepper. Remarkable alkaloids in this group include solenopsin, cynapine, lobeline, and coniine. Piperidine alkaloids exhibit versatile pharmacological effects, including analgesic, anti-inflammatory, antimicrobial, and insecticidal activity, making them useful for medicinal chemistry and ecological interactions. These alkaloids are commonly discovered in plants from the Lobeliaceae family, especially *Lobelia inflata*, which holds the alkaloid lobeline (Bottger et al., 2018).

5.2.1.5 Tropane Alkaloids

Tropane alkaloids are obtained from the amino acid ornithine and are distinguished by their unique structure, featuring a skeleton known as N-methyl-8-azabicyclo [3,2,1]octane (Ranjitha & Sudha, 2015). More than 250 alkaloids fall into this category and are mostly found in angiosperms from various families, which include Erythroxylaceae, Proteaceae, Solanaceae, Convolvulaceae, and Brassicaceae. These alkaloids have potential pharmacological properties, including being anti-cholinergic, and are commonly used in drugs. Medically, they are used for their mydriatic, anti-spasmodic, and anesthetic properties, though their toxicity requires mindful dosage management (Mao et al., 2014).

5.2.1.6 Pyrrolidine Alkaloids

These alkaloids are obtained from the amino acid L-ornithine and sometimes originate from lysine and arginine with added malonate units. They are identified by the ubiquity of rings in their structure. They are a subclass of alkaloids distinguished by a pyrrolidine ring, a five-membered heterocyclic structure holding a nitrogen atom. Examples of pyrrolidine alkaloids are putrescine, hygrine, and cuscohygrine (Kukula-Koch & Widelski, 2017). Pyrrolidine alkaloids are well known for their neuroactive, antimicrobial, and insecticidal activities, making them pharmacologically and agriculturally useful. Their structural versatility gives a broad range of applications in medicine and research. Also, they are found naturally in various plants and fungi and often have significant biological activity.

5.2.1.7 Isoquinoline Alkaloids

These alkaloids are biosynthesized mostly from the amino acid tyrosine and are mainly found in plants. These are found in families such as Papaveraceae, Berberidaceae, Fumariaceae, Ranunculaceae, and Annonaceae (Hussain et al., 2018). Structurally, isoquinoline alkaloids can be classified into two groups: simple isoquinolines, which feature a benzene ring fused to a pyridine ring, and benzylisoquinolines, which include an additional aromatic ring (Baros et al., 2012). Isoquinoline alkaloids have a wide range of pharmacological activities, such as analgesic, antimicrobial, and anticancerous properties, making them crucial in medicine and the pharmaceutical field.

5.2.1.8 Indole Alkaloids

Indole alkaloids are a versatile group of N-containing compounds differentiated by the presence of an indole ring system. They feature a bicycle structure that consists of a six-membered benzene ring fused with a five-membered pyrrole structure containing nitrogen, which provides basicity to the compound (Hamid et al., 2017). The precursor for these alkaloids is the amino acid tryptophan (Sagi et al., 2016). Indole alkaloids have numerous pharmacological activities like anticancer, antihypertensive, Central Nervous System modulating effects, and antimalarial, which make them key for therapeutic applications. Their different structural diversity and biological importance continue to attract interest in drug discovery and chemical synthesis in the research field.

5.2.1.9 Imidazole Alkaloids

Imidazole alkaloids consist of an imidazole ring with a five-membered heterocyclic structure containing two nitrogen atoms. These alkaloids are sourced from L-histidine and include compounds like histidine, histamine, pilocarpine, and pilosine. They are mainly derived from the Rutaceae and Cactaceae families (Aniszewski, 2015). Imidazole alkaloids have been recognized for their antimicrobial, anti-inflammatory, and numerous other pharmacological activities, which make them important in the medicinal field.

5.2.1.10 Quinolizidine Alkaloids

These compounds are obtained from the amino acid L-lysine. They are characterized by the fusion of two six-membered rings that share a nitrogen atom, with structural variations ranging from simple to complex. The quinolizidine ring structure provides these structures with various biological activities, making them useful in medicinal chemistry and drug development. They are mainly present in the Leguminosae family, particularly within the genus *Lupinus* (Bunsupa et al., 2012). These alkaloids are known for their pharmacological activities, which include neurotoxic, anti-inflammatory, and antimicrobial properties. Additionally, they have specific toxic behavior at higher doses.

5.2.2 Non-Heterocyclic Alkaloids

Non-heterocyclic alkaloids are chemically identified as nitrogen-containing and situated outside the ring, existing as a side chain rather than within the heterocyclic system. They can be obtained from amino acids or any biogenic amines. Examples involve mescaline, ephedrine, and colchicine, though they are relatively rare in nature. Protoalkaloids can originate from L-tyrosine and L-tryptophan, which convert into phenylethylamine and terpenoid indole, respectively (Almeida et al., 2017). Mescaline, a notable phenylethylamine alkaloid, originates from *Lophophora williamsii*, commonly called peyote. In contrast, monoterpenoid indole alkaloids constitute a sizable group, with approximately 3000 identified across families like Apocynaceae, Loganiaceae, and Rubiaceae (Pan et al., 2016).

5.3 THERAPEUTIC POTENTIAL OF ALKALOIDS

The basic properties of alkaloids enable them to form salts with both mineral acids (such as hydrochlorides, sulfates, and nitrates) and organic acids (like tartrates, sulfamates, and maleates). These alkaloid salts are typically dissolved in water and dilute alcohols but are usually insoluble in organic solvents. Crystallized salts are relatively stable and commonly used in commercial preparations. The solubility of alkaloids and their salts plays a significant role in pharmaceutical applications, as it affects their formulation and effectiveness (Bruneton, 1999). In medicine, plant-based alkaloids are mostly recognized for their potential in treating chronic diseases like cancer, diabetes, and neurological disorders, which provides an excellent and reliable source of medication (Adamski et al., 2020). Due to their diversity, alkaloids have numerous pharmacological applications in therapy, including analgesic (e.g., morphine), antiasthmatic (e.g., ephedrine), anticancer (e.g., vincristine), antihypertensive (e.g., reserpine), antipyretic (e.g., quinine), and antihyperglycemic (e.g., piperine) (Ng et al., 2015). This chapter will explore the diverse pharmacological activities of plant alkaloids, which are naturally occurring compounds known for their potential in biological activities.

5.3.1 Antioxidant Property

Alkaloids are recognized for their antioxidant activity, considering their various mechanisms, including free radical scavenging and the donation of electrons or a hydrogen atom. One notable example is a quinoline alkaloid originating from the aleurone base of *O. sativa cv*. This compound gives a strong antioxidant property, specifically when tested with the radical substrate 2,2-diphenyl-1-picrylhydrazyl (DPPH), mostly used to examine radical scavenging capability (Siddique & Husain, 2018). Indoles and their derivatives have significant antioxidant activity by effectively neutralizing radicals like 2,2'-azino-bis(3-ethylbenzothiazoline-6-sulphonic acid) (ABTS). These compounds show outstanding radical scavenging activity at physiological pH levels, highlighting their potential for therapeutic use in neutralizing oxidative stress (Reddy & Kumar, 2017).

Additionally, norditerpene alkaloids like linearilobin, linearilin, and browniine isolated from the roots of *Delphinium linearilobum* have also been shown to possess antioxidant activity. These alkaloids have been evaluated using both DPPH and metal chelation assays, confirming their role in neutralizing free radicals and binding metal ions, which can further contribute to their antioxidant property (Khan & Ghosh, 2020). Overall, the wide range of alkaloids and their antioxidant mechanisms make them valuable subjects for research in developing new therapeutic agents to manage oxidative stress-related problems. Their ability to neutralize free radicals and donate electrons or hydrogen makes them promising candidates for further study and potential application in clinical applications (Yang et al., 2006).

5.3.2 Anticancer Property

One of the primary challenges in cancer treatment is the advancement of multidrug resistance to chemotherapy agents, mostly due to the efflux of these drugs by p-glycoprotein (Joshi et al., 2017). The alkaloids vinblastine and vincristine, sourced from Apocynaceae, are well-known for their role in treating leukemia and Hodgkin's disease, as shown in Table 5.1. These compounds have chemopreventive characteristics by disrupting the microtubules that are amenable to the makeup of the mitotic spindle during cell division (Johnson et al., 1963). By breaking down or preventing the formation of these microtubules, they impede the division and maturation of cancer cells, thereby reducing cancer occurrence (Pradeep & Kuttan, 2004). Vinblastine and its derivatives work as anticancer agents by targeting similar binding sites, achieving efficacy against cancerous cells through a dual-binding process that interacts with both alpha and beta tubulin (Zhang et al., 2019). Similarly, Vincristine is a naturally occurring vinca alkaloid and organic molecule that functions as an antimicrotubule agent. It inhibits mitosis by arresting cells in metaphase (Schiller et al., 2007; Almagro et al., 2015). The antitumor effects of vinblastine and vincristine result from their binding to intracellular tubulin, which inhibits DNA repair and also RNA synthesis by blocking the DNA-dependent RNA polymerase enzyme. These drugs also disrupt cellular functions like glycolysis, respiration, nucleic acid, and protein synthesis (Bates & Eastman, 2017). Because of their hydrophobic nature when uncharged, vinblastine and vincristine can integrate into lipid bilayers easily and alter the structure as well as function of the membrane (Beck et al., 2000).

TABLE 5.1
Anticancer Alkaloids and Their Mechanisms

Alkaloid	Part Used	Potential	Mechanism of Action	References
Vinblastine	Stem, root, leaf	Neck cancer, breast cancer, head cancer	Binds to tubulin and inhibits microtubule formation, acting as an anti-mitotic agent.	(Meyers, 2007)
Vincristine	Stem, root, leaf	Lymphoblastic leukemia	Binds to the tubulin dimer, preventing the formation of microtubule structures, thereby acting as a mitotic inhibitor and anti-mitotic agent.	(Farnsworth, 1985)
Vindesine	Stem, root, leaf	Lung cancer melanoma	Anti-mitotic	(Arora et al., 2010)
Camptothecin	Whole tree	Ovarian and small-cell lung cancer	Inhibits Topoisomerase I, an enzyme essential for DNA replication	(Sadre et al., 2016)
Noscapine	Isolated from opium poppy	Breast and prostate cancer	Binds tubulin and disrupts microtubule dynamics, leading to cell arrest	(Yanran et al., 2018)
Evodiamine	Stem	Colon and lung cancer	Induces apoptosis and inhibits cell division and metastasis.	(Longfei et al., 2024)
Lycorine	Stem, leaf	Lung and breast cancer	Inhibits protein synthesis and promotes apoptosis	(Khalifa et al., 2018)

5.3.3 Anti-Inflammatory Activity

The inflammatory mechanism is the body's response to the invasion of an infectious agent, antigen, or cellular damage. It is the well-known indicator of disease and a crucial biological process (Kulinsky, 2007). The inflammatory response begins with the indication of signals, which can originate from infection, followed by the liberation of chemicals from migrating cells and tissues, which are commonly known as mediators (Blancas-Flores et al., 2010). Various experimental models were used to examine inflammation in the research of inflammation. The standard approach for examining the anti-inflammatory properties of compounds is to test them in both animal and biochemical inflammation models (Ekstein & Schachter, 2010). Isoquinoline, quinoline, and indole alkaloids are the most extensively researched classes for anti-inflammation. Surrounded by isoquinolines, berberine is the most widely used compound, showing effectiveness across nearly all experimental models (Kuo et al., 2004). Indole alkaloids, which include brucine and brucine-N-oxide, show notable analgesic and anti-inflammatory effects (Yin et al., 2003). Caulerpin, a marine alkaloid extracted from the lipid fraction of the algae *Caulerpa racemosa*, demonstrated anti-inflammatory potential in mice when administered orally (Liu et al., 2009). The plurality of the alkaloids shown in Figure 5.3 holds significant potential as anti-inflammatory agents, with some displaying remarkable activity.

5.3.4 Neurodegenerative Disorders (NDD)

NDDs affect the nervous system and are typically distinguished by the continuous loss of neurons (Kamagata et al., 2014). Recent research has shown that changes in the physicochemical properties of certain proteins can lead to their assemblage in the brain, which contributes to neuronal degeneration (Kovacs, 2014). Alzheimer's disease (AD) is a leading neurodegenerative disorder represented by the progressive decline in memory, learning, and other cognitive functions (Yan & Vassar, 2014). It is marked by the formation of abnormal plaques and neurofibrillary tangles, which contribute to significant neuronal loss (Schmitt et al., 2014). Parkinson's disease (PD) is a common

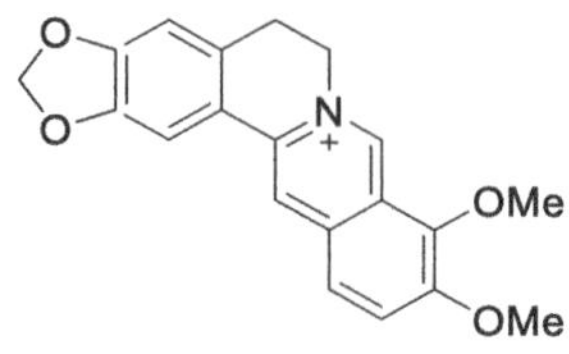

Berbine

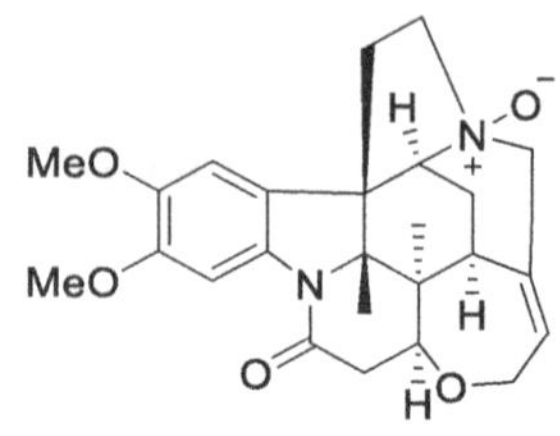

Brucine-N-oxide

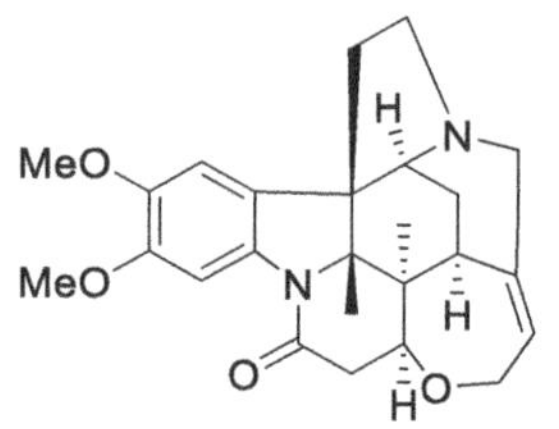

Brucine

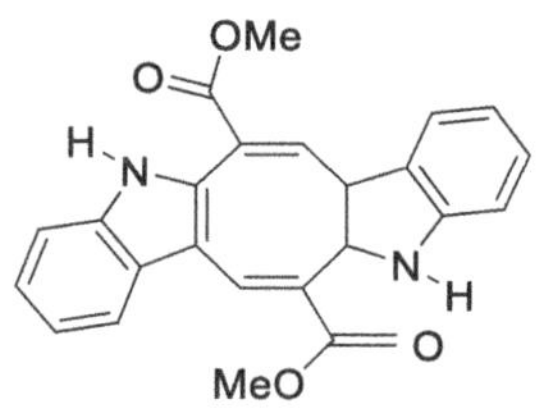

Cauleprin

FIGURE 5.3 Common structures of anti-inflammatory alkaloids.

neurodegenerative disorder that is identified by the loss of dopaminergic neurons in the substantia nigra pars compacta (SNpc) (Barnello et al., 2015). This neuronal degeneration leads to various symptoms such as tremors, rigidity, bradyphrenia (slowed thinking), bradykinesia, and postural instability (Schrag et al., 2015). Huntington's disease (HD) is a progressive neurodegenerative disorder with a worldwide incidence of 5–10 cases per 100,000 people. It is an autosomal dominant genetic disorder that leads to the death of neurons, specifically in the cerebral cortex and striatum. HD significantly reduces lifetime and causes serious disability in affected individuals (Vieweg, 2017). Plant-based compounds like alkaloids and flavonoids have been used since ancient times against neurodegenerative diseases. They have shown significant potential as therapeutic agents in reducing key aspects of disorders such as Alzheimer's disease and Huntington's disease (Choudhary et al., 2014).

Research suggests that berberine (BBR) has meaningful clinical use in Alzheimer's disease (AD) due to its potential to decrease amyloid-beta (Aβ) levels by inhibiting BACE-1, the enzyme responsible for cleaving amyloid precursor protein (APP) and initiating Aβ production (Han et al., 2012). Additionally, BBR has developed its capacity to alter behavioral disturbances by preventing hippocampal neurodegeneration and reducing BACE-1 activity (Kim et al., 2014). Likewise, autophagy has been known as a therapeutic target for Huntington's disease (HD), and BBR has been found to upregulate the autophagic function, which aids in the clearance of misfolded proteins, which is a key factor in the advancement of HD (Zhang et al., 2016). Furthermore, morphine plays a vital role in AD treatment by binding to mu-opioid receptors (MOR) in the central nervous system, increasing GABA levels in brain synapses. It also offers neuroprotection against oxidative stress. Morphine has been used to protect against the toxicity of intracellular amyloid beta (iAβ) in primary neuronal cultures of rats as well as humans (Cui et al., 2011).

5.3.5 Antimalarial Activity

Malaria has been cured with various medications, which include quinine, mefloquine, chloroquine, artemisinin, and their derivatives, shown in Figure 5.4. Quinine is an alkaloidal drug that was the first antimalarial agent isolated from the *Cinchona* tree (Ashok et al., 2013). Various alkaloids have been

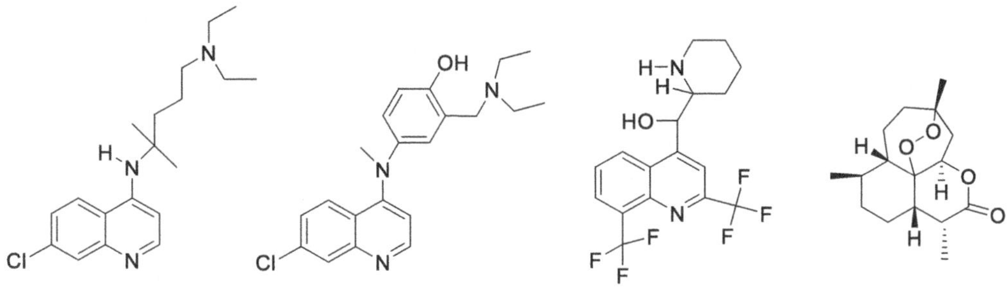

Chloroquine Amodiaquine Mefloquine Artemisinin

FIGURE 5.4 Some common antimalarial drugs.

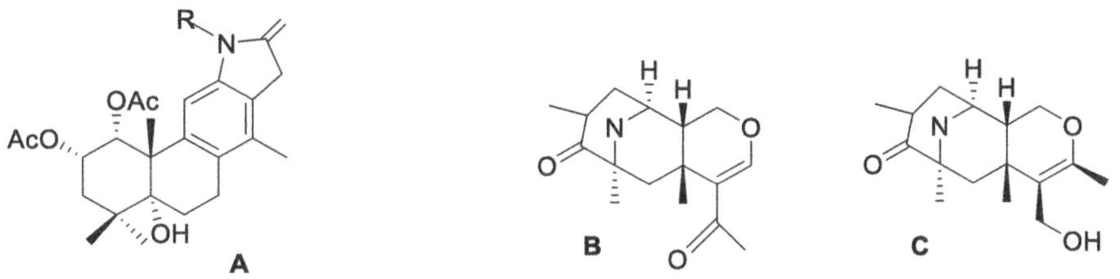

1.R=H
2.R=C_2H_4OH

Terpenoidal Alkaloid

Indole Alkaloids

FIGURE 5.5 Terpenoidal and indole alkaloids.

extracted from various biological sources and have been reported to have strong antimalarial activity. Chloroquine was developed in the 19th century as a synthetic derivative of quinine (Wiesner et al., 2003). It was potent, affordable, and less toxic, making it the preferred malaria treatment drug for many years (Onori, 1984). Mefloquine, structurally similar to quinine, was introduced to treat chloroquine-resistant malaria (Kain et al., 2001). Artemisinin, a natural compound derived from the sweet wormwood plant (*Artemisia annua*), and its semisynthetic derivatives are highly effective antimalarial drugs, especially in regions where resistance to other antimalarials has evolved (Mugittu et al., 2006).

Various categories of phytochemicals contribute to the antimalarial properties of plants, involving alkaloids, terpenes, steroids, and flavonoids. For example, Ma et al. isolated two new diterpene alkaloids from the seeds of *Caesalpinia minax* (Ma et al., 2014). These compounds feature a tetra-cyclic cassane-type diterpenoid structure with a γ-lactam ring and have antiplasmodial activity. Several indole alkaloids have been recently extracted and found to exhibit strong antiplasmodial activity. A chemical examination of *Alstonia macrophylla* bark led to the discovery of two new nitrogen containing compounds 1, has shown antiplasmodial activity against *P. falciparum* (Cheenpracha et al., 2013) (Figure 5.5).

5.3.6 Cardiovascular Activity

Alkaloids have a broad range of pharmacological activities, involving significant effects on the cardiovascular system. Most of the alkaloids influence the heart, blood, and vessels directly or indirectly (Kwan & Achike, 2002). These are often used in the treatment of high blood pressure, angina, and heart failure (Dawes, 1952). Numerous plant-derived alkaloids have demonstrated significant effects on the cardiovascular system. For instance, plants belonging to the genus *Tabernaemontana*, which are rich in various alkaloids, have been stated to exhibit antihypertensive properties (Podolsky, 2023). A significant number of alkaloids exhibit pharmacological effects on the heart, influencing its various functions. However, in many cases, these cardiac effects are secondary to their primary pharmacological activity or occur as toxic effects at higher doses. This section of the chapter focuses specifically on those alkaloids whose primary mode of action is directed at the heart, impacting its functions directly and significantly (Criscitiello & O'Rourke, 1965). From a therapeutic perspective, clinical cardiac diseases can be categorized into three main groups:

1. **Arrhythmias** (Abnormal heart condition that leads from irregular electrical impulses in the heart, causing the heart to beat fast or too slow)
2. **Congestive Heart Failure** (The heart's ability to pump blood is compromised, leading to fluid buildup in the lungs, causes shortage of breath, swelling in the legs, etc.)
3. **Coronary Insufficiency** (Coronary arteries fail to supply adequate oxygen-rich blood to the heart muscle due to atherosclerosis, causing symptoms like pain in the chest and increased risk of heart attack)

Quinidine is currently the preferred drug for treating atrial arrhythmias. Other Cinchona alkaloids, cryptopine-like compounds, sparteine and its derivatives, and Rauwolfia alkaloids, also hold significant interest in managing these conditions (Nixon et al., 2013). Digitalis-like glycosides form the cornerstone of treatment for congestive heart failure, and erythrophleum alkaloids have shown potential as promising therapeutic agents for this condition (McCawley, 1955). Pharmacological testing of cardioactive alkaloids yields three key types of facts: (1) changes in the normal physiological functions of the heart, (2) the ability to restore artificial cardiac arrhythmias to normal sinus rhythm, and (3) the effectiveness in compensating for experimentally induced congestive heart failure (Scherf & Romano, 1948). We will discuss the mechanism of some cardiovascular alkaloids.

5.3.6.1 Quinine

Early studies by Chirone (1875) demonstrated that quinine (Figure 5.6) has a direct effect on heart muscle: (a) the quinine-treated heart could not be restored either electrically or mechanically; (b) stimulation of the nerves leading to the heart did not produce the expected response in the quinine-treated heart. Another group of researchers (McCawley, 1955) also noted that quinine reduces the excitability of the heart. Quinine would help restore a regular rhythm in the heart that is beating irregularly.

Later, Grant and Iliescu (Grant & Iliescu, 1922) chose to assess quinine's potency using an indirect approach. They administered 0.6-g doses of quinine to patients with atrial fibrillation. These ineffective but "safer" doses did not exclude the arrhythmia but led to a reduction in the atrial rate and an increase in the ventricular rate. Once again, quinine was found to be less effective than quinidine. Quinine reduced the typical atrial rate by 86 beats per minute and increased the ventricular rate by 15 beats per minute, whereas quinidine produced reductions of 161 beats/min in the atrial rate and an increase of 45 beats/min in the ventricular rate.

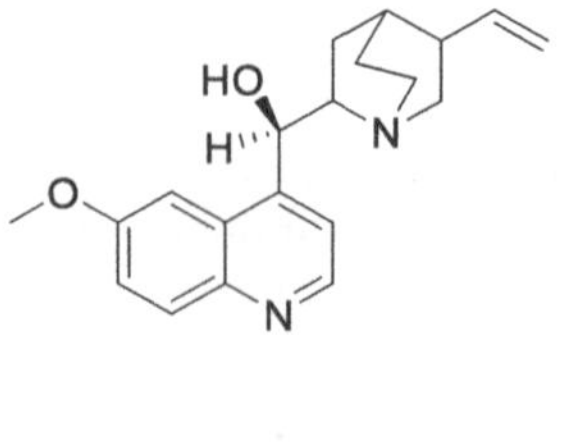

FIGURE 5.6 Structure of quinine.

5.3.6.2 Quinidine

When quinidine (Figure 5.7) is perfused throughout an isolated frog heart at a concentration of 1:20,000, it slows the heart rate and reduces the amplitude of contractions. Due to the extended diastolic pause, quinidine causes the heart to dilate. In cases where the frog heart develops an irregular beat, quinidine can restore a normal rhythm (Singh et al., 2020). Quinidine extends the atrial refractory period at low, clinically achievable concentrations.

Frey's (1918) work summarizes much of the current knowledge on quinidine therapy for atrial fibrillation. He successfully restored normal sinus rhythm in 6 out of 10 patients whose atrial fibrillation was caused by conditions such as atherosclerosis, mitral stenosis or insufficiency, rheumatic fever, or congestive heart failure. These patients also exhibited other arrhythmias, including atrial flutter and ventricular extrasystoles.

5.3.6.3 Hydroquinidine

The finding depicts that commercially available quinidine could contain up to 20% dihydroquinidine (Figure 5.8), which led to an in-depth study of the latter alkaloid. In dihydroquinidine, the double bond in the vinyl side chain is saturated with two hydrogen atoms (Zhang et al., 2014).

Van Dongen and Sanchez (1937) compared the anti-fibrillatory effects of hydroquinidine with specially purified quinidine, suggesting that the anti-fibrillatory activity of commercially available quinidine is mainly assigned to its dihydroquinidine content. Using rabbits as test subjects, they found that dihydroquinidine raised the electrical stimulus thresholds necessary to induce premature systoles, tachycardia, atrial fibrillation, and ventricular fibrillation. It also shortened the protraction of fibrillation after the electrical stimulus. Dihydroquinidine was effective in preventing heterotropic rhythms, particularly ventricular arrhythmias, induced by intravenous barium chloride. In contrast, de Boer (de Boer, 1990) reported different findings. In his study comparing hydroquinidine and quinidine on electrically induced arrhythmias in cats, both quinidine and dihydroquinidine only prevented arrhythmias at high doses. Hydroquinidine and quinidine were found to be similarly effective in this context (Geetha & Ramachandran, 2021).

5.3.6.4 Protopine

Protopine, a plant alkaloid, generally slows down the heart rate (bradycardia) in various animals, including frogs and rabbits (Brunton, 2006) observed that protopine caused slow and incomplete systole, along with prolonged diastole. At higher doses, it can cause irregular heartbeats and even

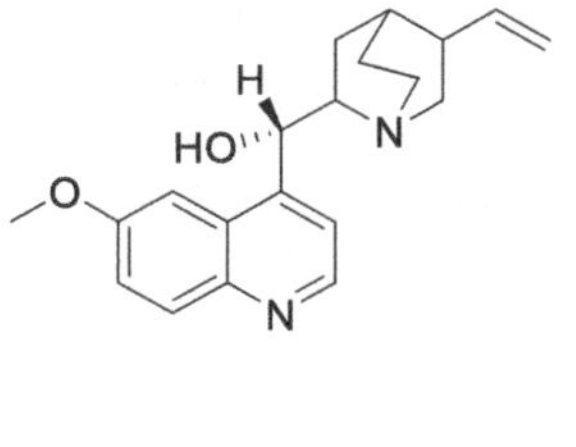

FIGURE 5.7 Structure of quinidine.

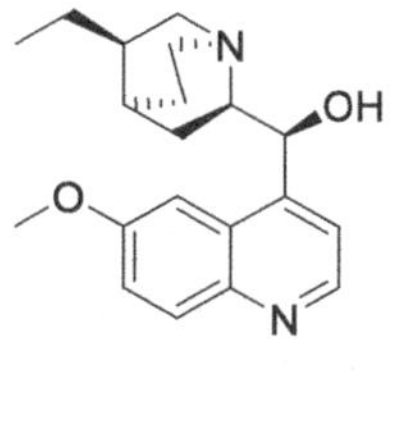

FIGURE 5.8 Structure of hydroquinidine.

ventricular fibrillation. The slowing effect of protopine is not easily reversed by atropine, a drug that typically counteracts bradycardia. Its impact on heart rate is significant, particularly in shock conditions where the heart's reflex responses are usually accelerated. Honda and Nagasaki (Honda & Nagasaki, 1908) studied protopine from Macleaya (Macleaya cordata) and observed its ability to slow the heart rate of frogs, a response that was not inverted by atropine. In paralyzed rabbits, they present that protopine still induced bradycardia, despite the blood pressure reaching shock levels, where one would typically expect cardiac acceleration due to reflexes.

5.3.7 Other Activities

Plant alkaloids, characterized by their unique and complex chemical structures, play a crucial role in antifungal, antimicrobial, and analgesic activities, making them valuable candidates for therapeutic applications. Their antifungal properties are well-established, with compounds like berberine, an isoquinoline alkaloid, disrupting fungal cell walls and interfering with enzymes critical for fungal growth. Another example, tomatidine, inhibits ergosterol synthesis in fungi like *Candida* and *Aspergillus* species, leading to fungal cell membrane disruption (Kaue et al., 2014).

Actions on the immunology system are also studied for these alkaloids, including inhibition and induction of gene expression, anti-proliferative, anti-system complement, anti-inflammatory, and apoptosis. At the Digestive system level, some related actions were studied, such as Anti-diarrhetic, electrolyte transport inhibition, and Anti-ulcer activities (Leitao Da-Cunha et al., 2005). Similar to apomorphine, buprenorphine, a semisynthetic opioid obtained from thebaine, a naturally occurring alkaloid in *P. somniferum*—is widely used as an alternative therapy in the treatment of opioid dependence. However, it underwent extensive first-pass metabolism, producing low oral bioavailability (Salminen et al., 2011).

In the realm of antimicrobial activity, alkaloids such as sanguinarine bind to bacterial DNA, inhibiting essential metabolic pathways in both Gram-positive and Gram-negative bacteria. **Quinine**, a well-known alkaloid from *Cinchona* bark, features a quinoline ring fused to a bicyclic system. It disrupts bacterial metabolism, demonstrating efficacy in treating bacterial infections alongside its historical use for malaria (Rosa et al., 2017). Alkaloids have also been studied for their antiviral potential, with sanguinarine showing promise against viral pathogens like human papillomavirus and herpes simplex virus (Imenshahidi & Hosseinzadeh, 2016).

Moreover, the analgesic activities of plant alkaloids are exemplified by morphine, a complicated benzylisoquinoline alkaloid that contains five fused rings. Morphine binds to opioid receptors in the CNS and effectively relieves chronic pain. Other alkaloids like berberine and piperine demonstrate analgesic potential by reducing inflammation and modulating neurotransmitter activity, such as GABA and opioid systems. These compounds inhibit pro-inflammatory cytokines and affect pain perception pathways, providing natural alternatives to pain relief (Cushnie & Lamb, 2005).

The versatile structural features of plant alkaloids enable their multi-faceted bio-applications, which include antifungal, antimicrobial, and analgesic properties, which are key in developing new therapies (Gupta et al., 2004). Their ability to address problems like drug resistance in fungi and bacteria, as well as provide pain relief, highlights their key role in medicine. Continued ongoing research into the structural diversity of alkaloids and their mechanisms of action will likely discover new applications in therapeutic potentials (Gutiérrez et al., 2008).

5.4 CONCLUSION

In conclusion, plant alkaloids are a versatile category of naturally sourced compounds with excellent pharmacological activities that provide a significant therapeutic potential. Throughout this chapter, we discussed the different types of alkaloids, like quinoline, indole, isoquinoline, and steroidal alkaloids, which have a broad range of biological uses. These include antimicrobial, antifungal, antimalarial, analgesic, anti-inflammatory, cardiovascular, and anticancer, making alkaloids highly important for the development in the medical field. The cardiovascular importance of alkaloids like quinidine is equally noteworthy. These compounds play a key role in maintaining heart rhythm and regulating high blood pressure. Alkaloids also have significant neuroprotective properties, with agents like caffeine and galantamine being used for cognitive enhancement and the treatment of NDD, like Alzheimer's disorder. Historically, alkaloids like quinine, berberine, and morphine have played an important role in traditional and clinical medicine and also provide an effective treatment for conditions like pain, malaria, and bacterial infections (Kaur et al., 2014). Their mechanisms of action, which range from enzyme inhibition to receptor interaction and modulation of biochemical pathways, underline their diversity as therapeutic agents. Additionally, the trending issue of drug-resistant pathogens has renewed interest in the antimicrobial and antifungal potential of plant alkaloids, which makes them crucial resources for addressing modern medical challenges. In summary, the pharmacological potential of alkaloids is diverse, and ongoing research continues to discover their immense applications. With their ability to treat a wide range of diseases, alkaloids remain an important part of both traditional healing practices and the advancement in medical science. As natural products, alkaloids offer a sustainable and valuable resource for addressing the growing demand for effective and improved medical uses. As new alkaloids are discovered and studied, their importance in drug development will only continue to grow.

ACKNOWLEDGMENT

The authors are thankful to the Department of Chemistry, Chandigarh University, India, for providing library access to conduct this review.

CONFLICT OF INTEREST

There is no conflict of interest between any of the authors.

COMPETING INTERESTS

The authors declare that they have no competing interest.

DECLARATION OF GENERATIVE AI AND AI-ASSISTED TECHNOLOGIES IN THE WRITING PROCESS

During the preparation of this work, the book chapter author(s) have not taken any assistance from AI, and the book chapter author(s) reviewed and edited the content as needed and take(s) full responsibility for the content of the publication.

REFERENCES

Adamski, Z., Blythe, L. L., Milella, L., & Bufo, S. A. (2020). Biological activities of alkaloids: From toxicology to pharmacology. *Toxins*, 12(4), 210.

Almagro, L., Fernández-Pérez, F., & Pedreño, M. A. (2015). Indole alkaloids from Catharanthus roseus: Bioproduction and their effect on human health. *Molecules*, 20(2), 2973–3000.

Alves de Almeida, A. C., de-Faria, F. M., Dunder, R. J., Manzo, L. P. B., Souza-Brito, A. R. M., & Luiz-Ferreira, A. (2017). Recent trends in pharmacological activity of alkaloids in animal colitis: Potential use for inflammatory bowel disease. *Evidence-Based Complementary and Alternative Medicine*, 2017, 8528210. https://doi.org/10.1155/2017/8528210

Aniszewski, T. (ed.) (2015) *Alkaloids: Chemistry, Biology, Ecology, and Applications* (p. 496). Elsevier.

Arora, R., Malhotra, P., Mathur, A. K., Mathur, A., Govil, C., & Ahuja, P. (2010). Anticancer alkaloids of Catharanthus roseus: Transition from traditional to modern medicine. In R. Arora (Ed.), *Herbal Medicine: A Cancer Chemopreventive and Therapeutic Perspective* (pp. 292–310). Jaypee Brothers Medical Publishers Pvt Ltd.

Ashok, P., Ganguly, S., & Murugesan, S. (2013). Review on in-vitro antimalarial activity of natural β-carboline alkaloid. *Mini Reviews in Medicinal Chemistry*, 13(12), 1778–1791.

Badri, S., Basu, V. R., Basha, S. K., Chandra, K., & Anasuya, D. (2019). A review on pharmacological activities of alkaloids. *World Journal of Current Medical and Pharmaceutical Research*, 1(6), 230–234. https://doi.org/10.37022/WJCMPR.2019.01068

Baranello, R. J., Bharani, K. L., Padmaraju, V., Chopra, N., Lahiri, D. K., Greig, N. H. et al. (2015). Amyloid-beta protein clearance and degradation (ABCD) pathways and their role in Alzheimer's disease. *Current Alzheimer Research*, 12, 32–46.

Baros, S., Karsayová, M., Jomová, K., Gáspár, A., & Valko, M. (2012). Free radical scavenging capacity of Papaver *somniferum* L. and determination of pharmacologically active alkaloids using capillary electrophoresis. *Journal of Microbiology, Biotechnology and Food Sciences*, 1, 725.

Bates, D., & Eastman, A. (2017). Microtubule destabilising agents: Far more than just antimitotic anticancer drugs. *British Journal of Clinical Pharmacology*, 83(2), 255–268.

Bauer, I., & Knölker, H.-J. (2012). Synthesis of pyrrole and carbazole alkaloids. In H.-J. Knölker (Ed.), *Alkaloid Synthesis* (pp 203–253). Springer-Verlag.

Bauer, K., & Loew, H. (1937). The chemistry of alkaloids. *Chemische Berichte*, 70, 2044–2061.

Beaudoin, G. A., & Facchini, P. J. (2014). Benzylisoquinoline alkaloid biosynthesis in opium poppy. *Planta*, 240(1), 19–32.

Beck, W. T., Cass, C. E., & Houghton, P. J. (2000). Microtubule-targeting anticancer drugs derived from plants and microbes: Vinca alkaloids, taxanes, and epothilones. In *Holland-Frei Cancer Medicine* (5th ed.). BC Decker.

Blancas-Flores, G., Almanza-Pérez, J. C., López-Roa, R. I., Alarcón-Aguilar, F. J., García-Macedo, R., & Cruz, M. (2010). Obesity as an inflammatory process. *Boletín Médico del Hospital Infantil de México*, 67, 88–96.

Böttger, A., Vothknecht, U., Bolle, C., & Alkaloids, W. A. (2018). *Lessons on Caffeine* (pp 179–203). Cannabis & Co. Springer.

Bruneton, J. (1999). *Pharmacognosy, Phytochemistry, Medicinal Plants* (2nd ed., p. 1136). Lavoisier.

Brunton, L. L., Lazo, J. S., & Parker, K. L. (2006). *The Pharmacological Basis of Therapeutics. Goodman Gilman's*, 11th ed. McGraw Hill, New York, pp. 727–733.

Bunsupa, S., Yamazaki, M., & Saito, K. (2012). Quinolizidine alkaloid biosynthesis: Recent advances and future prospects. *Frontiers in Plant Science*, 3, 239.

Cheenpracha, S., Ritthiwigrom, T., & Laphookhieo, S. (2013). Alstoniaphyllines A-C, unusual nitrogenous derivatives from the bark of Alstonia macrophylla. *Journal of Natural Products*, 76(4), 723–726.

Choudhury, B., Saytode, P., & Shah, V. (2014). Neurodegenerative disorders: Past, present and future. *International Journal of Pharmaceutical Technology and Biotechnology*, 5(2), 14–28.

Chirone, G. (1875). On the direct action of quinine on the heart muscle. *Archives Italiennes de Biologie*, 4, 289–302.

Cragg, G. M., & Newman, D. J. (2005). Plants as a source of anticancer agents. *Journal of Ethnopharmacology*, 100(1–2), 72–83.

Criscitiello, M. G., & O'Rourke, R. A. (1965). Documentation of transient arrhythmias by constant cardiac monitoring. *The American Journal of Cardiology*, 16(6), 779–788.

Cui, J., Wang, Y., Dong, Q., Wu, S., Xiao, X., Hu, J., et al. (2011). Morphine protects against intracellular amyloid toxicity by inducing estradiol release and upregulation of Hsp70. *The Journal of Neuroscience*, 31(45), 16227–16240.

Cushnie, T. P. T., & Lamb, A. J. (2005). Antimicrobial activity of flavonoids. *International Journal of Antimicrobial Agents*, 26(5), 343–356.

Dang, T. T., Onoyovwi, A., Farrow, S. C., et al. (2012). Biochemical genomics for gene discovery in benzylisoquinoline alkaloid biosynthesis in opium poppy and related species. *Methods in Enzymology*, 515, 231–266.

Dawes, G. S. (1952). Experimental cardiac arrhythmias and quinidine-like drugs. *Pharmacological Reviews*, 4(1), 43–84.

de Boer, A. G., van Hoogdalem, E. J., & Breimer, D. D. (1990). Improvement of drug absorption through enhancers. *European Journal of Drug Metabolism and Pharmacokinetics*, 15(2), 155–157.

Debnath, B., Singh, W. S., Das, M., Goswami, S., Singh, M. K., Maiti, D., & Manna, K. (2018). Role of plant alkaloids on human health: A review of biological activities. *Materials Today Chemistry*, 9, 56–72.

Dias, D. A., Urban, S., & Roessner, U. (2012). A historical overview of natural products in drug discovery. *Metabolites*, 2(2), 303–336.

Ekstein, D., & Schachter, S. C. (2010). Natural products in epilepsy—the present situation and perspectives for the future. *Pharmaceuticals*, 3, 1426–1445.

Estévez, V., Villacampa, M., & Menéndez, J. C. (2014). Recent advances in the synthesis of pyrroles by multicomponent reactions. *Chemical Society Reviews*, 43(13), 4633–4657.

Farnsworth, N. (1985). Plants and modern medicine: Where science and folklore meets. In *World Health Forum* (pp. 76–80).

Frey, E. (1918). Zur Behandlung des Vorhofflimmerns mit Chinidin. *Deutsche Medizinische Wochenschrift*, 44, 1025–1028.

Geetha, R. G., & Ramachandran, S. (2021). Recent advances in the anti-inflammatory activity of plant-derived alkaloid rhynchophylline in neurological and cardiovascular diseases. *Pharmaceutics*, 13(8), 1170.

Goel, P., Alam, O., Naim, M. J., Nawaz, F., Iqbal, M., & Alam, M. I. (2018). Recent advancement of piperidine moiety in treatment of cancer – A review. *European Journal of Medicinal Chemistry*, 157, 480–502. https://doi.org/10.1016/j.ejmech.2018.08.017

Gordon, M. R., & Rath, L. (1994). Synthetic alkaloids: Progress and challenges. *Journal of Medicinal Chemistry*, 37(25), 4232–4245.

Grant, R. T., & Iliescu, C. C. (1922). *Heart*, 9, 289.

Gupta, M., Mazumder, U. K., & Kumar, R. S. (2004). Antimicrobial activity of Alstonia macrophylla leaf extract. *Indian Journal Of Pharmacology*, 36(1), 38–40.

Gutiérrez, R. M., Mitchell, S., & Solis, R. V. (2008). Psidium guajava: A review of its traditional uses, phytochemistry, and pharmacology. *Journal of Ethnopharmacology*, 117(1), 1–27.

Hamid, H. A., Ramli, A. N. M., & Yusoff, M. M. (2017). Indole alkaloids from plants as potential leads for antidepressant drugs: A mini review. *Frontiers in Pharmacology*, 8, 96. https://doi.org/10.3389/fphar.2017.00096

Han, A. M., Heo, H., & Kwon, Y. K. (2012). Berberine promotes axonal regeneration in injured nerves of the peripheral nervous system. *Journal of Medicinal Food*, 15(4), 413–417.

Harborne, J. B. (1993). *Introduction to Ecological Biochemistry*. Elsevier Academic Press.

Henning, C. P. (2013). Compuestos secundarios nitrogenados: Alcaloides. In J. Ringuelet, & S. Viña (Eds.), *Productos Naturales Vegetales* (p. 18). Editorial de la Universidad de la Plata.

Heravi, M. M., et al. (2021). Synthesis of indole derivatives as prevalent moieties present in selected alkaloids. *RSC Advances*, 11(53), 33540–33612.

Honda, G., & Nagasaki, S. (1908). *Arch. Intern. Pharmacodynamie*, 18, 265.

Hussain, G., Rasul, A., Anwar, H., Aziz, N., Razzaq, A., Wei, W., Ali, M., Li, J., & Li, X. (2018). Role of plant derived alkaloids and their mechanism in neurodegenerative disorders. *International Journal of Biological Sciences*, 14(3), 341.

Imenshahidi, M., & Hosseinzadeh, H. (2016). Berberis vulgaris and berberine: An update review. *Phytotherapy Research*, 30(11), 1745–1764.

Johnson, I. S., Armstrong, J. G., Gorman, M., & Burnett, J. P. (1963). The vinca alkaloids: A new class of oncolytic agents. *Cancer Research*, 23(8), 1390–1427.

Joshi, P., Vishwakarma, R. A., & Bharate, S. B. (2017). Natural alkaloids as P-gp inhibitors for multidrug resistance reversal in cancer. *European Journal of Medicinal Chemistry*, 138, 273–292.

Kain, K. C., Shanks, G. D., & Keystone, J. S. (2001). Malaria chemoprophylaxis in the age of drug resistance. I. Currently recommended drug regimens, *Clinical Infectious Diseases*, 33(2), 226–234.

Kamagata, K., Tomiyama, H., Hatano, T., Motoi, Y., Abe, O., Shimoji, K., et al. (2014). A preliminary diffusional kurtosis imaging study of Parkinson disease: Comparison with conventional diffusion tensor imaging. *Neuroradiology*, 56(3), 251–258.

Kauê, A. P., da Silva, J. F. M., de Oliveira, A. L., & Pires, R. H. (2014). Antifungal activity of tomatidine against *Candida* and *Aspergillus* species: Mechanistic insights into membrane disruption. *Journal of Applied Microbiology*, 117(2), 511–520.

Kaur, R., & Arora, S. (2015). Alkaloids-important therapeutic secondary metabolites of plant origin. *Journal of Critical Reviews*, 2(3), 1–8.

Kaur, S., & Mondal, P. (2014). Antibacterial and antifungal properties of some natural products: A review. *Journal of Pharmacognosy and Phytochemistry*, 3(1), 115–119.

Khalifa, M. F., et al. (2018). An overview on the chemical and biological aspects of lycorine alkaloid. *Journal of Advanced Biomedical and Pharmaceutical Sciences*, 1(2), 41–49.

Khan, M. N., & Ghosh, N. (2020). Norditerpene alkaloids: Their antioxidant and anticancer properties. *Molecules*, 25(6), 1430.

Kim, M., Cho, K. I. H. O., Shin, M. S., Lee, J. M., Cho, H. S., Kim, C. J. U., et al. (2014). Berberine prevents nigrostriatal dopaminergic neuronal loss and suppresses hippocampal apoptosis in mice with Parkinson's disease. *International Journal of Molecular Medicine*, 33(4), 870–878.

Kovacs, G. G. (2014). Current concepts of neurodegenerative diseases. *EMJ Neurology*, 1, 78–86.

Kukula-Koch, W. A., & Widelski, J. (2017). Chapter 9 - Alkaloids. In S. Badal, & R. Delgoda (Eds.), *Pharmacognosy* (pp. 163–198). Academic Press.

Kulinsky, V. I. (2007). Biochemical aspects of inflammation. *Biochemistry*, 72, 733–746.

Kuo, C. L., Chi, C. W., & Liu, T. Y. (2004). The anti-inflammatory potential of berberine in vitro and in vivo. *Cancer Letters*, 203, 127–137.

Kwan, C. Y., & Achike, F. I. (2002). Tetrandrine and related bis-benzylisoquinoline alkaloids from medicinal herbs: Cardiovascular effects and mechanisms of action. *Acta Pharmacologica Sinica*, 23(12), 1057–1068.

Leitao Da-Cunha, E. V., Fechine, I. M., Guides, D. N., et al. Protoberberine alkaloids. In *The Alkaloids; Chemistry and Biology* (1st ed., 2005), p. 62. Elsevier Academic Press, ISBN: 0-12-469562-0.

Longfei, L., Wei, Z., Yifan, H., Jing, W., & Qiang, L. (2024). Evodiamine suppresses colon and lung cancer progression by inducing apoptosis and inhibiting cell proliferation and metastasis. *Phytomedicine*, 123, 155291.

Lin, L. et al. (2024). Evodiamine: A extremely potential drug development candidate of alkaloids from *Evodia rutaecarpa*. *International Journal of Nanomedicine*, 9843–9870.

Liu, Y., Morgan, J. B., Coothankandaswamy, V., Liu, R., Jekabson, M. B., Mahdi, F., Nagle, D. G., & Zhou, Y. D. (2009). The Caulerpa pigment caulerpin inhibits HIF-1 activation and mitochondrial respiration. *Journal of Natural Products*, 72, 2104–2109.

Ma, G., Sun, Z., Sun, Z. et al. (2014). Antimalarial diterpene alkaloids from the seeds of Caesalpinia minax. *Fitoterapia*, 95, 234–239.

Majid, A. (2021). Alkaloids: Classification, biosynthesis and pharmacological activities. *Journal of Pharmacognosy and Phytochemistry*, 10(2), 45–54.

Mao, Z., Huang, S., Gao, L., Wang, A., & Huang, P. (2014). A novel and versatile method for the enantioselective syntheses of tropane alkaloids. *Science China. Chemistry*, 57(2), 252–264. https://doi.org/10.1007/s11426-013-4998-2

McCawley, E. L. (1955). Cardioactive alkaloids. *The Alkaloids: Chemistry and Physiology*, 5, 79–107.

Meyers, M. A. (2007). *Happy Accidents: Serendipity in Modern Medical Breakthroughs*. Arcade Publishing.

Mugittu, K., Genton, B., Mshinda, H., & Beck, H. P. (2006). Molecular monitoring of Plasmodium falciparum resistance to artemisinin in Tanzania. *Malaria Journal*, 5(1), 126.

Ng, Y. P., Or, T. C. T., and Ip, N. Y. (2015). Plant alkaloids as drug leads for Alzheimer's disease. *Neurochemistry International*, 89, 260–270.

Nixon, G. L., Moss, D. M., Shone, A. E., Lalloo, D. G., Fisher, N., O'Neill, P. M., … & Biagini, G. A. (2013). Antimalarial pharmacology and therapeutics of atovaquone. *Journal of Antimicrobial Chemotherapy*, 68(5), 977–985.

Onori, E. (1984). The problem of Plasmodium falciparum drug resistance in Africa south of the Sahara. *Bulletin of the World Health Organization*, 62, 55–62.

Pan, Q., Mustafa, N. R., Tang, K., Choi, Y. H., & Verpoorte, R. (2016). Monoterpenoid indole alkaloids biosynthesis and its regulation in Catharanthus roseus: A literature review from genes to metabolites. *Phytochemistry Reviews*, 15(2), 221–250.

Podolsky, S. H. (2023). The (Harry) Gold standard: Angina, suggestion and the path to the "double-blind test" and clinical pharmacology. Part 3: The double blind and the rise of 'clinical pharmacology'. *Journal of the Royal Society of Medicine*, 116(9), 307–313.

Pradeep, C. R., & Kuttan, G. (2004). Piperine is a potent inhibitor of nuclear factor-κB (NF-κB), c-Fos, CREB, ATF-2 and proinflammatory cytokine gene expression in B16F-10 melanoma cells. *International Immunopharmacology*, 4(14), 1795–1803.

Ranjitha, D., & Sudha, K. (2015). Alkaloids in foods. *International Journal of Pharmaceutical, Chemical & Biological Sciences*, 5(4), 896–906.

Reddy, B. S., & Kumar, P. S. (2017). Indole Alkaloids and Their Antioxidant Activity. *Phytochemistry Reviews*, 16(1), 1–15.

Rosa, M. S., et al. (2017). Antifungal activity of alkaloids: Perspectives on their use as antifungal agents. *Frontiers in Microbiology*, 8, 575.

Sadre, R., et al. (2016). Metabolite diversity in alkaloid biosynthesis: A multilane (diastereomer) highway for camptothecin synthesis in Camptotheca acuminata. *The Plant Cell*, 28(8), 1926–1944.

Sagi, S., Avula, B., Wang, Y.-H., & Khan, I. A. (2016). Quantification and characterization of alkaloids from roots of Rauwolfia serpentina using ultra-high performance liquid chromatography-photo diode array-mass spectrometry. *Analytical and Bioanalytical Chemistry*, 408(1), 177–190.

Salminen, K. A., Meyer, A., Jerabkova, L., et al. (2011). Inhibition of human drug metabolizing cytochrome P450 enzymes by plant isoquinoline alkaloids. *Phytomedicine*, 18, 533–538.

Scherf, D., & Romano, F. J. (1948). Extrasystoles in groups. *American Heart Journal*, 35(1), 81–93.

Schiller, G. J., Damon, L. E., Coutre, S. E., Hsu, P., Bhat, G., & Douer, D. (2007). High-dose vincristine sulfate liposome injection, for advanced, relapsed, or refractory Philadelphia chromosome-negative acute lymphoblastic leukemia in an adolescent and young adult subgroup of a phase 2 clinical trial. *Journal of Adolescent and Young Adult Oncology*, 7(5), 546–552.

Schmid, M., & Keller, W. (2013). Modern applications of alkaloids in medicine. *Pharmacological Reviews*, 65(1), 23–47.

Schmitt, F., Hussain, G., Dupuis, L., Loeffler, J.-P., Henriques, A. (2014). A plural role for lipids in motor neuron diseases: Energy, signaling and structure. *Frontiers in Cellular Neuroscience*, 8(February), 25.

Schrag, A., Horsfall, L., Walters, K., Noyce, A., & Petersen, I. (2015). Prediagnostic presentations of Parkinson's disease in primary care: A case-control study. *Lancet Neurology*, 14(1), 57–64.

Seigler, D. S., and Seigler, D. S. (1998). Introduction to alkaloids. *Plant Secondary Metabolism*, 506–512.

Sertürner, F. (1804). Über das Morphium des Opiums. *Annalen der Pharmacie*, 3, 137–146.

Shakil, A. (1998). Isolation and structural elucidation of chemical constituents from fumaria indica, ferula oopoda and withania somnifer. Doctorate Thesis. International Center for Chemical Research, University of Karachi, 10–25.

Shi, Q. I. U., et al. (2014). Natural alkaloids: Basic aspects, biological roles, and future perspectives. *Chinese Journal of Natural Medicines*, 12(6), 401–406.

Siddique, Y., & Husain, S. (2018). Antioxidant and antimicrobial activities of quinoline alkaloids: A review. *Journal of Drug Delivery and Therapeutics*, 8(3), 22–29.

Silva Teles, M. M. R., Vieira Pinheiro, A. A., Da Silva Dias, C., Fechine Tavares, J., Barbosa Filho, J. M., & Leitão Da Cunha, E. V. (2019). Chapter Three - Alkaloids of the Lauraceae. In H.-J. Knölker (Ed.), *The Alkaloids: Chemistry and Biology* (pp 147–304). Academic Press.

Singh, P., Kalunke, R. M., Shukla, A., Tzfadia, O., Thulasiram, H. V., & Giri, A. P. (2020). Biosynthesis and tissue-specific partitioning of camphor and eugenol in Ocimum kilimandscharicum. *Phytochemistry*, 177, 112451.

Talapatra, S. K., & Talapatra, B. (2015). Alkaloids. General introduction. In *Chemistry of Plant Natural Products: Stereochemistry, Conformation, Synthesis, Biology, and Medicine* (pp. 717–724). Springer.

Van Dongen, K., & Sanchez, J. A. (1937). On the antifibrillatory action of quinidine and hydroquinidine. *Archives of International Pharmacodynamics and Therapy*, 56, 1–14.

Van Rensburg, M. G. J. V., & Van Zyl, L. S. P. (2013). Alkaloids: Classification, biosynthesis, and pharmacological properties. *Journal: Natural Products and Bioprospecting*. https://doi.org/10.1007/s13659-013-0167-1

Vieweg, S. (2017). *The Role of the Polyglutamine and N-Terminal Domains in Regulating the Aggregation and Structural Properties of Huntingtin Exon 1*. EPFL.

Weina, M., Man, Z., & Dongdong, Z. (2017). Berberine inhibits the proliferation and migration of breast cancer ZR-75-30 cells by targeting Ephrin-B2. *Phytomedicine*, 25, 45–51.

Wiesner, J., Ortmann, R., Jomaa, H., & Schlitzer, M. (2003). New antimalarial drugs. *Angewandte Chemie International Edition*, 42(43), 5274–5293.

William, E. (2009). Pharmacopoeial and related drugs of biological origin. In *Trease and Evans' Pharmacognosy* (pp. 353–416). WB Saunders Co. Ltd.

Wink, M. (2008). Ecological roles of alkaloids. *Modern Alkaloids*, 3–52.

Wink, M. (2010). Functions and biotechnology of plant secondary metabolites. Introduction, Chapter 1. *Annual Plant Reviews*, 39, 1–20.

Yanran, L., Zhang, Y., Wang, X., & Chen, Z. (2018). Noscapine as a potential anticancer agent: Molecular mechanisms and therapeutic applications. *Biomedicine & Pharmacotherapy*, 103, 1578–1586.

Yan, R., & Vassar, R. (2014). Targeting the β-secretase BACE1 for Alzheimer's disease therapy. *The Lancet Neurology*, 13, 319–329.

Yang, C. W., Chen, W. L., Wu, P. L., Tseng, H. Y., & Lee, S. J. (2006). Anti-inflammatory mechanisms of phenanthroindolizidine alkaloids. *Molecular Pharmacology*, 69(3), 749–758.

Yin, W., Wang, T. S., Yin, F. Z., & Cai, B. C. (2003). Analgesic and anti-inflammatory properties of brucine and brucine-N-oxide extracted from seeds of Strychnos nuxvomica. *Journal of Ethnopharmacology*, 88, 205–214.

Zhang, J. X., Yang, Z. R., Wu, D. D., Song, J., Guo, X. F., Wang, J., & Dong, W. G. (2014). Suppressive effect of sinomenine combined with 5-fluorouracil on colon carcinoma cell growth. *Asian Pacific Journal of Cancer Prevention*, 15(16), 6737–6743.

Zhang, Q., Bian, H., Guo, L., & Zhu, H. (2016). Pharmacologic preconditioning with berberine attenuating ischemia-induced apoptosis and promoting autophagy in neuron. *American Journal of Translational Research*, 8(2), 1197–1207.

Zhang, T., Chen, W., Jiang, X., Liu, L., Wei, K., Du, H., Wang, H., & Li, J. (2019). Anticancer effects and underlying mechanism of Colchicine on human gastric cancer cell lines in vitro and in vivo. *Bioscience Reports*, 39(1), BSR20181802.

Zhao, Z., Xiao, J., Wang, J., Dong, W., Peng, Z., & An, D. (2015). Anti-inflammatory effects of novel sinomenine derivatives. *International Immunopharmacology*, 29(2), 354–360.

Zuo, F., Nakamura, N., Akao, T., & Hattori, M. (2006). Pharmacokinetics of berberine and its main metabolites in conventional and pseudo germ-free rats determined by liquid chromatography/ion trap mass spectrometry. *Drug Metabolism and Disposition*, 34(12), 2064–2072.

6 Natural Anticancer Alkaloids
Sources, Potential Therapeutic Effects, and Mechanistic Approaches

Esraa A. Elhawary, Orchid A. Mahmoud, Heba A. S. El-Nashar, Omayma A. Eldahshan, and Abdel Nasser B. Singab

6.1 INTRODUCTION

Cancer is one of the leading causes of death worldwide, and disturbances in the autophagy process often accompany its development. Autophagy, a catabolic process included in several physiological processes, is crucial for cell growth and survival. It is an intracellular lysosomal/vacuolar degradation system. In this system, lysosomal hydrolases degrade the inner cytoplasmic cell membrane, and the products are released back into the cytoplasm. Indole alkaloids are natural products extensively found in nature and have been proven to possess various pharmacological activities. In recent years, pharmacological studies have proposed another potential of indole alkaloids, autophagy regulation (Luo et al., 2022).

Plant-based drugs have evolved as propitious preventive chemo options both in developing and developed nations. The plant secondary metabolites, including alkaloids, have proven efficacy and acceptability for cancer treatment (Dhyani et al., 2022).

Alkaloids, derived from plant sources, are basic, containing one or more nitrogen atoms (usually in a heterocyclic ring) and usually have a marked physiological action on man or other animals. Alkaloids have potent anticancer activity against various cancers (Mohan et al., 2012).

Cancer, a complex multifactorial disease, results from alterations in many physiological and biochemical functions. Over the last few decades, it has become obvious that cancer cells can acquire multidrug resistance to conventional anticancer drugs, leading to tumor relapse. Thus, there is an urge to discover new and effective anticancer drugs. Natural-derived products from various sources, including plants, have served as a primary source of anticancer drugs (Tilaoui et al., 2021). This book chapter includes a summary of different natural sources for anticancer alkaloids, *viz.*, plants, marine, and endophytes, in addition to an emphasis on their mechanisms of action.

6.2 METHODOLOGY

This study examines the pharmacological properties, research techniques, and active compounds isolated from natural sources, including medicinal plants, marine, and endophytic fungi. Different research engines were used for literature analysis guided by pertinent keyword exploration in Clarivate Analytics, Scopus, Web of Science, Science Direct, Google Scholar, MDPI, PubMed, and Google Academic.

DOI: 10.1201/9781003539179-6

6.3 MEDICINAL PLANT-DERIVED ANTICANCER ALKALOIDS

6.3.1 *Papaver somniferum* L.

HeLa (Human Cervix Carcinoma), HT29 (Human Colorectal Adenocarcinoma), C6 (Rat Brain Tumor Cells), and Vero (African Green Monkey Kidney) cell lines were used to explore the cytotoxic and antiproliferative properties of *Papaver somniferum* L. extracts. Particularly at high concentrations, *P. somniferum* L. extracts rich in alkaloids displayed antiproliferative effects on a variety of cancer cell lines. Extracts from *P. somniferum* L. degrade tumor cell lines' cellular membranes at high doses. Interestingly, the findings of the LDH test showed that *P. somniferum* L. had no cytotoxicity on cells at intermediate doses (Güler et al., 2016).

6.3.2 *Lantana camara* L.

Positively charged alkaloids were separated from *the Lantana camara* leaves' methanol extract using magnetic nanoparticles (MNPs). HPLC was used to isolate the highest peak of the alkaloid fraction (HPAF) from the crude alkaloid. The antiproliferative effects of crude alkaloids (CA) and HPAF were evaluated against endothelial cells (EA.hy926), a reference cell line, and cancer cell lines (MCF-7, HCT-116, and HeLa). The rat aortic ring assay was used to investigate antiangiogenic characteristics. While HPAF showed relatively low cytotoxicity against the HCT-116 cell line (IC_{50} = 8.38 µg/mL), it exhibited a strong anticancer impact against the MCF-7 and HeLa cell lines (IC_{50} = 0.027 µg/mL and 5.90 µg/mL, respectively). Additionally, the CA showed a mild cytotoxic activity against colon cancer HCT-116 cells and a strong anticancer effect against MCF-7 and HeLa. Cationic alkaloids were completely harmless on the normal test cell line; however, they showed specific antiproliferative action against HeLa. At non-toxic dosages, HPAF showed excellent antiangiogenic efficacy. Furthermore, cationic alkaloids had strong antiangiogenic properties (Al-Hakeim et al., 2021).

6.3.3 *Solanum aculeastrum*

Using spectroscopic methods, two steroid glycosides were identified as tomatidine and solasodine after being extracted from *Solanum aculeastrum* berries. These chemicals' effects on cell development were examined using the cancer cell lines HeLa, MCF7, and HT29. The IC_{50} of the combined compounds was lower than that of solasodine and unaffected by that of tomatidine, confirming that both solasodine and tomatidine had the strongest inhibitory impact on HeLa cells. Nevertheless, in HT29 and MCF7 cells, the combined IC_{50} values of the two compounds were likewise lower than those of the individual compounds. Following a 24-hour exposure, solasodine and tomatidine both reduced cell growth by obstructing the cell cycle in the G_0=G_1 phase, increasing from 55.6% to 64.2% and 66.8%, respectively. Flow cytometry labeled with Annexin V–FITC=PI revealed extremely low apoptotic indices for the compounds (Koduru et al., 2007).

6.3.4 *Opuntia polyacantha*

To obtain the complete alkaloid extract, leaves of *Opuntia polyacantha* were extracted individually using 80% methanol and chloroform at pH 2 and pH 10. The chloroform part was then dried. The reagents developed by Hager, Dragendorff, and Mayer qualitatively identified the whole alkaloid. HCS investigated the cytotoxic effect of the alkaloid extract of *O. polyacantha* in one cultured cellular model (MCF7 breast cancer cell line). Because *O. polyacantha* induces apoptosis, it inhibits the development of MCF7 cells. Through nuclear condensation, cell membrane permeability, Mitochondrial Membrane Potential (MMP) disruption, fragmentation, and cytochrome c release from mitochondria into the cytoplasm, the alkaloid extract of *O. polyacantha* can decrease the survival of breast cancer cells. Thus, *O. polyacantha* is considered a possible MCF7 inhibitor compared to the positive control, doxorubicin (Al-Shukri et al., 2020).

6.3.5 *Voacanga globosa*

Spirobisindole alkaloids from the plant *Voacanga globosa* were studied for their cytotoxic and antiproliferative properties, as well as the most potent alkaloid's pro-apoptotic properties. The alkaloids vobtusine lactone, globospiramine, and deoxyvobtusine that had previously been identified were tested using the MTT and Cell Titer Blue assays *in vitro*. The most physiologically active of the aforementioned alkaloids was globospiramine, which demonstrated cytotoxic and antiproliferative activities against the examined cell lines (KB3.1, MCF-7, A431, A549, L929, PC-3, and SKOV-3). Globospiramine boosted the cleaved caspase-3 and PARP expression, two markers in cancer cells that experienced caspase-dependent death, according to Western blot examination (Manzano et al., 2024).

6.3.6 *Annona squamosa*

Annona squamosa yielded two alkaloids that were identified as benzylisoquinoline alkaloids. Coclaurine was previously isolated from *A. squamosa* L., while the other alkaloid, 6,7-dimethoxy-1-(α-hydroxy-4-methoxybenzyl)-2-methyl-1,2,3,4-tetrahydroisoquinoline, was isolated from natural sources for the first time. Additionally, the extracted alkaloids were evaluated on several cell lines (HepG-2, MCF-7, and HCT-116) to discover their anti-cancer properties (Al-Ghazzawi, 2019).

6.3.7 *Citrullus colocynthis*

One of the families with the greatest commercial significance is Cucurbitaceae. *Citrullus colocynthis* (L.) is one species in the family that has a lot of bioactive secondary metabolites. Several plant extracts from *C. colocynthis* demonstrated cytotoxicity to some cancer cells. HPLC/QTOF/MS stands for High-Performance Liquid Chromatography/Quadrupole Time of Flight/Mass Spectrometry. This technique was used to examine the chemicals in the fruit extract of ethyl acetate (EtOAc). MIAPaCa-2 and A-431 cell lines were used to test for anticancer activities. The DPPH test was used to measure antioxidant activity. To understand the molecular basis for the hs anticancer action, molecular modeling was utilized. *Via* comparing the compounds' fragmentation pattern in positive mode with the reference spectra and published research, 24 compounds were inferred. A-431 and MIAPaCa-2 cancer cells responded well to the *C. colocynthis* EtOAc extract, but normal (healthy) cells showed little to no adverse effects. In contrast to the oxidative possibility of the reference (caffeic acid, 98%), the extracts of EtOAc and BuOH demonstrated strong antioxidant activity (86% and 76%, respectively) for the DPPH experiment. Cucurbita-5(10),6,23-triene-3β,25-diol was one of the main cucurbitacin compounds that LC/MS inferred from the EtOAc extract. Docking tests and MD simulations conducted during this work indicated that Cucurbita-5(10),6,23-triene-3β,25-diol's inhibition of EGFR was responsible for the anti-pancreatic cancer action of *C. colocynthis* extract (Mandour et al., 2023).

6.3.8 *Gnidia glauca*

Through various mechanisms, including the induction of apoptosis (as demonstrated by the TUNEL assay), modulation of tubulin polymerization, preservation of mitochondrial function (designated by the JC-1 assay), and inhibition of DNA (deoxyribonucleic acid) polymerase α and β activities, *Gnidia glauca* has shown significant anticancer potential against MCF-7 cells. The alkaloid extract from *G. glauca* leaves was examined for possible anticancer effects. After being cleaned and allowed to dry in the shade, the powdered *G. glauca* leaves were defatted, then extracted using 10% acetic acid in ethanol, and finally treated to alkaloid extraction. Utilizing human breast cancer (MCF-7) cells, the effects of the partly pure *G. glauca* leaf alkaloid extract on tubulin polymerization, DNA polymerase activity, MMP, and apoptosis were assessed by flow cytometry. It was found that the extract had a concentration-dependent impact on microtubule assembly (15.125–250 μg/mL), suggesting the presence of alkaloids that act as spindle poisons. With an IC_{50} value of 144.51 μg/mL, *G. glauca* leaf alkaloid extract was also discovered to influence the MMP and inhibit DNA polymerase α and β activities in a dose-dependent manner, which may interfere with DNA replication and repair activities. According to the TUNEL assay, leaf alkaloid extract displayed the capacity to cause DNA damage of 53.6%, which is marginally less than that of the common medication camptothecin (64.94%). Furthermore, phase arrests in sub-G_0/G_1, S, and G_2/M were revealed by the GgLAE (IC50 144.51 μg/mL), which demonstrated a considerable suppression of MCF-7 cell proliferation after 24 hours. *G. glauca* is considered a natural source of interesting phytochemical therapeutic candidates for additional assessment for pre-clinical and clinical investigations, according to these findings, which imply that *G. glauca* leaf alkaloid extract includes alkaloids with anticancer effects with several targets (Valleti et al., 2024).

6.3.9 *Cyclea peltata*

Phaeanthine, a bisbenzylisoquinoline alkaloid that was extracted from *Cyclea peltata* (Lam) Hook. F and Thoms' rhizomes were tested for their anticancer effects. Against HeLa cervical cancer cells, it demonstrated selective cytotoxicity, with an IC_{50} of 8.11 ± 0.04 μM. A docking score of −5.023 indicates that phaeanthine exhibited preferential binding to the antiapoptotic protein Akt in subsequent research using *in silico* molecular docking. It is interesting to note that the compound's *in vitro* evaluation explicitly linked it to mitochondria-mediated apoptosis by suppressing the production of Akt and p-Akt, as well as additional antiapoptotic proteins MCl-1, IGF-2, and XIAP. Intercellular cytochrome c dynamics and mitochondrial membrane polarization confirmed the innate mechanism of the apoptotic process in the complementary in vitro evaluation (Valsan et al., 2023).

6.3.10 *Alphonsea sclerocarpa*

An apomorphine alkaloid that was extracted from the medicinal plant *Alphonsea sclerocarpa* was found to have anticancer and antioxidant properties. The alkaloid Crebanine was isolated and purified using preparative HPLC techniques and conventional column chromatography. The DPPH and FRAP assays were implemented to evaluate the antioxidant activity. Antioxidant activity was dose-dependent, meaning that as concentration rose, so did activity. The MTT assay was utilized to perform the cytotoxicity test. After a 24-h incubation period, K562 (chronic myeloid leukemia blood cancer) cells exhibited remarkable cytotoxic action, with a 37.545 percent inhibition at 500 μm concentration (Suman Joshi et al., 2018).

6.3.11 *Catharanthus roseus* (L.)

This plant is also known as "The Madagascar periwinkle". In the warmer regions of the world, it is a common ornamental plant that is found in homes and gardens. The tropical and subtropical plant *Catharanthus roseus* (L.) G. Don, also referred to as "an anticancer drug yielding plant", is a member of the Apocynaceae family. The amazing qualities of the plant are highlighted in this review. It is regarded as a milestone in cancer therapy because alkaloids like vinblastine and vincristine, which are mostly found in the aerial sections of *C. roseus*, are utilized for the treatment of different human tumors. In addition, it exhibits anti-diarrheal, anti-ulcer, anti-bacterial, anti-oxidant, and anti-diabetic qualities. *C. roseus* leaves and stems contain a vast number of phytochemical components. Since it has been determined to be endangered, conservation is also a concern (Sain & Sharma, 2013).

Using Hoagland's nutritional solution, *C. roseus* seedlings were subjected to variable potassium (K^+) concentrations (1.5, 3.16, 15, and 30 mmol) and two forms (K_2SO_4 and KNO_3). On plant days 68 and 78, spraying ascorbic acid (AsA) on the leaves was performed twice at two doses (750 and 1,500 mg/L). Measurements were made of the leaves' H_2O_2 content, vinblastine, vincristine, tryptophan levels, the gene expression of D_4H and DAT, and peroxidase activity. Compared to 30% and 20% with K_2SO_4, potassium in KNO_3 form raised vinblastine (60%) and vincristine (50%). Vinblastine and vincristine prevent microtubules from assembling and, as a result, stop the polymerization-induced metaphase. Using an excess of K^+, the gene expression was three times higher in KNO_3 and 2.5 times higher in K_2SO_4. Vinblastine (20%) and vincristine (16%) increased with foliar administration of 750 mg/L AsA. In contrast to the control plant, the extra effects of both AsA doses on D4H and DAT expression were roughly 30% and 60%, respectively. Tryptophan dropped 2.5 times more than K^+ and 35% because of AsA's exterior. With AsA therapy, H_2O_2 dropped, and peroxidase activity rose. The levels of vinblastine, vincristine, tryptophan, and gene expression were positively correlated with K^+ and AsA (Sahi et al., 2022).

6.3.12 *Pinus sabiniana* L., *Phoenix dactylifera* L., and *Ferocactus* sp. L.

Pinus sabiniana L., *Phoenix dactylifera* L., and *Ferocactus sp.* L. total alkaloid extracts were evaluated for their *in vitro* cytotoxic activities against the non-tumorigenic fetal hepatic cell line (WRL-68) and the breast cancer cell line Michigan Cancer Foundation-7 (MCF-7). Plant powder was made from the leaves of *P. sabiniana* L., the pollen grains of *P. dactylifera* L., and the leaves of *Ferocactus* sp. L. The 3-[4,5-dimethylthiazoyl]-2,5-diphenyltetrazolium bromide assay was utilized to assess the cytotoxic activity. Compared to the other plants, *P. sabiniana* L. extract exhibited the highest total alkaloid concentration (164.62 ± 2.8 mg/100 g dry weight of plant). The total alkaloids of Ferocactus sp decreased the viability of the cells in both cell lines. L. and *P. dactylifera* L. The greatest reduction was observed at 400 µg/mL: 46 ± 2.20% (MCF-7) and 56.2 ± 2.2% (WRL-68) for *Ferocactus* sp. L., then 56.2 ± 2.2% (MCF-7) and 57.5 ± 3.2% (WRL-68) for *P. dactylifera* L. At the same concentration, the *P. sabiniana* alkaloids had minimal effect on the two cell lines, MCF-7 and WRL-68, with respective values of 89.3 ± 3.44% and 90.16 ± 2.7% (Al-Marzook & Omran, 2017).

6.3.13 *Vinca alkaloids*

Vinca is the source of a novel phenanthroindolizidine alkaloid (PA) derivative (YPC-10157), whose anticancer activity was assessed and its mode of action clarified. YPC-10157's anticancer efficacy was assessed against several human cancer cell lines (*in vitro*) and against mouse xenograft models (*in vivo*). Caspases-3/7 activity was used to measure cell apoptosis. An *in vitro* cell-free translation assay system was used to evaluate the impact on protein production. YPC-10157 caused apoptosis and showed a significant suppression of cell growth *in vitro*. In human colon cancer and leukemia xenograft models, YPC-10157 demonstrated potent anticancer activity *in vivo*. Furthermore, there was a substantial correlation between the inhibitory activity of YPC-10157 and its derivatives on protein synthesis and cell proliferation (Ali et al., 2012; Ueno et al., 2014).

6.3.14 *Solanaceae glycoalkaloids*

Solanaceae glycoalkaloids (GAs) are plant-based natural chemicals that are utilized extensively in traditional medicine to treat a wide range of illnesses. *Solanum* plants contain solanine, chaconine, solasonine, solamargine, and tomatine. Among other things, GAs have notable anticancer characteristics, such as a potent growth-inhibiting action on cancer cells. Through a variety of molecular routes, this action may cause tumor cells to undergo apoptosis. Intense research is being done on the molecular mechanisms of GA activity since a better understanding of these mechanisms may result in the creation of novel cancer treatments. Because protooncogenes, suppressors, and genes that control and repair apoptosis constitute the genetic basis for the development of neoplasms, compounds with antineoplastic qualities may influence the levels of these genes' expression or the levels of their expression products. Therapeutic bioactive molecules can be applied separately or in collaboration with other drugs to improve the effectiveness of cancer therapy; they can act on the cell through different mechanisms at different stages of carcinogenesis, inducing apoptosis, obstructing cell proliferation and migration, and inhibiting angiogenesis (Winkiel et al., 2022).

6.4 MARINE-DERIVED ANTICANCER COMPOUNDS

6.4.1 Marine Sponge

In addition to being abundant in natural products, marine sponges have produced a limitless number of bioactive metabolites. An interesting illustration of the chemical diversity of secondary metabolites produced by marine invertebrates, bromopyrrole alkaloids are a good example and exhibit a wide range of biological activity. Investigations into the synthesis and isolation of this structural class have produced a number of bromopyrrole alkaloids that may be lead hits (Mahamed et al., 2023).

6.4.2 Actinomycetes

Over the past 50 years, extensive research has been conducted on the naturally occurring, physiologically active 1,2,3,4-tetrahydroisoquinoline–quinone (THIQ) family members that have been identified from *Actinomycetes* and marine species. Among these, marine natural products and their reduced compounds—like ecteinascidins and renieramycins—have garnered attention because of their strong anticancer characteristics, remarkable architectures, and limited natural availability. We have created an intriguing THIQ natural product chemistry and medicinal chemistry based on our understanding of saframycin antibiotics chemistry, including their isolation, characterization, transformation, partial synthesis, and total synthesis, as part of our quest for novel anticancer metabolites (Saito, 2021).

6.4.3 *Eudistoma viride*

Ascidians are classified as marine invertebrates that produce a wide range of bioactive compounds. MTT assay was used to examine the ascidian *Eudistoma viride*, which is found along India's southeast coast, for its *in vitro* cytotoxicity effect against HeLa cells. With an IC_{50} value of 53 μg/mL, the crude methanol extract of *E. viride* exhibited dose-dependent cytotoxicity. It reduced cell viability to 9.2% and was more effective at 100 μg/mL than the standard cyclohexamide (1 μg/mL). At 10 μg/mL, ECF-8 showed the most pronounced cytotoxic activity among the nine chromatographically separated fractions. With 39.8% viable cells in contrast to 51% in cyclohexamide-treated cells at a 1 μg/mL concentration, the HPLC fraction EHF-21 of ECF-8 demonstrated remarkable dose- and time-dependent cytotoxicity; the IC_{50} was 0.49 μg/mL. Treated HeLa cells with 0.5 μg/mL EHF-21 showed signs of apoptosis, including condensation of chromatin, blebbing of the membrane, cell shrinkage, and formation of an apoptotic body. Changes in light scatter were seen in the cell size and granularity analysis, which is a marker of apoptotic cell death. Treated HeLa cells with fraction EHF-21 at 1 μg/mL exhibited a striking arrest of cells in the G_0/G_1, S, and G_2/M phases, as well as an increase in the sub-G_0/G_1 population, which suggested a rise in the population of apoptotic cells, according to the cell cycle analysis. The sub-G_1 region's statistical analysis revealed that apoptosis was induced in a dose-dependent manner. EHF-21-treated HeLa cells also showed signs of DNA fragmentation. The brominated indole alkaloid Eudistomin H, which is the active EHF-21 portion, caused HeLa cells to undergo apoptosis (Rajesh & Annappan, 2015).

6.4.4 Genus *Aaptos* (Demospongiae, Suberitida, Suberitidae)

As a prospective scaffold for the synthesis of anticancer, antibacterial, and antioxidant drugs, aaptamine (8,9-dimethoxy-1H-benzo[de][1,6]-naphthyridine), an alkaloid derived from sea sponges of the genus *Aaptos* (family Suberitidae), has garnered a lot of attention. Through a variety of signaling pathways, such as the modulation of PTEN/PI3K/Akt and CDK2/4, the inhibition of cyclin D_1/E in cell cycle arrest, and inhibiting cancer cell proliferation and apoptosis induction, aptamine exhibits encouraging anticancer efficacy. Aptamine's distinct chemical structure presents prospects for structural alterations meant to strengthen its anticancer properties. There is a lot of potential for resolving several issues related to cancer treatment by investigating aaptamine as a scaffold in the creation of innovative therapeutic agents (Tailor et al., 2024).

6.4.5 *Palmaria palmata*

Marine natural products include terpenes, polyketides, bioengineered alkaloids, high molecular weight organic sugars, and marine animal peptides. Additionally, land-based bacteria, plants, animals, and invertebrates are used to produce anticancer medications. While polysaccharides activate the innate immune system and cause apoptosis in human leukemia cells and pancreatic islet cancer cells, polyphenols, which are present in edible seaweeds like *Palmaria palmata*, can reduce the division and multiplication of cancer cells. Marine-derived alkaloids have the ability to stop the growth of cancer cells. Numerous plant species contain peptides that have been validated to have cytotoxic effects on different human cell lines, including those from bladder, lung, breast, and pancreatic malignancies. Issues with marine medication approval include the potential for a revival of various metabolites as a result of environmental factors and the scarcity of lead compounds (Rana et al., 2024).

6.4.6 *Monanchora pulchra*

Monanchoxymycalin C (MomC), a novel pentacyclic guanidine alkaloid, was isolated from *Monanchora pulchra*, a marine sponge. Here, a human prostate cancer (PCa) model was used to examine anticancer activity and the mechanism of action for the first time. At low micromolar concentrations, MomC was effective in all PCa cell lines and caused an uncommon non-apoptotic, caspase-independent cell death. One of the main molecular mechanisms of MomC's anticancer effect, according to kinase activity screening, is the activation of mitogen-activated protein kinase (MAPK) c-Jun N-terminal protein kinase (JNK1/2). Functional experiments showed that JNK1/2 was specifically and selectively activated before other processes linked to cell death were induced. The cytotoxic action of the marine compound was counteracted by pretreatment with the JNK-inhibitor SP600125, which inhibited JNK1/2. Cytotoxic ROS were upregulated as a result of MomC. However, unlike various ROS-inducing drugs, co-treatment with olaparib, a PARP-inhibitor, showed opposing outcomes, suggesting that MomC action requires an active PARP. It is interesting to note that while there was no evidence of direct control of p38 or ERK1/2, MomC's effectiveness required active p38 kinase, and its cytotoxicity increased when ERK1/2 was inhibited (Dyshlovoy et al., 2020a).

In another study, *M. pulchra*, which was collected in Northwestern Pacific waters, yielded the new bicyclic guanidine alkaloid Urupocidin C (Ur-C) and the previously identified Urupocidin A (UrA). Mass spectra and 1D and 2D NMR spectroscopy were used to clarify Ur-C's distinct structure. In contrast to non-malignant cells, we found that both alkaloids showed promising selectivity for human PCa cells, including highly drug-resistant lines. Marine-derived chemicals were able to cause caspase-mediated cell death and G_1- and S-cell cycle arrest in cancer cells. Accordingly, administration of the isolated alkaloids caused mitochondrial membrane permeabilization, which in turn caused the liberation of cytotoxic mitochondrial proteins into the cellular cytoplasm, the upregulation of ROS, the activation of caspase-9 and -3, PARP cleavage, DNA fragmentation, and apoptosis. Additionally, when Ur-C and Ur-A were coupled with the clinically licensed olaparib, synergistic effects were noted. Lastly, when combined with the androgen receptor inhibitor enzalutamide and docetaxel, which are both used in PCa therapy, these alkaloids showed additive effects (Dyshlovoy et al., 2020b).

6.4.7 *Gibraltar strait ascidians*

A range of natural compounds with different biosynthetic origins has been characterized as a result of the chemical analysis of a few Gibraltar Strait ascidians. Spectroscopic techniques were used to clarify the structures, and in some instances, chiral auxiliaries based on the Mosher method were used to ascertain the absolute stereochemistry. Key elements of the structural elucidation are described, with special attention paid to the novel compounds isolated from *Phallusia fumigata, Botryllus leachi, Stolonica socialis*, and *Pseudodistoma obscurum* ascidians. For the prediction a possible antitumor application, the cytotoxicity of the new compounds was evaluated against cell lines P-388 (suspension culture of a lymphoid neoplasm from a DBA/2 mouse), A-549 (monolayer culture of a human lung carcinoma), HT-29 (monolayer culture of a human colon carcinoma), and MEL-28 (monolayer culture of a human melanoma) (Zubía et al., 2003).

6.4.8 *Ecteinascidia turbinate*

The ecteinascidins, lurbinectedin, and trabectedin have unique modes of action and a good toxicity profile, making them highly intriguing antineoplastic medicines. These medications create adducts in the minor groove of DNA, which cause DSBs and SBBs and set off a chain of events that leads to apoptosis and cell cycle arrest. Additionally, interactions with the tumor microenvironment, limitations in cytokine and chemokine release, and a decrease in the extent of tumor-associated macrophages occur. For patients with liposarcoma or leiomyosarcoma that is incurable or metastatic and who have previously received an anthracycline-based treatment, the FDA has approved trameptin. Furthermore, trabectedin has been approved in the European Union in combination with pegylated liposomal doxorubicin (PLD) for the treatment of platinum-sensitive recurrent ovarian cancer. Lurbinectedin has been licensed by the FDA for patients with metastatic small-cell lung cancer after platinum-based chemotherapy or with the progression of the disease (Gadducci & Cosio, 2022).

6.4.9 *Corticium niger*

Selective cytotoxic activity against human colon cancer cell lines was demonstrated by a subset of extracts, according to the data from bioinformatic analysis of the NCI-60 cell cytotoxicity screen. Fractionation of a colon cancer selective extract using bioassay guidance from a Philippine marine sponge collection. In addition to two recognized plakinamine-class chemicals, *Corticium niger* yielded two novel steroidal alkaloids, plakinamines N and O. By interpreting the integrated MS and NMR spectroscopic data, the structures of these substances were clarified. Plakinamines N, O, and J showed improved inhibitory impacts against all cell lines of the colon with mean IC_{50} values of 11.5, 2.4, and 1.4 μM, respectively (Sunassee et al., 2014).

6.5 ENDOPHYTIC FUNGI-DERIVED ANTICANCER COMPOUNDS

6.5.1 *Millingtonia hortensis*

Throughout South Asia, *Millingtonia hortensis* is a significant medicinal herb. The current study included seven endophytic fungi from *Millingtonia hortensis* L. that have been previously reported. Five of them were discovered to generate taxol. The antibacterial and antifungal properties of Taxol against fungi and humans were evaluated using the agar well diffusion method. Fungal taxol successfully inhibited every pathogen that was tested. The grouping of bacterial cells against the chemicals (MID_6 and MID_9) was examined using flow cytometry in light of the notable antibacterial activity. In contrast to the control, MID9 displayed a considerable degree of cell death. The cytotoxic impacts of fungal taxol against the human liver cancer cell line $HepG_2$ were investigated using the MTT test. After 24 h of treatment, there was a noticeable reduction in cell viability (0.05–0.5 μM), and the IC_{50} dosage was determined to be 0.25 μM. Changes in morphology were also noted (Pandy et al., 2023).

6.5.2 *Xestospongia* sp.

Although the underlying mechanism is yet unknown, Renieramycin M has demonstrated encouraging anticancer efficacy against specific cancer cell lines. The blue sponge *Xestospongia* sp. was used to isolate Renieramycin M. Renieramycin M's anticancer and antimetastatic properties were examined in human non-small cell lung cancer cells. Anti-apoptotic MCL-1 and BCL-2 proteins were down-regulated as a result of p53 activation brought on by Renieramycin M therapy, but pro-apoptotic BAX protein levels remained unchanged. The subtoxic levels of renieramycin M dramatically reduced the cancer cells' capacity for invasion and migration. Furthermore, this substance demonstrated a substantial inhibitory effect on the cells' anchorage-independent growth. Through a p53-dependent mechanism, Renieramycin M caused lung cancer cells to undergo apoptosis, and the substance may prevent lung cancer cells from proliferating and spreading (Halim et al., 2011).

6.5.3 *Jatropha heynei*

To find the diversity indices and colonization frequency of endophytic fungi, the root, fruit, and leaf samples of *Jatropha heynei* were incubated using water agar, Czapek Dox agar, potato dextrose agar, and malt extract agar. Based on their physical traits, 33 endophytic fungal species belonging to 20 genera were identified after recovering from 5,400 segments of *J. heynei*. Leaf species were the most abundant, followed by root and fruit species. The season and various plant components have an impact on the diversity indices. The endophytic fungus *Penicillium citrinum*'s metabolites were examined *in vitro* for cytotoxic, antioxidant, and antibacterial properties. The *P. citrinum* culture filtrate (CF) extract exhibited strong antibacterial activity against *Staphylococcus aureus* with an inhibition zone of 16.32 ± 0.5 mm and *Pseudomonas syringae* with an inhibition zone of 7.65 ± 1 mm. The cyclic voltammetry approach, which is based on metabolite redox potential detection, was used to determine the antioxidant potential. The A549 and MCF-7 cell lines were both significantly cytotoxically affected by the CF extract at a concentration of 500 μg/mL, with IC_{50} values of 280.7 and 283.0 μg/mL, respectively. Twenty-one bioactive compounds were identified in CF extract *via* Orbitrap high-resolution liquid chromatography mass spectroscopy (OHRLCMS); the main ones are trigonelline, artemisinin, spectinomycin, sulfamethazine, psoralidin, nicotinic acid, kanosamine, and 8-hydroxyquinoline, among several bioactive compounds (Ashoka & Shivanna, 2023).

6.5.4 *Phaleria macrocarpa*

It is recognized that *Phaleria macrocarpa* (Scheff) Boerl contains active metabolites that may have anticancer effects. It has been reported that certain endophytes can generate secondary metabolites that resemble those of their parent plants. *Via* sorting out endophytic fungi that produced secondary metabolites from portions of *P. macrocarpa*, which were obtained from the Yogyakarta area, endophytes that produced such compounds were investigated. HeLa, MCF_7, $T_{47}D$, and normal Vero cells were used in the anticancer bioassay, which was conducted *in vitro*. Using flow cytometry, the mechanisms of apoptosis and the regulation of the cell cycle in potential extracts were identified. Identification of fungi that produce potentially cytotoxic extracts was done. Then, LC-HRMS (Liquid Chromatography-High Resolution Mass Spectrometry) was applied to assess the component content. The leaves' endophytic extract (code PC-L_2) exhibited the best cytotoxic activity when linked to other extracts. It showed the optimal effect on MCF-7 cells (IC_{50} = 110.66 µg/mL), according to the results of the six extracts that were taken from the stem, leaf, and fruit portions. The *Clonostachys wenpingii* strain DUCC5606 is closely related to the fungi that are produced by PC-L2 extract. Apoptosis was triggered by this extract, but not by the cell cycle modification method (Agustina et al., 2024).

6.5.5 Entrophospora Infrequents

Nothapodytes foetida inner bark was used to extract a novel endophytic fungus that, in shaking flask conditions, generated camptothecin in Sabouraud broth (SB). Fungal endophytes may be an alternate source of manufacture for camptothecin and its related chemicals, which are currently extracted from entire plants. The impact of various nutritional combinations and precursors (citral, geraniol, leucine, mevalonic acid, tryptophan, and tryptamine) on the endophytic fungus *Entrophospora infrequents*'s buildup of camptothecin was noted. The precursors (tryptophan and leucine, tryptophan and citral, tryptophan and mevalonic acid, and tryptophan and geraniol) were supplied either alone or in combination. In Sabouraud medium, the greatest concentration of camptothecin was found to be 503 ± 25 µg/100g dry cell mass. Tryptophan and leucine raised the amount of camptothecin in the medium by 2.5 times, whereas tryptophan production differed considerably from other treatments. Additionally, research was conducted to investigate the impact of the fungus camptothecin on the morphology of human cancer cell lines. The size of the treated cells decreased, as did the nucleus's condensation and the protoplasmic extensions. Apoptotic cells exhibit all of these traits (Amna et al., 2012).

6.5.6 *Nigrospora sphaerica, phyllostict*a sp., and *Diaporthe phaseolorum*

Thirteen native herbs' leaves yielded 30 endophytic fungal isolates. Morphological traits and ITS rDNA sequence phylogenetic studies were used to identify most of the isolates. The majority of the fungal extracts inhibited bacteria (Gram-positive and Gram-negative) with diverse degrees of inhibition zone ranging from 6.5 to 25.0 mm, according to quantitative screening using the disc diffusion assay. *Pseudomonas aeruginosa* was not inhibited. With inhibition zones ranging from 17.0 to 24.6 mm, respectively, all endophytic fungi crude extracts in this study revealed encouraging bioactivities against reference strains of *Salmonella typhimurium* (ATCC 14028) and *Bacillus cereus* (ATCC 11778). When tested against Gram-positive bacteria, the ethyl acetate extract displayed MBC values ranging from 62.50 to 500.00 µg/mL and MIC values between 62.50 and 125.00 µg/mL. For Gram-negative bacteria, on the other hand, values of MBC ranged from 250.0–1000.00 µg/mL, while MIC values were between 125.00 and 250.00 µg/mL. When the cytotoxic effect of fungal isolates was evaluated against four cancer cell lines, the results demonstrated that isolates of *Phyllosticta* sp., *Nigrospora sphaerica*, and *Diaporthe phaseolorum* had excellent antiproliferative effects against all cancer cell lines, with IC_{50} values ranging from 1.946 to 16.96 µg/mL. According to this study, endophytic fungi may be a new source of antibacterial and anticancer compounds (Gazaem et al., 2025).

6.5.7 Endophytes from *Kandelia candel, Rhizophora stylosa,* and *Rhizophoraceae*

The leaves, roots, and stems of three different species, namely, *Kandela candel, Rhizophora stylosa*, and *Rhizophoraceae*, were collected from the Zhanjiang region, and endophytic fungi were separated, purified, and identified from these plants. The isolated endophytic fungus outcomes on the development of A549 and NCI-H460 lung cancer cells were examined using the MTT test. The impact of endophytic fungus on the angiogenesis of lung cancer *in vitro* caused by the HPV-16 E7 oncoprotein was noted. According to the current study, 28 endophytic fungal strains were separated, refined, and identified from these different kinds of mangrove plants. Ten strains of endophytic fungus markedly inhibited the growth of NCI-H460 and A549 cells. The A549 cells exhibited the following average inhibitory rates: 64.4%, 59.5%, 81.9%, 43.9%, 58.3%, 56.2%, 48.3%, 42.4%, 93.0%, and 49.7%. The NCI-H460 cells exhibited average inhibitory rates of 41.2%, 49.3%, 82.7%, 40.7%, 53.9%, 52.6%, 56.8%, 64.3%, 91.0%, and 45.6%, in that order. In particular, three endophytic fungal strains significantly reduced the angiogenesis of lung cancer caused by the HPV-16 E7 oncoprotein *in vitro*. These results aid in the continued screening of endophytic fungi originating from mangroves for possible chemotherapeutic drugs (Liu et al., 2017b).

6.5.8 *Lasiodiplodia theobromae*

Among the endophytic fungi were *Curvularia lunata* Bv4 purified from *Bambusa vulgaris, Lasiodiplodia theobromae* purified from *Cola acuminata*, and *C. lunata* Eg7 purified from *Elaeis guineensis*. The internal transcribed spacer (ITS-rDNA) sequence was used to identify the fungi. Ethyl acetate was used to extract the secondary metabolites after the fungi underwent solid-state fermentation. The SYBR green I-based fluorescence test was used to evaluate extracts for antiplasmodial activity *in vitro* using *Plasmodium falciparum* (chloroquine-resistant strain) DD2. Using the tetrazolium-based colorimetric MTT test, the extracts' cytotoxicity toward human red blood cells and Jurkat (leukemia) cells was evaluated. Fungal extracts were investigated *via* gas chromatography-mass spectrometry (GC–MS) to ascertain their components. The extract of *L. theobromae* was not detrimental to erythrocytes (CC_{50} > 100 µg/mL) and had the highest antiplasmodial efficacy against chloroquine-resistant *P. falciparum* (IC_{50} = 5.4 µg/mL). The cytotoxic activity of all three fungal extracts on Jukart cell lines was mild (CC_{50} > 100 µg/mL). Forty significant bioactive chemicals, including isobutyl nonyl ester, oxalic acid, hexadecanoic acid, and 2,4-*di-tert*-butylphenol, were found in the three endophytic fungal extracts according to GC–MS analysis. Auspicious sources for bioactive chemicals that may be further assessed as potential medications for the treatment of malaria caused by *P. falciparum*-resistant strains were the endophytic fungi from the medicinal plants in this investigation (Nwobodo et al., 2024).

6.5.9 *Aspergillus neoniger*

Some species, like the medical herb *Ficus carica*, which has been shown to have anticancer properties, have not had their endophytes studied. Because the endophyte *Aspergillus neoniger* demonstrated anticancer properties in initial screening tests, it was chosen for the study of its metabolites. Asperazine and asperazine A were identified by spectroscopic examination after the endophytic extract underwent bioactivity-guided chromatographic fractionation. Asperazine showed moderate antiproliferative impacts against K-562 and HUVEC cell lines (GI_{50} = 24.8 and 31.5 µg/mL, respectively) and moderate cytotoxicity against HeLa cell lines (CC_{50}=18.4 µg/mL). However, Asperazine A exhibited no antibacterial action and only weak cytotoxic activity against HeLa cell lines (CC_{50}=34.6 µg/mL) and weak cytostatic activity against HUVEC and K-562 cell lines (GI_{50}=40.7 and 50.2 µg/mL, respectively). These findings offer a new source of anticancer compounds with therapeutic promise and imply that *A. neoniger* contributed to the host plant's documented anticancer action (Abdou et al., 2021). Different cytotoxic alkaloids are illustrated in Table 6.1. Chemical structures of natural alkaloids with anticancer activities are displayed in Figures 6.1 and 6.2.

TABLE 6.1
Summary of Cytotoxic Activities of Some Plants, Marines, and Fungi

Alkaloid Name	Cell Lines	Experiment Results	Ref.
P. sabiniana L. leaf extract	MCF-7 WRL-68	Cell viability at 400 μg/mL: 89.3 ± 3.44% Cell viability at 400 μg/mL: 90.16 ± 2.7%	(Al-Marzook & Omran, 2017)
Ferocactus sp. L. leaf extract	MCF-7 WRL-68	Cell viability reduction at 400 μg/mL: 46 ± 2.20% Cell viability reduction at 400 μg/mL: 56.2 ± 2.2%	
P. dactylifera L. pollen grain extract	MCF-7 WRL-68	Cell viability reduction at 400 μg/mL: 56.2 ± 2.2% Cell viability reduction at 400 μg/mL: 57.5 ± 3.2%	
L. camara leaf extract	HCT-116 MCF-7 HeLa	IC_{50} = 8.38 μg/mL IC_{50} = 0.027 μg/mL IC_{50} = 5.90 μg/mL	(Al-Hakeim et al., 2021).
S. aculeastrum berries (tomatidine and solasodine)	Hela	Inhibitory effect by obstructing the cell cycle in the G_0=G_1 phase, increasing from 55.6% to 64.2% and 66.8%, respectively.	(Koduru et al., 2007)
O. polyacantha extract	MCF-7	The survival of breast cancer cells compared to doxorubicin	(Al-Shukri et al., 2020)
V. globosa (Globospiramine)	(L929, KB3.1, A431, MCF-7, A549, PC-3, and SKOV-3)	The expression of cleaved caspase-3 and PARP	(MANZANO et al., 2024)
YS306 and YS206	$HepG_2$ HCT116 and HT29	Cell distribution at the G0/G1 and S phases Cell cycle arrest at the G2/M phase	(Liu et al., 2017a)
C. colocynthis	MIAPaCa-2 and A-431	Cytotoxic effect against the two cell lines	(Mandour et al., 2023)
G. glauca leaf alkaloid extract	MCF-7	Phase arrests in sub G_0/G_1, S, and G_2/M IC_{50} 144.51 μg/mL	(Valleti et al., 2024)
C. peltate rhizome extract (Pheanthine)	Hela	Cytotoxic with an IC_{50} value of 8.11 ± 0.04 μM	(Valsan et al., 2023)
A. sclerocarpa extract (Crebanine)	K562	Cytotoxic (I% = 37.545 %) at 500 μm concentration	(Suman Joshi et al., 2018)
E. viride extract	Hela	Cytotoxic with an IC_{50} value of 53 μg/mL	(Rajesh and Annappan, 2015)
C. niger extract (Plakinamine N, O, and J)	NCI-60	Cytotoxic with IC_{50} values of 11.5, 2.4, and 1.4 μM, respectively.	(Sunassee et al., 2014)
M. hortensis extract (Taxol)	Hep-G2	Cytotoxic with an IC_{50} value of 0.25 μM.	(Pandy et al., 2023)
P. citrinum extract	A549 and MCF-7	Cytotoxic with IC_{50} values of 280.7 and 283.0 μg/mL, respectively	(Ashoka & Shivanna, 2023)
C. wenpingii extract	MCF-7	Cytotoxic with an IC_{50} value of 110.66 μg/mL	(Agustina et al., 2024)
Asperazine	HUVEC and K-562	Cytotoxic with GI_{50} values of 31.5 and 24.8 μg/mL, respectively.	(Abdou et al., 2021)
Asperazine A	HUVEC and K-562	Cytotoxic with GI_{50} values of 40.7 and 50.2 μg/mL, respectively.	(Abdou et al., 2021)

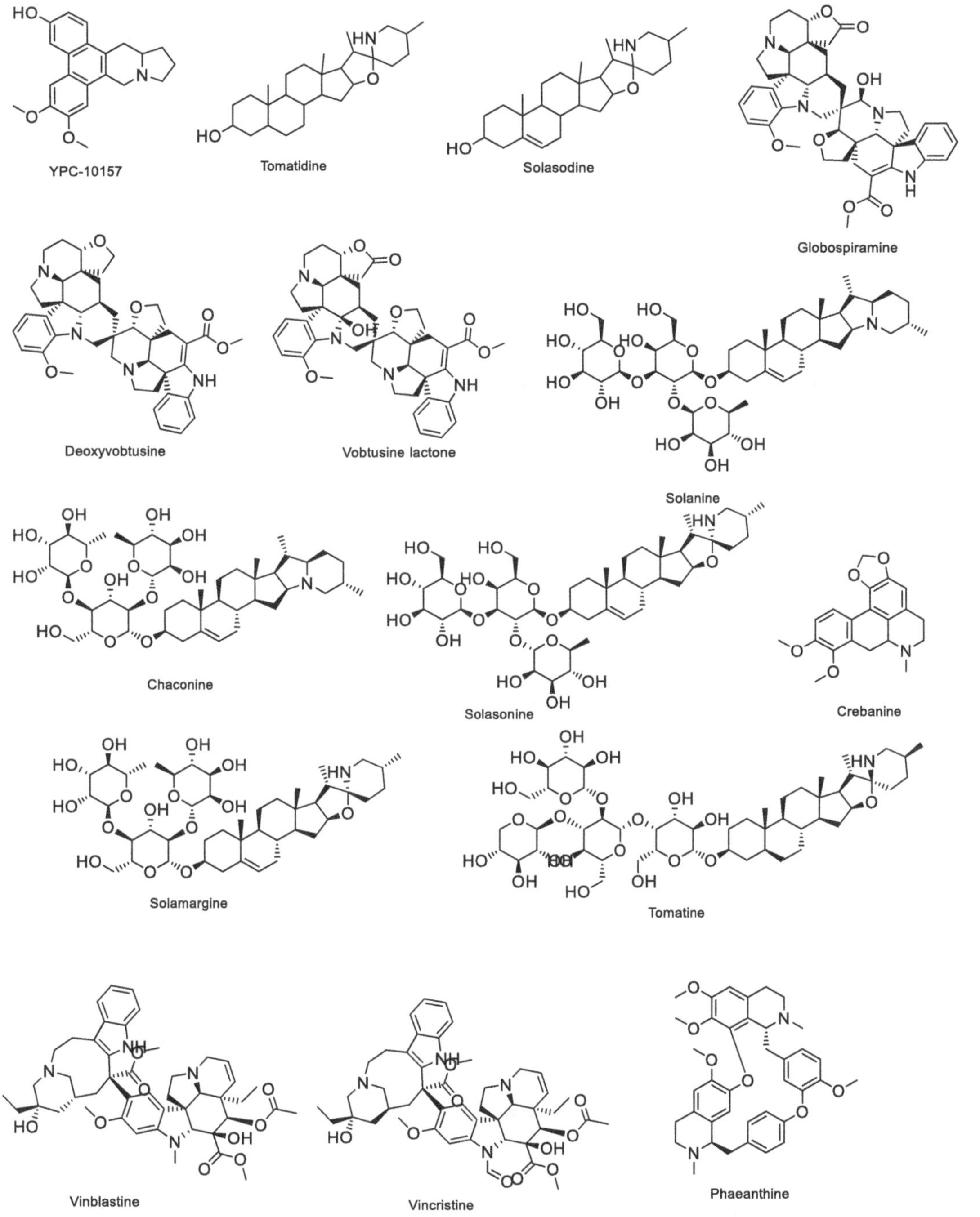

FIGURE 6.1 Chemical structures of natural alkaloids with anticancer activities. *(Continued)*

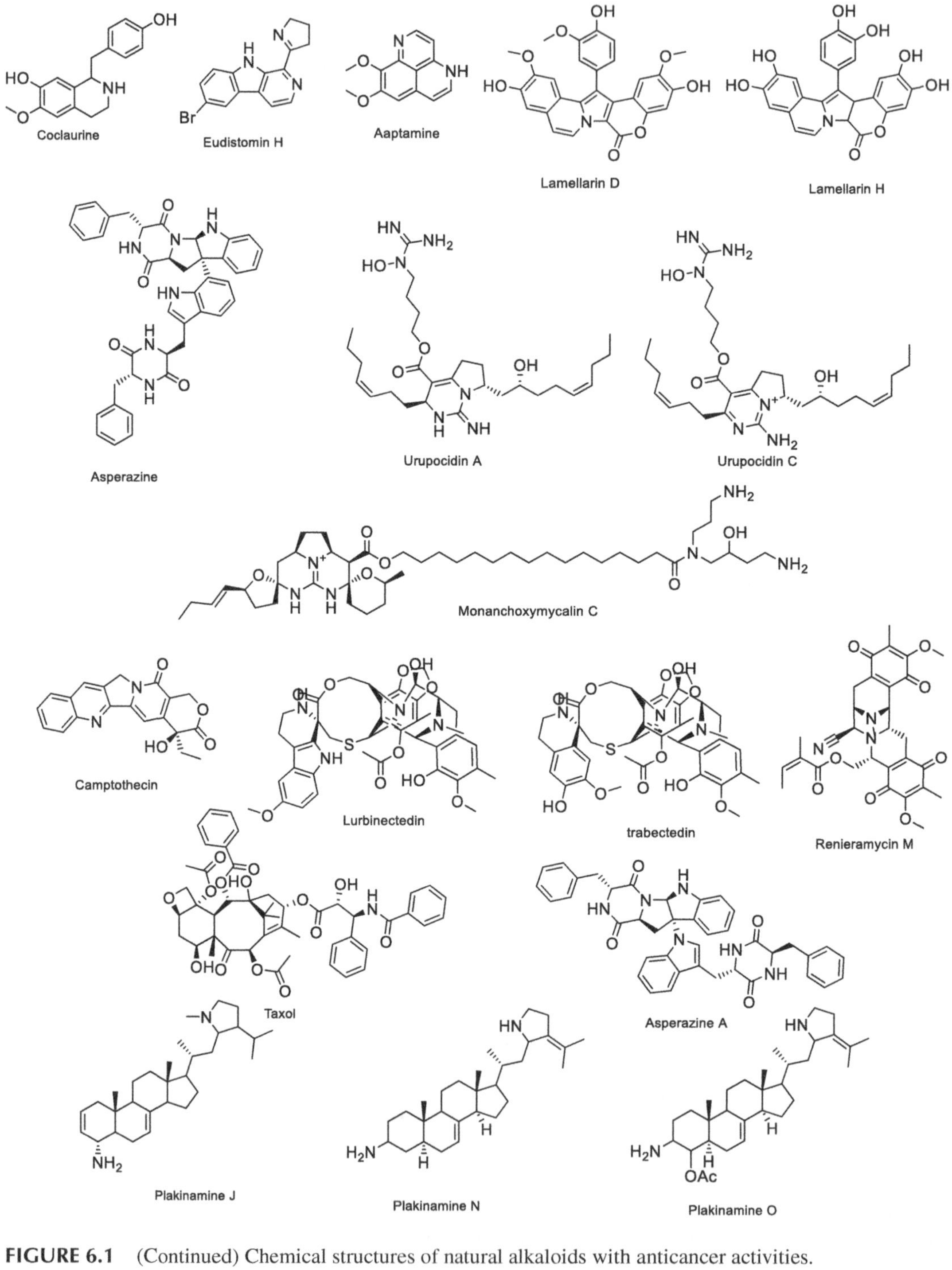

FIGURE 6.1 (Continued) Chemical structures of natural alkaloids with anticancer activities.

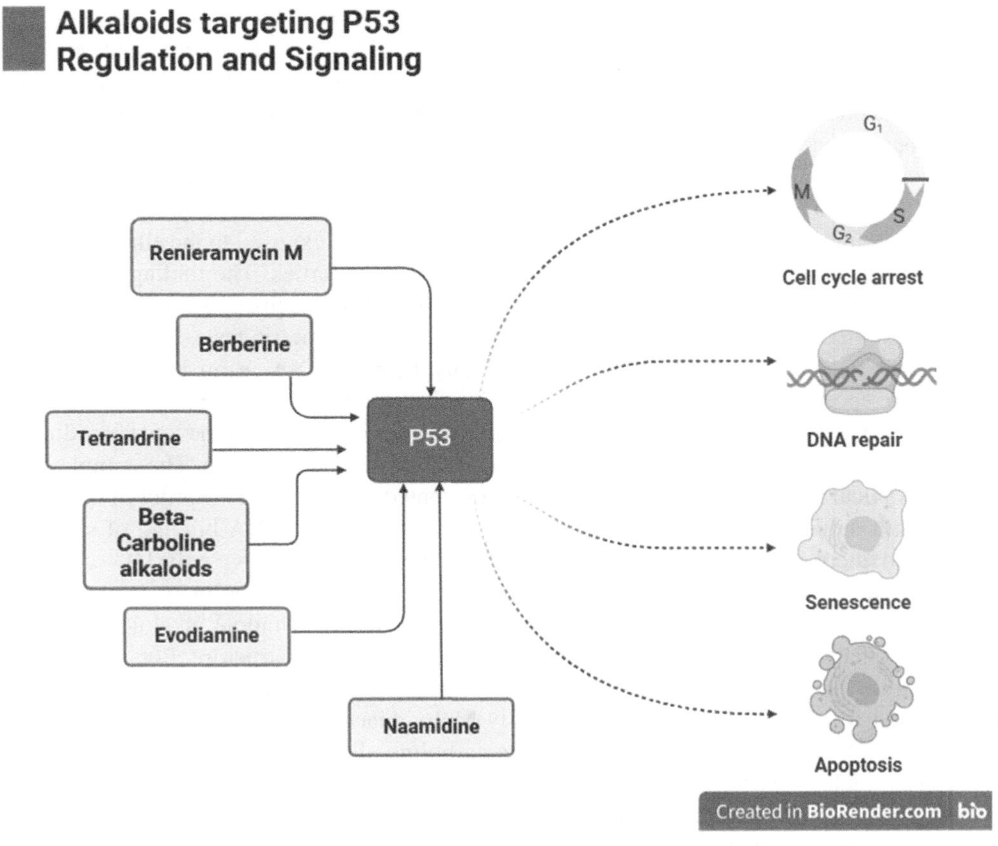

FIGURE 6.2 Alkaloids targeting the P53 signaling pathway.

6.6 MECHANISMS OF ANTICANCER ACTIVITY OF NATURAL ALKALOIDS

6.6.1 Targeting Nucleic Acids

The broad outline of information transfer within the cell is represented by the sequence DNA/RNA → mRNA → protein. It is possible to think of the processes of differentiation and carcinogenesis as modifications of this information flow. The cancer cell has a genetic flaw in the regular system that regulates cell division. Given that nucleic acids are the chemical building blocks of cell heredity, it makes sense to assume that a change in nucleic acids occurs during the carcinogenesis process. Many therapeutic chemicals target cells because DNA functions as the long-term reserve of the cell's genetic information. Ribonucleic acids (RNA) are equally important.

Numerous botanical groups, including the Papaveraceae, Berberidaceae, Menispermaceae, Fumariaceae, and many more, contain berberine, an isoquinoline alkaloid. There are numerous publications on berberine's anticancer properties against various cell lines. By causing DNA damage and the death of p53-dependent PCa cells, berberine was able to affect cell cycle suppression and apoptosis in human leukemia HL-60, Balb/c 3 T3 cells, and human osteosarcoma cells in a dose-dependent way. Numerous research groups have explored the interaction between berberine and DNA since the 1950s. Berberine favors AT sequences by the partial intercalative method of binding, according to Maiti and colleagues Bhadra et al., who used multi-spectroscopic and calorimetric techniques to study the interlinkage of berberine with various natural and synthesized DNAs.

Palmatine readily attaches to DNA and causes singlet oxygen (1O2) production, which in turn causes DNA photooxidation. When applied to HL-60 leukemic cells, palmatine exhibits strong anticancer activity. Bailly and colleagues reported on palmatine's DNA interaction as a possible bioreceptor for this alkaloid. Both intercalation and external stacking of the alkaloid on complexation were proposed by this study, which was based on absorption and fluorescence on natural DNA. DNA is photooxidized by palmatine through the production of singlet oxygen.

The Papaveraceae family's benzophenanthridine isoquinoline alkaloids, such as sanguinarine, a naturally occurring alkaloid derived from *Sanguinaria canadensis* L. and *Chelidonium majus* L., produce a wide range of pharmacological and metabolic effects, including prevention of cancer. Numerous cell lines have been shown to undergo apoptosis when exposed to the alkaloid. A number of different mechanisms has explained its antiproliferative properties. The findings unequivocally link sanguinarine's cytotoxicity to its capacity to intercalate DNA, causing strand breaks and other DNA-damaging consequences. Through spectroscopic and viscometric tests, Maiti and associates were the first to demonstrate this alkaloid's capacity to bind DNA in an intercalative manner. According to reports, sanguinarine has a 105 M−1 binding affinity for DNA. Sanguinarine's charged iminium form was the DNA-binding form. Giri and Kumar contributed to the understanding of the interlinkage between sanguinarine and single- and double-stranded poly(A). The complex's cooperative CD optical and DSC melting profiles were presented as unambiguous proof of the alkaloid's self-structured induction of poly(A). Das et al. explored the triplex RNA binding of sanguinarine and proposed interpolation of the alkaloid to the complex poly(U).poly(A)*poly(U).

Farouk et al. (2008), and numerous others have examined *beta*-carbolines, another family of natural alkaloids, which are a broad range of synthetic and natural indole alkaloids with varying degrees of aromaticity and a well-known tricyclic pyrido-indole ring structure. Beta-carboline alkaloids, which are used conventionally as herbal medicines in North Africa and the Middle East, were first identified from *Banisteriopsis caapi* (family Malpighiaceae) and *Peganum harmala* (family Zygophyllaceae). Additionally, a variety of *beta*-carboline alkaloids were identified from marine invertebrates, such as hydroids, soft coral, tunicates, and various sponges.

Beta-carboline type alkaloids, such as harmane, harmine, harman, norharman, and harmol, the partially or totally saturated compounds are called dihydro-beta-carbolines, such as harmalol, harmalan, harmaline, and tetrahydro-beta-carbolines, or THBCs, respectively. The protonated form of harmalol demonstrated a robust, intercalative way of binding when Sarkar and Bhadra examined both harmalol's structural forms and their interactions with CT DNA.

They also looked at the sequence-specific binding of harmalol, emphasizing hetero GC and then hetero AT. Profiles of ITC demonstrated enthalpy-driven and entropy-driven binding with DNA sequences, as well as cooperative and noncooperative binding. With an IC_{50} value of 14 μM, the alkaloid was also observed to exhibit maximal cytotoxicity against HepG2.

Complete structural and thermodynamic profiles of the binding of isoquinoline alkaloids with G-quadruplex, including sanguinarine, palmatine, berberine, and coralyne (a synthetic isoquinoline alkaloid), were presented by Bhadra and Kumar. The findings emphasize the significance of certain structural components, especially planarity, in small compounds for maintaining quadruplex DNA structure and creating more effective quadruplex-targeted medicinal treatments.

6.6.2 Targeting Enzymes such as Topoisomerases I and II, DNA Polymerase, Telomerase, and RNA Polymerase Associated with Nucleic Acids

Many anticancer medications currently in clinical use target DNA topoisomerases, which are essential enzymes for DNA replication and repair. Among the marine natural alkaloids that have been shown to block topoisomerases I and II are granulatimide, ascididemin, lissoclinidine B, lamellarin D, and polycarpines. As topoisomerase inhibitors, topotecan and irinotecan are FDA-approved anticancer medications that target the ovary, colon, and lung. They are synthetic derivatives of camptothecin, a naturally occurring alkaloid. Moreover, some of the widely recognized topoisomerase inhibitors include dicentrinone, an aporphine isolated from Ocotea leucoxylon, yohimbine, an indole purified from *Rauwolfia serpentine*, beta-carboline alkaloids like harmine, the quinoline alkaloid lunacridine purified from *Lunasia amara*, and indoquinoline alkaloid cryptolepine isolated from *Cryptolepis sanguinolenta*, among many others.

By targeting topoisomerases, eribulin mesylate, an analogue of the macrocyclic polyether halichondrin B, an alkaloid obtained from marine sponges, has advanced to phases I and II of clinical trials for the treatment of breast cancer. Neoamphimedine, another naturally occurring substance that was separated from *Xestospongia* sp., likewise has a distinct mode of activity as an inhibitor of topoisomerase *in vitro* and *in vivo* settings. Other than these, telomere targeting and inhibition of telomerase are equally feasible, efficient, and alternative ways to stop cell proliferation effectively. This allows researchers to create strong ligand derivatives to start more accurate diagnostic research for biomedical applications. A protein called telomerase is implicated in cancer and is found in 85–90% of cases.

Making use of this information, a novel medication for human cancer therapy that inhibits telomerase action *via* stabilizing the telomeric DNA quadruplex structure is highly significant. Among the natural compounds that act as telomerase inhibitors are berberine isolated from *Coptis chinensis* and *Berberis vulgaris*, and the apomorphine alkaloid boldine from *Lindera aggregate* and *Peumus boldus*. Furthermore, it was shown that by binding with the human telomeric G-quadruplex structure, chelerythrine, which was extracted from *C. majus* L., likewise inhibits telomerase action and causes cancer cell death. It is worth mentioning that direct targeting of DNA polymerase through anticancer natural alkaloids has not yet been investigated. DNA-damaging agents (vide supra) to inhibit processive DNA polymerases are considered the major strategy, though trabectedin was reported to cause the breakdown of RNA polymerase II.

6.6.3 Targeting Kinases

It is anticipated that the most promising approach to the development and spread of tumors will be to target oncogene kinases of signaling pathways.

The family of kinase enzymes known as protein kinase C (PKC) controls the activity of other proteins through the phosphorylation of hydroxyl groups of threonine and serine amino acid residues.

It has been demonstrated that chelerythrine has a potent antitumor impact on a variety of malignancies, including those of the breast, colon, and prostate. In human osteosarcoma cells, chelerythrine upregulates downstream kinases (p90RSK) and activates the JNK/p38-MAPK (c-Jun N-terminal kinase-mitogen-activated protein kinase), and in HeLa cells, the MEK signaling pathways. In HL-60 cells, it also results in bimodal cell death and G1 phase cell cycle arrest. Additionally, chelerythrine induces apoptosis in tumor cells and functions as a selective PKC inhibitor.

The cell cycle depends on the actions of a group of serine-threonine protein kinases designated as cyclin-dependent kinases (CDKs). There is a great need for rigorous screening to find natural alkaloids that exhibit excellent selectivity and proficiency as CDK inhibitors. These compounds function through antiproliferation action that arrests cells in G1 and/or G2/M phases of the cell cycle, such as berberine, evodiamine, a bioactive quinolone alkaloid extracted from *Evodia rutaecarpa*, piperine, harmalol, harmaline, and many more. The marine sponge Stylotella aurantium yielded two alkaloids: hemenialdisine and debromohymenialdisine. They act through the inhibition of CDKs *via* exhibiting competitive inhibition at the ATP-binding site. Fascaplysin, a different marine natural product that was extracted from a marine sponge, binds to the kinase ATP site and inhibits CDK4 specifically. Its IC_{50} value is 0.35 μM. Fascaplysin inhibits angiogenesis and endothelial cell growth.

Meridianins A–G are another class of marine indole alkaloids that are joined to an aminopyrimidine ring. They were discovered to preferentially inhibit CDK1, CDK5, and other protein kinases implicated in cancer and various diseases after being extracted from the marine tunicate *Aplidium meridianum*. Meridianins A–G function as ATP competitive inhibitors through binding to the protein kinases' ATP pocket, just like the alkaloids previously described.

Through the MAPK signaling pathway, sanguinarine has also been shown to increase ROS (reactive oxygen species), which in turn causes apoptosis of human lens epithelial cells.

Another possible cancer treatment target is the insulin-like growth factor-1 receptor (IGF-1R). They play a vital function in the invasion, survival, and proliferation of tumor cells. Solid tumors, for instance, ovarian carcinoma, small cell lung cancer, and non-small cell lung cancer, can be effectively treated with IGF-1R inhibitors. Among the principal ways *via* which IGF-1R signaling is transduced are the RAS/RAF/MAP kinase pathway and the phosphoinositide-3 kinase (PI3K)/Akt pathway. By inducing apoptosis and inhibiting invasion, angiogenesis, and metastasis in a range of cancer cell lines at the micro- and nanomolar levels as well as in the Balb-c/nude mouse model, evodiamine has anticancer properties. Phosphatidylinositol 3-kinase/Akt/caspase has been shown to trigger cell death in response to evodiamine, which also affects tubulin polymerization. The human osteosarcoma cell line, U20S cell, is susceptible to dose-dependent inhibition of proliferation and apoptosis caused by the isoquinoline alkaloid berberine. Berberine suppresses PI3K/Akt activation, which raises the PARP expression and Bax, and decreases the expression of Bcl-2 and caspase 3. Berberine's cytotoxic efficacy depends on the cell type and treatment time and is significantly lower than that of other anticancer medications; its IC_{50} ranges from 10 μM to 100 μM. It is believed that the indole alkaloids shearinine A, D, and E, which are obtained from *Penicillium janthinellum*, block endothelial growth factor (EGF) in HL-60.

A serine/threonine protein kinase, which has been extensively implicated in vital cell biology processes, glycogen synthase kinase-3 beta (GSK-3β) is a promising multipurpose kinase for the treatment of cancer. After being extracted from the sponge Acanthostrongylophora, the complex alkaloid manzamine was found to act as a non-competitive, specific ATP inhibitor. It is bound to GSK-3β (IC_{50} value of 10.2 μM). By inhibiting the expression of Ki-67, CD31, p-Brk, and p-FAK in tumor samples, phenylmethylene hydantoin (PMH-1), which was extracted from the sponge *Hemimycale arabica*, was proposed to be able to dock into the GSK-3β binding pocket and manage the growth of invasive breast cancer based on molecular modeling.

Tetrandrine, another naturally occurring bisbenzylisoquinoline alkaloid, was isolated from *Stephania tetrandra* roots and has been shown to exhibit several therapeutic uses, including anticancer activities in micromolar concentrations. Tetrandrine's anticancer properties may be attributed to GSK-3β activation, production of reactive oxygen species (ROS), p38 MAPK activation, and suppression of Wnt/beta-catenin signaling. Additionally, tetrandrine upregulates p27, p53, p21, and Fas; downregulates cyclins, CDKs, and Akt phosphorylation; modifies Bcl-2 family members; and triggers caspases.

6.6.4 Targeting Tubulin Polymerization

Hemosterlin, extracted from the marine sponge *Hemiasterella minor*, binds to tubulin and inhibits its polymerization, resulting in mitotic catastrophe and finally apoptosis at nanomolar concentrations. This results in antiproliferative activities against murine leukemia p388 cells, human glioblastoma U373, mammary carcinoma MCF-7, and human ovarian carcinoma HEY. However, hemiasterlins have a significant level of toxicity *in vivo*; thus, there is a need to synthesize novel analogues with improved therapeutic efficacy. Sanguinarine has also been shown to interfere with the kinetics of microtubular assembly by binding to tubulin. Halichondrin B, a marine alkaloid, was also declared to display tubulin polymerization impact.

6.6.5 Targeting Apoptosis

The best strategies to stop tumor growth and avoid malignancies are to increase apoptosis in tumors and inhibit cell proliferation. The capacity to cause cancer cells to undergo apoptosis is regarded as one of the key characteristics of an anticancer medication with excellent anticancer potential. Apoptosis can be brought on by intrinsic mechanisms regulated by mitochondria or extrinsic pathways regulated by receptors.

Hepatic carcinoma cells (HepG2) are more susceptible to a greater extent to apoptosis than uterine carcinoma cells (HeLa) because beta-carboline alkaloids have been shown to cause fragmentation of nucleic acids, ultramorphological changes of cellular organelles, reduced potential of mitochondrial membrane, upregulation of p53 and caspase-3, triggering ROS (reactive oxygen species), and a marked increase in G2/M population. Smac/DIABLO is used by the steroidal alkaloid cephalostatin 1 to trigger apoptosis. One or more apoptosis-inhibitory protein members are neutralized by the mitochondrial protein Smac/DIABLO. Smac passes into the cytosol from mitochondria when UV or γ-irradiation causes apoptosis.

Briofilin, another steroidal alkaloid, increases the expression of the Bax protein and the division of poly (ADP-ribose) polymerase-1 (PARP-1) in HeLa cells while suppressing the initiation of Bcl-2 and caspase-3. It has been documented that solumargine causes apoptosis by upregulating TNFR-1 (tumor necrosis factor receptor-1). A multipurpose class of pro-inflammatory cytokines, TNF initiates signaling pathways for cellular differentiation, death, inflammatory response, and cell survival. When Bax is present, the isoquinoline alkaloid sanguinarine causes intrinsic apoptosis. The transactivation of Bax from cytosol to mitochondria requires the weaker connection of STRAP (serine-threonine kinase receptor-associated protein) and MELK (maternal embryonic leucine zipper kinase).

Cytochrome c is released from the mitochondria into the cytosol because of MOMP (mitochondrial outer membrane permeabilization), which is brought on by the buildup of Bax in the mitochondria. The antiapoptotic proteins Bcl-2, XIAP, Bclxl, and cyclin D are downregulated in multiple myeloma cells treated with sanguin. Additionally, it increases the expression of the pro-apoptotic protein Bax. By blocking JAK2/STATs signaling (Janus kinases, JAK, and signal transducer and activator of transcription proteins, STAT), it essentially causes multiple myeloma cells to undergo apoptosis. It remarkably elevates breast cancer cells' sensitivity to ligand-mediated apoptosis that is induced by TNF-related apoptosis. It has been declared that the combination of piperlongumine and piperine, which are derived from *Piper longum L.* and *Piper nigrum L.*, respectively, suppresses STAT3 phosphorylation and regulates downstream pathways for apoptosis induction in breast cancer cells.

Evodiamine has been shown to quickly increase ROS and cause mitochondrial depolarization, which triggers apoptosis. Evodiamine increases Bax expression and decreases the expression of Bcl-2 in certain cancer cells, causing death, both caspase-dependent and caspase-independent. Matrine, a prime bioactive metabolite extracted from *Sophora flavescens* Ait and *S. japonica*, was assessed in both *in vitro* studies, including 4 T1, H22, MNNG/HOS, and BxPC-3 cells, and *in vivo* studies such as the BALB/c mice model. It impacts proteins associated with cell proliferation or death, including Fas, E2F-1, Bcl-2, Bax, and Fas-L.

At 50–100 mg/kg, matrine reduces the volumes of pancreatic tumors and suppresses the growth of MNNG/HOS xenografts. Furthermore, it also uses apoptosis to stop the growth of other kinds of cancer cells, including K567 cells, SGC-7901 cells, and hepatoma G2 cells. Another example of a substance that targets apoptosis is napidine, an imidazole alkaloid originating from sponges that was found in *Leucetta chagosensis*. By disrupting mitochondrial potential, naamidine triggers caspase-dependent apoptosis in tumor cells through intrinsic and extrinsic cell death mechanisms.

By stabilizing the p21 CDK inhibitor (CDK2), stimulating active caspases 8 and 9, and causing the splitting of PARP, the cell cycle is arrested at the G1 phase. Additionally, in both in vitro and in vivo models, naamidine caused the effector caspase 3 cleavage through a p53-independent manner.

6.6.6 Targeting Transcription Factors Controlling the Expression of Cancer Genes

A protein known as a transcription factor attaches itself to specific DNA sequences, controlling the expression of genes by either stimulating or inhibiting transcription, which is crucial for the initiation and spread of cancer. Hypoxia-inducible factor 1 (HIF-1), a potential tumor factor linked to metabolism, the proliferation of tumor cells, apoptosis, and angiogenesis, is considered to be one of

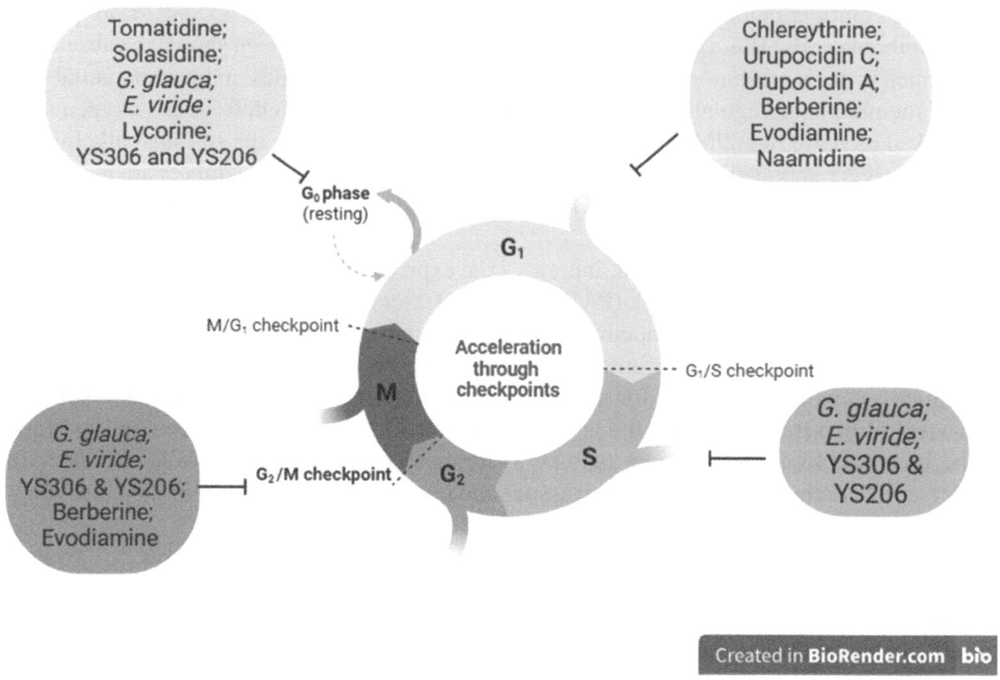

FIGURE 6.3 Different alkaloids and plants targeting cell cycle arrest at different phases in cancer.

the primary effective targets for treating a variety of cancers. Pyrroloiminoquinone rings containing alkaloids were isolated from the sponge *Latrunculia sp.*, which were found to be unprecedented inhibitors of HIF-1α/p300. These alkaloids disrupted the interaction between HIF-1α and p300, effectively inhibiting the growth of prostatic carcinoma and HCT 116 cell lines.

According to reports, the alkaloid berberine inhibits MDM2-DAXX-HAUSP interactions, which in turn causes MDM2 self-ubiquitination and destruction. The human MDM2 gene encodes the mouse double minute 2 homolog. Furthermore, it is a p53 tumor inhibitor negative regulator. One effective method for reactivating p53 is for MDM2 to interfere with the regulatory processes. Another inducible transcription factor, NF-κB, is a dimer protein that regulates inflammatory, immunological, and carcinogenic responses. It has also emerged as a key molecular target in therapeutic development. NF-κB is a segment of the Rel family that, in addition, includes c-Rel, p50 (NF-κB1), RelA (p65), RelB, and p52.

According to reports, evodiamine causes cancer cell death via activating the Fas ligand (Fas-L)/NF-κB signaling pathways. The activation of NF-κB caused by interleukin-1, TNF, phorbol ester, and okadaic acid is also strongly suppressed by sanguinarine. The signal transducer and activator of transcription 3 activation (STAT-3) is also effectively inhibited; Bcl-2 family members are modulated; caspases are activated, and death receptor 5 (DR-5) is upregulated; CDKs, cyclins, MMP-2, and MMP-9 are downregulated; and p21, p27, and phosphorylation of p53 are upregulated. Alkaloids targeting the P53 signaling pathway are illustrated in Figure 6.3.

6.6.7 Targeting Histone Deacetylases to Cause Genomic Instability

Among the classes of epigenetic enzymes are histone deacetylases (HDACs), which control gene expression *via* suppressing histone deacetylation and transcription. They enable the histones to more firmly enclose the nucleic acid through the removal of acetyl groups from the N-acetyl lysine amino acids on the histone. Tumor cell genomic instability is caused by the dysregulation

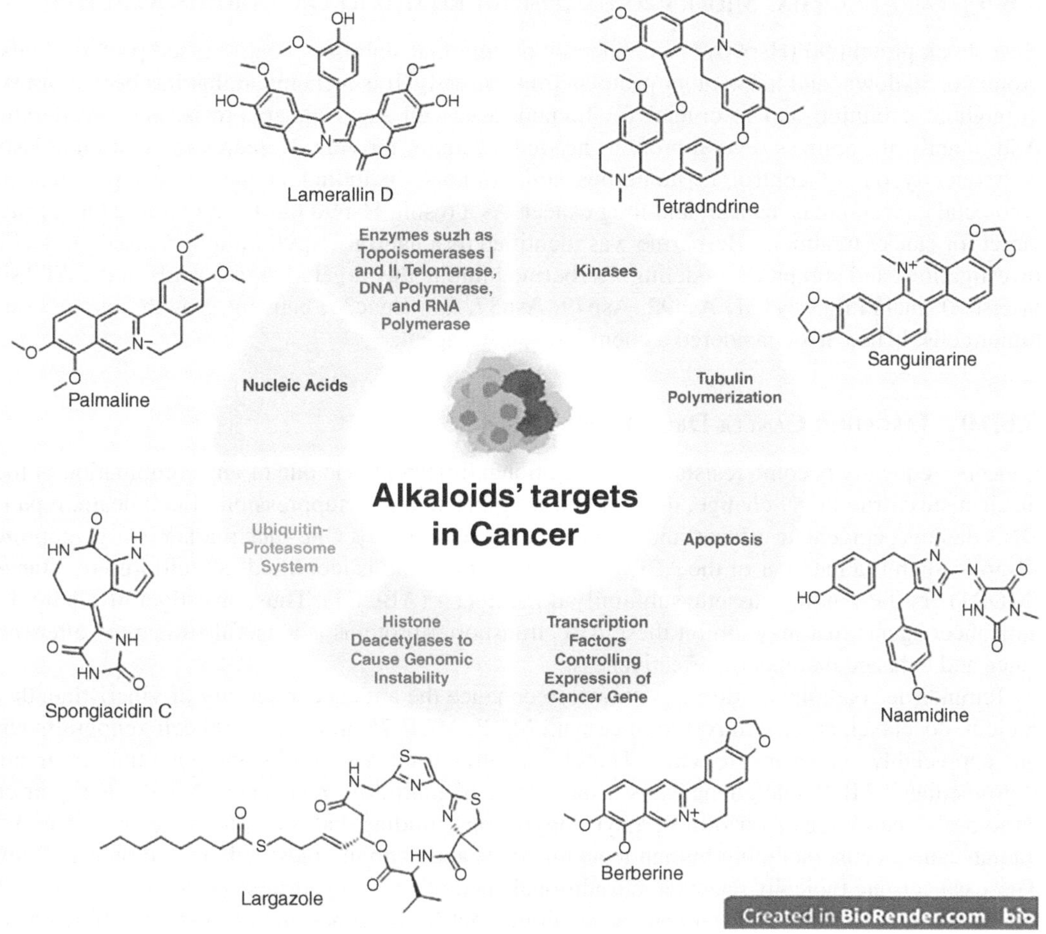

FIGURE 6.4 Summary of the most important alkaloids' targets in cancer.

of DNA methylation in addition to the acetylation of lysine residues on histone tails. HDACs are therefore likely therapeutic targets for the therapy of cancer. By altering histone deacetylase, berberine was reported to cause apoptosis and growth arrest in the A549 cell line. Lycorine, a poisonous crystalline alkaloid from the Amaryllidaceae family, is another naturally occurring plant alkaloid that inhibits HDAC in K562 cells and induces cell cycle arrest in the G0/G1 phase. Different alkaloids and plants targeting cell cycle arrest at different phases in cancer are displayed in Figure 6.4. Largazole, a macrocyclic depsipeptide alkaloid purified from cyanobacteria, also affects the suppression of histone deacetylases (HDAC) and displays anticancer activity.

6.6.8 Targeting Ubiquitin–Proteasome System

One tightly controlled mechanism for intracellular protein turnover and degradation is the ubiquitin–proteasome system (UPS). A group of enzymes, including ubiquitin-activating enzyme (E1), ubiquitin-conjugating enzyme (E2), and ubiquitin–protein ligase (E3), work together to mark proteins for proteasomal breakdown. In recent studies, one of the therapeutic approaches is targeting the UPS for cancer treatment because it is involved in the regulation of significant carcinogenesis pathways. With an IC_{50} of 3.8 μM, the pyrrole alkaloid Spongiacidin C, which is extracted from the sea sponge *Stylissa massa*, was reported to be the first natural inhibitor of USP7.

6.6.9 Targeting Heat Shock Protein (Hsp90) Related to Oncoprotein Maturity

Heat shock protein 90 (Hsp90) is a chaperone protein that stabilizes heat-stressed proteins, aids in protein breakdown, and helps other proteins fold correctly. It is a chemical that has been conserved throughout evolution and is crucial for homeostasis, cell survival, apoptosis, and proliferation. Additionally, it stabilizes several proteins needed for tumor thriving. Research indicates that Hsp90 is frequently out of control in numerous solid tumors, including hematological malignancies, colorectal cancer, breast cancer, and lung cancer. As a result, Hsp90 has been identified as a crucial target for cancer treatment. Berberine was identified as a natural Hsp90 inhibitor based on docking investigations and computer modeling. Berberine inhibits the targeted amino acids in the ATP sites in Hsp90, including Gly121, Asn92, Asp79, Asn37, and Leuc34, causing growth suppression of tumor cells; hence, it's considered a chemopreventive agent.

6.6.10 Targeting Cancer Drug Resistance

Cancers frequently become resistant to conventional treatments via one or any combination of these mechanisms: drug target change, drug efflux, drug inactivation, suppression of cell death, repair of DNA damage, epigenetic effects, and intrinsic cell heterogeneity. One such marker is P-glycoprotein (P-gp), which is a member of the ABC transporter family and is identified as multidrug resistance 1 (MDR1) or ATP-binding cassette subfamily B member 1 (ABCB1). Thus, investigating alkaloids as anticancer agents that may inhibit these ABC transporter proteins is a useful strategy to alter resistance and enhance therapeutic effectiveness.

Tetrandrine coadministration was shown to enhance the anticancer activity of vincristine, doxorubicin, docetaxel, and paclitaxel in mice with MDR, MCF-7/adr, or KBv200 cell xenografts without appreciably increasing toxicity. Therefore, tetrandrine has significant potential for treating P-gp-mediated MDR malignancies as a modulator of multidrug resistance (MDR). Kirkpatrickia variolosa is the source of variolin B, a pyridopyrrolopyrimidine that is cytotoxic against a variety of human cancer cells, including human leukemia cells and ovarian, intestinal, and colon carcinomas. These cancers are typically resistant to traditional chemotherapeutic drugs because of their high Pgp expression levels. Moreover, anticancer alkaloids that have been shown to exhibit MDR reversal properties include harmine, piperine, and piperidine.

6.6.11 Targeting Carbohydrate Metabolism

Along with a host of other biological and pathological processes, carbohydrates are recognized to be important in the spread of cancer. Increased metabolism and accumulated glucose promote a number of cancer-specific traits. Warburg's effect is the process by which a significant amount of glucose can be transformed into lactic acid in cancer cells, independent of oxygen availability. Abnormal glycolytic metabolism and hyperactivated glucose uptake are thought to be characteristics of cancer cells. Overexpressed extracellular tumor-specific carbohydrate antigens (TACAs) on tumor cell surfaces are regarded as biosignatures for cancer diagnosis and have been given priority in the creation of new anticancer medications. Nevertheless, little is now known about the precise regulation of cancer cells' glucose metabolism.

Since cancer cells' overexpression of glucose transporters (GLUTs) reduces their higher glucose consumption compared to normal cells, focused chemotherapy has frequently been made possible. As a result, numerous natural alkaloids have been chosen as anticancer agents. It has been demonstrated that berberine, a naturally occurring alkaloid with previously documented tumor-selective anticancer properties, improves glucose absorption in metabolic organs and cells. By blocking the intake of glucose and the transcription of glucose metabolic genes, GLUT1 (glucose transporter 1), LDHA (lactate dehydrogenase A), and HK2 (hexokinase 2), Berberine was reported to regulate the proliferation of colon cancer cell lines, HCT116 and KM12C.

The study also showed that berberine inhibits the production of HIF-1α (hypoxia-inducible factor 1-alpha), a popular transcription factor essential for cancer cells with uncontrolled glucose metabolism, in the HCT116 cell line. Additionally, the treatment of colon cancer cells by berberine involves inhibiting the synthesis of HIF-1α proteins that are dependent on mTOR (mammalian target of rapamycin), which is a unique strategy for berberine's establishment as an anticancer therapeutic compound.

Certain groups, such as glycoconjugated prodrugs and 2-D-glucose paclitaxel, have conjugated sugars at various locations and target GLUT-1. Following cellular absorption, the hydrolytic enzymes cleave these prodrugs, releasing the active medications that regulate tumor cells.

By creating a nuclear ternary complex with DNA and GAPDH (glyceraldehyde 3-phosphate dehydrogenase), saframycin A, another naturally occurring alkaloid that was extracted from marine tunicates, was shown to possess a strong anticancer impact. Glyceraldehyde 3-phosphate is converted to 1,3-diphosphoglycerate by the glycolytic enzyme GAPDH. APE1 (human apurinic and apyrimidinic endonuclease), a crucial enzyme taking part in the redox control of several transcription factors and abasic sites' repair in damaged DNA, is also physically associated with GAPDH. This reactivates the endonuclease activity of the oxidized APE1 to cleave abasic DNA sites by converting it to the reduced form.

6.6.12 Targeting Lipid Metabolism

According to recent research, lipid metabolism problems contribute significantly to carcinogenesis by taking advantage of several aberrant gene and protein expression patterns as well as the dysregulation of signaling pathways and cytokines. Diseases, including hyperlipidemia, obesity, and cancer, are caused by metabolic abnormalities that arise when the balance between the anabolism and catabolism of lipids is upset. Natural alkaloids that were discovered to be interesting possibilities for metabolic illnesses to address lipid metabolism-related anomalies include piperine, betaine, berberine, nuciferine, and different nigellamines and dolabellane derivatives. High-fat diets and endogenous fatty acid consumption are examples of lipid intake. These factors increase VEGF (vascular EGF) and encourage angiogenesis by attracting M2 macrophages, which leads to the formation of cancer cells.

A diet high in fat raises levels of microRNA-130a, IGF (insulin-like growth factor), and several pro-inflammatory cytokines that promote the growth of cancer cells, including interleukins IL-1α, IL-1β, IL-6, TNF-α, EMR1 (macrophage markers like EGF-like module-containing mucin-like hormone receptor-like1), CR4 (complement receptor 4), and TLR-4 (toll-like receptor). A high-fat diet also speeds up the growth of cancer cells by activating the MCP-1/CCR2 signaling pathway. Fatty acids, LPA (lysophosphatidic acid), and PUFAs (ω-3 polyunsaturated fatty acids) are examples of endogenous intake.

The proliferation and invasion of mammalian malignant cells are accelerated by decreased expression of the mRNA encoding LPA5, one of the G protein-coupled receptors. PUFAs significantly raise EDPs (epoxydocosapentaenoic acids) levels, a metabolite of ω-3PUFA developed by cytochrome P450 enzymes in tumor cells. This, in turn, lowers the expression of pro-oncogenic genes such as C-myc, Axin2, and C-jun in cancer tissues, which inhibits the growth of cancer. The B-scavenger receptor, CD36 (cluster of differentiation 36), one of the cell surface fatty acid receptors that influences cancer metastasis, is also expressed more when fatty acids are consumed.

By activating AMPK (AMP-dependent protein kinase) in the liver and muscle tissues, berberine helps with fatty liver and hyperlipidemia. Increasing fatty acid oxidation requires AMPK activation. Additionally, it controls lipid metabolism by inhibiting 5-tetradecyloxy-2-furoic acid (TOFA) and fatty acid synthetase, which causes cancer cells to undergo apoptosis. Changes in hepatic and muscle gene expression processes, which promote fatty acid oxidation and decrease lipogenesis or hyperlipidemia, are brought about by the berberine-dependent reduction of dyslipidemia in obesity. By preventing the formation of cholesteryl ester, piperine has been shown to reduce the accumulation of lipid droplets in mouse macrophages.

It raises serum levels of high-density lipoprotein (HDL) and dramatically lowers serum lipid concentrations in obese rats to levels close to normal. Another substance that is present in a variety of foods, such as spinach, beets, and numerous whole grains, is betaine. Through their involvement in mitochondrial oxidative demethylation, betaine and choline help to maintain lipid homeostasis in obese individuals. Through the PPARα/PPARγ coactivator-1α pathway, nuciferine, which is present in the leaves of *Nelumbo nucifera*, enhances dyslipidemia in vivo. In vivo and in vitro, oxymatrine, which is mostly contained in Sophora roots from Sophora japonica (kushen), has been shown to reduce hepatic lipid metabolism by controlling the expression of microRNA-182 in nonalcoholic fatty liver disease. Different targets of alkaloids for cancer treatment are displayed in Figure 6.4.

6.7 CONCLUSIONS AND FUTURE REMARKS

Many natural alkaloids can act as chemoprotective agents against various types of cancer. They can be isolated from medicinal plants, marines, and endophytic fungi. Mechanistically, different targets were identified, such as nucleic acids, enzymes related to cancer progression, such as topoisomerase, telomerase, polymerases, kinases, apoptosis, transcription factors, histone deacetylases, UPS, carbohydrates, and lipid metabolism. The diversity of nature implies that chemical leads can be developed that interact with most therapeutic targets. Interestingly, the mechanism of anticancer action of various natural alkaloids is still uninvestigated. This chapter provides a solid foundation that enables oncologists and pharmaceutical scientists to manipulate natural anticancer candidates for further experiments, including pharmacodynamics, pharmacokinetics, *in vivo* studies, and clinical trials. Additional studies are needed to investigate the anticancer efficacy of combinations of more than two alkaloids or combinations with commercial chemotherapeutic drugs. Moreover, many of these alkaloids should be investigated against different cancer types to expand their activity spectrum. Further bioavailability experiments should be conducted to determine the efficacy of the alkaloids and address pharmacokinetic concerns. Thus, new, safe, and efficacious chemotherapeutics are highly needed to be developed into pharmaceutical formulations.

COMPETING INTERESTS

The authors declare that they have no competing interest.

DECLARATION OF GENERATIVE AI AND AI-ASSISTED TECHNOLOGIES IN THE WRITING PROCESS

During the preparation of this work, the book chapter author(s) have not taken any assistance from AI, and the book chapter author(s) reviewed and edited the content as needed and take(s) full responsibility for the content of the publication.

REFERENCES

Abdou, R., Alqahtani, A. M., & Attia, G. H. (2021). Anticancer natural products from Aspergillus neoniger, an endophyte of Ficus carica. *Bulletin of the National Research Centre*, 45, 1–6.

Agustina, I., Lay, C. S., Astuti, P., & Hertiani, T. (2024). Cytotoxicity screening of endophytic fungi from Phaleria macrocarpa (scheff) boerl collected in Yogyakarta district. *Indonesian Journal of Pharmacy*, 219–226.

Al-Ghazzawi, A. M. (2019). Anti-cancer activity of new benzyl isoquinoline alkaloid from Saudi plant Annona squamosa. *BMC Chemistry*, 13, 1–6.

Al-Hakeim, H. K., Al-Zabibah, R. S., Alzihari, H. F., Almensoori, A. K., Al-Zubaidi, H. A., & Hassan, L. E. A. H. (2021). Anticancer and antiangiogenic activities of alkaloids isolated from Lantana camara by adsorption on the magnetic nanoparticles. *Karbala International Journal of Modern Science*, 7(1), 11.

Al-Marzook, F. A. & Omran, R. (2017). Cytotoxic activity of alkaloid extracts of different plants against breast cancer cell line. *Asian Journal of Pharmaceutical Research*, 10(7), 168–171.

Al-Shukri, A. F., Al-Marzook, F. A., Al-Hammadi, N. A., & Mutlag, I. H. (2020). Antitumor activity of alkaloids extract from opuntia polyacantha plant using high content screening technique (HCS). *Pharmacophore*, 11(1), 129–135.

Ali, R., Mirza, Z., Ashraf, G. M., Kamal, M. A., Ansari, S. A., Damanhouri, G. A., Abuzenadah, A. M., Chaudhary, A. G., & Sheikh, I. A. (2012). New anticancer agents: Recent developments in tumor therapy. *Anticancer Research*, 32(7), 2999–3005.

Amna, T., Amina, M., Sharma, P., Puri, S., Al-Youssef, H. M., Al-Taweel, A. M., & Qazi, G. (2012). Effect of precursors feeding and media manipulation on production of novel anticancer pro-drug camptothecin from endophytic fungus. *Brazilian Journal of Microbiology*, 43, 1476–1489.

Ashoka, G. B. & Shivanna, M. B. (2023). Antibacterial, antioxidant, and anticancer activities of Penicillium citrinum Thom. endophytic in Jatropha heynei. *Journal of Applied Pharmaceutical Science*, 13(3), 196–207.

Dhyani, P., Quispe, C., Sharma, E., Bahukhandi, A., Sati, P., Attri, D. C., Szopa, A., Sharifi-Rad, J., Docea, A. O., & Mardare, I. (2022). Anticancer potential of alkaloids: A key emphasis to colchicine, vinblastine, vincristine, vindesine, vinorelbine and vincamine. *Cancer Cell International*, 22(1), 206.

Dyshlovoy, S. A., Kaune, M., Kriegs, M., Hauschild, J., Busenbender, T., Shubina, L. K., Makarieva, T. N., Hoffer, K., Bokemeyer, C., & Graefen, M. (2020a). Marine alkaloid monanchoxymycalin C: A new specific activator of JNK1/2 kinase with anticancer properties. *Scientific Reports*, 10(1), 13178.

Dyshlovoy, S. A., Kudryashova, E. K., Kaune, M., Makarieva, T. N., Shubina, L. K., Busenbender, T., Denisenko, V. A., Popov, R. S., Hauschild, J., & Fedorov, S. N. (2020b). Urupocidin C: A new marine guanidine alkaloid which selectively kills prostate cancer cells via mitochondria targeting. *Scientific Reports*, 10(1), 9764.

Farouk, L., Laroubi, A., Aboufatima, R., Benharref, A., & Chait, A. (2008). Evaluation of the analgesic effect of alkaloid extract of Peganum harmala L.: Possible mechanisms involved. *Journal of Ethnopharmacology*, 115(3), 449–454.

Gadducci, A., & Cosio, S. (2022). Trabectedin and lurbinectedin: Mechanisms of action, clinical impact, and future perspectives in uterine and soft tissue sarcoma, ovarian carcinoma, and endometrial carcinoma. *Frontiers in Oncology*, 12, 914342.

Gazaem, M. A. H., Ismail, N. H., Sadiran, S. H., Abd Mutalib, N., Alsayadi, M. M. S., Alkhudhayri, D. A., & Syed Mohamad, S. A. (2025). Secondary metabolites of endophytic fungi from Malaysian local herbs as potential antimicrobial and anticancer agents. *Journal of Medicinal and Pharmaceutical Chemistry Research*, 7(7), 1435–1452.

Güler, D. A., Aydın, A., Koyuncu, M., & Parmaksız, İ. (2016). Anticancer activity of Papaver somniferum. *Journal of the Turkish Chemical Society Section A: Chemistry*, 3(3), 349–366.

Halim, H., Chunhacha, P., Suwanborirux, K., & Chanvorachote, P. (2011). Anticancer and antimetastatic activities of renieramycin M, a marine tetrahydroisoquinoline alkaloid, in human non-small cell lung cancer cells. *Anticancer Research*, 31(1), 193–201.

Koduru, S., Grierson, D., Van de Venter, M., & Afolayan, A. (2007). Anticancer activity of steroid alkaloids isolated from Solanum aculeastrum. *Pharmaceutical Biology*, 45(8), 613–618.

Liu, J., He, Y., Zhang, D., Cai, Y., Zhang, C., Zhang, P., Zhu, H., Xu, N., & Liang, S. (2017a). In vitro anticancer effects of two novel phenanthroindolizidine alkaloid compounds on human colon and liver cancer cells. *Molecular Medicine Reports*, 16(3), 2595–2603.

Liu, X., Wu, X., Ma, Y., Zhang, W., Hu, L., Feng, X., Li, X., & Tang, X. (2017b). Endophytic fungi from mangrove inhibit lung cancer cell growth and angiogenesis in vitro. *Oncology Reports*, 37(3), 1793–1803.

Luo, M.-L., Huang, W., Zhu, H.-P., Peng, C., Zhao, Q., & Han, B. (2022). Advances in indole-containing alkaloids as potential anticancer agents by regulating autophagy. *Biomedicine & Pharmacotherapy*, 149, 112827.

Mahamed, S., Motal, R., Govender, T., Dlamini, N., Khuboni, K., Hadeb, Z., Shaik, B. B., Moodley, K., Mohite, S. B., & Karpoormath, R. (2023). A concise review on marine bromopyrrole alkaloids as anticancer agents. *Bioorganic & Medicinal Chemistry Letters*, 80, 129102.

Mandour, Y. M., Refaat, E., & Hassanein, H. D. (2023). Anticancer activity, phytochemical investigation and molecular docking insights of Citrullus colocynthis (L.) fruits. *Scientific Reports*, 13(1), 20038.

Manzano, J. A., Abellanosa, E. A., Yen, C.-H., Macabeo, A. P., & Austriaco, N. P. (2024). *Cytotoxic, Antiproliferative and Pro-apoptotic Activities of Spirobisindole Alkaloids from the Philippine Medicinal Plant Voacanga globosa.* ChemRxiv; doi:10.26434/chemrxiv-2024-g19r6

Mohan, K., Jeyachandran, R., & Deepa, R. (2012). Alkaloids as anticancer agents. *Annals of Phytomedicine*, 1(1), 46–53.

Nwobodo, D. C., Okoye, N. N., Sifir Mudkhur, M., Ikem, J. C., Eze, P. M., Okoye, F. B. C., Saki, M., & Esimone, C. O. (2024). In vitro antiplasmodial and anticancer analyses of endophytic fungal extracts isolated from selected Nigerian medicinal plants. *Scientific Reports*, 14(1), 19765.

Pandy, R., Kumar, S. S., Suresh, P., Annaraj, J., Pandi, M., Vellasamy, S., & Sagadevan, S. (2023). Screening and characterization of fungal Taxol-producing endophytic fungi for evaluation of antimicrobial and anticancer activities. *Open Chemistry*, 21(1), 20220344.

Rajesh, R. P. & Annappan, M. (2015). Anticancer effects of brominated indole alkaloid Eudistomin H from marine ascidian Eudistoma viride against cervical cancer cells (HeLa). *Anticancer Research*, 35(1), 283–293.

Rana, A., Mondal, M., Rana, M., & Pargaien, A. V. (2024). Marine bioactives in cancer prevention and treatment. *International Journal of Pharmaceutical Quality Assurance*, 15(3), 1820–1826.

Sahi, N., Mostajeran, A., & Ghanadian, M. (2022). Changing in the production of anticancer drugs (vinblastine and vincristine) in Catharanthus roseus (L.) G. Don by potassium and ascorbic acid treatments. *Plant, Soil & Environment*, 68(1).

Sain, M. & Sharma, V. (2013). Catharanthus roseus (An anti-cancerous drug yielding plant)-A review of potential therapeutic properties. *International Journal of Pure & Applied Bioscience*, 1(6), 139–142.

Saito, N. (2021). Chemical research on antitumor isoquinoline marine natural products and related compounds. *Chemical and Pharmaceutical Bulletin*, 69(2), 155–177.

Suman Joshi, D., Chander Amgoth, S. N. S., Padmavathi, C., Madhavi, J., & Krishna Satya, A. (2018). Antioxidant and anticancer activities of an Aporphine alkaloid isolated from Alphonsea sclerocarpa. *The Journal of Pharmacology*, 7(1): 51–55.

Sunassee, S. N., Ransom, T., Henrich, C. J., Beutler, J. A., Covell, D. G., McMahon, J. B., & Gustafson, K. R. (2014). Steroidal alkaloids from the marine sponge Corticium niger that inhibit growth of human colon carcinoma cells. *Journal of Natural Products*, 77(11), 2475–2480.

Tailor, N. K., Deswal, G., & Grewal, A. S. (2024). Aaptamine: A versatile marine alkaloid for antioxidant, antibacterial, and anticancer therapeutics. *Chemistry*, 6(4), 677–694.

Tilaoui, M., Ait Mouse, H., & Zyad, A. (2021). Update and new insights on future cancer drug candidates from plant-based alkaloids. *Frontiers in Pharmacology*, 12, 719694.

Ueno, S., Yamazaki, R., Ikeda, T., Yaegashi, T., & Matsuzaki, T. (2014). Antitumor effect of a novel phenanthroindolizidine alkaloid derivative through inhibition of protein synthesis. *Anticancer Research*, 34(7), 3391–3397.

Valleti, P. V., Kumar, V., Ramayanam, P. K., Gopalappa, R., Vijendra Dittekoppa, P., Cm, A., Sillanpaa, M., & Al-Farraj, S. (2024). Multifaceted anticancer potential of Gnidia glauca (Fresen.) Gilg leaf alkaloids: Impact on multiple cellular targets. *ACS Omega*, 9(8), 9615–9624.

Valsan, A., Meenu, M. T., Murali, V. P., Malgija, B., Joseph, A. G., Nisha, P., Radhakrishnan, K. V., & Maiti, K. K. (2023). Exploration of phaeanthine: A bisbenzylisoquinoline alkaloid induces anticancer effect in cervical cancer cells involving mitochondria-mediated apoptosis. *ACS Omega*, 8(16), 14799–14813.

Winkiel, M. J., Chowański, S., & Słocińska, M. (2022). Anticancer activity of glycoalkaloids from Solanum plants: A review. *Frontiers in Pharmacology*, 13, 979451.

Zubía, E., Ortega, M. J., & Salvá, J. (2003). Antitumor potential of natural products from marine ascidians of the Gibraltar Strait: A survey. *Ciencias Marinas*, 29(2), 251–260.

7 Alkaloids with Antiparasitic Activity

From Traditional to Modern Medicine

Noureddine Chaachouay and Nawal Zemmouli

7.1 INTRODUCTION

Parasitic infections are a crucial public health condition globally caused by protozoan parasites and intestinal helminths, particularly in developed countries, and are considered a primary cause of disease and illness (Despommier et al., 2012). Among such infections, one of the most disastrous is malaria, the leading cause of death in many areas, with Plasmodia leading to the deadliest, where *falciparum malaria* is the most fatal. Malaria is estimated to have caused 247 million cases and 619,000 deaths globally in 2022, the World Health Organization says, with the hardest-hit being sub-Saharan Africa (Zoloth, 2023). Several other parasitic infections, such as leishmaniasis, schistosomiasis, and Chagas disease, also significantly contribute to the burden of morbidity and mortality, particularly in poor populations with poor access to healthcare (Bhutta et al., 2014). The development of drug-resistant parasite strains increases the complexity of treatment challenges, leading to a pressing demand for new clinical drug agents.

Plants continue to be useful as remedies for parasitic diseases. The use of plants as remedies for parasitic infection is longstanding in many cultures (Chaachouay et al., 2022). In South America, for example, native people used *Cinchona* tree bark, abundant with quinine alkaloids, to cure fevers – a principle that also formed the basis of contemporary antimalarial medicines (Milliken et al., 2021). Also, in traditional Chinese medicine, *Artemisia annua* was used to relieve symptoms of malaria, from which the malaria drug artemisinin was derived. For centuries, African indigenous healers have used plants such as *Strychnos* species, with indole alkaloids, for their antiparasitic activities (Kingston & Cassera, 2022). These local medicinal uses demonstrate the depth of wisdom in indigenous medicine and its value in orienting modern drug development.

The development from ethnomedicine to modern pharmacology has been associated with discovering and defining the bioactive compounds present in medicinal plants (Chaachouay et al., 2021; Süntar, 2020). Quinine, isolated from the bark of the *Cinchona* tree in the early 19th century, was the first effective antimalarial drug (Yeka et al., 2009). Its success led to the production of synthetic compounds such as chloroquine and mefloquine. Artemisinin, also isolated from *Artemisia annua*, spawned artemisinin-based combination therapies, which have become malaria's treatment of choice (Njuguna et al., 2012). These landmarks demonstrate that natural products provide essential frameworks for creating new medications (Chaachouay & Zidane, 2024).

A diverse collection of alkaloid compounds that contain nitrogen exists as a large structural group, and scientific researchers have studied them for their ability to work as broad-spectrum antiparasitic agents (Bhambhani et al., 2021; Chaachouay et al., 2023). Parasite metabolism interference, nucleic acid synthesis inhibition, and cellular structure disruption constitute alkaloids' three

DOI: 10.1201/9781003539179-7

main operational modes. For example, berberine, an isoquinoline alkaloid, is known to possess anti-*Leishmania* activity, and likewise, the steroidal alkaloid conessine from *Holarrhena pubescens* is known to have anti-*Plasmodium falciparum* action (Gutierrez, 2007). The range of alkaloid compounds across chemical space represents new opportunities to find potential antiparasitic drug starts.

Scientific development of drugs based on alkaloids faces multiple obstacles because they present toxicity risks and exhibit poor distribution within the body, alongside extensive challenges in large-scale synthetic methods (Daley & Cordell, 2021). New high-throughput screening and metabolomics methods have enabled researchers to detect potential novel alkaloids during the past few years (Fattorusso & Taglialatela-Scafati, 2007). Microbial fermentation and plant tissue culture are sustainable biotechnology systems for alkaloid production. The complex chemical group known as alkaloids demonstrates potential as drug candidates targeting parasitic diseases (Srivastava & Srivastava, 2013). New, effective, and available therapies can be developed through the combination of traditional knowledge and modern scientific methods. This chapter aims to discuss the antiparasitic activity of alkaloids from ancient medicinal applications to their contributions to contemporary drug discovery. It discusses their mode of action, leading examples, and prospects in antiparasitic therapy.

7.2 CLASSIFICATION OF ALKALOIDS WITH ANTIPARASITIC PROPERTIES

Scientists investigate alkaloids extensively because of their extensive antiparasitic applications as natural products containing nitrogen-based molecular structures (Cushnie et al., 2014). The isoquinoline alkaloids represent a vital chemical group that researchers obtain from various medical plants (Plazas et al., 2022). Studies show that antiparasitic and *Leishmania* species-active compounds exist in the natural products berberine and palmatine. The parasitic activity of berberine against *Leishmania* results from inhibiting parasite mitochondrial function and DNA synthesis (Cicero & Baggioni, 2016). The antiparasitic properties of indole alkaloids can be attributed to compounds such as reserpine and ajmalicine. Research indicates that reserpine demonstrates effectiveness against both Trypanosoma species responsible for sleeping sickness and Chagas disease, according to Osorio et al. (2008). The toxic mechanisms of these alkaloids are developed by disrupting the neurotransmitter storage system and release processes essential for parasite survival.

Quinoline and quinolizidine alkaloids represent an essential group of antiparasitic substances. The antimalarial medication quinine and related compounds belong to the quinoline alkaloids since they block malaria parasites from breaking down hemoglobin (Kingston & Cassera, 2022). Plants from the *Sophora* species contain active antiparasitic compounds matrine and oxymatrine, which effectively combat the fish parasite *Cryptocaryon irritans*. The alkaloids interrupt the parasite's cellular integrity and bioenergetic processes (Goto et al., 2015). Steroidal and terpenoid alkaloids serve as well-known antiparasite compounds. The antiparasitic effect of solasodine and tomatidine combines two beneficial features of inhibiting microorganisms while reducing inflammation, according to research by Delbrouck et al. (2023). Experimental trials confirmed the antimicrobial properties of chonemorphine as extracted from the tropical plant *Chonemorpha fragrans*, which people traditionally employ to treat intestinal infections (Devi et al., 2019).

Most alkaloid families receive scholarly attention for their antiparasitic properties and select distinguished classes. Science publications indicate that bisindole alkaloids named sungucine and isosungucine, which were identified from *Strychnos icaja*, demonstrated antiparasitic properties that target *Plasmodium falciparum* as reported by Beaufils et al. (2021) and Frédérich et al. (2001). These compounds disrupt parasites' replicative and survival life cycle as their primary mechanism of action. Their diverse chemical nature and biological activity make these alkaloids potentially good starting materials for antiparasitic drug development. Performing additional research on the biosynthesis of antiparasitic compounds alongside structure–activity examinations and pharmacodynamic evaluations will help develop the antiparasitic drug discovery field (Table 7.1).

TABLE 7.1
Alkaloid Class, Representative Compounds, and Key Characteristics for Each Group

Alkaloid Class	Representative Compounds	Natural Sources	Target Parasites	References
Isoquinoline Alkaloids	Berberine, Palmatine, Jatrorrhizine, Haemanthamine, Lycorine, Narciclasine	*Berberis* spp., *Enantia chlorantha, Annickia kummeriae, Hippeastrum aulicum, Narcissus* spp.	*Plasmodium falciparum, Leishmania donovani, Trypanosoma cruzi*	(Bentley, 1998; Bisset, 1985; Plazas et al., 2022; Shamma, 2012)
Indole Alkaloids	Reserpine, Ajmalicine, Alstonisine, Strychnobaillonine, Harmaline	*Rauvolfia* spp., *Alstonia macrophylla, Strychnos icaja, Peganum harmala*	*Trypanosoma brucei, Plasmodium falciparum, Toxoplasma gondii*	(Aniszewski, 2007; Choudhary et al., 2010, 2010; Frédérich et al., 2001; Holmstedt et al., 1980)
Quinoline and Quinolizidine Alkaloids	Quinine, Chloroquine, Mefloquine, Matrine, Oxymatrine, Conessine	*Cinchona* spp., *Sophora* spp., *Holarrhena antidysenterica*	*Plasmodium falciparum, Cryptocaryon irritans*	(Aniszewski, 2007; Bentley, 1998; Bharti et al., 2018; Bisset, 1985)
Steroidal and Terpenoid Alkaloids	Solasodine, Tomatidine, Conessine, Mokluangin A, Irehline	*Solanum* spp., *Holarrhena pubescens, Chonemorpha fragrans*	*Plasmodium falciparum, Leishmania donovani*	(Almadiy & Nenaah, 2018; Amoa Onguéné et al., 2013; Delbrouck et al., 2023; El Bazaoui et al., 2011; Huang et al., 2022)
Other Notable Classes	Harmaline, Piperine, Manzamine A, Longamide B, Dibromopalau'amine	*Peganum harmala, Piper nigrum*, Marine sponges (*Agelas* spp.)	*Toxoplasma gondii, Plasmodium falciparum, Leishmania donovani, Trypanosoma brucei*	(Aniszewski, 2007; Bisset, 1985; Cushnie et al., 2014; Sharifi et al., 2024)

7.3 MECHANISMS OF ANTIPARASITIC ACTION

Alkaloids have powerful antiparasitic properties, the main mechanisms of which are parasite metabolism and replication blockade. It has been shown that some alkaloids, such as quinine and quinidine, used to treat malaria and berberine, inhibit key metabolic pathways in parasites, like glycolysis and mitochondrial function, leading to a block in energy supply and inhibition of parasite growth (Shukla et al., 2025). Furthermore, nucleic acid synthesis disruption is observed in the potential inhibition of DNA topoisomerases of *Plasmodium falciparum*, leading to parasite replication inhibition (Attasart et al., 2016). Such mechanisms emphasize the potential of alkaloids as promising antiparasitic agents by blocking vital biochemical events (Figure 7.1).

Another important mechanism is that parasitic membranes and organelles are broken down and lack structural integrity, resulting in cell death. Studies have demonstrated that alkaloids such as sanguinarine and matrine disrupt membrane permeability in *Leishmania* spp., leading to cytotoxic effects and impairing organelle function (De Souza Silva et al., 2024; Ornelas-Cruces et al., 2025). In addition, investigations on *Trypanosoma cruzi* also found that alkaloids could disrupt mitochondrial membranes, leading to apoptosis-like cell death (Romanelli et al., 2022; Rosenkranz & Wink, 2008). The results demonstrate alkaloids' ability to directly impact the integrity of parasites' cells, resulting in a broad-spectrum antiparasitic effect.

In addition, alkaloids may manipulate host immune responses to aid in parasite clearance. Some alkaloids (capsaicin and piperine) have immunostimulant properties, activating macrophage

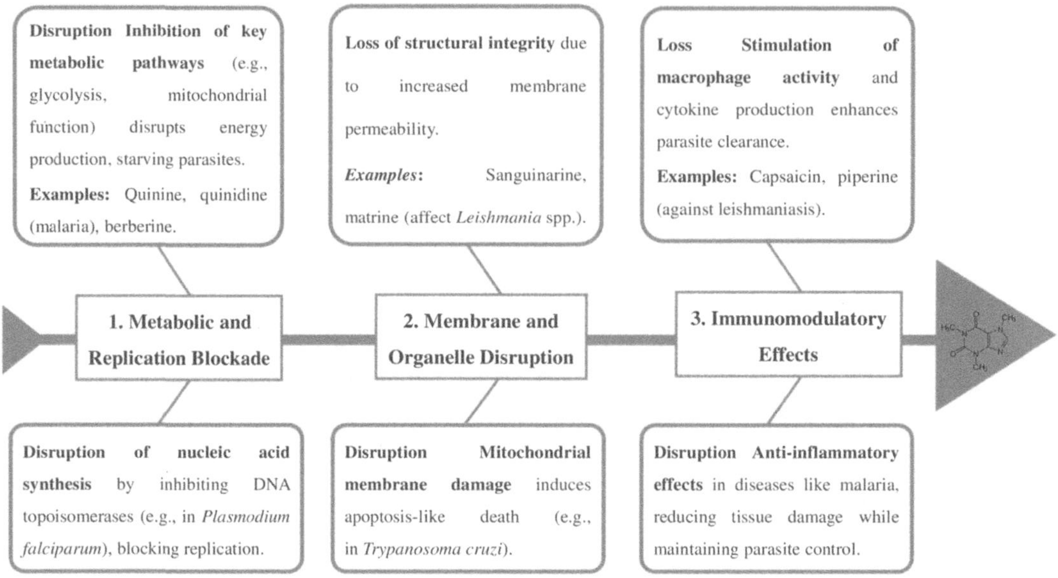

FIGURE 7.1 Mechanisms of antiparasitic action of alkaloids.

activation and the production of cytokines, which help in the host of systemic infection (leishmaniasis) (Garnier & Shahidi, 2021). Other studies suggest alkaloids could down-regulate inflammation in diseases like malaria, thereby limiting tissue damage and maintaining effective parasite control (Amoa Onguéné et al., 2013; Osorio et al., 2008). By inhibiting parasite-specific pathways and modifying the immune system in the host, alkaloids present a multi-source-targeting strategy for antiparasitic purposes.

7.4 ALKALOIDS IN TRADITIONAL MEDICINE

The traditional systems of medicine of all parts of the world have long documented the use of preparations from alkaloid-containing plants. Indigenous peoples have historically used plants ranging from *Cinchona* spp., used for the treatment of malaria, to *Berberis* spp. (Arayne et al., 2007; Canales et al., 2020; Diputación de Málaga, 2022; Imenshahidi & Hosseinzadeh, 2016; Song, 2009). Intestinal parasites use bark, showing an empirical knowledge of their curative properties. Such traditional usages are frequently confirmed by contemporary pharmacological investigations that support the antiparasitic activity of plant alkaloids. Ethnobotany research, by providing an important base for discovering bioactive compounds, narrows the dichotomy between traditional knowledge and scientific inquiry (Benkhnigue et al., 2023; Benlarbi et al., 2023; Orch et al., 2021).

Several plants that contain highly antiparasitic alkaloids have been investigated for their antiparasitic activity. For example, *Uncaria tomentosa*, which is used in Amazonian medicine and known to contain oxindole alkaloids active against *Leishmania* and *Plasmodium* species ("Antileishmanial Potentials of Phytochemicals", 2021; Shahrajabian et al., 2023; Sharifi et al., 2024). In addition, *Argemone mexicana*, used in the Ayurveda system of medicine, and Mexican folk medicine, is known for producing benzylisoquinoline alkaloids, potent antihelminthic and antimalarial compounds (Brahmachari et al., 2013). These examples demonstrate how ethnomedicine can lead to new antiparasitic drug discovery.

Incorporating traditional information into the scientific study has accelerated the research progress of plant alkaloids as antiparasitic drugs. Ethnopharmacology has led to the isolation and enhancement of bioactive plant-derived compounds, including artemisinin, obtained from the plant

Artemisia annua, traditionally used in Chinese medicine for fever and malaria (Van Agtmael et al., 1999). Cooperation between traditional healers and scientists, in addition, has helped to develop sustainable sourcing of medicinal plants in combination with the safeguarding of traditional knowledge. By integrating ancient knowledge with modern drug discovery, researchers can exploit alkaloids to the fullest as market-friendly antiparasitic agents.

7.5 MODERN DRUG DISCOVERY AND DEVELOPMENT

Isolation and characterization of bioactive alkaloids are the initial and decisive steps toward modern antiparasitic drug discovery. More advanced chromatographic methods, such as HPLC-MS and NMR spectroscopy, allowed the isolation and purification of some new alkaloids that are highly active against parasites such as *Plasmodium* and *Leishmania* (Ahmad & Nadeem, 2023; Brkljača & Urban, 2011; Du et al., 2018). High-throughput screening methods with bioassay-guided fractionation have significantly increased the speed of discovery, as evidenced by the discovery of anti-trypanosomal alkaloids from *Tabernaemontana* spp. (Nzama, 2023). Such approaches permit the accurate determination of molecular structures and initial antiparasitic assessment, which can form the basis of further drug discovery efforts.

Upon identification of bioactive alkaloids, semi-synthesis, and structure–activity relationship research tailors such alkaloids to achieve the maximal therapeutic benefit. Classical medicinal chemistry methodology has already proven to work well, i.e., to increase the bioavailability and decrease toxicity of natural alkaloids while maintaining their antiparasitic efficacy. For example, derivatives of quinolizidine alkaloids with structural modifications have resulted in compounds with enhanced antimalarial activity against drug-resistant strains of *Plasmodium falciparum* (Mishra et al., 2025). Computer-aided drug design and molecular docking also contribute to a better knowledge of the effect of some structural features on binding to the parasitic targets and assist in rational analog development.

Despite encouraging results in preclinical studies, there are obstacles to translating laboratory findings into clinical practice. Information from preclinical animal studies must show efficacy and safety before testing in humans may be allowed, which frequently reveals unforeseen pharmacokinetic restrictions or toxicities. However, even if these clinical trials are successful, several challenges may arise in commercializing alkaloid-based drugs, including large-scale production, regulatory approval, and identifying an optimal market niche to ensure patient benefits from these new drugs. Furthermore, many alkaloids possess complex chemical structures, leading to expensive synthetic processes, which trigger research on semi-synthetic methods or alternative sourcing to provide sustainable and cheaper therapies.

7.6 SAFETY, TOXICITY, AND PHARMACOKINETICS

Antiparasitic alkaloids have varied toxicological profiles and must be assessed for potential therapeutic use (Amoa Onguéné et al., 2013). Although numerous alkaloids isolated from plants, such as quinine and berberine, have shown selective toxicity against parasites, they also have toxic effects on the host when used at higher doses, such as cardiotoxicity, hepatotoxicity, and neurotoxicity. Pharmacokinetic studies indicate substantial difficulties with the absorption, distribution, metabolism, and excretion of drugs in this class; as such, many of the alkaloids demonstrate poor oral bioavailability or rapid hepatic metabolism. For example, quinine has a small therapeutic window, so it must be dosed carefully to achieve a balance of efficacy against malaria versus the risk of cinchonism (nausea, tinnitus, visual disturbance). These observations emphasize the need for extensive preclinical toxicological evaluations and the determination of safe dosing schedules for antiparasitic alkaloids (Warrell et al., 2017).

Researchers have exploited numerous optimization strategies to minimize adverse effects and simultaneously enhance therapeutic efficacy. Structural changes (e.g., artemisinin derivatives with better pharmacokinetic properties) have been made to increase drug safety and decrease toxicity. Furthermore, new approaches for drug delivery, such as nanoparticle encapsulation and liposomal formulation, enhance alkaloid bioavailability and minimize the off-target response of the alkaloids, respectively. Pharmacogenomic methods can also detect patient-specific metabolic differences contributing to drug response and provide personalized drug dosing strategies (TumkurSattar et al., 2023). With these strategies, the therapeutic applications of antiparasitic alkaloids can be maximized, and patient safety is ensured.

7.7 FUTURE PERSPECTIVES AND RESEARCH DIRECTIONS

One approach that holds promise to treat drug-resistant parasite strains with minimal toxicity is the creation of combination therapies based on alkaloids. Indeed, new findings have demonstrated the effects of the synergistic use of plant-derived alkaloids with traditional antiparasitic agents, for example, artemisinin combination therapies for malaria, that have increased treatment efficacy to new levels (Oliveira et al., 2009; Rasoanaivo et al., 2011). Furthermore, progress in biotechnology, metabolic engineering, and plant cell culture systems is changing how rare alkaloids can be sustainably produced at scale (Diamond & Desgagné-Penix, 2016). These advances overcome challenges within the supply chain and open the door to newly explored alkaloids, which act through unique mechanisms, targeting, for example, parasite epigenetics or quorum-sensing pathways.

Further investigation is also warranted to better understand novel molecular targets and create a supportive policy for translational research. New parasite-specific vulnerabilities, such as essential enzymes and signaling pathways found exclusively in protozoans, are identified through high-throughput screening and computational modeling (Müller & Hemphill, 2016). However, developing all these findings in the clinic was time-consuming and would need substantial investment channels and international cooperation to surmount the regulatory and commercial barriers. Public-private partnerships and incentives for neglected tropical disease research are key in expediting the development of new-generation alkaloid-based antiparasitics to provide these therapies ethically and equitably to endemic populations.

7.8 CONCLUSION

The comprehensive analysis described in this chapter reveals a high interest in alkaloids as efficient antiparasitic compounds. A critical analysis shows that alkaloids have varied modes of action, such as interrupting parasite metabolism, membrane integrity, host immunomodulation, and being active against drug-resistant strains, among other important findings. Empiricism aside, traditional medicine has offered invaluable leads in exploring natural product cargoes, and current pharmacological research also supports the antimalarial efficacy of alkaloid-containing plants, such as *Cinchona* and *Artemisia*, thus blurring the horizons between avant-garde and traditional approaches. In addition, the evolution of drug design has paved the way for the rational design of these compounds, enabling control over their structure and optimization by novel drug delivery systems to overcome their previous problems of bioavailability and toxicity. The significance of alkaloids as antiparasitic agents is likely to become increasingly important, particularly due to the rise in antimicrobial resistance. The advent of alkaloid combinational therapy and biotechnological approaches, including metabolic engineering and nanotechnology, provides new opportunities for improved efficacy and sustainable production. Recent findings on new targets, parasite-specific enzymes, and epigenetic modulators also broaden these compounds' activity spectrum for therapeutic use. However, achieving this potential requires sustained research on all levels, from ethnobotanical surveys to clinical studies, exploiting the pharmacological diversity of alkaloids. To enhance the efficacy of antiparasitic alkaloids, researchers and future government support must prioritize collaborative efforts that merge traditional knowledge with scientific bases. Recommendations are to build international collaborations to benefit drug development initiatives, especially for neglected tropical diseases, and to help develop policies encouraging affordable therapeutic development. In addition, translational research should emphasize pharmacokinetic tailoring and extensive safety evaluation to justify clinical application. When innovation is combined with equitable access strategies, the synergies derived from alkaloids will help treat global parasitic disease burdens, improving health outcomes in endemic parts of the world.

COMPETING INTERESTS

The authors declare that they have no competing interest.

DECLARATION OF GENERATIVE AI AND AI-ASSISTED TECHNOLOGIES IN THE WRITING PROCESS

During the preparation of this work, the book chapter author(s) have not taken any assistance from AI, and the book chapter author(s) reviewed and edited the content as needed and take(s) full responsibility for the content of the publication.

LIST OF ABBREVIATIONS

DNA	Deoxyribonucleic Acid
HPLC	High-Performance Liquid Chromatography
MS	Mass Spectrometry
NMR Spectroscopy	Nuclear Magnetic Resonance Spectroscopy

REFERENCES

Ahmad, F., & Nadeem, H. (2023). Mass spectroscopy as an analytical tool to harness the production of secondary plant metabolites: The way forward for drug discovery. In G. C. Pereira (Éd.), *Gene, Drug, and Tissue Engineering* (p. 77–103). Springer US. https://doi.org/10.1007/978-1-0716-2716-7_5

Almadiy, A. A., & Nenaah, G. E. (2018). Ecofriendly synthesis of silver nanoparticles using potato steroidal alkaloids and their activity against phytopathogenic fungi. *Brazilian Archives of Biology and Technology*, *61*, e18180013.

Amoa Onguéné, P., Ntie-Kang, F., Lifongo, L. L., Ndom, J. C., Sippl, W., & Mbaze, L. M. (2013). The potential of anti-malarial compounds derived from African medicinal plants. Part I: A pharmacological evaluation of alkaloids and terpenoids. *Malaria Journal*, *12*(1), 449. https://doi.org/10.1186/1475-2875-12-449

Aniszewski, T. (2007). *Alkaloids-Secrets of Life: Aklaloid Chemistry, Biological Significance, Applications and Ecological Role*. Elsevier. https://books.google.fr/books?hl=fr&lr=&id=a1Z6oJL-dgMC&oi=fnd&pg=PP1&dq=Indole+Alkaloids+Rauvolfia+spp.,+Alstonia+macrophylla,+Strychnos+icaja,+Peganum+harmala&ots=Ms7jkUNlJC&sig=u-KqjgL_wvOlARhjW4wMBwFHJxc

Antileishmanial Potentials of Phytochemicals. (2021). In S. M. Ezzat, M. A. Salem, & A. Zayed (eds.), *Neglected Tropical Diseases and Phytochemicals in Drug Discovery* (1st éd., pp. 183–198). Wiley. https://doi.org/10.1002/9781119617143.ch6

Arayne, M. S., Sultana, N., & Bahadur, S. S. (2007). The berberis story: Berberis vulgaris in therapeutics. *Pakistan Journal of Pharmaceutical Sciences*, *20*(1), 83–92.

Attasart, P., Boonma, S., Sunintaboon, P., Tanwilai, D., Pothikasikorn, J., & Noonpakdee, W. T. (2016). Inhibition of Plasmodium falciparum proliferation in vitro by double-stranded RNA nanoparticle against malaria topoisomerase II. *Experimental Parasitology*, *164*, 84–90.

Beaufils, N., Boucher, J., Peydecastaing, J., Rigal, L., Vilarem, G., Villette, M.-J., Candy, L., & Pontalier, P.-Y. (2021). The effect of time and temperature on the extraction of xylose and total phenolic compounds with pressurized hot water from hardwood species used for pulp and paper production in the South of France. *Bioresource Technology Reports*, *16*, 100832.

Benkhnigue, O., Khamar, H., Bussmann, R. W., Chaachouay, N., & Zidane, L. (2023). Ethnobotanical and ethnopharmacological study of medicinal plants used in treating some liver diseases in the Al-Haouz Rehamna region (Morocco). *Ethnobotany Research and Applications*, *25*, 1–32.

Benlarbi, F., Mimoune, N., Chaachouay, N., Souttou, K., Saidi, R., Mokhtar, M. R., Kaidi, R., & Benaissa, M. H. (2023). Ethnobotanical survey of the traditional antiparasitic use of medicinal plants in humans and animals in Laghouat (Southern Algeria). *Veterinary World*, 357–368. https://doi.org/10.14202/vetworld.2023.357-368

Bentley, K. W. (1998). *Isoquinoline Alkaloids* (Vol. 1). CRC Press. https://books.google.fr/books?hl=fr&lr=&id=MDHnf0MbgGcC&oi=fnd&pg=PR11&dq=Isoquinoline+Alkaloids&ots=g77PctIgFv&sig=2m8yNEOfEVn0XahNrefmhHjP07A

Bhambhani, S., Kondhare, K. R., & Giri, A. P. (2021). Diversity in chemical structures and biological properties of plant alkaloids. *Molecules*, *26*(11), 3374.

Bharti, N., Mohapatra, P. P., & Singh, S. (2018). Plant-based antiamoebic natural products: Literature review and recent developments. *Frontiers in Natural Product Chemistry*, *4*, 1–40.

Bhutta, Z. A., Sommerfeld, J., Lassi, Z. S., Salam, R. A., & Das, J. K. (2014). Global burden, distribution, and interventions for infectious diseases of poverty. *Infectious Diseases of Poverty*, *3*(1). https://doi.org/10.1186/2049-9957-3-21

Bisset, N. G. (1985). Plants as a source of isoquinoline alkaloids. In *The Chemistry and Biology of Isoquinoline Alkaloids* (pp. 1–22). Springer. https://link.springer.com/content/pdf/10.1007/978-3-642-70128-3_1?pdf=chapter%20toc

Brahmachari, G., Gorai, D., & Roy, R. (2013). Argemone mexicana: Chemical and pharmacological aspects. *Revista Brasileira de Farmacognosia*, *23*, 559–567.

Brkljača, R., & Urban, S. (2011). Recent advancements in HPLC-NMR and applications for natural product profiling and identification. *Journal of Liquid Chromatography & Related Technologies*, *34*(13), 1063–1076.

Canales, N. A., Hansen, T. N. G., Cornett, C., Walker, K., Driver, F., Antonelli, A., Maldonado, C., Nesbitt, M., Barnes, C. J., & Rønsted, N. (2020). Historical chemical annotations of Cinchona bark collections are comparable to results from current day high-pressure liquid chromatography technologies. *Journal of Ethnopharmacology*, *249*, 112375.

Chaachouay, N., Azeroual, A., Benkhnigue, O., & Zidane, L. (2023). Alkaloids: A suitable precursor for nanomaterials synthesis, and their various applications. In A. Husen (Éd.), *Secondary Metabolites Based Green Synthesis of Nanomaterials and Their Applications* (pp. 23–48). Springer Nature. https://doi.org/10.1007/978-981-99-0927-8_2

Chaachouay, N., Azeroual, A., Douira, A., & Zidane, L. (2022). Ethnoveterinary practices of medicinal plants among the Zemmour and Zayane tribes, Middle Atlas, Morocco. *South African Journal of Botany*, *151*, 826–840.

Chaachouay, N., Douira, A., & Zidane, L. (2021). COVID-19, prevention and treatment with herbal medicine in the herbal markets of Salé Prefecture, North-Western Morocco. *European Journal of Integrative Medicine*, *42*, 101285.

Chaachouay, N., & Zidane, L. (2024). Plant-derived natural products: A source for drug discovery and development. *Drugs and Drug Candidates*, *3*(1), 184–207.

Choudhary, S. P., Bhardwaj, R., Gupta, B. D., Dutt, P., Gupta, R. K., Biondi, S., & Kanwar, M. (2010). Epibrassinolide induces changes in indole-3-acetic acid, abscisic acid and polyamine concentrations and enhances antioxidant potential of radish seedlings under copper stress. *Physiologia Plantarum*. https://doi.org/10.1111/j.1399-3054.2010.01403.x

Cicero, A. F. G., & Baggioni, A. (2016). Berberine and its role in chronic disease. In S. C. Gupta, S. Prasad, & B. B. Aggarwal (Éds.), *Anti-inflammatory Nutraceuticals and Chronic Diseases* (Vol. 928, pp. 27–45). Springer International Publishing. https://doi.org/10.1007/978-3-319-41334-1_2

Cushnie, T. T., Cushnie, B., & Lamb, A. J. (2014). Alkaloids: An overview of their antibacterial, antibiotic-enhancing and antivirulence activities. *International Journal of Antimicrobial Agents*, *44*(5), 377–386.

Daley, S., & Cordell, G. A. (2021). Alkaloids in contemporary drug discovery to meet global disease needs. *Molecules*, *26*(13), 3800.

De Souza Silva, M. S., Dos Santos, M. L. M. F., Da Silva, A. M., França, W. W. M., Araújo, S. B., Da Silva, R. L., Do Nascimento, W. R. C., Da Silva Santos, N. P., Da Cruz Filho, I. J., De Azevedo Albuquerque, M. C. P., De Araújo, H. D. A., & De Lima Aires, A. (2024). Sanguinarine: An alkaloid with promising in vitro and in vivo antiparasitic activity against different developmental stages of Schistosoma mansoni and in silico pharmacokinetic properties (ADMET). *Parasitology Research*, *123*(2). https://doi.org/10.1007/s00436-024-08153-w

Delbrouck, J. A., Desgagné, M., Comeau, C., Bouarab, K., Malouin, F., & Boudreault, P.-L. (2023). The therapeutic value of Solanum steroidal (Glyco) alkaloids: A 10-year comprehensive review. *Molecules*, *28*(13), 4957.

Despommier, D. D., Gwadz, R. W., & Hotez, P. J. (2012). *Parasitic Diseases*. Springer Science & Business Media. https://books.google.fr/books?hl=fr&lr=&id=Eo_fBwAAQBAJ&oi=fnd&pg=PA1&dq=Parasitic+infections+are+a+crucial+public+health+condition+globally+caused+by+protozoan+parasites+and+intestinal+helminths,+particularly+in+developed+countries,+and+are+considered+a+primary+cause+of+disease+and+illness.&ots=K6XrRDkOr7&sig=qbc7dBPAo3GnyCyfF7H18hj5RPk

Devi, N., Gupta, A. K., & Prajapati, S. K. (2019). Prospects of traditionally important Apocynaceae plants of India in cancer remediation. *Journal of Drug Delivery and Therapeutics*, *9*(1), 293–302.

Diamond, A., & Desgagné-Penix, I. (2016). Metabolic engineering for the production of plant isoquinoline alkaloids. *Plant Biotechnology Journal*, *14*(6), 1319–1328. https://doi.org/10.1111/pbi.12494

Diputación de Málaga. (2022). *Vinette Thunberg (Berberis hispanica)*. https://www.malaga.es/fr/turismo/naturaleza/lis_cd-13051/agracejo-andaluz-berberis-hispanica

Du, W., Jin, L., Li, L., Wang, W., Zeng, S., Jiang, H., & Zhou, H. (2018). Development and validation of a HPLC-ESI-MS/MS method for simultaneous quantification of fourteen alkaloids in mouse plasma after oral administration of the extract of corydalis yanhusuo tuber: Application to pharmacokinetic study. *Molecules*, *23*(4), 4. https://doi.org/10.3390/molecules23040714

El Bazaoui, A., Bellimam, M. A., & Soulaymani, A. (2011). Nine new tropane alkaloids from Datura stramonium L. identified by GC/MS. *Fitoterapia*, *82*(2), 193–197.

Fattorusso, E., & Taglialatela-Scafati, O. (2007). *Modern Alkaloids: Structure, Isolation, Synthesis, and Biology*. John Wiley & Sons.

Frédérich, M., De Pauw, M.-C., Prosperi, C., Tits, M., Brandt, V., Penelle, J., Hayette, M.-P., DeMol, P., & Angenot, L. (2001). Strychnogucines A and B, two new antiplasmodial bisindole alkaloids from *Strychnos icaja*. *Journal of Natural Products*, *64*(1), 12–16. https://doi.org/10.1021/np000285t

Garnier, A., & Shahidi, F. (2021). Spices and herbs as immune enhancers and anti-inflammatory agents: A review. *Journal of Food Bioactives*, *14*. http://www.isnff-jfb.com/index.php/JFB/article/view/225

Goto, T., Hirazawa, N., Takaishi, Y., & Kashiwada, Y. (2015). Antiparasitic effect of matrine and oxymatrine (quinolizidine alkaloids) on the ciliate *Cryptocaryon irritans* in the red sea bream *Pagrus major*. *Aquaculture*, *437*, 339–343. https://doi.org/10.1016/j.aquaculture.2014.12.026

Gutierrez, R. M. P. (2007). *Handbook of Compounds with Antiprotozoal Activity Isolated from Plants*. Nova Publishers. https://books.google.fr/books?hl=fr&lr=&id=IhTac3zPdRoC&oi=fnd&pg=PP9&dq=berberine,+an+isoquinoline+alkaloid,++is+known+to+possess%E2%80%82anti+Leishman+ia+activity,+and+likewise,+the+steroidal+alkaloid+conessine+from+Holarrhena+pubescens+is+known+to+have+anti+Plasmodium+falciparum+action&ots=_-mHAzSyCS&sig=1xXs5ytSaGFG7iYWeTFbJJO2MZI

Holmstedt, B., Lindgren, J. E., Plowman, T., Rivier, L., Schultes, R. E., & Tovar, O. (1980). Indole alkaloids in Amazonian Myristicaceae: Field and laboratory research. *Botanical Museum Leaflets, Harvard University*, *28*(3), 215–234.

Huang, W., Wang, Y., Tian, W., Cui, X., Tu, P., Li, J., Shi, S., & Liu, X. (2022). Biosynthesis investigations of terpenoid, alkaloid, and flavonoid antimicrobial agents derived from medicinal plants. *Antibiotics*, *11*(10), 1380.

Imenshahidi, M., & Hosseinzadeh, H. (2016). Berberis vulgaris and berberine: An update review. *Phytotherapy Research*, *30*(11), 1745–1764.

Kingston, D. G. I., & Cassera, M. B. (2022). Antimalarial natural products. In A. D. Kinghorn, H. Falk, S. Gibbons, Y. Asakawa, J.-K. Liu, & V. M. Dirsch (Éds.), *Antimalarial Natural Products* (pp. 1–106). Springer International Publishing. https://doi.org/10.1007/978-3-030-89873-1_1

Milliken, W., Walker, B. E., Howes, M.-J. R., Forest, F., & Nic Lughadha, E. (2021). Plants used traditionally as antimalarials in Latin America: Mining the tree of life for potential new medicines. *Journal of Ethnopharmacology*, *279*, 114221.

Mishra, R., Da Cunha Xavier, J., Kumar, N., Krishna, G., Dhakad, P. K., Dos Santos, H. S., Bandeira, P. N., Rodrigues, T. H. S., Gondim, D. R., Ribeiro, W. H. F., Da Silva, D. S., Teixeira, A. M. R., Pereira, W. F., Marinho, E. S., & Sucheta. (2025). Exploring quinoline derivatives: Their antimalarial efficacy and structural features. *Medicinal Chemistry*, *21*(2), 96–121. https://doi.org/10.2174/0115734064318361240827072124

Müller, J., & Hemphill, A. (2016). Drug target identification in protozoan parasites. *Expert Opinion on Drug Discovery*, *11*(8), 815–824. https://doi.org/10.1080/17460441.2016.1195945

Njuguna, N. M., Ongarora, D. S. B., & Chibale, K. (2012). Artemisinin derivatives: A patent review (2006–present). *Expert Opinion on Therapeutic Patents*, *22*(10), 1179–1203. https://doi.org/10.1517/13543776.2012.724063

Nzama, V. N. (2023). *Antiplasmodial and Antigonococcal Bioactivity of Tabernaemontana Elegans Stapf. (Apocynaceae) Stem Bark Fractions* [Master's Thesis, University of Pretoria (South Africa)]. https://search.proquest.com/openview/a4acf159fabad21576c2b12208618df51?ps://search.1?proquest.com/o1?penview/a4acf159fabad21576c2b12208618df5/1?1?pq-origsite=gscholar&cbl=2026366&diss=y

Oliveira, A. B., Dolabela, M. F., Braga, F. C., Jácome, R. L., Varotti, F. P., & Póvoa, M. M. (2009). Plant-derived antimalarial agents: New leads and efficient phythomedicines. Part I. Alkaloids. *Anais da Academia Brasileira de Ciências*, *81*, 715–740.

Orch, H., Chaachouay, N., Douiri, E. M., Faiz, N., Zidane, L., & Douira, A. (2021). Use of medicinal plants in dermato-cosmetology: An ethnobotanical study among the population of Izarene. *Jordan Journal of Pharmaceutical Sciences*. https://scholar.google.com/citations?view_op=view_citation&hl=fr&user=DMXnc0oAAAAJ&citation_for_view=DMXnc0oAAAAJ:7PzlFSSx8tAC

Ornelas-Cruces, M., Escalona-Montaño, A. R., Salaiza-Suazo, N., Sifontes-Rodríguez, S., & Aguirre-García, M. M. (2025). The potential role of sanguinarine as an inhibitor of Leishmania PP2C in the induction of apoptosis. *Acta Parasitologica*, *70*(1). https://doi.org/10.1007/s11686-024-00977-x

Osorio, E. J., Robledo, S. M., & Bastida, J. (2008). Alkaloids with antiprotozoal activity. *The Alkaloids: Chemistry and Biology*, *66*, 113–190.

Plazas, E., Avila, M. M. C., Muñoz, D. R., & Cuca, S. L. E. (2022). Natural isoquinoline alkaloids: Pharmacological features and multi-target potential for complex diseases. *Pharmacological Research*, *177*, 106126. https://doi.org/10.1016/j.phrs.2022.106126

Rasoanaivo, P., Wright, C. W., Willcox, M. L., & Gilbert, B. (2011). Whole plant extracts versus single compounds for the treatment of malaria: Synergy and positive interactions. *Malaria Journal*, *10*(S1). https://doi.org/10.1186/1475-2875-10-s1-s4

Romanelli, M. M., Amaral, M., Thevenard, F., Santa Cruz, L. M., Regasini, L. O., Migotto, A. E., Lago, J. H. G., & Tempone, A. G. (2022). Mitochondrial Imbalance of *Trypanosoma cruzi* Induced by the Marine Alkaloid 6-Bromo-2′-de-*N*-Methylaplysinopsin. *ACS Omega*, *7*(32), 28561–28570. https://doi.org/10.1021/acsomega.2c03395

Rosenkranz, V., & Wink, M. (2008). Alkaloids induce programmed cell death in bloodstream forms of trypanosomes (Trypanosoma b. Brucei). *Molecules*, *13*(10), 2462–2473.

Shahrajabian, M. H., Cheng, Q., & Sun, W. (2023). The importance of neglected and underutilized medicinal plants from South America in modern pharmaceutical sciences. *Letters in Drug Design & Discovery*, *20*(11), 1688–1706. https://doi.org/10.2174/1570180819666220512113812

Shamma, M. (2012). *The Isoquinoline Alkaloids Chemistry and Pharmacology* (Vol. 25). Elsevier. https://books.google.fr/books?hl=fr&lr=&id=ACvOK3ph3wMC&oi=fnd&pg=PP1&dq=Isoquinoline+Alkaloids&ots=MkY70LkgF3&sig=QlFzL-B76JneZsqb6nRddSsJpZ4

Sharifi, F., Mohamadi, N., Abiri, A., Sharifi, I., Harandi, M. F., & Sharififar, F. (2024). Nature-derived alkaloids as a promising bioactive compound in drug discovery to meet global leishmania needs. *Letters in Drug Design & Discovery*, *21*(1), 34–61. https://doi.org/10.2174/1570180820666230220141636

Shukla, S., Kushwah, S., & Mani, A. (2025). Targeting malaria's Achilles' heels: A review of plasmodium life cycle vulnerabilities for drug discovery. *Current Topics in Medicinal Chemistry*, *25*. https://doi.org/10.2174/0115680266348099250108065838

Song, C. E. (2009). *Cinchona Alkaloids in Synthesis and Catalysis: Ligands, Immobilization and Organocatalysis* (pp. 1–525). Wiley Online Library. https://doi.org/10.1002/9783527628179

Srivastava, S., & Srivastava, A. K. (2013). Biotechnology and genetic engineering for alkaloid production. In G. K. Ramawat, & J.-M. Mérillon (Eds.), *Natural Products: Phytochemistry, Botany and Metabolism of Alkaloids, Phenolics and Terpenes* (pp. 213–250). Springer.

Süntar, I. (2020). Importance of ethnopharmacological studies in drug discovery: Role of medicinal plants. *Phytochemistry Reviews*, *19*(5), 1199–1209. https://doi.org/10.1007/s11101-019-09629-9

TumkurSattar, A. H., Abushoumi, F., Al Dobayan, Z., & Al-Ghamdi, F. (2023). Investigate how genetic variations impact drug response and explore the integration of pharmacogenomics into clinical practice. *European Journal of Cardiovascular Medicine*, *13*(4). https://search.ebscohost.com/login.aspx?direct=true&profile=ehost&scope=site&authtype=crawler&jrnl=20424884&AN=174826044&h=QB0OwoPmu2cvFf9BOG6zf5WH6vf8%2B9qtP3s8RgB4%2BscDUSiv9lRLWJn%2FnZFUqOszfCBuN9tPREmqOzmmFXeR7g%3D%3D&crl=c

Van Agtmael, M. A., Eggelte, T. A., & Van Boxtel, C. J. (1999). Artemisinin drugs in the treatment of malaria: From medicinal herb to registered medication. *Trends in Pharmacological Sciences*, *20*(5), 199–205.

Warrell, D. A., Watkins, W. M., & Winstanley, P. A. (2017). Treatment and prevention of malaria. In *Essential Malariology* (4th ed., pp. 268–312). CRC Press. https://www.taylorfrancis.com/chapters/edit/10.1201/9780203756621-12/treatment-prevention-malaria-david-warrell-william-watkins-peter-winstanley

Yeka, A., Achan, J., D'Alessandro, U., & Talisuna, A. O. (2009). Quinine monotherapy for treating uncomplicated malaria in the era of artemisinin-based combination therapy: An appropriate public health policy? *The Lancet Infectious Diseases*, *9*(7), 448–452.

Zoloth, L. (2023). *May We Make the World? Gene Drives, Malaria, and the Future of Nature*. MIT Press. https://books.google.fr/books?hl=fr&lr=&id=bJO0EAAAQBAJ&oi=fnd&pg=PR11&dq=Malaria+is%E2%80%82estimated+to+have+caused+247+million+cases+and+619,000+deaths+globally+in+2022,+the+World+Health+Organization+says,+with+hardest-hit+sub-Saharan+Africa&ots=OS5nRdOo6y&sig=w0JCXIFGiscW1d13p53ImjrSUvw

8 Alkaloids in Cardiovascular Health and Disease Management

Shankul Kumar, Vedant Kumar Prajapati, Arvind Kumar Patel, Shashi Ranjan Singh, and Sumitra Nain

8.1 INTRODUCTION

8.1.1 Overview of Cardiovascular Diseases

In recent decades, the global health prospect has been increasingly dominated by cardiovascular diseases (CVDs), also known as non-communicable diseases (NCDs). These diseases associated with hypertension and hyperlipidemia have high mortality and morbidity. Since ancient times, medicinal plants have played a significant role in traditional medical systems, providing a natural and comprehensive approach to well-being. Medicinal plants are rich in bioactive substances with a variety of pharmacological characteristics, including polyphenols, flavonoids, alkaloids, and terpenoids. Increasingly, contemporary scientific studies are confirming the effectiveness of these substances by demonstrating how they can alter various physiological pathways associated with CVDs (Flora & Nayak, 2019).

Many therapeutic plants include alkaloids, a class of chemical compounds found in nature that are mainly composed of basic nitrogen atoms. Both conventional and contemporary medicine have significantly benefited from the use of alkaloids. Alkaloids are produced by a variety of taxa, including fungi, bacteria, plants, and animals (Manske, 1965). Through acid–base extraction, they can be separated from the crude extracts of these organisms. Because these bioactive substances have a wide range of pharmacological characteristics, alkaloidal medicinal plants are essential for managing and treating a number of CVDs.

Peripheral artery disease, hypertension, coronary artery disease, and stroke are among the conditions collectively referred to as "CVDs" that affect the heart and blood arteries. These conditions not only result in major health issues, including heart attacks, strokes, heart failure, and arrhythmias, but they also degrade quality of life and often lead to long-term damage. Beyond just physical health, CVD influences mental health and places a heavy financial strain on people, families, and healthcare systems around the globe. A multimodal strategy is needed to protect oneself from these crippling illnesses, which includes controlling risk factors and leading a healthy lifestyle (Ghodeshwar et al., 2023). Limit saturated fats, trans fats, and sodium while still maintaining a well-balanced diet that includes fruits, vegetables, whole grains, and lean proteins. Essential nutrients include these foods. Heart and blood vessel health are maintained by moderate-intensity exercise and regular physical activity. Another important step in reducing CVDs is reducing alcohol consumption and avoiding smoking (Pallazola et al.,2019). Also important for cardiovascular wellness is the presence of supportive social networks, sufficient sleep, and the ability to relax. Most importantly, an individual with risk factors – diabetes, high blood pressure, or high cholesterol levels – should have a constant review and proper attention by the physicians. Regular consultation with physicians would expedite

DOI: 10.1201/9781003539179-8

timely intervention, as symptoms signaling cardiovascular problems could be captured early. However, the sustainability of healthcare systems is facing a considerable challenge since the prevalence of CVD among older populations in the LMICs has dramatically increased (Santos, 2022).

According to the WHO, CVDs account for an estimated 17.9 million deaths annually or 32% of all deaths globally. Of these, 85% are due to heart attacks and strokes (World Health Organization, 2017). Many causes particular to LMICs, including poor public health infrastructure, limited access to high-quality healthcare, and a greater prevalence of risk factors such as smoking, diabetes, and hypertension, all play a role in this alarming trend (Vasudeva et al., 2024).

8.1.2 Cardiovascular Biomarker

Cardiovascular markers are biochemicals secreted in the blood under stress or injury in the cardiac environment. These biomarkers have made a substantial contribution to the diagnosis, treatment, and management of CVDs, including atherosclerosis, heart failure, and myocardial infarction or heart attack. The most well-known of these are cardiac biomarkers, which are crucial for the early diagnosis, risk assessment, and treatment of various disorders. These include troponins, C-reactive protein, CKMB, BNP, NT-proBNP, galectin, adiponectin, IL-6, TNF-α, miRNAs, oxylipins, and LDL/HDL (Wang & Pencina, 2017).

8.1.2.1 Clinical Relevance of Cardiovascular Biomarker

Cardiovascular biomarkers play a crucial role in the diagnosis (troponins and CK-MB), prognosis (BNP and NT-proBNP), and management of various heart conditions, providing valuable insights into patient risk stratification and treatment efficacy. These biomarkers enable clinicians to identify patients at high risk for cardiovascular events, allowing for timely interventions that may improve outcomes and reduce healthcare costs. Emerging research continues to explore novel biomarkers that could enhance our understanding of CVDs, potentially leading to the development of targeted therapies and personalized medicine strategies (Dhingra & Vasan, 2017).

8.1.3 Molecular Imaging for Cardiovascular Diseases

Molecular imaging is a sophisticated diagnostic method that offers non-invasive visualization, characterization, and quantification of biological processes at the molecular and cellular levels in living organisms. It is similar to chest X-rays, ECGs, ultrasounds, CT scans, SPECT scans, PET scans, and MRI scans. In CVDs, it has important applications that enable early detection, tracking the disease's course, and assessing the effectiveness of treatment (Sharma & Sharma, 2020).

8.2 SEPARATION OF ALKALOIDS FROM PLANT SOURCES

Alkaloids are naturally occurring nitrogen-containing compounds with significant pharmacological activities. Alkaloid extraction methods can be broadly classified into conventional and modern techniques. Conventional methods, such as acid–base extraction and Soxhlet extraction, are widely used but have notable drawbacks. They involve long extraction times, high solvent consumption, and lower selectivity. Acid–base extraction typically uses acidified solvents to extract alkaloid salts, which are then converted to free bases using a base and extracted with organic solvents. Soxhlet extraction allows for the continuous washing of plant material with hot solvent, but it can lead to the thermal degradation of compounds (Harborne, 1998).

In contrast, modern methods offer improved efficiency, reduced environmental impact, and better selectivity. Ultrasound-assisted extraction (UAE) uses ultrasonic waves to rupture cell walls, enhancing the release of alkaloids at lower temperatures. Microwave-assisted extraction (MAE) rapidly heats plant material, speeding up extraction while using less solvent. Supercritical Fluid Extraction (SFE) utilizes CO_2 under high pressure for selective, clean extraction, while Pressurized Liquid Extraction (PLE) or Accelerated Solvent Extraction (ASE) offers fast, automated processing with minimal solvent use. Although modern methods require higher initial investment and optimization, they offer faster, eco-friendly, and more efficient alternatives. Overall, modern techniques represent a significant advancement in alkaloid extraction, particularly in pharmaceutical and analytical applications (Chemat et al., 2012).

8.3 CLASSIFICATION OF ALKALOIDS

Alkaloids are nitrogen-containing organic compounds that are mostly found in plants. They are categorized according to their chemical structure (e.g., indole, pyridine, tropane), biological origin (e.g., particular plant families like Solanaceae or Papaveraceae), or pharmacological effects (e.g., analgesics like morphine, stimulants like caffeine, and antimalarials like quinine; Cordell, 1981) (Table 8.1). Because of their therapeutic qualities, alkaloids help plants defend against herbivores and microbes, but their potential toxicity must be used with caution.

TABLE 8.1
Chemical Classification of Alkaloids Present in Medicinal Plants

S. No.	Type	Basic Ring	Example
1	Pyrrole and Pyrrolidine	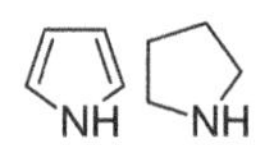	Hygrine, Coca species
2	Pyridine and Piperidine	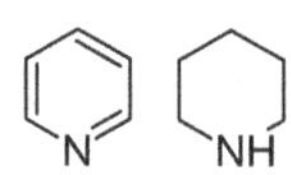	Arecoline, Anabasine, Lobeline
3	Pyrrolizidine	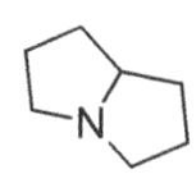	Echimidine
4	Tropane	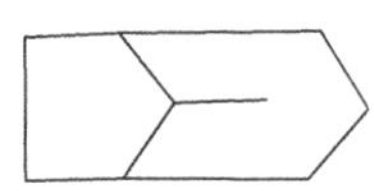	Atropine, Hyoscine, Hyocyamine
5	Quinoline	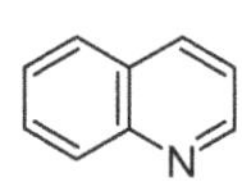	Quinine, Quinidine, cinchonine, Camptothecin
6	Isoquinoline	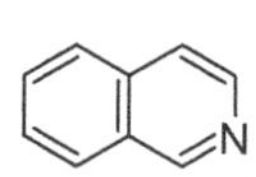	Berberine, Emetine, papaverine
7	Phenanthrene	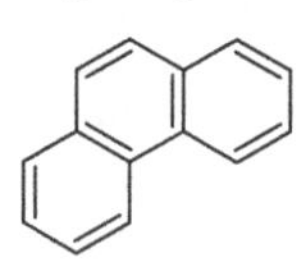	Morphine, Codeine
8	Aporphine	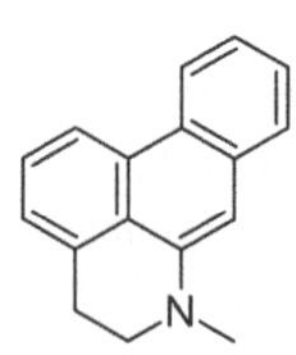	Lysicamine, litsericinone
9	Indole	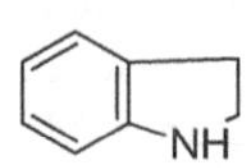	Ergometrine, Reserpine, Strychnine

(Continued)

TABLE 8.1 (Continued)
Chemical Classification of Alkaloids Present in Medicinal Plants

S. No.	Type	Basic Ring	Example
10	Imidazole	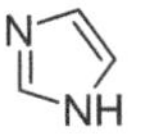	Pilocarpine, pilosine
11	Norlupinane	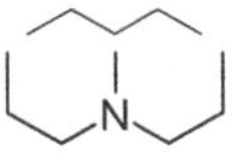	Cystisine, laburnine
12	purine	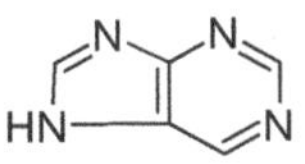	Caffeine, Theobromine
13	Steroidal	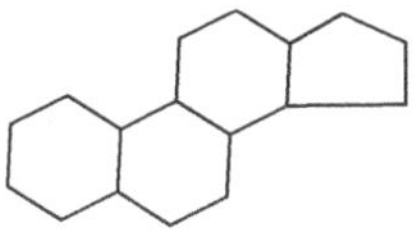	Solanidine, Conessine
14	Diterpene	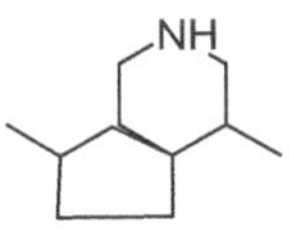	Aconitine, Aconine
15	Alkylamine	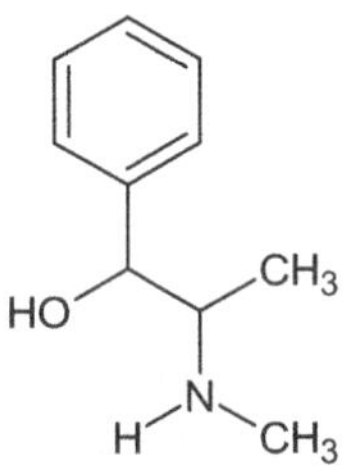	Ephedrine, colchicine

8.4 PHARMACOLOGICAL ACTION OF ALKALOIDS

Alkaloids are a diverse group of naturally occurring secondary metabolites obtained from different parts of plant having a wide range of pharmacological activities, including applications in the management of CVDs: Bark (cinchona), Stem (ephedra), Leaves (vinca), Bulb (common snowdrop), Latex (opium), Fruit (Black pepper), Root (rauwolfia). Among the most well-known alkaloids, morphine has been and is being used in medicine. This alkaloid is a strong narcotic used to alleviate pain, but because of its addictive properties, its usefulness is limited (Chisholm, 2015). Codeine, a methyl ether derivative of morphine, is found naturally in opium poppies alongside morphine. It has been demonstrated to have good analgesic properties and to be comparatively nonaddictive. These alkaloids stimulate the heart or respiratory system. Atropine is an alkaloid used as a medication in various clinical settings. For instance, bradycardia (low heart rate) is treated with an injection of atropine. An alkaloid derived from toxic curare, tubocurarine is used as a muscle relaxant during surgery. Many forms of cancer are treated with vincristine and vinblastine as chemotherapeutic agents. One of the most potent local anesthetics is cocaine, an alkaloid found in *Erythroxylum coca*. Blood vessel constrictors include ephedrine, which was obtained from Ephedra species, and ergonovine, an alkaloid derived from the fungus *Claviceps purpurea*. Ephedrine is also used to treat bronchial asthma and to ease the symptoms of common colds, sinusitis, and hay fever. Quinine, a potent antimalarial agent, is now frequently substituted with synthetic medications, which are less harmful and more effective. Quinidine, another alkaloid from the Cinchona species, is used in medicine to treat arrhythmias, or abnormal heartbeats. Another alkaloid that has been used for a long time to treat severe gout attacks is colchicine, which is found in plants of the Liliaceae family. Lobeline, extracted from Lobelia inflata, is another alkaloid used in medicine with multiple modes of action. Reserpine is one of the indole alkaloids obtained from the Rauwolfia plant and used as an antihypertensive agent. Caffeine is one of the purine alkaloids present in tea or coffee and is used as a diuretic. Vasacine and vasacinone are two of the quinazoline alkaloids obtained from the Vasaka plant and are clinically used as an expectorant and smooth muscle relaxant. Strychnine and brucine are active constituents of Nux vomica used as bitter stomachic agents (Kurek, 2019).

8.4.1 Pharmacological Class of Drug Acting on CVDs

Alkaloids have a variety of pharmacological effects on CVDs, including Antihypertensive, Antiarrhythmic, Hypolipidemic, Antithrombotic, Cardioprotective, and Diuretic.

8.4.1.1 Antihypertensive Action

Alkaloids are naturally occurring compounds that have significant antihypertensive effects. Several alkaloids exert blood pressure-lowering actions through different mechanisms. One of the most well-known is reserpine, derived from *Rauwolfia serpentina*, which works by depleting catecholamines (norepinephrine, dopamine) from sympathetic nerve endings. This reduces sympathetic tone, leading to decreased heart rate and peripheral vascular resistance (Sneader, 2005; Wainwright, 2013).

Another example is berberine, found in plants like Berberis, which exhibits vasodilatory effects by promoting nitric oxide (NO) synthesis and inhibiting calcium channels in vascular smooth muscle. This results in vessel relaxation and a drop-in blood pressure. Some alkaloids, such as ajmaline and serpentine, also modulate cardiac and vascular functions, contributing to antihypertensive action (Zhang et al., 2011).

Additionally, certain alkaloids block adrenergic receptors, particularly alpha-1 receptors, reducing vasoconstriction and improving blood flow. Others may act as diuretics, aiding in the excretion of excess sodium and water, thereby lowering blood volume and pressure (Johansen et al.,1991).

While many alkaloids demonstrate promising antihypertensive properties, their clinical use is limited by potential side effects such as sedation or depression (as seen with reserpine). However, ongoing research aims to develop safer alkaloid-based antihypertensives with improved efficacy and tolerability (Mancia et al.,2007).

8.4.1.2 Antiarrhythmic Action

Alkaloids possess notable antiarrhythmic properties by acting on cardiac ion channels and autonomic regulation. One prominent example is ajmaline, an indole alkaloid derived from Rauwolfia serpentina. Ajmaline functions as a class Ia antiarrhythmic agent, blocking sodium channels in cardiac muscle fibers. This slows the depolarization phase of the action potential, prolongs the refractory period, and stabilizes the cardiac rhythm, making it effective in treating atrial and ventricular arrhythmias (Tripathi, 2013; Nattel & Carlsson, 2006).

Another alkaloid, quinidine (from Cinchona bark), also acts as a class Ia agent with similar sodium channel-blocking effects. It also has anticholinergic properties and mild potassium channel-blocking effects, further contributing to rhythm control.

These alkaloids help suppress ectopic pacemaker activity, reduce conduction velocity, and prevent re-entry circuits responsible for arrhythmias. While effective, their use requires careful monitoring due to potential side effects, such as QT prolongation or proarrhythmic risks. Nonetheless, alkaloids continue to be valuable in antiarrhythmic pharmacotherapy (Chiale & Schmidberg, 2004).

8.4.1.3 Hypolipidemic Action

Alkaloids exhibit hypolipidaemic (lipid-lowering) activity through multiple mechanisms that influence lipid metabolism. One of the most studied is berberine, an isoquinoline alkaloid found in plants like Berberis species. Berberine reduces serum cholesterol and triglyceride levels by upregulating LDL receptor expression in the liver, enhancing the clearance of low-density lipoprotein (LDL) cholesterol from the bloodstream. It also activates AMP-activated protein kinase (AMPK), a key regulator of lipid and glucose metabolism, leading to decreased lipid synthesis (Kong et al., 2004).

Another example is piperine, found in black pepper (*Piper nigrum*), which has been shown to reduce total cholesterol and improve HDL/LDL ratios by modulating enzymes involved in lipid metabolism. Similarly, alkaloids like caffeine and theobromine may mildly influence lipid breakdown through stimulation of lipolysis (Vijayakumar et al., 2004).

These natural compounds hold promising therapeutic potential for managing hyperlipidemia and associated cardiovascular risks; however, further clinical studies are necessary to establish their efficacy and safety in long-term use.

8.4.1.4 Antithrombotic Action

Alkaloids demonstrate significant antithrombotic activity by interfering with platelet aggregation, coagulation pathways, and fibrinolysis. One key example is berberine, an isoquinoline alkaloid, which inhibits platelet activation and aggregation by modulating intracellular calcium levels and suppressing the synthesis of thromboxane A_2. This reduces clot formation and improves blood flow (Kang et al., 2002).

Similarly, alkaloids like tetrandrine and rutaecarpine exhibit calcium channel-blocking properties, which help inhibit platelet aggregation and vascular smooth muscle contraction. These actions contribute to reduced thrombus formation and better vascular protection (Huang et al., 2007).

Additionally, some alkaloids affect endothelial function by promoting the release of nitric oxide (NO), leading to vasodilation and reduced platelet adhesion. Certain indole alkaloids, such as harmine, have shown potential in modulating clotting factors and inhibiting thrombin activity.

Due to these mechanisms, alkaloids hold promise as natural antithrombotic agents. However, further research and clinical validation are needed to optimize their therapeutic use and ensure safety in long-term cardiovascular management (Moncada et al., 1991).

8.4.1.5 Cardioprotective Action

Alkaloids exhibit cardioprotective effects through antioxidant, anti-inflammatory, anti-apoptotic, and vasodilatory mechanisms. Berberine, for instance, enhances myocardial energy metabolism and reduces oxidative stress, preserving heart tissue during ischemic events. Ajmaline stabilizes cardiac rhythm by modulating ion channels, reducing arrhythmic risks. Alkaloids like vincamine improve cerebral and cardiac blood flow via vasodilation. Some also enhance endothelial function and inhibit lipid peroxidation, preventing atherosclerosis. These actions collectively protect the myocardium from ischemia-reperfusion injury, hypertrophy, and progression of heart failure (Zhang et al., 2020).

8.4.1.6 Diuretic Action

Certain alkaloids promote diuresis by acting on renal tubules or influencing hormonal control of fluid balance. Caffeine and theobromine, methylxanthine alkaloids, increase renal blood flow and inhibit sodium and water reabsorption in the nephron, leading to increased urine output. They also mildly inhibit phosphodiesterase, raising cyclic AMP levels, which enhances diuretic action. Additionally, alkaloids may interact with aldosterone or vasopressin pathways, altering fluid retention. This diuretic effect helps manage hypertension, edema, and congestive heart failure by reducing plasma volume and cardiac workload (Ritchie, 1991).

8.5 ALKALOID-CONTAINING MEDICINAL PLANT USED IN CVDS

Some of the medicinal plant and their active constituents have possible mechanisms of action in cardiovascular health (Table 8.2).

TABLE 8.2
Alkaloids Present in Medicinal Plants Having Therapeutic Application in CVDs

Alkaloidal Constituent	Structure	Pharmacological Action	Reference
Berberine	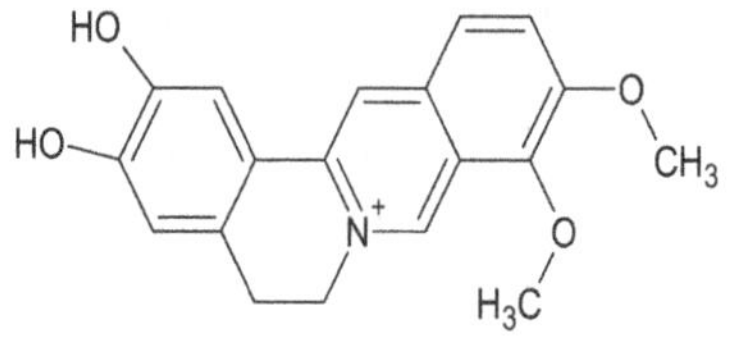	Anti diabetic, Antithrombotic	Utami et al. (2023)
Reserpine	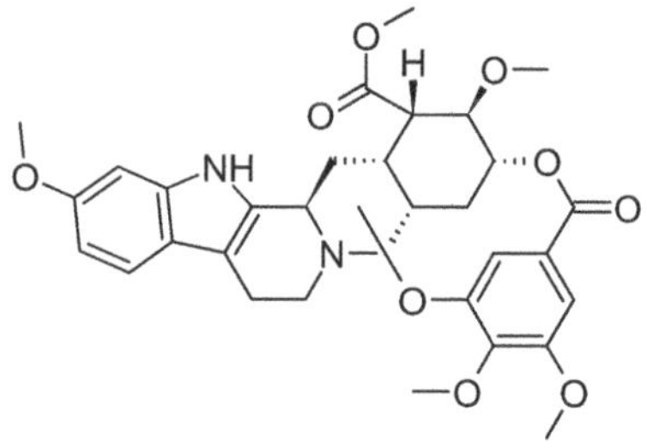	Antihypertensive	Mashour et al. (1998)
Ajmalcine	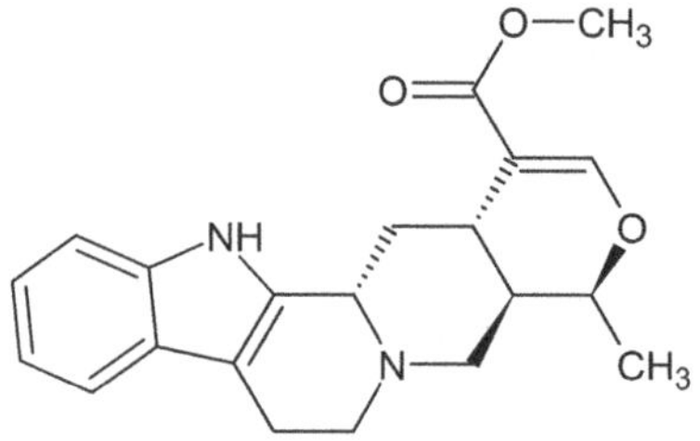	Antihypertensive	Kashyap et al. (2020)
Rescinnamine	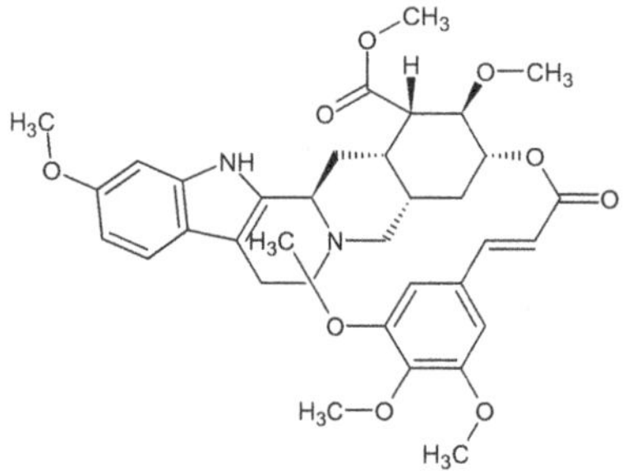	Antihypertensive	Winsor (1959)
Matrine	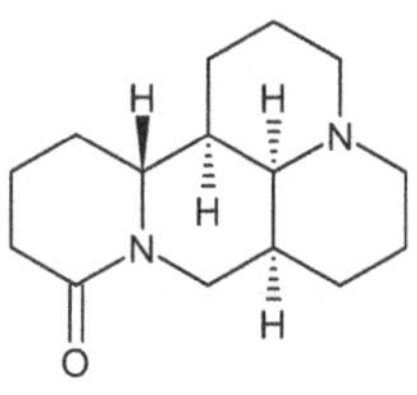	Cardioprotective	Shanjiang Chen et al. (2024)
Oxymatrine	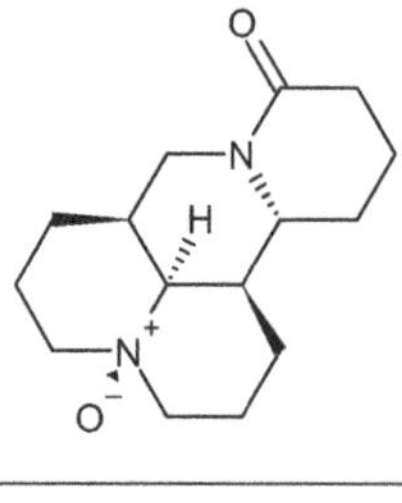	Cardioprotective	Shanjiang Chen et al. (2024)

(*Continued*)

TABLE 8.2 (Continued)
Alkaloids Present in Medicinal Plants Having Therapeutic Application in CVDs

Alkaloidal Constituent	Structure	Pharmacological Action	Reference
Deserpidine	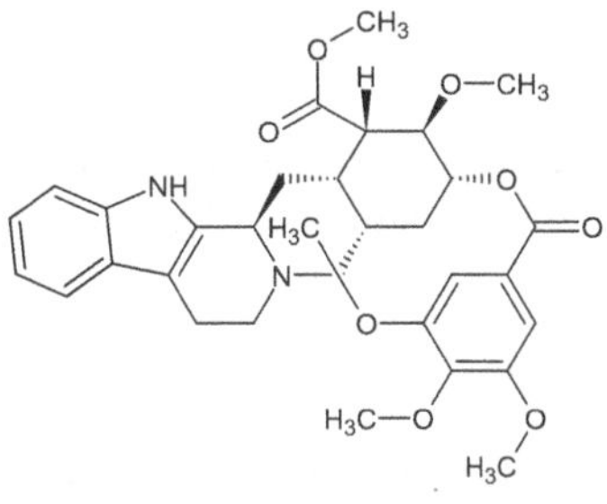	Antihypertensive	Winsor (1959)
Quinidine	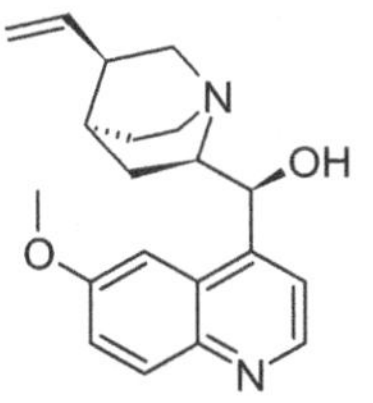	Antiarrhythmic	White (1992)
Protoveratrine	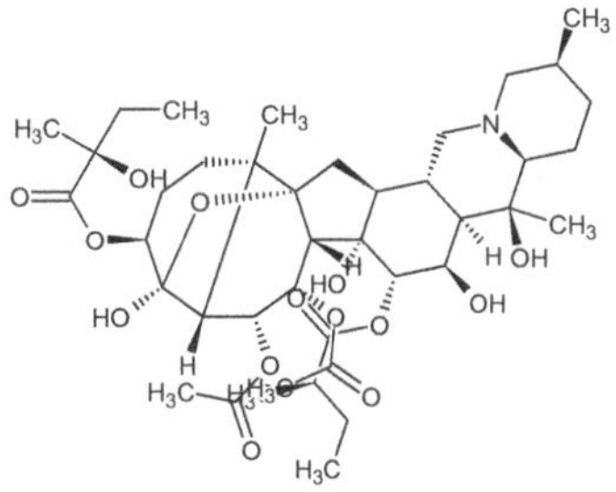	Antihypertensive	Currens et al. (1953)
Sophocarpine	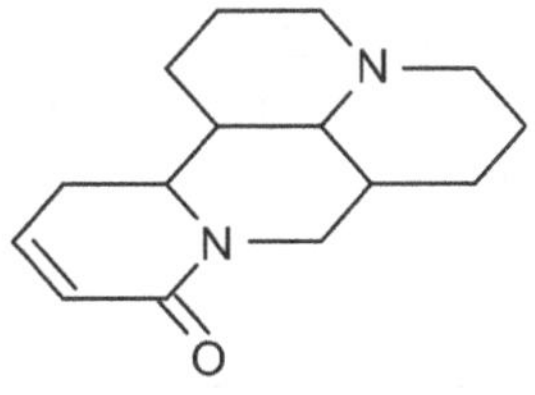	In Cardiovascular disease	Zhang et al. (2017)
Sanguiranine	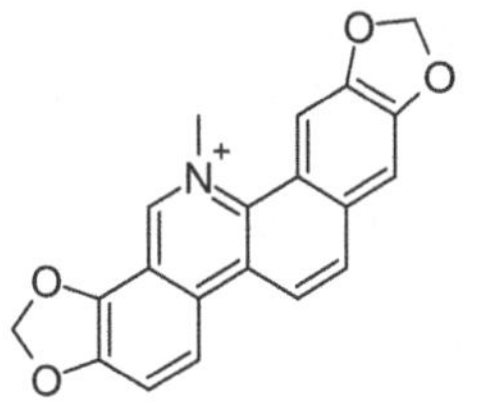	In Cardiovascular disease	Ahsan et al. (2007)
Caffeine	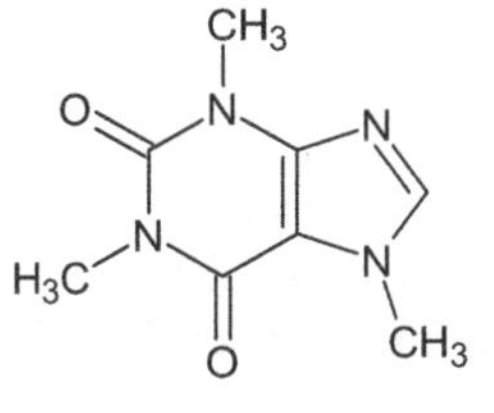	Diuretic	Chieng and Kistler (2022)

8.5.1 *Rauwolfia serpentina* (Reserpine, Ajmaline)

Reserpine lowers sympathetic tone, which lowers blood pressure, by depleting serotonin and catecholamines from central and peripheral nerve endings. Historically, antihypertensive medications have been used to treat hypertension (Mashour et al., 1998).

Ajmaline stabilizes cardiac electrical activity and prevents arrhythmias by blocking sodium channels. It is used to identify and treat ventricular arrhythmias and Brugada syndrome (Antzelevitch & Brugada, 2002).

8.5.2 *Camellia sinensis* (Theophylline)

Theophylline increases cyclic AMP levels by acting as a phosphodiesterase inhibitor, which causes vasodilation and enhances cardiac output. It is used to treat pulmonary hypertension and bronchodilate respiratory conditions linked to the heart (Barnes, 2003).

8.5.3 Ephedra Species (Ephedrine)

Ephedrine stimulates adrenergic receptors, leading to increased cardiac output and blood pressure. It is used in managing hypotension during anesthesia and as a cardiotonic agent (Morris, 1984).

8.5.4 *Claviceps purpurea* (Ergotamine, Dihydroergotamine)

Ergot alkaloids, including ergotamine and dihydroergotamine, alter vascular tone by acting as partial agonists or antagonists on serotonin and alpha-adrenergic receptors. It is used to treat vascular headaches that affect the heart as well as migraines (Tfelt-Hansen & Saxena, 2000).

8.5.5 *Sophora flavescens* (Matrine, Oxymatrine)

The quinolizidine alkaloids matrine and oxymatrine were among the main matrine-type alkaloids that were isolated from the medicinal plants of Sophora. Increasing research has shown that OMT exhibits a variety of positive pharmacological effects, including anti-inflammatory, anti-cancer, antidiabetic, and antiviral properties, as well as protective effects on the brain, liver, heart, lungs, vascular system, gastrointestinal tract, bones, kidneys, and skin (Duong et al., 2023).

8.5.6 *Sinomenium acutum* (Sinomenine)

Sinomenine exhibits endothelium-dependent vasorelaxation through the release of PGI2 and NO from the endothelium. An increase in endothelial cells' intracellular Ca^{++} concentration triggers NOS activation and PGI2 release (Nishida & Satoh, 2007).

8.5.7 *Nelumbo nucifera* (Neferine, Nuciferine)

The *Nelumbo nucifera* Gaertn seed embryo was used to extract the bisbenzyl isoquinoline alkaloid known as nerine. Its pharmacological actions are multifaceted, including anti-atherosclerotic, anti-hypertensive, anti-thrombotic, anti-diabetic, and vascular protective properties, as well as anti-arrhythmic effects (Liu et al., 2018).

The aporphine alkaloid nuciferine, mostly found in lotus plants (Nelumbo nucifera), is of interest due to its wide range of pharmacological actions, especially regarding CVDs. Due to its lipid-lowering, vasorelaxant, anti-inflammatory, and antioxidant properties, nuciferine may be considered as a medicinal agent for treating CVDs (Xie et al., 2024).

8.5.8 *Tabernaemontana divaricate* (Villocrine-A)

The indole alkaloid villocarine A is extracted and isolated from plants like *T. divaricate* and exhibits antioxidative, anti-inflammatory, and vaso-modulatory properties, which are crucial for the treatment of CVDs and are associated with its cardiovascular benefits (Lee & Park, 2020).

8.5.9 *Stephaniae tetrandrae* (Tetrandrine, Fangochinoline)

Two bisbenzylisoquinoline alkaloids, tetrandrine and fangchinoline, were isolated from the roots of *Stephania tetrandra* and other closely related species. It has a variety of pharmacological actions, making it a prospective therapeutic agent for the treatment of CVDs. In several CVD situations, tetrandrine may improve outcomes by addressing key processes, including inflammation, oxidative stress, and calcium dysregulation (Yu et al., 2004; Chen & Xu, 2021).

8.6 CONCLUSION

Alkaloids form a large category of naturally occurring compounds that carry significant therapeutic potential in the management of CVDs. Their pharmacological activities, including antioxidant, anti-inflammatory, vasodilatory, and lipid-regulating effects, address key pathophysiological processes, such as oxidative stress and endothelial dysfunction, inflammation, hypertension, and atherosclerosis. Through their multifaceted mechanisms, alkaloids such as reserpine, ajmaline, theophylline, ephedrine, ergotamine, dihydroergotamine, matrine, oxymatrine, sinomenine, neferine, nuciferine, villocarine A, tetrandrine, and fangchinoline have demonstrated therapeutic efficacy in various cardiovascular contexts. Specific alkaloids, including neferine, nuciferine, and villocarine A, have been reported to have preclinical and clinical evidence supporting their role in protecting cardiovascular health and arresting disease progression.

Alkaloids modulate the signaling pathways involved in CVD, enhance the bioavailability of nitric oxide, and suppress inflammatory mediators. However, further research will be necessary to translate their potential into clinically approved therapies, including ascertaining their safety, efficacy, and optimal formulations. Harnessing the pharmacological properties of alkaloids offers a natural and innovative approach to combat the global burden of CVDs.

COMPETING INTERESTS

The authors declare that they have no competing interest.

DECLARATION OF GENERATIVE AI AND AI-ASSISTED TECHNOLOGIES IN THE WRITING PROCESS

During the preparation of this work, the book chapter author(s) have not taken any assistance from AI, and the book chapter author(s) reviewed and edited the content as needed and take(s) full responsibility for the content of the publication.

REFERENCES

Ahsan, H., Reagan-Shaw, S., Breur, J., and Ahmad, N. (2007). Sanguinarine induces apoptosis of human pancreatic carcinoma AsPC-1 and BxPC-3 cells via modulations in Bcl-2 family proteins. *Cancer Letters*, *249*(2), 198–208.

Antzelevitch, C., and Brugada, P. (2002). Fever and Brugada syndrome. *Pacing and Clinical Electrophysiology*, *25*(10), 1537–1539.

Barnes, P. J. (2003). Theophylline: New perspectives for an old drug. *American Journal of Respiratory and Critical Care Medicine*, *167*(6), 813–818.

Chemat, F., Abert-Vian, M., & Cravotto, G. (2012). Green extraction of natural products: Concept and principles. *International Journal of Molecular Sciences*, *13*(7), 8615–8627. https://doi.org/10.3390/ijms13078615

Chen, H., and Xu,W. (2021). Antioxidant and anti-inflammatory effects of fangchinoline in vascular endothelial dysfunction. *Journal of Traditional and Complementary Medicine*, *11*(2), 134–141. https://doi.org/10.1016/j.jtcme.2021.06.004

Chen, S., Wu, S., & Lin, B. (2024). The potential therapeutic value of the natural plant compounds matrine and oxymatrine in cardiovascular diseases. *Frontiers in Cardiovascular Medicine, 11*. https://www.frontiersin.org/journals/cardiovascular-medicine/articles/10.3389/fcvm.2024.1417672

Chiale, P. A., andSchmidberg, J. (2004). Antiarrhythmic properties of quinidine: Clinical applications. *Cardiovascular Drugs and Therapy, 18*(5), 399–409.

Chieng, D., and Kistler, P. M. (2022). Coffee and tea on cardiovascular disease (CVD) prevention, *Trends in Cardiovascular Medicine, 32*(7):399–405. https://doi.org/10.1016/j.tcm.2021.08.004

Chisholm, H. (2015). *Encyclopedia Britannica, A Dictionary of Arts, Sciences, Literature and General Information*. Vol. 22. Sagwan Press. ISBN-13: 9781340141998.

Cordell, G. A. (1981). *Introduction to Alkaloids: A Biogenetic Approach*. Wiley-Inter Science.

Currens, J.H., Myers, G.S., and White, P.D. (1953). The use of protoveratrine in the treatment of hypertensive vascular disease,*American Heart Journal,46*(4):576–592. https://doi.org/10.1016/0002-8703(53)90068-4

Dhingra, R., and Vasan, R.S.(2017). Biomarkers in cardiovascular disease: Statistical assessment and section on key novel heart failure biomarkers. *Trends in Cardiovascular Medicine,27*(2), 123–133. https://doi.org/10.1016/j.tcm.2016.07.005

Duong, C. D., Vu, T. N., & Ngo, T. V. N. (2023). Applying a modified technology acceptance model to explain higher education students' usage of ChatGPT: A serial multiple mediation model with knowledge sharing as a moderator. *The International Journal of Management Education, 21*(3), 100883. https://doi.org/10.1016/j.ijme.2023.100883

Flora, G.D., and Nayak, M.K. (2019). A brief review of cardiovascular diseases, associated risk factors and current treatment regimes. *Current Pharmaceutical Design,25*, 4063–4084.

Ghodeshwar, G.K., Dube, A., and Khobragade, D. (2023). Impact of lifestyle modifications on cardiovascular health: A narrative review. *Cureus, 15*, e42616.

Harborne, J. B. (1998). *Phytochemical Methods: A Guide to Modern Techniques of Plant Analysis*. Springer. A comprehensive book detailing phytochemical techniques, including alkaloid extraction and isolation.

Huang, Y., Li, Y., Zhang, L., et al. (2007). Tetrandrine, a Ca^{2+} channel blocker, prevents thromboxane A_2-induced vascular contraction and platelet aggregation. *European Journal of Pharmacology, 566*(1–3), 153–159.https://doi.org/10.1016/j.ejphar.2007.02.021

Johansen, T. E. B., et al. (1991). Alpha-adrenoceptor antagonists in the treatment of hypertension. *Drugs, 41*(3), 365–376.

Kang, D. G., Oh, H., Sohn, E. J., and Lee, H. S. (2002). Anti-hypertensive and anti-thrombotic effects of berberine. *Biological & Pharmaceutical Bulletin, 25*(6), 788–792.

Kashyap, P., KalaiselvanV, KumarR, and KumarS. (2020). Ajmalicine and reserpine: indole alkaloids as multi-target directed ligands towards factors implicated in Alzheimer's disease. *Molecules,25*(7), 1609. https://doi.org/10.3390/molecules25071609

Kong, W. J., et al. (2004). Berberine is a novel cholesterol-lowering drug working through a unique mechanism distinct from statins. *Nature Medicine, 10*(12), 1344–1351. https://doi.org/10.1038/nm1135

Kurek, J. (2019). Introductory chapter: Alkaloids - Their importance in nature and for human life [Internet]. In *Alkaloids - Their Importance in Nature and Human Life*. IntechOpen. https://doi.org/10.5772/intechopen.85400

Lee, S. J., and Park, K. (2020). Alkaloids in vascular health: Focus on Villocarine A. *Phytotherapy Research, 34*(3), 523–530. https://doi.org/10.1002/ptr.6501

Liu, X., Wang, W., and Li, P. (2018). Antioxidant and anti-inflammatory effects of neferine in cardiovascular diseases. *Phytomedicine, 46*, 25–34. https://doi.org/10.1016/j.phymed.2018.07.010

Mancia, G., Grassi, G., andZanchetti, A. (2007). New-onset diabetes and antihypertensive drugs. *Journal of Hypertension, 25*(1), 3–12.

Manske, R.H.F. (1965), *The alkaloids: Chemistry and Physiology*.Academic Press Inc.

Mashour, N. H., Lin,G. I., and Frishman, W. H. (1998). Herbal medicine for the treatment of cardiovascular disease. *Archives of Internal Medicine, 158*(20), 2225–2234.

Moncada, S., Palmer, R. M., and Higgs, E. A. (1991). Nitric oxide: Physiology, pathophysiology, and pharmacology. *Pharmacological Reviews, 43*(2), 109–142.

Morris, J. (1984). Ephedrine in the treatment of circulatory collapse. *Journal of the Royal Society of Medicine, 77*(6), 477–478.

Nattel, S., and Carlsson, L. (2006). The use of ajmaline in the diagnosis of arrhythmias. *European Heart Journal, 27*(11), 1269–1270. https://doi.org/10.1093/eurheartj/ehi885

Nishida, S., and Satoh, H. (2007). Cardiovascular pharmacology of sinomenine: the mechanical and electropharmacological actions. *Drug Target Insights,2*, 97–104.

Pallazola, V.A., Davis, D.M., Whelton, S.P., Cardoso, R., Latina, J.M., Michos, E.D., Sarkar, S., Blumenthal, R.S., Arnett, D.K., and Stone, N.J. (2019). A clinician's guide to healthy eating for cardiovascular disease prevention. *Mayo Clinic Proceedings: Innovations, Quality & Outcomes*, *3*, 251–267.

Ritchie, J. M. (1991). Central nervous system stimulants II: The xanthines. In A. G. Gilman et al. (Eds.), *Goodman & Gilman's The Pharmacological Basis ofTherapeutics* (8th ed., pp. 592–607). Pergamon Press.

Santos, L. (2022). The impact of nutrition and lifestyle modification on health. *European Journal of Internal Medicine*,*97*, 18–25.

Sharma, R., and Sharma, P. (2020). Molecular imaging of cardiovascular diseases: The evolving paradigm. *Journal of Nuclear Cardiology*, *27*(6), 2051–2066.

Sneader, W. (2005). *Drug Discovery: A History*. Wiley.

Tfelt-Hansen P., and SaxenaP. R. (2000). Ergot alkaloids in the treatment of migraine: Pharmacological basis. *Journal of Neurology*, *247*(4), 95–108.

Tripathi, K. D. (2013). *Essentials of Medical Pharmacology*(7th ed.).Jaypee Brothers Medical Publishers.

Utami, A.R., Maksum, I.P., and Deawati, Y. (2023). Berberine and its study as an antidiabetic compound. *Biology (Basel)*,*12*(7), 973. https://doi.org/10.3390/biology12070973

Vasudeva, R., Mehta, H., Chan, W. C., Jiwani, S., Yarlagadda, S. G., Acharya, P., et al. (2024). Nationwide trends and outcomes of percutaneous coronary intervention for stable ischemic heart disease in end-stage kidney disease: A longitudinal study. *Annals of Translational Medicine*, *12*(6), 111.

Vijayakumar, R. S., Surya, D., and Nalini, N. (2004). Hypolipidemic effect of piperine in experimentally induced hyperlipidemic rats. *Phytotherapy Research*, *18*(7), 614–618. https://doi.org/10.1002/ptr.1507

Wainwright, M. (2013). *Miracle Cure: The Story of Penicillin and the Golden Age of Antibiotics*. Blackwell Science.

Wang, T. J., and Pencina, M. J. (2017). Cardiovascular disease risk assessment: Insights from CRP and BNP. *Nature Reviews Cardiology*, *14*, 195–205.

White, N. J. (1992). Antimalarial pharmacokinetics and treatment regimens. *British Journal of Clinical Pharmacology*, *34*(1), 1–10.

Winsor, R. (1959). Comparative effects of various Rauwolfia alkaloids in hypertension.*Diseases of the Chest*, *35*(4):415–421. https://doi.org/10.1378/chest.35.4.415

World Health Organization. (2017). Cardiovascular-Diseases @ www.who.int. Available from: https://www.who.int/health-topics/cardiovascular-diseases?utm_source=chatgpt.com

Xie, W., Chen, S., Wang, W., Qin, X., Kong, C., Wang, D. (2024). Nuciferine reduces vascular leakage and improves cardiac function in acute myocardial infarction by regulating the PI3K/AKT pathway. *Scientific Reports*,*14*(1), 7086. https://doi.org/10.1038/s41598-024-57595-w

Yu, X.C., Wu, S., Chen, C.F., Pang, K. T., and Wong, T. M. (2004). Antihypertensive and anti-arrhythmic effects of an extract of Radix Stephaniae Tetrandrae in the rat. *Journal of Pharmacy and Pharmacology*, *56*(1), 115–122. https://doi.org/10.1211/0022357022458

Zhang, J.-X., Feng, Y., Zhang, Y., Liu, Y., Li, S.-D., & Yang, M.-H. (2017). Hemorheology index changes in a rat acute blood stasis model: A systematic review and meta-analysis. *African Journal of Traditional, Complementary and Alternative Medicines*, *14*(4), 96–107. https://doi.org/10.21010/ajtcam.v14i4.12

Zhang, Y., et al. (2011). Berberine lowers blood glucose and improves insulin sensitivity. *Metabolism*, *59*(2), 285–292.

Zhang, Y., et al. (2020). Cardioprotective effect of berberine on ischemia/reperfusion injury via AMPK pathway. *International Journal of Molecular Sciences*, *21*(15), 5238.

9 Alkaloids and Neurological Disorders

Neuroprotective and Neurotherapeutic Potential

Kiran Chauhan, Abhishek Gupta, Deepti Singh Vashishth, and Piyush Bhalla

9.1 INTRODUCTION

Neurological disorders are a wide range of ailments that impact the central and peripheral nervous systems, involving nerves, the brain, and the spinal cord. They are mainly caused by infections, trauma, genetic mutations, neurodegenerative processes, and environmental factors (Dimopoulou et al., 2025). Alzheimer's disease (AD), epilepsy, Parkinson's disease (PD), stroke, and multiple sclerosis are some common neurological disorders that exhibit different clinical symptoms affecting daily life (Fan et al., 2025). Several of these conditions are associated with oxidative stress, protein misfolding, neuroinflammation, and disruptions in neurotransmission, which ultimately lead to motor impairments and progressive cognitive impairment (Fan et al., 2025). Currently, many synthetic drugs are in use to treat various neurological disorders by targeting specific neurotransmitter systems, enzymes, or receptors within the central nervous system (CNS). These drugs work to restore the balance of neurotransmitters, reduce neuroinflammation, or protect neuronal cells (such as MAO inhibitors used in PD). However, despite their benefits, these synthetic drugs also come with various side effects ranging from mild to severe, including cognitive impairments, motor disturbances like dyskinesia, hepatotoxicity, cardiovascular effects, sexual dysfunction, insomnia, and neurotoxicity. To combat these side effects, there is a rise in global demand for safer and more sustainable therapeutic options to treat natural disorders, and researchers have shifted their focus to drugs of natural origin (Kumar et al., 2024).

Alkaloids are one such group of naturally occurring compounds characterized by the presence of nitrogen in their structure. They are commonly found in various parts of plants as well as in fungi, bacteria, and even certain marine species. Based on their chemical structure, alkaloids are classified into different classes, including indole, pyrrolizidine, tropane, and isoquinoline alkaloids, each of which distinctly interacts with biological systems (El-Kashak et al., 2025). For many years, alkaloids have attracted considerable attention due to their broad pharmacological properties. They are reported to exhibit various biological properties, including anti-inflammatory, antimicrobial, anticancer, and neuroprotective properties. Some well-known alkaloids like morphine, quinine, and vinblastine have shown remarkable therapeutic potential with their ability to modulate enzymes, influence intracellular pathways, and bind to specific receptors, increasing their importance in natural drug discovery (Cabrera-Pardo et al., 2025; Gou et al., 2025; Wei et al., 2025). Huperzine A, berberine, and galantamine are some common alkaloids from various medicinal plants reported to have neuroprotective properties that inhibit neuroinflammatory pathways, lower oxidative stress, and alter neurotransmitter activity (Yadav et al., 2024).

DOI: 10.1201/9781003539179-9

The role of alkaloids in the treatment of neurodegenerative disorders is predicted to grow steadily. Recent advances in drug-delivery approaches, as well as strategies for bioavailability improvement, have also increased the therapeutic promise that these alkaloids can offer (Singh et al., 2025). This chapter, thus, explores the reported role of plant-derived alkaloids in the treatment of various neurodegenerative disorders, looking into the mechanism by which alkaloids act as well as discussing recent developments on bioavailability and novel drug-delivery systems that can impact the pharmacological actions of these alkaloids. Overall, the chapter addresses the growing role of natural alkaloids in drug discovery and neuropharmacology.

9.2 MECHANISMS OF NEUROPROTECTION BY ALKALOIDS

9.2.1 Antioxidant Activity

Oxidative stress can lead to neuronal injury and functional decline and, therefore, is one of the main factors behind various neurodegenerative disorders. Almost all plant-derived products, including alkaloids, exhibit antioxidant properties to some extent by scavenging reactive oxygen species (ROS), modulating the activity of antioxidant enzymes, and limiting lipid peroxidation (Selim et al., 2025).

Enhancement of the endogenous antioxidant system is one of the major mechanisms by which alkaloids mediate neuroprotection. Abundant reports have demonstrated that superoxide dismutase (SOD), hydrogen peroxidase (CAT), and glutathione peroxidase (GPx) are essential enzymes that prevent oxidative stress in neural cells (Nahar et al., 2025). The alkaloids such as berberine have been shown to activate the nuclear factor erythroid 2–related factor 2 (Nrf2) pathways, which induce the transcription of the antioxidant response element, thus, inducing cellular resistance to oxidative stress (Touhtouh et al., 2025).

Alkaloids can directly scavenge free radicals, thereby preventing neurotoxicity induced by ROS quite effectively. Indole alkaloids from *Vinca minor* and isoquinoline alkaloids from *Berberis* species significantly mitigate oxidative stress in various neuronal models by neutralizing radicals (Cabrera-Pardo et al., 2025). Alkaloids exhibit neuroprotective properties by modulating the function of mitochondria fairly significantly. Mitochondrial dysfunction manifests aberrantly in neurodegeneration, and tetrahydroberberine has been shown to preserve mitochondrial membrane potential, thereby reducing apoptotic signaling (Li et al., 2025).

9.2.2 Anti-Inflammatory Effects

Neuroinflammation plays a pivotal albeit somewhat enigmatic role in the pathogenesis of various neurodegenerative disorders, including AD and PD, quite frequently. Activated microglia and astrocytes mediate it primarily, resulting in excessive release of pro-inflammatory cytokines like TNF-α, IL-6, and IL-1β, which cause neuronal damage (ALNasser et al., 2025). Alkaloids have gained considerable attention lately owing mainly to their propensity to exert profoundly neuroprotective effects by heavily modulating various neuroinflammatory pathways.

Alkaloids reduce neuroinflammation essentially by inhibiting nuclear factor kappa B (NF-κB), a key transcription factor regulating the production of inflammatory cytokines. Berberine has been shown quite effectively to suppress NF-κB activation, thereby reducing expression of pro-inflammatory markers in various neurodegenerative conditions (Cabrera-Pardo et al., 2025). Tetrandrine, a bisbenzylisoquinoline alkaloid, has significantly demonstrated anti-inflammatory effects by inhibiting microglial activation and downregulating NF-κB signaling, as reported by Yang et al. (2025).

Alkaloids heavily influence the mitogen-activated protein kinase pathway, which mediates neuronal apoptosis triggered by inflammation. Research indicates huperzine A and galantamine tweak MAPK signaling significantly, preventing excessive inflammatory mediator release and bolstering neuronal survival (de Lima et al., 2025). Alkaloids can skew microglial polarization from pro-inflammatory M1 phenotype towards neuroprotective M2 phenotype, thereby contributing heavily to reduced inflammation and rampant neuronal repair (Levaya et al., 2025). Alkaloids like vincamine show pronounced vasodilatory effects and anti-inflammatory properties, thereby enhancing blood flow within cerebral regions noticeably. According to Touhtouh et al. (2025), neuroinflammation plays a significant role in aggravating secondary brain damage, particularly in patients who have suffered a stroke.

9.2.3 Neurotransmitter Modulation

Neurotransmitter modulation plays a critical role in maintaining cognitive function, motor control, and mood regulation within the human brain. Disruptions in neurotransmitter balance are frequently associated with neurodegenerative disorders such as AD and PD (Touati et al., 2024). Many alkaloids have shown the ability to regulate neurotransmitter activity. These alkaloids usually act by modulation of cholinergic pathways. Alkaloids like galantamine and huperzine A are reported to inhibit acetylcholinesterase (AChE) by preventing the breakdown of acetylcholine (Arumugam, 2024). Since deficiency of acetylcholine is associated with AD, this mechanism offers particular therapeutic value.

Besides modulating cholinergic pathways, many alkaloids also act by influencing dopaminergic signaling, which is necessary for motor control and emotional regulation. Alkaloids like harmine have exhibited monoamine oxidase inhibiting potential, which increases dopamine levels (Xie et al., 2025). Some iboga alkaloids are also reported to modulate dopamine signaling along with gamma-aminobutyric acid (GABA) neurotransmission, which helps in lowering neurotoxicity (Pratihar et al., 2025). Alkaloids are also reported to modulate serotonergic signaling pathways. Some alkaloids from *Sceletium tortuosum* are reported to increase serotonin levels by acting as serotonin reuptake inhibitors, promoting neurogenesis (Hanif & Gul, 2025). Alkaloids also work by affecting glutamatergic signaling, which is a major pathway behind excitotoxicity that can damage neurons. Some alkaloids from *Dendrobium nobile* are reported to aid in maintaining the calcium balance and prevent the release of excess glutamate, thereby reducing excitotoxicity (Touati et al., 2024).

9.2.4 Inhibition of Protein Aggregation

Protein aggregation manifests notably in various neurodegenerative disorders, including AD, PD, and Huntington's disease. Misfolded proteins, such as amyloid-beta and tau, accumulate in AD, and alpha-synuclein accumulates in PD, causing neuronal toxicity and cell death (GK & Priyadarshini, 2024). Recent research highlighted the potential of natural alkaloids in preventing protein aggregation, thereby offering some serious neuroprotective benefits.

Alkaloids directly interfere with amyloid fibril formation, thereby inhibiting protein aggregation through a primary, yet obscure, mechanism. Berberine binds tightly to misfolded proteins, thereby preventing oligomerization and subsequently reducing cytotoxicity quite significantly in neuronal cells. Caffeine and berberine together largely inhibit phosphorylated α-synuclein aggregation, heavily implicated in PD progression, quite effectively (Waheeb et al., 2025). Alkaloids provide neuroprotection by modulating protein degradation pathways under cellular stress. Certain alkaloids enhance autophagic clearance and proteasomal degradation of aggregated proteins, thereby reducing toxic accumulation within neurons. Harmine, a β-carboline alkaloid, has been shown to activate lysosomal degradation pathways, preventing tau protein aggregation in various models of AD. In addition, alkaloids influence chaperone proteins that are critical for protein folding and stabilization. Some plant-derived alkaloids increase the expression of heat shock proteins, which facilitate the refolding of misfolded proteins and inhibit aggregation (Arumugam, 2024).

9.2.5 Enhancement of Neurogenesis and Neuronal Survival Pathways

The generation of new neurons and their survival is important for brain plasticity and recovery from neurodegenerative disorders. Many alkaloids have been found to activate several molecular pathways that promote neurogenesis and support neuronal survival, including boosting neurotrophic factors (Numakawa & Odaka, 2021). Alkaloids notably enhance neurogenesis mainly by increasing brain-derived neurotrophic factor (BDNF) levels in various brain regions. BDNF plays a key role in neuron survival and growth, especially during differentiation. Studies have shown that alkaloids like huperzine A and berberine increase BDNF signaling, which in turn improves neuronal survival and synaptic plasticity significantly (de Lima et al., 2025). Harmine and other β-carboline alkaloids have also been shown to increase levels of BDNF, improving synaptic plasticity and helping with memory formation (Xie et al., 2025). Alkaloids also activate important intracellular pathways involved in neuronal survival, such as PI3K/Akt and MAPK/ERK signaling, downstream. These pathways regulate cell proliferation, guide differentiation, and help neurons resist apoptosis effectively. Alkaloids like galantamine and harmine are reported to stimulate these pathways, leading to enhanced neuroprotection and noticeable cognitive improvements (Chaurasiya et al., 2022).

9.2.6 Mitochondrial Protection and Energy Metabolism Enhancement

Mitochondria play an important role in the maintenance of various neuronal functions by regulating energy metabolism, redox balance, and apoptosis. The dysfunction in this process causes oxidative stress and neuronal cell death, leading to numerous neurodegenerative disorders, including AD and PD (de Lima et al., 2025). Alkaloids are reported to improve mitochondrial bioenergetics, adding to their neuroprotective behavior. Many alkaloids, like huperzine A and berberine, have been reported to show neuroprotective effects by increasing mitochondrial ATP production through oxidative phosphorylation (Wang et al., 2021). Certain alkaloids are also reported to activate the AMP-activated protein kinase (AMPK) pathway, which regulates cellular energy homeostasis and promotes mitochondrial biogenesis (Hani et al., 2024).

Reducing mitochondrial oxidative stress is another important way alkaloids protect neurons. Mitochondria are a major source of ROS, which damage cell components and trigger apoptosis in neurons under pathological conditions. Compounds like harmine and galantamine have been shown to boost mitochondrial antioxidant enzymes, including SOD and CAT, which help lower oxidative damage and support neuronal survival (Sun et al., 2016). Alkaloids also exhibit neuroprotective properties by preventing mitochondrial-mediated apoptosis. In many neurodegenerative disorders, increased mitochondrial permeability causes the release of various pro-apoptotic factors like cytochrome c, leading to caspase-dependent cell death. A few alkaloids, including berberine and vincamine, are reported to maintain neuronal integrity by modulating Bcl-2 family proteins and preventing caspase-3 activation (Ashok et al., 2017).

Lastly, alkaloids also improve mitochondrial dynamics and mitophagy, which are essential for keeping mitochondria healthy. When mitochondrial fission and fusion go wrong, damaged mitochondria accumulate and accelerate neurodegeneration over time. Alkaloids such as sinomenine and protopine are shown to enhance mitochondrial fusion and trigger mitophagy through AMPK/PGC1α signaling, which significantly improves mitochondrial function (Lu et al., 2024). They also support neuronal function by boosting glucose metabolism and energy production while directly protecting mitochondria. Geissoschizine methyl ether, for example, induces a Warburg-like effect in neurons, increasing glycolysis under oxidative stress conditions (Sun et al., 2016).

9.3 ALKALOIDS IN NEUROLOGICAL DISORDERS

Table 9.1 outlines several plant-based alkaloids, their natural sources, pharmacological profiles, effective dosages, and related references. The following subsections explore their specific roles in treating and managing specific neurological conditions.

TABLE 9.1
Some Key Alkaloids from Plants and Their Neuroprotective Effects

Alkaloid Name	Name of the Plant	Plant Part	Tested Activity Name	Dose Effective	Reference
Anatabine	*Nicotiana tabacum (Tobacco)*	Leaves	Anti-inflammatory, neuroprotective	1–2 mg in clinical studies	Gatto and Etcheverry (2014)
Berberine	*Berberis vulgaris*	Roots, Bark	Neuroprotective, anti-inflammatory	100–200 mg/kg in animal studies	Anbai et al. (2022)
	Berberis spp.	–	NMDA receptor antagonism, reduction of ROS, NF-κB, TNF-α, IL-1β	10–50 mg/kg (*in vivo*)	Song et al. (2020), Gawel et al. (2020), Sedaghat et al. (2017), Jiang et al. (2015), Cortes et al. (2015), Hussain et al. (2018)
Boldine	*Peumus boldus*	Leaves	Antioxidant, neuroprotective	50–200 mg/kg in animal studies	Sáez et al. (2024)
	Litsea cubeba	–	GABA receptor modulation, an increase in antioxidant enzyme levels	10–75 mg/kg (*in vivo*)	Moezi et al., (2019), de Souza et al. (2022), Akotkar et al. (2023)
Caffeine	*Coffea arabica*	Seeds (Beans)	Neuroprotective, cognitive enhancer	100–400 mg/day in humans	Newhouse et al. (2012)
	Coffea species (coffee beans)		Neuroprotective (AD, PD), antioxidant, cognitive enhancement, analgesic adjuvant	Effective at low–moderate doses (e.g., 250 μM or ~20–50 mg/kg in rodents)	Lima et al. (2025)
Chelerythrine	*Chelidonium majus*	Whole plant	Neuroprotective, anti-inflammatory	1–10 μM *in vitro*	Zhang et al. (2020)
Columbamine	*Coptis chinensis (Chinese Goldthread)*	Roots	AChE inhibitor	IC_{50} = 0.44 μM	Zhao et al. (2015)
Corynoline	*Corydalis bungeana*	Aerial parts	Neuroprotective, anti-inflammatory	10–50 μM *in vitro*	Yang et al. (2025)
Dendrobine	*Dendrobium nobile (Noble Dendrobium)*	Stems	Neuroprotective against Aβ-induced apoptosis	Not specified	Zhang et al. (2023)
Epiberberine	*Corydalis species*	Roots	AChE inhibitor	IC_{50} = 6.5 μM	Orhan et al. (2006)
Evodiamine	*Evodia rutaecarpa*	Fruits	Anti-inflammatory, neuroprotective	10–40 mg/kg in animal studies	Wang et al. (2024)
Galanthamine	*Galanthus nivalis*	Bulbs	Acetylcholinesterase inhibitor	Therapeutic dose: 16–24 mg/day in humans	Wilcock et al. (2000)
	Galanthus woronowii, Leucojum aestivum, Narcissus spp.	–	Cholinesterase Inhibition, Alzheimer's disease treatment	Clinically used dose varies	Heinrich and Teoh (2004), Heinrich (2010)

Harman, Harmol, Harmine, Harmalol, Harmaline	*Passiflora incarnata (Passion Flower)*	–	Serotonin reuptake inhibition, forced swimming test, open field test, spatial memory in water maze	50 µg/mL (*in vitro*), 45 mg/kg (*in vivo*), 30–300 mg/kg/day (*in vivo*)	Fiebich et al. (2011), Jawna-Zboińska et al. (2016)
Harmine	*Banisteriopsis caapi*	Stems	MAO-A inhibitor, neuroprotective	5–20 mg/kg in animal studies	Schwarz et al. (2003)
Huperzine A	*Huperzia serrata*	Whole plant	Acetylcholinesterase inhibitor	0.2–0.4 mg/day in humans	Rafii et al. (2011)
			Acetylcholinesterase inhibition, Alzheimer's disease treatment	0.2–0.8 mg/day	Ovais et al. (2018)
			Acetylcholinesterase inhibition, neuroprotective in AD, anti-apoptotic, BBB protective	Commonly effective at 0.1–0.5 mg/kg in rodents	Lima et al. (2025)
Isolongifolene	*Murraya koenigii (Curry Leaf Tree)*	Leaves	Antioxidant, neuroprotective against rotenone-induced mitochondrial dysfunction	EC_{50} = 77.34 µg/mL (DPPH assay); 5–45 mg/kg *in vivo*	Balakrishnan et al. (2021)
Isorhynchophylline	*Uncaria rhynchophylla*	Stems, Hooks	Neuroprotective, anti-inflammatory	10–50 mg/kg in animal studies	Li et al. (2021)
Jatrorrhizine	*Coptis chinensis (Chinese Goldthread)*	Roots	AChE inhibitor	IC_{50} = 0.80 µM	Zhao et al. (2015)
Kavalactones (Dihydromethysticin)	*Piper methysticum (Kava)*		Clinical trials, anxiety and depression studies	50–300 mg/day, 1.8 g tablet (3 times/day)	Malsch and Kieser (2001), Sarris et al. (2009)
Luteolin	*Lonicera japonica*	Flowers	Anti-inflammatory, neuroprotective	50–100 mg/kg in animal studies	Lee et al. (2023)
Magnoflorine	*Coptis chinensis (Chinese Goldthread)*	Rhizomes	Antioxidant, neuroprotective	Not specified	Zhang et al. (2022)
Mahanimbine	*Murraya koenigii (Curry Leaf Tree)*	Leaves	Neuroprotective against LPS-induced neuronal deficits, antioxidant, AChE inhibitor	IC_{50} = 0.03 mg/mL; 20 mg/kg *in vivo*	Azahan et al. (2019)
Mitragynine	*Mitragyna speciosa (Kratom)*		Forced swimming test, tail suspension test, apomorphine-induced climbing behavior, haloperidol-induced catalepsy, ketamine-induced social withdrawal	10–30 mg/kg (*in vivo*), 50–500 mg/kg (*in vivo*)	Idayu et al. (2011), Vijeepallam et al. (2016)

(*Continued*)

TABLE 9.1 (Continued)
Examples of Case Studies of Phytomedicine in Real-World Applications

Alkaloid Name	Name of the Plant	Plant Part	Tested Activity Name	Dose Effective	Reference
Morphine	*Papaver somniferum (Opium Poppy)*	Latex	Analgesic, neuroprotective	Not specified	Hu et al. (2007)
N-Methylasimilobine	*Nelumbo nucifera (Lotus)*	Leaves	AChE inhibitor	IC_{50} = 1.5 µg/mL	Yang et al. (2012)
Nicotine	*Nicotiana tabacum*	Leaves	Neuroprotective, cognitive enhancer	Variable; transdermal patches: 7–21 mg/day	Newhouse et al. (2012)
Nuciferine	*Nelumbo nucifera (Lotus)*	Leaves	Anti-inflammatory, antioxidant, neuroprotective	Not specified	Zhang et al. (2018)
Oxymatrine	*Sophora flavescens*	Roots	Neuroprotective, anti-inflammatory	25–100 mg/kg in animal studies	Yang et al. (2014)
Palmatine	*Coptis chinensis (Chinese Goldthread)*	Rhizomes	AChE inhibitor, antioxidant, neuroprotective	IC_{50} = 0.69 µM	Och et al. (2023)
Physostigmine	*Physostigma venenosum*	Seeds	Acetylcholinesterase inhibitor	0.5–2 mg IV in humans	Klein-Schwartz and Oderda (1987)
		–	Acetylcholinesterase inhibition, glaucoma, Alzheimer's disease	Not in clinical use due to short half-life and side effects	Mach et al. (2004), Howes and Perry (2011)
Piperlongumine	*Piper longum (Long Pepper)*	Fruits	Neuroprotective in Parkinson's disease model	120 mg/kg (oral in mice)	Liu et al. (2018)
Pseudo-berberine	*Corydalis species*	Roots	AChE inhibitor	IC_{50} = 4.3 µM	Grycová et al. (2007)
Pseudo-copsitine	*Corydalis species*	Roots	AChE inhibitor	IC_{50} = 4.5 µM	Zhang and Wei (2019)
Pseudo-corydaline	*Corydalis species*	Roots	AChE inhibitor	IC_{50} = 8.4 µM	Zhang et al. (2016)
Reserpine	*Rauvolfia serpentina*	Roots	Antipsychotic, antihypertensive	0.1–0.25 mg/day in humans	PVAMC (1982)
Rhynchophylline, Isorhynchophylline	*Uncaria rhynchophylla*	–	NMDA receptor inhibition, MAPK pathway modulation, reduction of pro-inflammatory cytokines, mitochondrial protection	1–30 µM (*in vitro*)	He et al. (2021), Hu et al. (2018), Geng et al., (2021), Mathien et al. (2021), Ho et al. (2014), Zhou et al. (2010), Ndagijimana et al. (2013)
Sanguinarine	*Sanguinaria canadensis (Bloodroot)*	Rhizomes	AChE inhibitor, neuroprotective	IC_{50} = 6.2 µM	Och et al. (2021)
Serpentine	*Catharanthus roseus (Madagascar Periwinkle)*	Whole plant	AChE inhibitor	IC_{50} = 0.775 µM	Pereira et al. (2010)

Sinomenine	*Sinomenium acutum*	Stems, Roots	Anti-inflammatory, neuroprotective	10–40 mg/kg in animal studies	Zhang et al. (2014)
		–	Inhibition of NLRP1 inflammasome, protection of hippocampal neurons from PTZ-induced damage	20–80 mg/kg (*in vivo*)	Li et al. (2023), Hong et al. (2022), Gao et al. (2018), Ramazi et al. (2020)
Solasodine	*Solanum nigrum (Black Nightshade)*	Berries	Neuroprotective, antioxidant	50 mg/kg *in vivo*	Kumar et al. (2014)
Stylopine	*Corydalis species*	Roots	AChE inhibitor	IC_{50} = 15.8 µM	Grycová et al. (2007)
Taspine	*Magnolia grandiflora (Southern Magnolia)*	Bark	AChE inhibitor	IC_{50} = 0.33 µM	Rollinger et al. (2006)
Tetrandrine	*Stephania tetrandra*	Roots	Neuroprotective, anti-inflammatory	30–60 mg/kg in animal studies	Yang et al. (2014)
		–	Reduction of P-gp expression, regulation of calcium channels	2 µM (*in vitro*)	Zhang et al. (2018), Chen et al. (2015)
Valepotriates	*Valeriana officinalis (Valerian)*	–	Forced swimming test, horizontal wire test	100–500 mg/kg	Hattesohl et al. (2008)

'-' indicates not known.

9.3.1 Alzheimer's Disease

Galanthamine, from the bulbs of *Galanthus nivalis* (snowdrop), is a reversible AChE inhibitor. It helps cholinergic function by preventing acetylcholine from breaking down, which is beneficial in managing AD. Studies have shown that taking galanthamine 16–24 mg/day can actually improve cognition in people with mild to moderate dementia (Wilcock et al., 2000). Huperzine A from *Huperzia serrata* works similarly, but it also has extra neuroprotective benefits. Importantly, it can cross the blood–brain barrier and binds strongly to AChE. Clinical trials suggest that huperzine A in oral doses of 0.2–0.4 mg/day is helpful for age-related memory problems (Rafii et al., 2011). Physostigmine from *Physostigma venenosum* seeds is mainly used intravenously for anticholinergic toxicity but has been looked at for AD too, at 0.5–2 mg doses (Klein-Schwartz & Oderda, 1987). There are several other alkaloids with neuroprotective properties. Dendrobine from *Dendrobium nobile* protects neurons from Aβ-induced apoptosis (Zhang et al., 2023). N-methylasimilobine from *Nelumbo nucifera* inhibits AChE with an IC_{50} of 1.5 μg/mL (Yang et al., 2012). Palmatine from *Coptis chinensis* not only inhibits AChE (IC_{50} = 0.69 μM) but also has antioxidant effects (Och et al., 2023). Mahanimbine from *Murraya koenigii* works on multiple targets, inhibits AChE (IC_{50} = 0.03 mg/mL), and has shown activity *in vivo* at 20 mg/kg (Azahan et al., 2019). Some other alkaloids like columbamine (IC_{50} = 0.44 μM), jatrorrhizine (0.80 μM), pseudo-copsitine (4.5 μM), stylopine (15.8 μM), and epiberberine (6.5 μM) also inhibit AChE to varying degrees, mostly coming from *Coptis chinensis* or *Corydalis* species (Zhao et al., 2015; Grycová et al., 2007; Orhan et al., 2006). Moreover, some alkaloids like pseudo-corydaline, pseudo-berberine, taspine (IC_{50} = 0.33 μM), serpentine (0.775 μM), and sanguinarine (6.2 μM) from plants such as *Magnolia grandiflora, Catharanthus roseus*, and *Sanguinaria canadensis* have also shown potential against AD (Rollinger et al., 2006; Pereira et al., 2010; Och et al., 2021).

9.3.2 Anti-Inflammatory Effects

Berberine, from *Berberis vulgaris*, is reported to exhibit anti-inflammatory and neuroprotective effects by modulating AMPK, NF-κB, and MAPK signaling pathways. In animal studies, doses of 100–200 mg/kg effectively reduced cognitive dysfunction and neuroinflammation (Anbai et al., 2022). Nuciferine, obtained from *Nelumbo nucifera* leaves, also demonstrates multiple neuroprotective actions, although specific dosages have not been reported (Zhang et al., 2018). Anatabine, from *Nicotiana tabacum*, has shown efficacy in inflammation-related neurological disorders at clinical doses of 1–2 mg (Gatto & Etcheverry, 2014). Caffeine, a well-known methylxanthine from *Coffea arabica*, protects neural tissues through adenosine receptor antagonism, with daily intake of 100–400 mg demonstrating benefits in models of cognitive decline (Newhouse et al., 2012).

Additional alkaloids with anti-inflammatory activity include tetrandrine (from *Stephania tetrandra*, 30–60 mg/kg), oxymatrine (from *Sophora flavescens*, 25–100 mg/kg), and sinomenine (from *Sinomenium acutum*, 10–40 mg/kg), which exert their effects via calcium channel modulation and NF-κB inhibition (Yang et al., 2014; Zhang et al., 2014). Flavonoid alkaloids have also shown promising anti-neuroinflammatory potential. Alkaloid luteolin from *Lonicera japonica* (50–100 mg/kg), corynoline from *Corydalis bungeana* (10–50 μM), chelerythrine from *Chelidonium majus* (1–10 μM), and evodiamine from *Evodia rutaecarpa* (10–40 mg/kg) have all been studied extensively for these effects (Lee et al., 2023; Yang et al., 2025; Zhang et al., 2020; Wang et al., 2024). Besides these, alkaloids isorhynchophylline and rhynchophylline, from *Uncaria rhynchophylla*, are reported to interact with NMDA receptors and affect both inflammatory and oxidative pathways. *In vivo* studies, at a dose of 10–50 mg/kg, and *in vitro* studies at a dose of 1–30 μM, have highlighted their potential in treating excitotoxicity and neuroinflammation (Li et al., 2021; He et al., 2021).

9.3.3 Parkinson's Disease

Nicotine, from the leaves of *Nicotiana tabacum*, works by acting on nicotinic acetylcholine receptors. Some studies suggest it can help with cognitive decline and neurodegeneration in Alzheimer's and Parkinson's models. Using transdermal patches at 7–21 mg/day is reported to improve things like working memory and attention (Newhouse et al., 2012).

Harmine, from *Banisteriopsis caapi*, is a monoamine oxidase-A (MAO-A) inhibitor. It also helps mitochondria work better and can increase neurotrophic factors. Preclinical animal studies have found it effective at 5–20 mg/kg doses (Schwarz et al., 2003). Piperlongumine, from *Piper longum*, protects neurons mainly because of its antioxidant nature. In Parkinson's mouse models, giving it orally at 120 mg/kg provided significant protection (Liu et al., 2018).

9.3.4 Antipsychotic Effects

Reserpine, obtained from *Rauvolfia serpentina*, has been used for a long time to treat high blood pressure and some psychiatric problems. It acts by blocking vesicular monoamine transporters (VMAT), which lowers neurotransmitters like dopamine and serotonin. Clinically, even at pretty low doses of 0.1–0.25 mg/day, reserpine has demonstrated significant efficacy, establishing it as an important compound in neuropharmacology (PVAMC, 1982).

9.3.5 Antioxidant Effects

Boldine, one of the main alkaloids from *Peumus boldus* leaves, has been extensively studied for its antioxidant and neuroprotective effects. In animals, at doses of 50–200 mg/kg, boldine has shown significant neuroprotective benefits (Sáez et al., 2024). Magnoflorine, from *Coptis chinensis*, also shows some neuroprotective and antioxidant activity, although comprehensive data regarding its precise potency is still limited (Zhang et al., 2022).

Solasodine, from *Solanum nigrum*, has been shown to demonstrate neuroprotective effects at 50 mg/kg *in vivo*. Moreover, isolongifolene, from *Murraya koenigii*, is reported to reduce mitochondrial problems and oxidative stress at doses of 5–45 mg/kg, with a DPPH IC_{50} of 77.34 μg/mL (Kumar et al., 2014; Balakrishnan et al., 2021).

9.3.6 Anxiety and Mood Disorders

Kavalactones, particularly dihydromethysticin from *Piper methysticum* (Kava), have demonstrated anxiolytic properties in clinical settings, with effective dosages ranging from 50–300 mg/day to up to 1.8 g/day in some trials (Malsch & Kieser, 2001; Sarris et al., 2009). Valepotriates from *Valeriana officinalis* (Valerian) were tested in behavioral assays like the forced swim test and found effective *in vivo* at 100–500 mg/kg (Hattesohl et al., 2008). Compounds harman, harmol, harmine, harmalol, and harmaline from *Passiflora incarnata* have demonstrated serotonin reuptake inhibition and enhanced cognition in rodent models. These compounds were active at both *in vitro* concentrations (~50 μg/mL) and *in vivo* doses (30–300 mg/kg/day) (Fiebich et al., 2011; Jawna-Zboińska et al., 2016). Mitragynine from *Mitragyna speciosa* (Kratom) also showed efficacy in several behavioral models of depression and psychosis at doses ranging from 10–500 mg/kg (Idayu et al., 2011; Vijeepallam et al., 2016).

Kavalactones, especially dihydromethysticin from *Piper methysticum* (Kava), have been shown to help with anxiety in clinical studies. The doses used are usually between 50 and 300 mg per day, though some studies went up to 1.8 g/day (Malsch & Kieser, 2001; Sarris et al., 2009). Valepotriates from *Valeriana officinalis* (Valerian) have been tested in animals using behavioral assays like the forced swim test and were found effective at 100–500 mg/kg (Hattesohl et al., 2008). A few compounds from *Passiflora incarnata*, namely harman, harmol, harmine, harmalol, and harmaline, have shown inhibition of serotonin reuptake and improved cognition in rodents. They were effective both *in vitro* at about 50 µg/mL and *in vivo* at 30–300 mg/kg/day (Fiebich et al., 2011; Jawna-Zboińska et al., 2016). Mitragynine from *Mitragyna speciosa* (Kratom) also showed some effects in behavioral models of depression and psychosis, with doses ranging from 10 to 500 mg/kg (Idayu et al., 2011; Vijeepallam et al., 2016).

9.4 FUTURE PROSPECTS AND CHALLENGES

Despite the considerable potential of plant-derived alkaloids in the treatment of various neurological disorders, several challenges and research gaps continue to limit their broader application. Firstly, it is challenging to get plant-based alkaloids to work for brain disorders, with the biggest problem being getting the drug to the right place. Many of these alkaloids are poorly absorbed orally because they break down too quickly, and they struggle to cross the blood–brain barrier, which makes them significantly less effective. To address this, researchers are looking at liposomes, nanoparticles, and nanoemulsions, which can basically act as tiny shields to protect the alkaloids, keep them in the system longer, and help them target the brain better, which should make them more useful for treating CNS disorders (Goswami et al., 2025). The second challenge is isolating the desired alkaloid form from the plant extract, ensuring purity and consistency. Various factors, including plant origin, environmental conditions, and harvest timing, greatly influence the chemical composition of these plant extracts. It becomes increasingly important to address this variability through a comprehensive study of the effects of geographical and seasonal variations on plant extracts, as well as to identify superior and desired genotypes. To fully harness the potential of the alkaloids and for quality control frameworks, it becomes more important to optimize the isolation process. Lastly, to enhance the neuroprotective behavior of the alkaloids, another promising strategy involves the design and synthesis of alkaloid analogs or derivatives. Structural modification of natural alkaloids can enhance selectivity, potency, metabolic stability, and brain penetration, while potentially reducing adverse effects. Structure–activity relationship (SAR) studies have already yielded analogs with improved efficacy in models of neurodegenerative and psychiatric disorders. Advances in computational chemistry and high-throughput screening already provide powerful tools to accelerate the identification of optimized alkaloid-based candidates with potential clinical applications.

9.5 CONCLUSION

Alkaloids demonstrate much potential for treating different neurological disorders, like Alzheimer's, Parkinson's, anxiety, depression, and also conditions with neuroinflammation. As discussed in the chapter, many alkaloids act in more than one way; some block AChE, some affect monoamine systems, and many of them have antioxidant or anti-inflammatory effects. They have shown effects in animal studies and even some clinical trials, clearly indicating their usefulness for neurotherapeutics. Even so, their full potential has not been released yet. There are numerous challenges, such as poor absorption when taken orally and insufficient solid clinical evidence, which make it challenging to use them more widely. On the plus side, some new approaches could help, like targeted drug delivery systems, making modified analogs, and high-throughput screening to find better candidates. To make alkaloids reliable for therapy, significantly more research is needed. Studies that look carefully at how they work and that are backed by strong, well-designed clinical trials are required. Such efforts are the only way to translate the neurotherapeutic potential of alkaloids into evidence-based interventions for neurological health.

COMPETING INTERESTS

The authors declare that they have no competing interest.

DECLARATION OF GENERATIVE AI AND AI-ASSISTED TECHNOLOGIES IN THE WRITING PROCESS

During the preparation of this work, the book chapter author(s) have not taken any assistance from AI, and the book chapter author(s) reviewed and edited the content as needed and take(s) full responsibility for the content of the publication.

REFERENCES

Akotkar, L., Aswar, U., Ganeshpurkar, A., Raj, R., and Pawar, A. (2023). An overview of chemistry, kinetics, toxicity and therapeutic potential of boldine in neurological disorders. *Neurochemical Research* 48(11), 3283–3295. doi:10.1007/s11064-023-03992-y

ALNasser, M. N., Alboraiy, G. M., and Alsowig, E. M. (2025). Cholinesterase inhibitors from plants and their potential in alzheimer's treatment: Systematic review. *Brain Sciences* 15(2), 215.

Anbai, S. M., Niazi, F., Siahposht-Khachaki, A., Sefidari, L., and Farzin, D. (2022). Neuroprotective effects of berberine after severe traumatic brain injury in male rats: The role of IL-1β and IL-10. *Journal Mazandaran University Medical Science* 31(204), 1–13.

Arumugam, P. (2024). Investigating the neuroprotective effects of alkaloids from Galanthus woronowii as a potential treatment for Alzheimer's disease. *Natural Product Research*. 1–2.

Ashok, B. S., Ajith, T. A., and Sivanesan, S. (2017). Hypoxia-inducible factors as neuroprotective agents in Alzheimer's disease. *Clinical and Experimental Pharmacology and Physiology* 44(3), 327–334.

Azahan, N. S. M., Mani, V., Ramasamy, K., Lim, S. M., Alhowail, A., Sajid, S., and Majeed, A. B. A. (2019). Neuroprotective potential of mahanimbine against lipopolysaccharides (LPS)-induced neuronal deficits on SK-N-SH cells and antioxidant potentials in ICR mice brain. *Journal of Pharmaceutical Research International* 31(2), 1–11.

Balakrishnan, R., Vijayraja, D., Mohankumar, T., Manimaran, D., Ganesan, P., Choi, D.-K., and Elangovan, N. (2021). Isolongifolene mitigates rotenone-induced dopamine depletion and motor deficits through anti-oxidative and anti-apoptotic effects in a rat model of Parkinson's disease. *Journal of Chemical Neuroanatomy*, 112, 101890.

Cabrera-Pardo, J. R., Pérez, R., Burgos, V., and Ortiz, L. (2025). 9-Hydroxyaristoquinolone: A new indole alkaloid with inhibitory activity of NF-κB in HMC-3 microglia cells. *International Journal of Molecular Sciences* 26(6), 2419.

Chaurasiya, N. D., Leon, F., and Tekwani, B. L. (2022). Natural product inhibitors of monoamine oxidases—Potential new drug leads for neuroprotection, neurological disorders, and neuroblastoma. *Molecules* 27(13), 4297.

Chen, Y. H., Xiao, X., Wang, C. C., Jiang, H. Y., Hong, Z., and Xu, G. X. (2015). Beneficial effect of tetrandrine on refractory epilepsy via suppressing P-glycoprotein. *The International Journal of Neuroscience* 125(9), 703–710. doi:10.3109/00207454.2014.966821

Cortes, N., Posada-Duque, R. A., Alvarez, R., Alzate, F., Berkov, S., Cardona-Gómez, G. P., et al. (2015). Neuroprotective activity and acetylcholinesterase inhibition of five Amaryllidaceae species: A comparative study. *Life Sciences* 122, 42–50. doi:10.1016/j.lfs.2014.12.011

de Lima, E. P., Laurindo, L. F., Catharin, V. C. S., Direito, R., Tanaka, M., Jasmin Santos German, I., Lamas, C. B., Guiguer, E. L., Araújo, A. C., Fiorini, A. M. R., and Barbalho, S. M. (2025). Polyphenols, alkaloids, and terpenoids against neurodegeneration: Evaluating the neuroprotective effects of phytocompounds through a comprehensive review of the current evidence. *Metabolites* 15(2), 124.

Dimopoulou, M., Kolonas, A., Stagos, D., and Gortzi, O. (2025). A review of the sustainability, chemical composition, bioactive compounds, antioxidant and antidiabetic activity, neuroprotective properties, and health benefits of microalgae. *MDPI Biomass* 5, 11.

El-Kashak, W. A., Essa, A. F., Abdelhameed, M. F., et al. (2025). *Unveiling the Neuroprotective Potential of Ipomoea Carnea Ethanol Extract via the Modulation of Tau and β-Secretase Pathways in AlCl3-Induced Memory Impairment*. Springer.

Fan, H., Wang, C., Liu, S., Li, Y., and Xu, Z. (2025). Therapeutic potential of natural products in stress granules: Underlying molecular mechanisms and future perspectives. *Current Pharmaceutical Analysis*, 21(2), 88–99.

Fiebich, B. L., Knörle, R., Appel, K., Kammler, T., and Weiss, G. (2011). Pharmacological studies in an herbal drug combination of St. John's Wort (Hypericum perforatum) and passion flower (Passiflora incarnata): In vitro and in vivo evidence of synergy between Hypericum and Passiflora in antidepressant pharmacological models. *Fitoterapia* 82, 474–480. doi: 10.1016/j.fitote.2010.12.006

Gao, B., Wu, Y., Yang, Y. J., Li, W. Z., Dong, K., Zhou, J., et al. (2018). Sinomenine exerts anticonvulsant profile and neuroprotective activity in pentylenetetrazole kindled rats: Involvement of inhibition of NLRP1 inflammasome. *Journal of Neuroinflammation* 15, 152. doi:10.1186/s12974-018-1199-0

Gatto, E. M., and Etcheverry, J. L. (2014). Effects of dietary supplementation with the Solanaceae plant alkaloid anatabine in patients with relapsing–remitting multiple sclerosis: A pilot study. *Clinical Pharmacology: Advances and Applications* 6, 1–5.

Gawel, K., Kukula-Koch, W., Nieoczym, D., Stepnik, K., van der Ent, W., Banono, N. S., et al. (2020). The influence of palmatine isolated from Berberis sibirica radix on pentylenetetrazole-induced seizures in zebrafish. *Cells* 9(5), 1233. doi: 10.3390/cells9051233

Geng, H., Chen, X. Q., and Wang, C. Z. (2021). Systematic elucidation of the pharmacological mechanisms of Rhynchophylline for treating epilepsy via network pharmacology. *BMC Complementary Medicine and Therapies* 21(1), 9. doi:10.1186/s12906-020-03178-x

Gk, P., and Priyadarshini, P. (2024). Crocetin inhibits Aβ aggregation in vitro and reduces ROS production and Aβ-mediated cellular toxicity in neuronal SHSY5Y cells. *Alzheimer's & Dementia* 20, e087283.

Goswami, U., Romero-Ben, E., and Cruz, J. S. (2025). Polymer-based nanocarriers to transport therapeutic biomacromolecules across the blood-brain barrier. *Acta Biomaterialia.*

Gou, G., Liu, L., Bao, W., Li, J., and Aisa, H. A. (2025). Dimeric Amide Alkaloid Enantiomers from Piper longum L. with anti-inflammatory and antidiabetic effects. *Journal of Agricultural and Food Chemistry* 73(11), 6955–6969.

Grycová, L., Dostál, J., & Marek, R. (2007). Quaternary protoberberine alkaloids. *Phytochemistry*, 68(2), 150–175.

Hani, U., Ghazwani, M., and Haider, N. (2024). *Bitter Yet Beneficial: The Dual Role of Dietary Alkaloids in Managing Diabetes and Enhancing Cognitive Function.* IUBMB Life.

Hanif, A., and Gul, A. (2025). Kanna (Sceletium tortuosum (L.) NE Br. Aizoaceae). In *Comprehensive Guide to Hallucinogenic Plants.* Taylor & Francis.

Hattesohl, M., Feistel, B., Sievers, H., Lehnfeld, R., Hegger, M., and Winterhoff, H. (2008). Extracts of Valeriana officinalis L. sl show anxiolytic and antidepressant effects but neither sedative nor myorelaxant properties. *Phytomedicine* 15, 2–15. doi: 10.1016/j.phymed.2007.11.027

He, L. Y., Hu, M. B., Li, R. L., Zhao, R., Fan, L. H., He, L., et al. (2021). Natural medicines for the treatment of epilepsy: Bioactive components, pharmacology and mechanism. *Frontiers in Pharmacology* 12, 604040. doi:10.3389/fphar.2021.604040

Heinrich, M. (2010). *Galanthamine from Galanthus and other Amaryllidaceae– Chemistry and Biology Based on Traditional Use, the Alkaloids: Chemistry and Biology.* Elsevier Academic Press, 157–165. doi: 10.1016/S1099-4831(10)06804-5

Heinrich, M., and Teoh, H. L. (2004). Galanthamine from snowdrop: the development of a modern drug against Alzheimer's disease from local Caucasian knowledge. *Journal of Ethnopharmacology* 92(2–3), 147–162. doi: 10.1016/j.jep.2004.02.012

Ho, T. Y., Tang, N. Y., Hsiang, C. Y., and Hsieh, C. L. (2014). Uncaria rhynchophylla and rhynchophylline improved kainic acid-induced epileptic seizures via IL-1β and brain-derived neurotrophic factor. *Phytomedicine* 21(6), 893–900. doi:10.1016/j.phymed.2014.01.011

Hong, H. X., Lu, X., Lu, Q., Huang, C., and Cui, Z. M. (2022). Potential therapeutic effects and pharmacological evidence of sinomenine in central nervous system disorders. *Frontiers in Pharmacology* 13, 1015035. doi:10.3389/fphar.2022.1015035

Howes, M.-J. R., and Perry, E. (2011). The role of phytochemicals in the treatment and prevention of dementia. *Drugs & Aging* 28(6), 439–468. doi: 10.2165/11591310-000000000-00000

Hu, S., Sheng, W. S., Lokensgard, J. R., and Peterson, P. K. (2007). Morphine inhibits microglia-induced neuron death. *The Journal of Immunology* 179(2), 1198–1206.

Hu, S. Q., Mak, S. H., Zuo, X. L., Li, H. T., Wang, Y. Q., and Han, Y. F. (2018). Neuroprotection against MPP+-Induced cytotoxicity through the activation of PI3-K/Akt/GSK3β/MEF2D signaling pathway by rhynchophylline, the major tetracyclic oxindole alkaloid isolated from Uncaria rhynchophylla. *Frontiers in Pharmacology* 9, 768. doi:10.3389/fphar.2018.00768

Hussain, G., Rasul, A., Anwar, H., Aziz, N., Razzaq, A., Wei, W., et al. (2018). Role of plant derived alkaloids and their mechanism in neurodegenerative disorders. *International Journal of Biological Sciences* 14 (3), 341–357. doi:10.7150/ijbs.23247

Idayu, N. F., Hidayat, M. T., Moklas, M., Sharida, F., Raudzah, A. N., Shamima, A., et al. (2011). Antidepressant-like effect of mitragynine isolated from Mitragyna speciosa Korth in mice model of depression. *Phytomedicine* 18, 402–407. doi: 10.1016/j.phymed.2010.08.011

Jawna-Zboińska, K., Blecharz-Klin, K., Joniec-Maciejak, I., Wawer, A., Pyrzanowska, J., Piechal, A., et al. (2016). Passiflora incarnata L. Improves spatial memory, reduces stress, and affects neurotransmission in rats. *Phytotherapy Research* 30, 781–789. doi: 10.1002/ptr.5578

Jiang, W., Wei, W., Gaertig, M. A., Li, S., and Li, X. J. (2015). Therapeutic effect of berberine on Huntington's disease transgenic mouse model. *PLoS One* 10, e0134142. doi:10.1371/journal.pone.0134142

Klein-Schwartz, T. W., and Oderda, G. M. (1987). Physostigmine: A review of therapeutic uses and toxicity. *Journal of Toxicology. Clinical Toxicology* 25(1–2):1–22. doi:10.3109/15563658708992658

Kumar, B. N. P., Srivastav, Y., Kumar, A., and Sirisha, S. (2024). Natural remedies for neurodegenerative disorders: Insights from pharmacognosy - A comprehensive review. *International Journal of Pharmaceutical and Life Sciences* 10(10), 187–199.

Kumar, M., Sharma, S., Vasudeva, N., and Joshi, A. (2014). Solasodine protects rat brain against ischemia/reperfusion injury-induced oxidative stress and apoptosis. *Iranian Journal of Basic Medical Sciences* 17(12), 1001–1007.

Lee, H. L., Kim, J. M., Go, M. J., et al. (2023). Protective Effect of lonicera japonica on $PM_{2.5}$-induced pulmonary damage in BALB/c mice via the TGF-β and NF-κB pathway. *Antioxidants* 12(4), 968. doi:10.3390/antiox12040968

Levaya, Y., Atazhanova, G., Gabe, V., and Badekova, K. (2025). A review of botany, phytochemistry, and biological activities of eight salvia species widespread in Kazakhstan. *Molecules*, 30(5), 1142.

Li, D., Zhong, Z. F., Ko, C. N., Tian, T. T., and Yang, C. (2023). From mundane to classic: Sinomenine as a multi-therapeutic agent. *British Journal of Pharmacology* 22. doi:10.1111/bph.16267

Li, X., Yu, H., Liu, R., Miao, J., Lv, J., Yang, S., and Zhu, Y. (2025). Activation of the Nrf2 pathway by tetrahydroberberine suppresses ferroptosis and enhances functional recovery following spinal cord injury. *Molecular Neurobiology*, 62, 8439–8456.

Li, X., Yuan, Y., Xu, H., et al. (2021). Isorhynchophylline ameliorates cerebral ischemia/reperfusion injury by inhibiting NLRP3 inflammasome activation and modulating gut microbiota. *Frontiers in Pharmacology* 12, 574793. doi:10.3389/fphar.2021.574793

Liu, J., Liu, W., Lu, Y., Tian, H., Duan, C., Lu, L., … and Zheng, P. (2018). Piperlongumine restores the balance of autophagy and apoptosis by increasing BCL2 phosphorylation in rotenone-induced Parkinson disease models. *Autophagy* 14(5), 845–861.

Lu, L., Pang, M., Chen, T., Hu, Y., and Chen, L. (2024). Protopine exerts neuroprotective effects on neonatal hypoxic-ischemic brain damage in rats via activation of the AMPK/PGC1α pathway. *Drug Design, Development and Therapy*, 18, 4975–4992.

Mach, M., Grubbs, R. D., Price, W. A., Paton, S. J., and Lucot, J. B. (2004). Behavioral changes after acetylcholinesterase inhibition with physostigmine in mice. *Pharmacology, Biochemistry, and Behavior* 79(3), 533–540. doi: 10.1016/j.pbb.2004.09.009

Malsch, U., and Kieser, M. (2001). Efficacy of kava-kava in the treatment of non-psychotic anxiety, following pretreatment with benzodiazepines. *Psychopharmacology* 157, 277–283. doi: 10.1007/s002130100792

Mathien, S., Tesnière, C., Meloche, S., and Gottesman, M. (2021). Regulation of mitogen-activated protein kinase signaling pathways by the ubiquitin-proteasome system and its pharmacological potential. *Pharmacological Reviews* 73(4), 263–296. doi:10.1124/pharmrev.120.000170

Moezi, L., Yahosseini, S., Jamshizadeh, A., and Pirsalami, F. (2019). Acute boldine treatment induces anticonvulsant effects in mice through its antioxidant activity. *Drug Research* 69(4), 227–233. doi:10.1055/a-0659-2478

Nahar, L., Habibi, E., Guo, M., and Charoensup, R. (2025). Natural products in neurodegenerative diseases: Recent advances and future outlook. *Frontiers in Pharmacology* 16, 1529194.

Ndagijimana, A., Wang, X. M., Pan, G. X., Zhang, F., Feng, H., and Olaleye, O. (2013). A review on indole alkaloids isolated from Uncaria rhynchophylla and their pharmacological studies. *Fitoterapia* 86, 35–47. doi:10.1016/j.fitote.2013.01.018

Newhouse, P. A., Potter, A., and Singh, A. (2012). Nicotine treatment of mild cognitive impairment: A 6-month double-blind pilot clinical trial. *Neurology* 78(2), 91–101. doi:10.1212/WNL.0b013e31823efcbb

Numakawa, T., and Odaka, H. (2021). Brain-derived neurotrophic factor signaling in the pathophysiology of Alzheimer's disease: Beneficial effects of flavonoids for neuroprotection. *International Journal of Molecular Sciences* 22(11), 5719.

Och, A., Komsta, Ł., Kołodziej, P., and Kocki, J. (2021). Application of HPLC-DAD for in vitro investigation of acetylcholinesterase inhibition activity of selected isoquinoline alkaloids from Sanguinaria canadensis extracts. *Molecules* 26(1), 176.

Och, A., Komsta, Ł., Kołodziej, P., and Kocki, J. (2023). In vitro and in silico of cholinesterases inhibition and in vitro and in vivo anti-melanoma activity investigations of extracts obtained from selected Berberis species. *Molecules* 28(3), 1234.

Orhan, G., Orhan, I., and Sener, B. (2006). Recent developments in natural and synthetic drug research for Alzheimer's disease. *Letters in Drug Design & Discovery* 3(4), 268–274.

Ovais, M., Zia, N., Ahmad, I., Khalil, A. T., Raza, A., Ayaz, M., et al. (2018). Phyto-therapeutic and nanomedicinal approach to cure alzheimer disease: Present status and future opportunities. *Frontiers in Aging Neuroscience* 10, 284. doi: 10.3389/fnagi.2018.00284

Participating Veterans Administration Medical Centers. (1982). Low doses vs standard dose of reserpine: A randomized, double-blind, multiclinic trial in patients taking chlorthalidone. *JAMA* 248(19), 2471–2477. doi:10.1001/jama.1982.03330190035027

Pereira, D. M., Ferreres, F., Oliveira, J. M. A., Gaspar, L., Faria, J., Valentão, P., Sottomayor, M., and Andrade, P. B. (2010). Pharmacological effects of Catharanthus roseus root alkaloids in acetylcholinesterase inhibition and cholinergic neurotransmission. *Phytomedicine* 17(8–9), 646–652. 10.1016/j.phymed.2010.01.012

Pratihar, S., Gupta, A., Bhattacharya, T., et al. (2025). Development of bioisosteric iboga-alkaloids as antinociceptive and anxiolytic agents with neuroprotective effects. bioRxiv.

Rafii, M. S., Walsh, S., Little, J. T., et al. (2011). A phase II trial of huperzine A in mild to moderate Alzheimer disease. *Neurology* 76(16), 1389–1397. doi:10.1212/WNL.0b013e318216eb7b

Ramazi, S., Fahanik-Babaei, J., Mohamadi-Zarch, S. M., Tashakori-Miyanroudi, M., Nourabadi, D., Nazari-Serenjeh, M., et al. (2020). Neuroprotective and anticonvulsant effects of sinomenine in kainate rat model of temporal lobe epilepsy: Involvement of oxidative stress, inflammation and pyroptosis. *Journal of Chemical Neuroanatomy* 108, 101800. doi:10.1016/j.jchemneu.2020.101800

Rollinger, J. M., Schuster, D., Baier, E., Ellmerer, E. P., Langer, T., Stuppner, H., and Prast, H. (2006). Taspine: Bioactivity-guided isolation and molecular ligand-target insight of a potent acetylcholinesterase inhibitor from Magnolia x soulangiana. *Journal of Natural Products*, 69(9), 1341–1346.

Sáez, J. C., Burrell, J. C., Cahill, C. M., Cullen, D. K., and Cardozo, C. P. (2024). Pharmacology of boldine: Summary of the field and update on recent advances. *Frontiers in Pharmacology* 15, 11427365.

Sarris, J., Kavanagh, D., Byrne, G., Bone, K., Adams, J., and Deed, G. (2009). The kava anxiety depression spectrum study (KADSS): A randomized, placebo-controlled crossover trial using an aqueous extract of Piper methysticum. *Psychopharmacology* 205, 399–407. doi: 10.1007/s00213-009-1549-9

Schwarz, M. J., Houghton, P. J., Rose, S., Jenner, P., and Lees, A. D. (2003). Activities of extract and constituents of Banisteriopsis caapi relevant to Parkinsonism. *Pharmacology Biochemistry and Behavior* 75(3), 627–633.

Sedaghat, R., Taab, Y., Kiasalari, Z., Afshin-Majd, S., Baluchnejadmojarad, T., and Roghani, M. (2017). Berberine ameliorates intrahippocampal kainate-induced status epilepticus and consequent epileptogenic process in the rat: underlying mechanisms. *Biomedicine & Pharmacotherapy* 87, 200–208. doi:10.1016/j.biopha.2016.12.109

Selim, Y. A., Kelany, N. A., and Elshafey, N. (2025). A promising strategy for treating Alzheimer's disease with Origanum majorana extraction. *The Egyptian Journal of Chemistry* 98, 813–821.

Singh, A. A., Katiyar, S., and Song, M. (2025). Phytochemicals targeting BDNF signaling for treating neurological disorders. *Brain Sciences* 15(3), 252.

Song, D. Y., Hao, J. Y., and Fan, D. M. (2020). Biological properties and clinical applications of berberine. *Frontiers in Medicine* 14(5), 564–582. doi:10.1007/s11684-019-0724-6

Sun, J., Ren, X., Qi, W., Yuan, D., and Simpkins, J. W. (2016). Geissoschizine methyl ether protects against oxidative stress-mediated cytotoxicity in neurons through the 'Neuronal Warburg Effect'. *Journal of Ethnopharmacology* 187, 249–258.

Touati, I., Boulaamane, Y., Britel, M. R., et al. (2024). Dendrobium nobile alkaloids modulate calcium dysregulation and neuroinflammation in Alzheimer's disease: A bioinformatic analysis. *Chinese Medicine Research* 12, 100495.

Touhtouh, J., Laghmari, M., Chraa, F., Benali, T., Ghanam, J., El Shazly, M., Goh, K.W., Bouyahya, A., Lee, L.H., Aanniz, T. and Hammani, K., 2025. Celtis genus (Cannabaceae): a comprehensive review of the ethnomedicinal use, food value, phytochemistry, biological activities, valuable compounds, and insight into mechanisms of action. *Journal of Agriculture and Food Research* 21, 101797.

Vijeepallam, K., Pandy, V., Kunasegaran, T., Murugan, D. D., and Naidu, M. (2016). Mitragyna speciosa leaf extract exhibits antipsychotic-like effect with the potential to alleviate positive and negative symptoms of psychosis in mice. *Frontiers in Pharmacology* 7, 464. doi: 10.3389/fphar.2016.00464

Waheeb, T. S., Abdulkader, M. A., and Ghareeb, D. A. (2025). *Neuroprotective Efficacy of Berberine and Caffeine Against Rotenone-Induced Neuroinflammatory and Oxidative Disturbances Associated with Parkinson's Disease*. Springer.

Wang, E., Chen, F., Xiao, J., et al. (2021). Neuroprotective phytochemicals in experimental ischemic stroke: Mechanisms and potential clinical applications. *Oxidative Medicine and Cellular Longevity* 2021(1), 6687386.

Wang, Z., Lai, C., Shen, B., et al. (2024). Effects of evodiamine on behavior and hippocampal neurons through inhibition of angiotensin-converting enzyme and modulation of the renin-angiotensin pathway in a mouse model of post-traumatic stress disorder. *Nutrients* 16(12), 1957. doi:10.3390/nu16121957

Wei, X., Jiao, P., Jiang, Z., Wang, C., Liu, Q., and Li, Y. (2025). Study on differential metabolite active ingredients in maize roots based on network pharmacology and metabolomics analysis. *Journal of Agricultural and Food Chemistry* 73(11), 6646–6658.

Wilcock, G. K., Lilienfeld, S., and Gaens, E. (2000). Galantamine International-1 study group. Efficacy and safety of galantamine in patients with mild to moderate Alzheimer's disease: Multicentre randomised controlled trial. *BMJ* 321(7274), 1445–1449. doi: 10.1136/bmj.321.7274.1445

Xie, Z., Liu, W., Dang, R., et al. (2025). Effects and mechanisms of harmine on ameliorating ethanol-induced memory impairment. *Journal of Ethnopharmacology* 337, 118789.

Yadav, V. K., Kumar, M., and Sharma, R. (2024). *Role of Plant-Derived Alkaloids in Neurodegenerative Disease Treatment*. Pharmedico Publishers.

Yang, Q., Chen, Q., Zhang, K. B., Liu, Y., and Zheng, J. C. (2025). Sinomenine alleviates neuroinflammation in chronic cerebral hypoperfusion by promoting M2 microglial polarization and inhibiting neuronal pyroptosis. *Acta Neuropathologica Communications* 13(1), 48.

Yang, Z.-D., Zhang, X., Du, J., Ma, Z.-J., Guo, F., Li, S., and Yao, X.-J. (2012). An aporphine alkaloid from Nelumbo nucifera as an acetylcholinesterase inhibitor and the primary investigation for structure–activity correlations. *Natural Product Research* 26(5), 387–392. 10.1080/14786419.2010.487188

Yang, Z. Y., Zhang, Y., Liu, C., et al. (2014). Antinociceptive effect of tetrandrine on LPS-induced hyperalgesia by inhibiting IKKβ phosphorylation. *PLoS One* 9(4), e94584. doi:10.1371/journal.pone.0094584

Zhang, C. C., Kong, Y. L., Zhang, M. S., Wu, Q., and Shi, J. S. (2023). Two new alkaloids from Dendrobium nobile Lindl. exhibited neuroprotective activity, and dendrobine alleviated $A\beta_{1-42}$-induced apoptosis by inhibiting CDK5 activation in PC12 cells. *Drug Development Research* 84(2), 262–274. doi:10.1002/ddr.22030

Zhang, H., and Wei, J. (2019). Current anti-Alzheimer's disease effect of natural products and their derivatives. *Journal of Integrative Neuroscience* 18(3), 229–234.

Zhang, J., Zhang, C., Zhang, J., et al. (2020). In vitro anti-inflammatory potency of sanguinarine and chelerythrine. *eFood* 1(6), 392–398. doi:10.2991/efood.k.210118.001

Zhang, L., Liu, Y., Lu, X., Zhang, Y., and Li, G. (2022). Magnoflorine attenuates cerebral ischemia-induced neuronal injury by inhibiting oxidative stress and autophagy. *Oxidative Medicine and Cellular Longevity* 2022, 9482485.

Zhang, Q., Chen, C., Wang, F.-Q., Li, C.-H., Zhang, Q.-H., and Zheng, Y.-T. (2016). Simultaneous screening and analysis of antiplatelet aggregation active alkaloids from Rhizoma Corydalis. *Pharmaceutical Biology* 54(12), 3022–3028.

Zhang, X., Wang, Y., Dong, H., et al. (2014). Antinociceptive effects of sinomenine in a rat model of neuropathic pain. *Scientific Reports* 4, 7270. doi:10.1038/srep07270

Zhang, Y., Liu, Y., Wang, T., Li, Q., and Wang, Z. (2018a). Nuciferine inhibits proinflammatory cytokines via the PPARs in LPS-induced RAW264.7 cells. *Molecules* 23(10), 2723.

Zhang, Z. X., Liu, T., Yu, M., Li, K. D., and Li, W. H. (2018b). The plant alkaloid tetrandrine inhibits metastasis via autophagy-dependent Wnt/β-catenin and metastatic tumor antigen 1 signaling in human liver cancer cells. *Journal of Experimental & Clinical Cancer Research* 37, 7. doi:10.1186/s13046-018-0678-6

Zhao, F., Li, R., Xiao, S., and Sun, Y. (2015). Alkaloids from Coptis chinensis exhibit acetylcholinesterase inhibitory activity. *Journal of Enzyme Inhibition and Medicinal Chemistry* 30(6), 865–868.

10 Alkaloids in Pain Management
Analgesic and Antinociceptive Properties

Anant and Shrejal Tiwari

10.1 INTRODUCTION

Pain is the defense system that produces alarm signals to protect the human body. The International Association for the Study of Pain (IASP) defines pain as an unpleasant emotional and sensory experience related to actual or potential tissue damage. Nociceptive signals are modulated throughout the CNS, with pain traveling dynamically from the periphery to the brain. However, the defense mechanism of chronic pain is still considered one of the most important health problems in human survival and a challenge for modern medicine. The antinociceptive properties of analgesics refer to their ability to reduce pain perception by interfering with the transmission of nociceptive signals and interacting with the physiological pathways involved in pain. Currently, chronic pain management primarily relies on anticonvulsants, tricyclic antidepressants, and opioids, with gabapentin and pregabalin as the first choices. However, their effectiveness is often unsatisfactory. Moreover, adverse reactions are observed at the neurological level when gabapentin and pregabalin are used; these negative reactions were found to be positively correlated with the dosages of the drug (Chen et al., 2014). While the exact analgesic mechanism of TCAs, possibly linked to the descending noradrenergic inhibitory pathway, is unclear, they are commonly used off-label for chronic pain (Bair et al., 2003; Obata, 2017). In most studies, TCAs offer only slight symptom relief and limited effectiveness (Moore et al., 2012). Furthermore, unpleasant anticholinergic side effects (such as sleepiness, tiredness, dry mouth, and constipation) frequently limit their use (Till and Jones, 2019). Accordingly, TCAs are not always effective in practice; instead, they seem to favor a subset of patients with co-occurring psychological disorders (Till and Jones, 2019).

Opioids are effective for severe and chronic pain and are found among the strong alkaloids offered by nature. Today, powerful analgesic and antinociceptive agents are primarily natural alkaloids derived from plants, such as morphine, codeine, and oxycodone. These compounds act through multiple mechanisms, such as inhibiting inflammatory mediators, modulating neurotransmitters, and interacting with opioid, adrenergic, and serotonergic receptors. Opioid alkaloids, such as morphine and codeine, are well-known analgesics derived from the *Papaver somniferum*, or opium poppy. These alkaloids bind to opioid receptors in the central nervous system (CNS), blocking pain and producing analgesia. Another important group includes tropane alkaloids, like scopolamine and atropine from *Atropa belladonna*, which act by inhibiting muscarinic receptors and reducing pain perception. Others, such as berberine from *Berberis* species and ajmalicine from *Rauvolfia serpentina*, influence the adrenergic system, leading to pain relief. Additionally, alkaloids like quinine from *Cinchona* species and mitragynine from *Mitragyna speciosa* (kratom) exhibit antinociceptive activity through modulation of calcium and potassium ion channels, which affects nerve signal transmission but with less respiratory depression compared to morphine, making it a subject of interest for pain management. Furthermore, capsaicin (from chili peppers, *Capsicum* species) exerts analgesic effects by interfering with ion channels involved in pain signal transmission. Capsaicin

DOI: 10.1201/9781003539179-10

desensitizes TRPV1 receptors, reducing the sensation of pain over time. Due to their powerful analgesic properties, plant alkaloids have been widely used in traditional and modern medicine for pain relief. However, some of them, particularly opioid alkaloids, can cause tolerance, dependence, and side effects, necessitating careful usage and further research to develop safer, non-addictive alternatives.

10.2 ALKALOIDS: STRUCTURE, CLASSIFICATION, AND MECHANISMS OF ACTION

Alkaloids, as their name suggests, are found in about 20% of plants and mainly serve to defend the plant. Alkaloids are naturally occurring nitrogen-containing compounds produced by plants, fungi, and some animals as secondary metabolites. They possess various pharmacological activities, including analgesic, antipyretic, anti-inflammatory, and neuroprotective effects. Certain plants, including *Woodfordia fruticosa, Adhatoda vasica, Chenopodium ambrosioides, Viburnum cotinifolium, Vitex negundo, Peganum harmala*, and *Broussonetia papyrifera*, have been scientifically studied for their antinociceptive properties and the presence of alkaloids useful in pain management. Alkaloids are classified based on their chemical structure and biosynthetic origin. They typically contain a heterocyclic ring with nitrogen, which influences their physiological activity. The alkaloid must have a mechanism to deal with inflammation associated with injury, which can increase the sensitivity of nociceptors, leading to enhanced pain stimuli. Furthermore, it should have analgesic properties to relieve pain by interacting with pain pathways and receptors in the body, which involve binding with opioid receptors (morphine), TRPV1 channels (evodiamine), or NMDA receptors (chelerythrine), inhibiting inflammatory mediators, and modulating ion channels. Along with these properties, the alkaloid must have minimal side effects, dependence, and development of tolerance. It should be stable and have a synergistic effect when used in combination with other analgesics (Olofinsan et al., 2023).

10.2.1 Opioid Alkaloids

Natural alkaloids extracted from the resin of opium poppy come under this category. They include morphine, codeine, thebaine, and other semi-synthetic alkaloids produced using these, such as oxycodone. Hydromorphone is an opioid analgesic for severe pain, but it can cause life-threatening respiratory depression at high doses or when combined with other depressants. Morphine can also cause constipation, nausea, vomiting, and sedation. Long-term use can lead to tolerance, dependence, and addiction. It is crucial to use these under strict medical supervision to mitigate risks and ensure safe and effective pain management.

10.2.1.1 Morphine

An alkaloid harvested from the *P. somniferum* is fundamentally used as an analgesic for chronic pain management, including nociceptive pain. Like any other opioid, they have morphine and have an affinity to d-, k-, and m-opioid receptors. It strongly activates mu opioid receptors in both the CNS and PNS to control pain. This binding activates $G_{i/o}$ proteins and inhibits the adenylate cyclase enzyme, leading to a decrease in the levels of cyclic adenosine monophosphate (cAMP). Morphine inhibits the release of pain neurotransmitters, such as substance P and glutamate, in the spinal cord. It activates mu-opioid receptors (MOPr), which open GIRK channels through Gα proteins, allowing potassium (K+) to leave the neuron while reducing calcium (Ca^{2+}) influx via Gβγ proteins. This hyperpolarization prevents neurons from transmitting pain signals. Morphine furthermore stimulates the release of endogenous opioids (endorphins), inhibits the release of neurotransmitter GABA, and enhances pain inhibition. Treatment of morphine for a longer duration may cause activation of TLR4 receptors on astrocytes and microglia in the PAG, inducing several signaling cascades.

Long-term use of opioids for chronic pain can lead to side effects that diminish their ability to reduce GABAergic interneuron activity. This includes increased levels of cAMP and heightened calcium channel production, preventing opioid binding at MOPr from suppressing GABA release. Additionally, prolonged morphine treatment can upregulate Toll-like receptor 4 (TLR4) on astrocytes and microglia in the periaqueductal gray (PAG), increasing cytokine release and shifting MOPr coupling from Gi/o to Gs, resulting in morphine tolerance.

10.2.1.2 Codeine and Thebaine

The opioids are extracted from the poppy. Codeine first gets metabolized into morphine in the liver by enzyme CYP2D6 and then helps in pain management by regulating the activation of the descending pain pathway, hyperpolarization of neurons, and inhibition of pain signal transmission by reducing neurotransmitters like substance P and glutamate in the spinal cord, which can enhance endorphin release. This process also involves binding to MOPr in the CNS, decreasing cAMP production. The signal transmission helps in pain management. Codeine enhances pain relief when combined with other analgesics like acetaminophen or ibuprofen.

Whereas thebaine, also known as paramorphine or methyl enol ether, is the main alkaloid extracted from the Persian poppy (*Papaver bracteatum*). Due to its toxic effect in high doses, which can lead to convulsions similar to strychnine poisoning, its use is inhibited, although it is not used directly as a therapeutic, but is the main ingredient for the synthesis of semi-synthetic opioids.

10.2.2 Non-Opioid Alkaloids

Alkaloids that fall under this class include chemical components that are extracted from plant material other than *Opium* (*P. somniferum*). There are no alkaloids extracted from other plants, such as coffee, tea, or cortex, that can serve the purpose. Along with showcasing analgesic and nociceptive properties, these alkaloids, when used long-term, have low chances for the development of tolerance and dependence, which is common in the case of opioid analgesics. Several common alkaloids used for neuropathic pain management are mentioned further.

10.2.2.1 Arecoline

Pyridine alkaloids are primarily found in the arecaece, prominently in areca catechu, which is popularly used as a masticatory product in the Indian subcontinent. Studies conducted on the areca family point out the pharmacological effect of arecoline, which includes anti-inflammatory, analgesic, and antioxidant effects, which are related to nociceptive pain in the human body. The therapeutic effect of arecoline, as discussed by Chen et al. (2024), is attributed to its agonist action at cholinergic receptors, particularly its binding to M2 muscarinic receptors in the central nervous system stimulating/inhibiting the synaptic transmission in ACC, mainly found in postsynaptic response at the physiological level, while at the behavioral level, studies indicate analgesic effects of arecoline. In studies using a rat model, arecoline inhibits carbachol-stimulated phosphoinositide turnover, leading to depolarization, opening of voltage-dependent Ca^{2+} channels, and neurotransmitter release. This suggests that arecoline acts as a partial agonist, which may explain its analgesic effects. However, it also has side effects, including stimulation of the parasympathetic nervous system and potential carcinogenicity (Chen et al., 2024; Kozlakidis et al., 2022).

10.2.2.2 Aloperine

Aloperine is a piperidine alkaloid derived from *Sophora alopecuroides*, a plant used in traditional medicine across Central and Western Asia. It is primarily utilized for managing inflammation and chronic pain. Research indicates that aloperine inhibits the TLR4-dependent inflammatory pathway in macrophages, showcasing its anti-inflammatory effects. It suppresses TNF-α, IL-17A, and IL-6 levels, reduces COX-2 and iNOS activity, and subsequently lowers nitric oxide (NO) generation. Aloperine also decreases the expression of KLF2, resulting in lower MCP-1, VCAM-1, IL-6, and

E-selectin, which may contribute to its potential anti-atherosclerotic properties. Additionally, an 80 mg/kg intraperitoneal dose effectively alleviates neuropathic pain linked to chronic constriction injury by inhibiting NF-κB and reducing IL-1β and IL-6 levels. While aloperine is widely used as an analgesic, further studies are needed to explore its potential for treating inflammation and neurodegenerative diseases (NDDs).

10.2.2.3 Berberine

An isoquinoline alkaloid from plants like *Hydrastis canadensis* and *Cortex phellodendri* is used in traditional Chinese medicine to treat inflammation. It has a therapeutic action against fever and protects myocardial activity (Yun et al., 2021). In addition, Yang et al. (2020) suggest it shows the ability to activate opioid receptors. Berberine targets various channels to relieve pain. The effective pain relief is achieved when the AMPK signaling pathway and expression of NRF2 are activated (Chakrabarti et al., 2018) while inhibiting HMGB1/TLR4/NF-κB, which helps in improving pain by reducing oxidative stress. Inhibiting the nuclear factor kappa B pathway reduces the expression of TNF-α, IL-1β, and IL-6, which are pro-inflammatory cytokines playing the lead role in inflammation and pain. Berberine has shown effective results in improving pain in models of chronic inflammation, diabetic neuropathy, cancer, and chemotherapy-induced pain. Additionally, it regulates dopamine and NMDA receptors. Recent investigation shows that berberine regulates dopamine receptor DR1 in the cerebral cortex of rats (Yoo et al., 2006). Furthermore, by regulating NMDA receptors and regulating activation of PKC-mediated phosphorylation of μ-opioid receptor, which helps in reducing morphine addiction and tolerance (Huang et al., 2019). Berberine's properties reduce oxidative stress and protect neurons from damage, alleviating neuropathic pain. Furthermore, it blocks calcium channels, reducing calcium ion influx and relaxing vascular smooth muscle cells, which helps in managing pain associated with vascular conditions. These combined actions make berberine a promising multifaceted agent in pain management.

10.2.2.4 Corynoxeine

Corynoxeine is a naturally occurring alkaloid from the psychoactive plant Kratom (*M. speciosa*) (Papsun et al., 2019). It serves as a potentially safer alternative to traditional opioids, offering effective pain management while reducing the risks of respiratory depression, constipation, and tolerance development. Corynoxeine's basic function is as an anti-inflammatory, used to manage inflammation and related pain in conditions like arthritis and autoimmune diseases. The inflammation is reduced by inhibiting the proliferation of vascular cells and blocking the ERK1/2 phosphorylation pathway and calcium channel, while activating the potassium channel, which promotes vascular dilation, muscle relaxation, and improves blood flow. Furthermore, it acts as a neuron protection by enhancing autophagy to clear the neurotoxic protein (Alpha-synuclein).

10.2.2.5 Chelerythrine

Chelerythrine, extracted from *Chelidonium majus*, is a benzophenanthridine alkaloid known for its various biological activities, including its role in pain management. It primarily works by inhibiting protein kinase C activity. Chelerythrine has been found to inhibit NR2B-containing NMDA receptors. This inhibition is independent of PKC activity and involves direct interaction with the NMDA receptor2. Furthermore, it is also associated with inflammation, chronic pain, and the downregulation of the secretion of TNF-α and matrix metalloproteinase-9 (MMP-9). This has been observed in various pain models, including the tail flick and formalin tests.

10.2.2.6 Conolidine

An indole alkaloid from the bark of the tropical shrub *Tabernaemontana divaricata* is used in traditional Chinese, Ayurvedic, and Thai medicine (Edinoff et al., 2021). Conolidine, a component of crepe jasmine, can serve as a natural alternative to traditional opioids. It interacts with the chemokine receptor (ACK3), regulating the availability of opioid peptides like dynorphin and enkephalin.

It prevents the binding of endogenous opioid peptides to traditional opioids. As a modulator of ACKR3, conolidine increases the availability of endogenous opioid peptides, inducing analgesia without triggering classical G protein signaling. This receptor is found in high concentrations in important opiate-related brain centers. This alkaloid blocks the $Ca_v2.2$ channels (Arita et al., 2019). This section highlights the importance of calcium channels in pain perception through neuron depolarization. Although research on this alkaloid is limited, early studies suggest its potential as a nociceptive analgesic. The first synthesis of the compound occurred in 2011 by Tarselli et al. (Tarselli et al. 2011), who demonstrated its effectiveness in reducing chemically induced and inflammation-related pain in an in vivo model. Conolidine injections were found to alleviate both primary and secondary pain responses, indicating their promise for managing acute and persistent pain.

10.2.2.7 Capsaicin

Capsaicin, a naturally occurring alkaloid, is the main ingredient of hot peppers, especially cayenne peppers. Similar to other naturally occurring alkaloids, capsaicin has been used for pain management for ages. Capsaicin is available in various forms, including creams, gels, lotions, and patches, with varying concentrations. When applied to the skin, capsaicin causes a burning sensation by activating TRPV1 receptors (Singh et al., 2018). The repeated application can further desensitize neurons and can help in chronic pain management. Capsaicin induces a loss of reactivity in neurons through calcium-induced desensitization and inactivation of capsaicin channels. According to Reyes-Esquido et al., capsaicin selectively activates TRPV1 (Transient Receptor Potential Vanilloid 1), a calcium-permeable cation channel in nociceptor terminals. This activation leads to a reduced response to unpleasant stimuli, making capsaicin useful in treating various pain types, including neuropathic pain, osteoarthritis, and musculoskeletal pain. A clinical trial by Bernstein et al. demonstrated its effectiveness: 32 patients aged 54–90 applied a 0.075% capsaicin cream, resulting in a 77% clinical improvement compared to 31% in the placebo group. Notably, 54% of capsaicin patients experienced a 40% reduction in pain, while only 6% in the placebo group reported similar relief after six weeks.

10.2.2.8 Evodiamine

A naturally occurring indole alkaloid found in the fruit of *Evodia rutaecarpa* has potential antinociceptive properties. Evodiamine activated the TRVP1 Channels, which desensitize the receptors and reduce pain perception. Moreover, evodiamine works as an anti-inflammatory and has been shown to reduce peripheral hypersensitivity in rat models through activation of TRPV1 Channels. Evodiamine inhibits the production of pro-inflammatory cytokines such as interleukin-1β (IL-1β), interleukin-6 (IL-6), and tumor necrosis factor-alpha (TNF-α). This modulation helps reduce inflammation and associated pain. It has been shown to be effective as a neuroprotective by maintaining mitochondrial antioxidant functions and reducing oxidative stress in dorsal root ganglia (DRG) tissues.

10.2.2.9 Lappaconitine (LA)

A traditional Chinese–Japanese medicine extracted from the roots of the aconitum spp. This versatile compound offers various pharmacological properties, including analgesic, anti-inflammatory, anti-tumor, anti-arrhythmic, and antiepileptic activities. When used as an analgesic, it offers a spectrum effect with minimum side effects, unlike opioids (like addiction and respiratory depression), making it a potential analgesic for long-term usage. LA hydrobromide is equally effective as pethidine and is seven times more potent than aminopyrine. Furthermore, this analgesic allows doctors to adjust the dose according to the severity of pain correctly. Ruan et al. suggest that there are divergent views on its mechanism as it downregulates the expression of the P2X3 receptor in DRG neurons, which are responsible for neuron sensitivity in neuropathic pain. Xiao et al. demonstrated that LA alleviates burn-induced pain by upregulating P2X3 receptors in the PAG region and activating the endogenous analgesic system. Additionally, LA enhances the expression of the gene to release dynorphin A (DynA) and activate the k-opioid receptor on postsynaptic nerves for analgesic effect

(Wands et al.), where voltage-gated sodium channels are another key target of LA, which inhibits NAV1.7 channels. However, further validation of this pathway is required. Research by Zhao et al. in 2011 indicates that lappaconitine, administered intraperitoneally at 4 mg/kg, inhibits nociception from chronic constriction injury (CCI). The study found increases in interstitial ATP (IATP) and Ia,b-meATP, as well as upregulation of P2X3 receptor expression in rat DRG neurons, suggesting that lappaconitine has antinociceptive effects through regulation of purinergic signaling in the DRG.

10.2.2.10 Liquiritigenin

Liquiritigenin found in liquorice (*Glycyrrhizae radix*) roots has demonstrated excellent potential, particularly for nociceptive chronic pain. Where nociceptive pain arises from noxious stimuli, liquiritigenin exerts its analgesic effects through multiple mechanisms. Liquiritigenin inhibits post-inflammatory mediators, such as prostaglandins and cytokines (TNF-α and IL-6), reducing receptor sensitization and providing pain relief. A 2014 study on a rat model of CCI-induced neuropathic pain suggested it acts as a TRPM3 blocker and has an anti-hyperalgesic effect. It also inhibits calcium signals in DRG neurons and affects serotonin and norepinephrine uptake, influencing pain modulation at the spinal cord. Additionally, liquiritigenin modulates NF-kB activity, which is involved in the inflammatory response. While traditionally used for pain relief, its mechanisms are still being studied.

10.2.2.11 Mitragynine

Mitragynine is an indole alkaloid derived from the psychoactive plant Kratom (*M. speciosa*), traditionally used to treat muscle pain, diarrhea, and hypertension (Papsun et al., 2019; Ilmie et al., 2015; Meireles et al., 2019). It acts as a partial agonist at MOPr, providing pain relief while minimizing adverse side effects typical of traditional opioids (Alford et al., 2025; Ellis et al., 2020). Additionally, mitragynine antagonizes kappa (κ)-opioid receptors (KOR), which helps stabilize mood and reduce dysphoria (Huang et al., 2019). It also acts as a partial agonist at 5-HT2A and 5-HT1A receptors, enhancing serotonin signaling for anxiety and mood improvement. Moreover, it has agonistic effects on alpha-2 adrenergic receptors, aiding in pain management and opioid withdrawal (Leon et al., 2021). Unlike conventional opioids, mitragynine does not recruit β-arrestin 2 at MOR, lowering the risks of respiratory depression and constipation, and it also interacts with cannabinoid receptors (CB1 and CB2) to modulate pain and inflammation (Obeng et al., 2019; Sun et al., 2024).

10.2.2.12 Oxymatrine

Oxymatrine (OMT), a quinolizidine alkaloid extracted from the roots of *Sophora flavescens*, is widely used in traditional Chinese medicine due to its neuroprotective and anti-inflammatory properties. Research has shown that OMT can reduce inflammatory and neuropathic pain in mice and is used in China to treat conditions like bacterial and atrophic vaginitis. However, its clinical use is limited by serious side effects, including hepatotoxicity and neurotoxicity (You et al., 2020). OMT inhibits pro-inflammatory cytokines such as TNF-α, IL-6, and IL-1β by modulating the NF-κB pathway. It may also prevent morphine tolerance by influencing P-glycoprotein (P-gp) expression (Li et al., 2023), thereby enhancing drug sensitivity and improving chemotherapy outcomes. Additionally, OMT reduces neuronal apoptosis by lowering Bcl-2 and C-caspase-3 levels while increasing Bax expression (Liu et al., 2014).

10.2.2.13 Caffeine

A methylxanthine derivative from *Coffea arabica* can stimulate brain activity, enhancing alertness by increasing cortical activity and metabolic energy in the brain while influencing acetylcholine levels. Caffeine-rich coffee is a popular beverage in India and around the world. Caffeine's pharmacological effects mainly stem from blocking adenosine A1, A2A, and A2B receptors, which helps reduce neurotoxicity. Additionally, studies show that caffeine has antioxidant and anti-inflammatory properties. Caffeine reduces Aβ deposition in the cortex and hippocampus while suppressing the levels of presenilin-1 (PS1) and β-secretase. Caffeine, when added as an adjunct to painkillers like ibuprofen and acetaminophen, adds small but important relief in cases of acute pain. Studies show that 100–130 mg caffeine, along with other pain relief medications, was a huge help in managing pain such as postoperative dental pain, postpartum pain, and headache.

10.2.2.14 Vinpocetine

An alkaloid similar to the extract of the medicinal plant *Vinca minor*, originally marketed as Cavinton. Vinpocetine found its use in neurological and cerebrovascular conditions; its efficacy in pain management for humans is based on its anti-inflammatory actions, vasodilation, and neuroprotection. While its biological activity can increase blood flow, glucose uptake, and microcirculation, various underlying mechanism helps in pain management. One of these mechanisms includes increased uptake of oxygen and enhanced glucose levels, which help in vasodilation and increased production of neuronal ATP. Activated NF-κB enhances the transcription of mRNA by stimulating RNA polymerase-like proteins and co-activators. Vinpocetine manages chemokines and cytokines, leading to further NF-κB activation. By preventing NF-κB from moving to the nucleus, it stops the inflammatory response, reducing cerebral infarction and delaying necrosis, thus acting as an anti-inflammatory agent. A daily dose of 60 mg is considered safe, with minimal adverse effects reported (US Food and Drug Administration, 2019). However, potential side effects include nausea, stomach pain, headache, nervousness, facial flushing, and a weakened immune system.

10.2.2.15 Nicotine

Tobacco (*Nicotiana tobaccum*) is the main source of nicotine, a pyridine alkaloid with complex mechanisms for managing nociceptive pain. Nicotine binds stereoselectively to nicotinic cholinergic receptors (nAChRs) in the central and peripheral nervous systems. When smoked, it is quickly absorbed into the bloodstream and reaches the brain. In the brain, nicotine diffuses into tissues and binds to nAChRs, which are ligand-gated ion channels. This binding allows cations like sodium and calcium to enter, activating voltage-dependent calcium channels and promoting the release of neurotransmitters, such as dopamine, norepinephrine, and serotonin, while inhibiting GABA release. The alpha-7 nAChR is notably involved in nicotine's analgesic effects by activating the descending pain modulatory pathway from the brain to the spinal cord. However, despite potential neuroprotective benefits in NDDs, nicotine also has toxic effects. It disrupts mitochondrial membranes, induces reactive oxygen species (ROS) production, and causes endoplasmic reticulum (ER) stress, leading to protein degradation and apoptosis.

10.2.2.16 Piperine

Piper longum, commonly known as long pepper, and *Piper nigrum*, known as black pepper, are the primary sources of piperine. Piperine (PIP) is a significant alkaloid found in both long pepper and black pepper, both belonging to the Piperaceae family. This compound is known for its anti-inflammatory and analgesic properties, as well as its various neuroprotective effects, which include antioxidant, antidepressant, and anticonvulsant activities. Piperine influences the perception of pain at multiple levels. While it activates the TRPV1 receptors, it desensitizes these receptors, reducing pain perception and the response to noxious stimuli. It also affects the release of neurotransmitters (serotonin, GABA, Norepinephrine) involved in pain transmission in the spinal cord. A few studies show that PIP has an antagonistic effect on the sodium/calcium channel responsible for pain management. Furthermore, it can inhibit the release of neuropeptides playing a key role in pain signal transmissions from peripheral nerves to the CNS.

10.2.2.17 Tetrahydropalmatine

Tetrahydropalmatine (THP), an isoquinoline alkaloid from Corydalis and Stephania, exhibits anxiolytic, anti-inflammatory, analgesic, and cardioprotective properties. Research indicates that THP protects nerve cells from ketamine-induced apoptosis by counteracting the decrease in acetylcholine (ACh) levels and reducing acetylcholinesterase (AChE) activity. THP also reverses memory impairments caused by D-galactose, decreasing malondialdehyde (MDA) and nitric oxide (NO) levels while increasing glutathione (GSH) and the activities of superoxide dismutase (SOD), catalase (CAT), and GSH peroxidase. Furthermore, THP decreases pro-inflammatory cytokines, including TNF-α, IL-1β, and IL-6, and enhances the expression of glial cell-derived neurotrophic factor (GDNF) in ketamine-treated mice. Additionally, THP inhibits TNF-α-induced adhesion of monocytes to endothelial cells by blocking the production of vascular cell adhesion molecule-1 (VCAM-1) and reducing NF-κB activity. These findings suggest THP's potential as a therapeutic agent for inflammation and neuroprotection, warranting further pharmacological studies.

10.2.2.18 Tetrandrine

Tetrandrine is a bisbenzylisoquinoline alkaloid extracted from the roots of *Stephania japonica, S. tetrandra*, and *S. Moore*, known for its anti-inflammatory properties. It inhibits pro-inflammatory mediators such as TNF-α, IL-6, and IL-1β by blocking the NF-kB signaling pathway in LPS-induced macrophages. Additionally, it reduces metalloproteinase expression and decreases prostaglandin E2 (PGE2) and nitric oxide (NO) production. In animal studies, tetrandrine lowered NO levels in serum and pancreatic tissue in rats with acute hemorrhagic necrotizing pancreatitis and inhibited NF-κB activation. It also showed improvement in memory and learning deficits in a rat model of Alzheimer's disease by reducing TNF-α and IL-1β expression. Recently, a phase 4 clinical trial has explored tetrandrine tablets for treating COVID-19 patients. Overall, tetrandrine appears to be a potent alkaloid for addressing inflammation and NDDs.

10.2.2.19 Sinomenine

Sinomenine is a benzyl alkaloid found in Chinese herbal medicine, primarily extracted from *Sinomenium acutum*. It is used to treat conditions such as chronic nephritis, rheumatoid arthritis, myocardial ischemia, and ankylosing spondylitis while also helping reduce foot swelling caused by substances like formaldehyde. Its anti-inflammatory effects result from inhibiting the c-Jun N-terminal kinases (JNKs) and NF-κB signaling pathways, which lowers the expression of cytokines like IL-1β and TNF-α. Sinomenine also shows potential in treating NDDs, as it reduces ROS and nitric oxide in Alzheimer's model studies. In rats with temporal lobe epilepsy, it decreased seizure intensity and other adverse effects. Clinical studies indicate that 101 out of 120 rheumatoid arthritis patients showed improvement when sinomenine was combined with methotrexate. Thus, sinomenine is an effective treatment for both inflammatory conditions and NDDs.

10.3 MECHANISMS OF ACTION OF ALKALOIDS IN PAIN MODULATION THROUGH INTERACTION WITH OPIOID RECEPTORS (μ, κ, δ)

The opioid system is one of several neurotransmitter systems that control the intricate process of pain modulation. Natural chemical substances called alkaloids, which are present in plants, have been employed extensively for their analgesic properties, mainly through their interaction with opioid receptors. The main target for pain alleviation is the μ-opioid receptor (MOR), which is also linked to tolerance and addiction. The KOR plays a role in sedation, the experience of dysphoria, and the relief of pain. The δ-opioid receptor (DOR) is involved in mood control and the modulation of chronic pain. Alkaloids may interact with these receptors by indirect control of opioid pathways, allosteric modulation, or direct agonism. While certain alkaloids, like mitragynine, work as partial agonists and lower the danger of dependency and respiratory depression, others, like morphine, behave as complete agonists at MOR.

10.3.1 μ-Opioid Receptor (MOR) Activation

The main receptor in charge of opioid-induced analgesia is MOR. MOR activation causes adenylate cyclase inhibition, which lowers cAMP levels and neuronal excitability. It also causes voltage-gated calcium channels (Ca^{2+}) to close, which inhibits neurotransmitter release, and potassium (K^+)

channels to open, which causes hyperpolarization and decreased neuronal firing. While mitragynine (found in kratom) functions as a partial MOR agonist, activating the receptor but causing less respiratory depression than morphine, alkaloids like morphine, codeine, and thebaine behave as complete MOR agonists. Although MOR activation is the most efficient pain management method, it is also linked to respiratory depression, opioid dependency, and tolerance. Researchers are now developing biased MOR agonists to distinguish between analgesia and side effects (Foss et al., 2020).

10.3.2 κ-Opioid Receptor (KOR) Modulation

One G-protein-coupled receptor (GPCR) that is essential for controlling pain, stress, mood, and addiction-related behaviors is the KOR. Dynorphins, endogenous opioid peptides that control nociception and emotional processing, are the main factors that activate it. KOR activation is associated with aversion, stress reaction, and dysphoria, in contrast to the MOR, which is in charge of the pleasurable and addictive effects of opioids like morphine. KOR agonists block pain transmission in peripheral sensory neurones and the spinal cord's dorsal horn to provide antinociceptive effects. Without causing cerebral side effects like dysphoria, KOR activation in the peripheral nervous system can significantly reduce visceral pain, which is influenced by the dynorphin/KOR system. The unpleasant consequences of KOR activation offset the rewarding qualities of dopamine release in the brain's mesolimbic pathway. Because of this, the KOR regulation may be a target for addiction therapy. Dysphoria and an increase in drug-seeking behaviors are caused by chronic stress, which increases the production of dynorphins, activates KOR, and lowers dopamine levels (Marchette et al., 2025).

10.3.3 δ-Opioid Receptor (DOR) Involvement

DOR plays a role in mood control, neuroprotection, and the modulation of chronic pain. Since DOR activation does not result in euphoria or dysphoria like MOR and KOR do. It is a desirable target for treating neuropathic pain (Gopalakrishnan et al., 2022). Instead of binding to the receptor directly, certain alkaloids, such as berberine and tetrahydropalmatine, increase DOR signaling by modifying ion channels and neurotransmitter release. Derivatives of mitragynine reduce pain by altering neurotransmission at mu-opioid and delta receptors. Mitragynine derivatives, in contrast to traditional opioids, selectively activate DOR, lowering the risk of addiction (Paul et al., 2023).

10.4 MECHANISMS OF ACTION OF ALKALOIDS IN PAIN MODULATION THROUGH NEUROTRANSMITTER MODULATION

Alkaloids are a broad family of naturally occurring substances that work by interacting with several neurotransmitter systems, such as glutamate (Glu), serotonin, dopamine, and GABA, to produce their analgesic (pain-relieving) actions. Alkaloids are promising therapeutic agents for pain treatment because of these interactions, which affect nociceptive transmission, synaptic plasticity, and CNS excitability.

10.4.1 Alkaloids and Serotonin (5-HT) Modulation

Serotonin is essential for the experience of pain, especially for the descending pain inhibition circuits. Alkaloids have the following effects on serotonin levels: By enhancing serotonin release, alkaloids such as mitragynine (found in Kratom) and harmaline (found in *P. harmala*) raise 5-HT availability, which improves the activation of descending serotonergic pathways that inhibit pain signals (Hussain et al., 2018). Tricyclic antidepressants, which are used to treat chronic pain, work similarly in this way. Certain alkaloids, such as indole alkaloids from Rauwolfia and Tabernanthe iboga, bind to 5-HT1A and 5-HT2A receptors to modulate pain sensitivity and emotional reactions to pain. This process is known as serotonin receptor binding (Mikkilineni et al., 2012).

10.4.2 Alkaloids and Dopamine (DA) Modulation

Dopamine has a role in reward systems and pain perception. Alkaloids use several mechanisms to affect dopamine activity, such as dopaminergic receptor activation done through alkaloids like ibogaine and morphine (opioid alkaloid), which increase dopamine release in the mesolimbic pathway and may lead to addiction, euphoria, and decreased pain perception (Fasakin et al., 2022). Dopamine Transporters (DAT) are inhibited by certain alkaloids that prevent dopamine from being reabsorbed, extending its activity in the synaptic cleft and improving analgesia. Cocaine and tropane alkaloids, for instance, use this process to influence how pain is perceived. Interaction with dopamine–glutamate balance through alkaloids like ibogaine and harmine modulates this balance, potentially reducing neuropathic pain, and dopamine inhibits glutamate release, reducing excitatory pain transmission (Dubey et al., 2018).

10.4.3 Alkaloids and GABAergic Modulation

The main inhibitory neurotransmitter is GABA, and a variety of alkaloids affect GABAergic transmission to induce sleepy and analgesic effects. Alkaloids such as muscimol (found in Amanita mushrooms) and huperzine A activate GABA receptors, acting as GABA agonists to improve inhibitory neurotransmission and lessen pain perception (Radulovic et al., 2013). Alkaloids that limit GABA reuptake, including gabaculine, a naturally occurring GABA-transaminase inhibitor, raise GABA levels, extending its calming effects and lowering the transmission of pain (Jiang et al., 2022).

10.4.4 Alkaloids and Glutamatergic Modulation

The primary excitatory neurotransmitter involved in pain transmission is glutamate. Alkaloids like ibogaine, voacangine, and mitragynine work as NMDA receptor antagonists to prevent excessive glutamate activity, which contributes to hyperalgesia and chronic pain (Akyuz et al., 2021). Glutamate release is decreased by some alkaloids that inhibit presynaptic glutamate release, which lessens pain sensitivity and excessive neural excitement. For instance, it has been demonstrated that berberine and huperzine A prevent glutamate-mediated excitotoxicity (Otimenyin and Ior, 2022). GABA–glutamate homeostasis is influenced by alkaloids, which maintain the balance between excitatory and inhibitory impulses to avoid exacerbating pain.

10.5 ALKALOIDS' MECHANISMS OF ACTION IN REGULATING ION CHANNELS TO MODULATE PAIN

Through their modulation of ion channels, which are crucial for neuronal excitability and pain signaling, alkaloids provide notable analgesic (pain-relieving) effects. Each of the three main ion channels—potassium (K^+), calcium (Ca^{2+}), and sodium (Na^+)—has a unique function in the transmission and regulation of pain (Figure 10.1).

10.5.1 Sodium (Na^+) Channel Modulation

In pain-transmitting neurons, voltage-gated sodium channels (Nav) are crucial for generating action potentials. Many alkaloids inhibit Nav1.7 and Nav 1.8 channels through alkaloids like Brucine (from *Strychnos nuxvomica*), which inhibits Nav 1.7 (pain-related sodium channel) and reduces the excitability of nociceptive neurons (Yu et al., 2019). Additionally, matrine (found in *Sophora* species) suppresses Nav1.8, which lessens neuropathic pain. When aconitine and similar alkaloids bind toxically to sodium channels, they cause depolarization and pain-like feelings that last for a long time. This explains why excessive dosages of certain alkaloids can be harmful (Ameri, 1998). Like

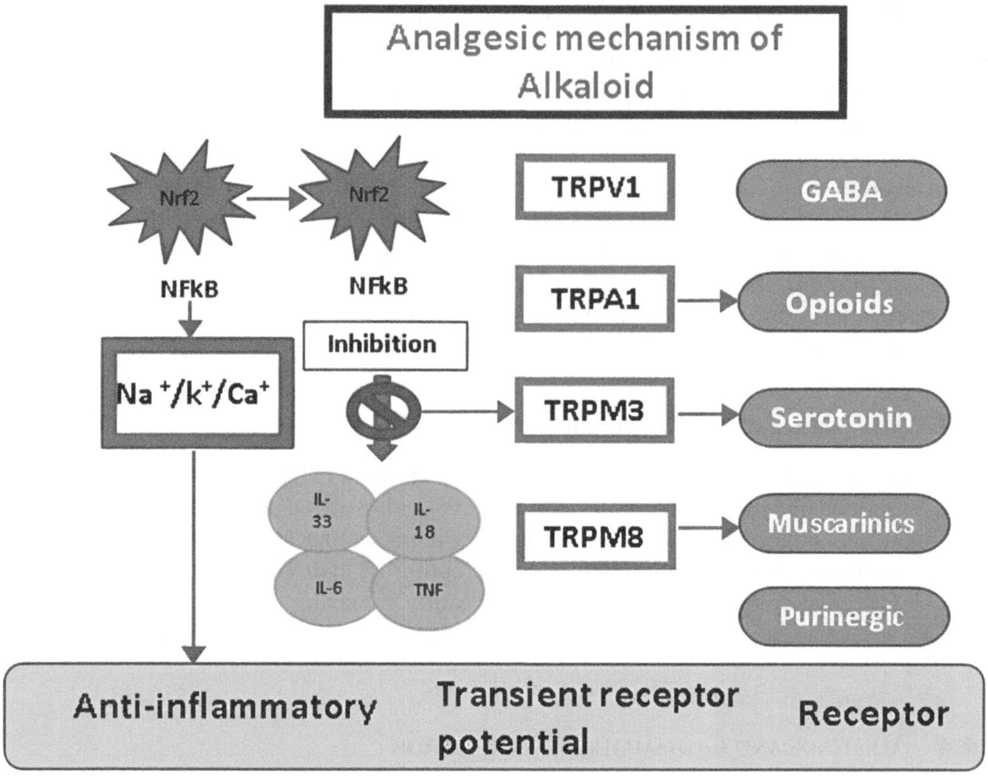

FIGURE 10.1 Analgesic mechanism of alkaloids.

lidocaine, alkaloids like cocaine block Nav channels, which lowers neuronal excitability and produces numbing effects (Friese et al., 1997).

10.5.2 Calcium (Ca^{2+}) Channel Modulation

In pain pathways, calcium channels control the release of neurotransmitters. Alkaloids can either increase or decrease synaptic transmission by interacting with these channels. Berberine (found in *Berberis species*) inhibits both N-type and L-type Ca^{2+} channels, blocking L-type calcium channels and lowering excessive neurotransmitter release and nociceptive signaling (Kochegarov, 2003). Alkaloids from *Corydalis yanhusuo*, such as tetrahydropalmatine, reduce chronic pain by blocking N-type calcium channels (Feng et al., 2024). T-type calcium channels are specifically modulated by some alkaloids, such as huperzine A, and may play a role in both neuroprotection and analgesia (Jiang et al., 2022) (Figure 10.2).

10.5.3 Potassium (K^+) Channel Modulation

Neurons are hyperpolarized by potassium channels, which lowers their excitability and muffles pain signals. These channels can be blocked or activated by alkaloids such as Papaverine, which comes from *P. somniferum* and has analgesic and vasodilatory actions via activating ATP-sensitive potassium (K ATP) channels (Bajaj et al., 2020). Certain alkaloids, such as cryptolepine, help reduce pain by activating Kv7 channels. At hazardous levels, isoteric acid (found in *Amanita muscaria*) causes neuronal hyperexcitability and pain sensitivity by blocking K-channels (Huang et al., 2019) (Figure 10.3).

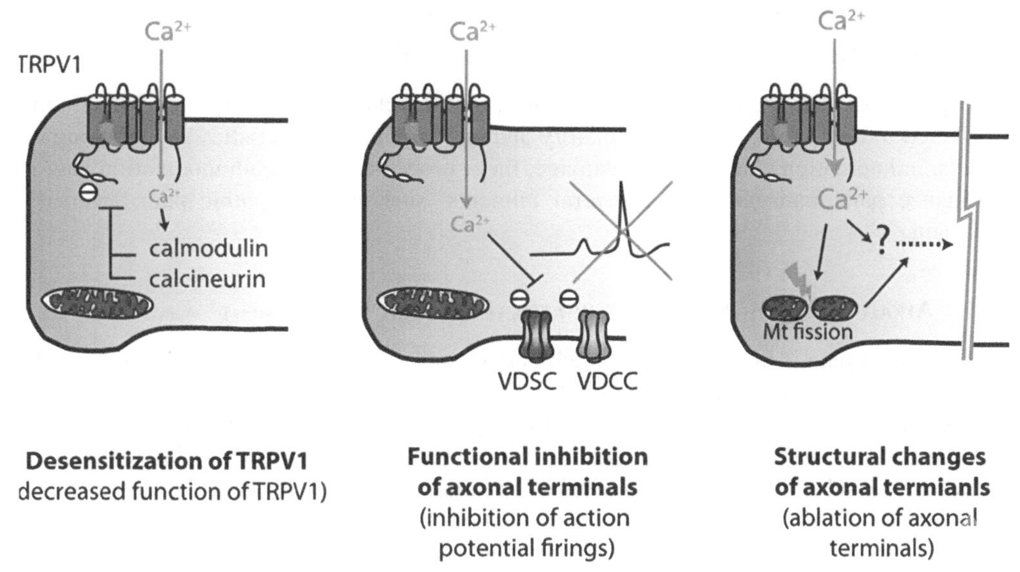

FIGURE 10.2 Ion modulation for pain management

(Chung & Campbell, 2016).

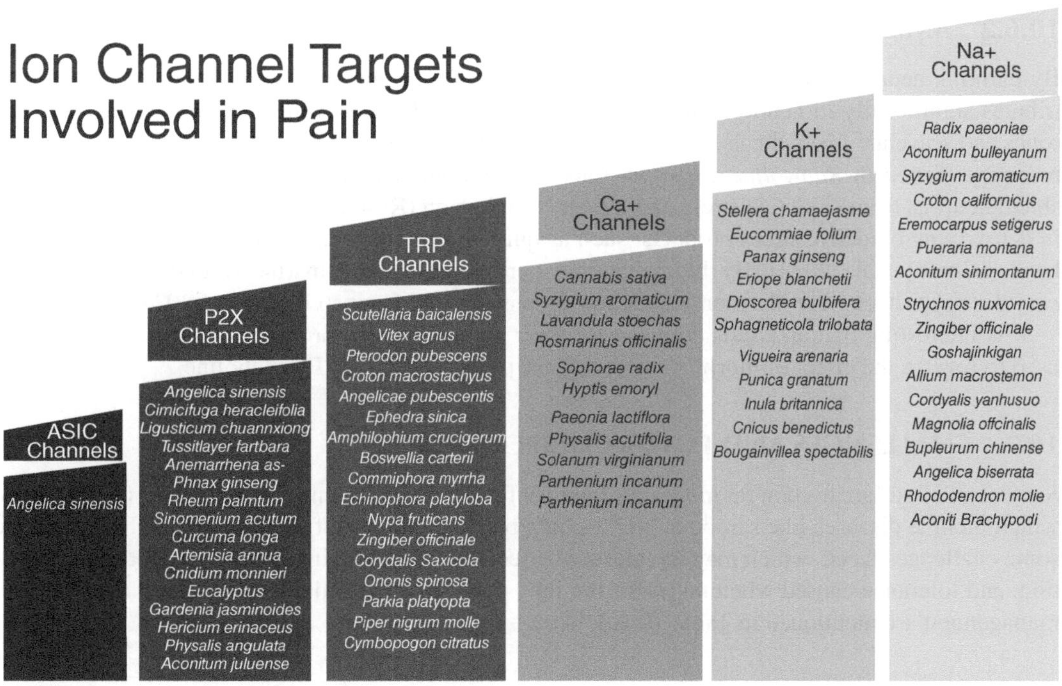

FIGURE 10.3 Alkaloids responsible for the modulation of ion channels.

10.6 ANTI-INFLAMMATORY AND ANTIOXIDANT PROPERTIES OF ALKALOIDS CONTRIBUTING TO ANALGESIA

The analgesic (pain-relieving) effects of alkaloids, a broad class of naturally occurring chemicals, are attributed to their strong anti-inflammatory and antioxidant characteristics. By lowering oxidative stress, inflammation, and neuronal damage, these processes aid in pain modulation and, in the end, enhance pain management for several ailments, such as neuropathic pain, arthritis, and chronic pain.

10.6.1 Alkaloids' Anti-Inflammatory Effects on Pain Management

A major cause of pain is inflammation, which is regulated by nuclear factor-kappa B (NF-κB) signaling, cyclooxygenase enzymes (COX-1, COX-2), and pro-inflammatory cytokines (TNF-α, IL-6, IL-1β). Alkaloids lower inflammation in several ways, e.g., inhibiting COX-2 and 5-lipoxygenase (LOX). Berberine (found in Berberis species) lowers the production of prostaglandin (PGE2) and leukotrienes, two substances that are essential for inflammatory pain (Dong et al., 2024). Evodiamine (found in *Evodia rutaecarpa*) reduces pain associated with chronic inflammation by suppressing NF-κB activation and downregulating inflammatory mediators, including TNF-α and IL-1β (Wang et al., 2024). *P. harmala*'s harmine alters the polarization of M1/M2 macrophages, fostering anti-inflammatory M2 macrophages that hasten tissue healing and pain alleviation (Chaubey and Singh, 2025). A major cause of pain is inflammation, which is regulated by nuclear factor-kappa B (NF-κB) signaling, cyclooxygenase enzymes (COX-1, COX-2), and pro-inflammatory cytokines (TNF-α, IL-6, IL-1β).

10.6.2 Antioxidant Effects of Alkaloids in Pain Modulation

By harming neurons and raising pain sensitivity (hyperalgesia and allodynia), oxidative stress contributes significantly to neuropathic and chronic pain. Alkaloids prevent oxidative damage to neurones and scavenge free radicals, therefore counteracting oxidative stress. As free radical scavengers, reserpine (found in *Rauwolfia serpentina*) and cryptolepine (found in *Cryptolepis sanguinolenta*) shield neurons from oxidative stress and lessen chronic pain (Rose et al., 2024). Alkaloids that assist in detoxifying oxidative stress products, such as quinine and colchicine, raise GSH, SOD, and CAT levels (Ghate et al., 2024). By blocking lipid peroxidation, sanguinarine (found in *Sanguinaria canadensis*) protects neuronal membranes and lessens pain sensitivity (Gupta et al., 2024) (Table 10.1).

It is interesting that alkaloids are used for their antinociceptive properties, but they can also be used to reverse the effect of morphine induced antinociception, e.g., Fang choline.

10.7 CHALLENGES AND FUTURE DIRECTIONS

Several alkaloids are used in rat models or humans for the management of nociceptive, chronic pain, which include diseases like sarcoma, NDDs, etc., but with the use of these plant alkaloid there are some challenges faced, which may include side effects or toxic drug interaction, dependence, addiction, and tolerance caused when taken for the long term. Several limitations of alkaloid-based pain management are mentioned in Table 10.2.

10.8 LEGAL AND REGULATORY ASPECTS OF ALKALOID USE IN MEDICINE

Although laws and regulations governing the use of alkaloids in medicine differ from one nation to the next, they all typically seek to guarantee controlled, safe, and effective use. Because of their potential for misuse and addiction, several alkaloids, including morphine and codeine, are categorized as restricted drugs. Laws such as India's Narcotic Drugs and Psychotropic Substances Act

TABLE 10.1
Summarized Plant Alkaloids and Their mechanism for the Treatment of Nociceptive Pain with References

Alkaloid	Plant Extract From	Mechanism Followed	References
Arecoline	*Areca catechu*	Agonist at muscarinic acetylcholine receptor, GABA inhibition	Chen et al. (2024)
Aloperine	*Sophora alopecuroides*	Inhibit the TLR4 pathway to upregulate NF-κB, IL-1β, and IL-6, associated with reduced ROS	Aryal et al. (2021)
Berberine and Berbamine	Berberis spp	Activation of AMP-activated protein kinase (AMPK) helps to reduce the expression of pro-inflammatory genes	Meng et al. (2022)
Bulleyaconitine A	*Aconitum bulleyanum*	Inhibit Nav1.3 and Nav1.7	Xie et al. (2018)
Brucine	*Strychnos nuxvomica*	Suppress voltage-dependent sodium channel	Yu et al. (2019)
Caffeine	*Coffea arabica*	Antagonism of adenosine receptors (A1 and A2A), as well as inhibition of cyclooxygenase activity	Grama et al. (2020)
Caulerpin	*Caulerpa racemosa*	Regulate α2-adrenoceptors and 5-HT3 receptors	Cavalcante-Silva et al. (2014)
chelerythrine	*Coptis chinensis*	Inhibit PKC and NR2B, NMDA receptors	Danielewski et al. (2025)
Discretamine	*Duguetia moricandiana*	Reduced production of nitric oxide (NO), interleukin-6 (IL-6), interleukin-1β (IL-1β), and tumor necrosis factor-alpha (TNF-α)	Da Silva et al. (2012)
Delavayine	*Incarvillea delavayi*	Acts as an anti-inflammatory	Nakamura et al. (2000)
Evodiamine	*Evodia rutaecarpa*	Regulates (TRPV1)	Iwaoka et al. (2016)
Eleagnine	*Chrysophyllum albidum*	Suppression of pro-inflammatory cytokines, cyclooxygenase-2, and nuclear factor-kappa B (NF-κB)	Ajayi et al. (2020)
Galantamine	*Galanthus woronowii*, *Leucojum aestivum*, and other members of the family Amaryllidaceae	enhances cholinergic neurotransmission by inhibiting acetylcholinesterase, pro-inflammatory signaling molecules (NF-κB p65), and inflation of the synapse-associated proteins	Kataria et al. (2024)
Geissospermine	*Geissospermum vellosii*	Anticholinesterase Activity and reduces associated inflammation	Lima et al. (2016)
Harmine	*Peganum harmala*	Inhibit NF-κB activation, downregulated TLR4, LRR, NOD and NLRP3	Liu et al. (2016)
Huperzine A	*Huperzia serrata*	Inhibits acetylcholinesterase and NMDA receptor, Zn channel	Yu et al. (2013)
Incarvillateine	*Incarvillea sinensis*	Activates adenosine receptors and inhibits cytokines	Kim et al. (2019)
Lobeline	*Lobelia inflata*	Release neurotransmitters like serotonin and dopamine and interact with nicotinic acetylcholine receptors	Kataria et al. (2024)

(*Continued*)

TABLE 10.1 (Continued)
Summarized Plant Alkaloids and Their mechanism for the Treatment of Nociceptive Pain with References

Alkaloid	Plant Extract From	Mechanism Followed	References
Lappaconitine	*Aconitum sinimontanum*	Inhibit Nav1.7, Modulation of P2X3 receptor channels	Ou et al. (2011)
Morphine	*Papaver somniferum*	Regulate mu-opioid receptors	Kaserer et al. (2020)
Montanine	Amaryllidaceae family	Reduced cytokines and central sensitization	Koutova et al. (2019)
Nicotine	Tobacco	Activation of Nicotinic Acetylcholine Receptors (nAChRs)	Luo et al. (2023)
Nantenine	*Nandina domestica*	Antagonist to 5-HT2A and Alpha-Adrenergic Receptor	Shi and Wu (2023)
Neoline	Goshanjimkigan	Inhibit Nav1.7	Nakatani et al. (2020)
Oxymatrine	*Sophora flavescentis*	Blocking NF-κB translocation, reducing the production of cytokines, TNF-α, and IL-17A	Liu et al. (2014)
Piperine	*Piper nigrum*	Inhibits interleukin-1β (IL-1β), interleukin-6 (IL-6), and tumor necrosis factor-alpha (TNF-α)	Mahboubi (2021)
Physotigmine	*Physostigma venenosum*	Reversible cholinesterase inhibitor, synergistic to opioids like morphine	Klivinyi et al. (2021)
Solanine	*Solanum trilobatum*	Inhibition of Acetylcholinesterase and regulates ion channels	Shi and Wu (2023)
Sinomenine	*Sinomenium acutum*	Inhibition of c-Jun N-terminal kinases (JNKs) and NF-κB signaling pathways, thus suppressing the mRNA expression of cytokines, such as IL-1β and TNF-α	Rao et al. (2017)
Ternanthranin	*Choisya ternata*	Activate potassium ion channel	Donald et al. (2021)
Tetrandrine	*Stephania japonica, S. tetrandra*, and *S. Moore*	Blocking the NF-kB signaling, IKKβ phosphorylation, and suppressing satellite glial cells (SGCs)	Qin et al. (2018)
Tetrahydropalmatine	*Stephania and Corydalis genus,*	THP lowered TNF-α, IL-1β, and IL-6 expression-suppressed iNOS and NF-κB protein activities, inhibit NAV1.7	Oh et al. (2010)
Tetramethylpyrazine	*Ligusticum wallichii*	Modulation of the P2X3 receptor channel	Wang et al. (2018)
Umbellatine	*Psychotria umbellata*	Partially activate the opioid receptor and inhibit the NMDA receptor	Both et al. (2002)
Vinpocetine	*Vinca minor*	Activation of nuclear factor kappa B (NF-κB), reduced glutathione, inhibits the production of pro-inflammatory cytokines (TNF-α) and (IL-1β)	Ruiz-Miyazawa et al. (2015)
Verticinone	Fritillaria sp.	Blocks ion channels and modulates central sensitization	Kumar et al. (2025)

TABLE 10.2
Several Limitations of Some Commonly Used Plant-Based Alkaloids for Pain Management

Alkaloid	Toxic Effects	References
Berberine	• Liver toxicity leading to elevated liver enzymes, jaundice, and abdominal pain. • It can interfere with medications, such as those for blood sugar regulation or heart conditions, by altering liver enzymes responsible for drug metabolism. • Gastrointestinal issues like diarrhea, constipation, gas, and upset stomach. • Adversely affect DNA synthesis and may exhibit cytotoxic effects.	Singh et al. (2018)
Capsaicin	• It can cause a severe burning sensation and irritation. • Inhalation of capsaicin (e.g., from pepper sprays) can lead to coughing, difficulty breathing, and throat irritation. • Consuming excessive amounts can result in stomach pain, nausea, vomiting, and diarrhea. • High doses may overstimulate nerve receptors, leading to desensitization or pain. • Some individuals may experience allergic responses, such as swelling or rash.	Johnson et al. (2007)
Lappaconitine	• Affects sodium channels in the heart, potentially leading to arrhythmias or other cardiac issues. • Symptoms like paresthesia (tingling sensations), muscle weakness, or seizures may occur in cases of overdose. • High doses can lead to nausea, vomiting, or abdominal discomfort. • Like other diterpenoid alkaloids, it may exhibit toxic effects on cells at high concentrations.	McCabe et al. (2022)
Mitragynine	• Act on opioid receptors, potentially leading to respiratory depression, especially in overdose cases. • High doses may cause seizures, psychosis, or sedation. • It may lead to irregular heart rhythms or other cardiovascular complications. • Nausea, vomiting, and constipation are common side effects.	Behonick et al. (2022)
Oxymatrine	• Liver toxicity, particularly at high doses. • Nausea, vomiting, and abdominal discomfort. • Dizziness or other mild neurological symptoms	Feng et al. (2024)
Caffeine	• High doses can lead to increased heart rate, palpitations, or even arrhythmias. • Overconsumption may cause anxiety, restlessness, insomnia, or, in severe cases, seizures. • Nausea, vomiting, or diarrhea. • Frequent use can result in dependence, which may cause withdrawal symptoms such as headaches, fatigue, and irritability. • An overdose can cause respiratory distress.	Meredith et al. (2013)
Vinpocetine	• It may cause embryo and fetal loss, reduced fetal weight, and developmental malformations in animal models. • Low blood pressure, rapid heartbeat, or other cardiovascular disturbances. • Dizziness, headache, and sleep disturbances. • Stomach discomfort and nausea. • May slow blood clotting, increasing the risk of bleeding, especially during surgery.	Hafez et al. 2025

(*Continued*)

TABLE 10.2 (Continued)
Several Limitations of Some Commonly Used Plant-Based Alkaloids for Pain Management

Alkaloid	Toxic Effects	References
Arecoline	• Linked to oral squamous cell carcinoma. • Chronic use can lead to oral submucous fibrosis, a precancerous condition. • Can overstimulate the CNS, potentially causing tremors, seizures, or other neurological symptoms. • Increased heart rate and blood pressure due to its parasympathomimetic properties. • Induces oxidative stress and DNA damage, contributing to its toxic effects	Huang et al. (2024)
Nicotine	• Increases heart rate and blood pressure, contributing to the risk of hypertension and cardiovascular diseases. • Dizziness, headaches, seizures, and, in severe cases, CNS depression. • Overdose may lead to nausea, vomiting, and abdominal pain. • High doses can result in respiratory distress or failure. • Highly addictive, and withdrawal symptoms include irritability, anxiety, and difficulty concentrating.	Schep et al. (2009)
Tetrahydropalmatine	• Liver impairment, particularly with long-term use or high doses. • Overuse can lead to drowsiness, dizziness, or extrapyramidal symptoms (movement disorders). • contraindicated in pregnant women due to potential risks to fetal development	Lai and Chan (1999)

control their manufacture, sale, and prescription. Alkaloid-based medications have to adhere to strict safety and quality requirements. The clearance procedure is supervised by regulatory agencies such as the FDA in the United States and the CDSCO in India. The Drugs and Cosmetics Act, which contains regulations for Ayurvedic, Siddha, and Unani medications, governs the use of alkaloids in traditional Indian systems such as Ayurveda. The European Medicines Agency (EMA) in Europe has regulations for alkaloids found in herbal medicines. Although alkaloid-containing drugs, especially opioids, have many therapeutic uses, they also carry the danger of addiction, toxicity, and abuse. Abuse-deterrent formulations (ADFs), alternative medicines, regulatory actions, and harm reduction techniques are some ways to lessen these hazards.

10.9 POTENTIAL STRATEGIES TO MITIGATE

- Creating formulations that hinder the abuse of drugs based on alkaloids. For instance, mixing them with substances that, when tampered with, negate their effects.
- Investigating medicines or treatments based on non-alkaloids to lessen reliance on alkaloids, particularly in the treatment of pain or other medical uses.
- Putting in place stringent dose recommendations and routine monitoring to stop abuse or overuse.
- Teach patients and medical professionals about the dangers of alkaloids and how to use them safely.
- Strategies for situations such as fescue toxicosis in cattle include substituting non-toxic plant kinds for toxic ones, reducing alkaloid consumption by combining it with other feed, or employing supplements to mitigate the effects.
- Enforcing laws to regulate the manufacture, sale, and use of compounds containing alkaloids.

10.10 CONCLUSION

Alkaloids have long been a valuable tool in managing pain, from ancient herbal remedies to modern medicine. These naturally occurring compounds offer powerful pain relief with fewer challenges than those faced while using poppy-based alkaloids. Through various mechanisms, such as interacting with opioid receptors, modulating neurotransmitters, and reducing inflammation, these plant-based alkaloids show potent antinociceptive and anti-inflammatory properties, which can be used for the management of chronic or acute pain without complicating the usage in the long term. Where opioid alkaloids like morphine and codeine remain essential for treating severe pain, their risks, such as addiction and tolerance, highlight the need for safer alternatives. Fortunately, non-opioid alkaloids like capsaicin, berberine, and mitragynine are showing promise, providing pain relief with fewer side effects. However, some can be toxic in high doses, cause dependency, or have unpredictable effects depending on the individual. Striking a balance between effectiveness and safety is crucial. Research advancement is required to refine these compounds, develop better drug formulations, and explore innovative ways to integrate them into pain management strategies. With this future direction, plant-based alkaloids could offer a future where pain relief is not only effective but also safer and more sustainable for those who need it most.

COMPETING INTERESTS

The authors declare that they have no competing interest.

DECLARATION OF GENERATIVE AI AND AI-ASSISTED TECHNOLOGIES IN THE WRITING PROCESS

During the preparation of this work, the book chapter author(s) have not taken any assistance from AI, and the book chapter author(s) reviewed and edited the content as needed and take(s) full responsibility for the content of the publication.

REFERENCES

Ajayi, A. M., Badaki, V. B., Ariyo, O. O., Ben-Azu, B., Asejeje, F. O., & Adedapo, A. D. (2020). *Chrysophyllum albidum* fruit peel attenuates nociceptive pain and inflammatory response in rodents by inhibition of pro-inflammatory cytokines and COX-2 expression through suppression of NF-κB activation. *Nutrition Research*, *77*, 73–84.

Akyuz, E., Polat, A. K., Eroglu, E., Kullu, I., Angelopoulou, E., & Paudel, Y. N. (2021). Revisiting the role of neurotransmitters in epilepsy: An updated review. *Life Sciences*, *265*, 118826.

Alford, A. S., Moreno, H. L., Benjamin, M. M., Dickinson, C. F., & Hamann, M. T. (2025). Exploring the therapeutic potential of mitragynine and corynoxeine: Kratom-derived indole and oxindole alkaloids for pain management. *Pharmaceuticals*, *18*(2), 222.

Ameri, A. (1998). The effects of Aconitum alkaloids on the central nervous system. *Progress in Neurobiology*, *56*(2), 211–235.

Arita, T., Asano, M., Kubota, K., Domon, Y., Machinaga, N., & Shimada, K. (2019). Discovery of conolidine derivative DS39201083 as a potent novel analgesic without mu opioid agonist activity. *Bioorganic & Medicinal Chemistry Letters*, *29*(15), 1938–1942.

Aryal, B., Raut, B. K., Bhattarai, S., Bhandari, S., Tandan, P., Gyawali, K., Sharma, K., Ranabhat, D., Thapa, R., Aryal, D., Ojha, A., Devkota, H. P., & Parajuli, N. (2021). Potential therapeutic applications of plant-derived alkaloids against inflammatory and neurodegenerative diseases. *Evidence-Based Complementary and Alternative Medicine*, *2022*(1), 7299778. https://doi.org/10.1155/2022/7299778

Bair, M. J., Robinson, R. L., Katon, W., & Kroenke, K. (2003). Depression and pain comorbidity: A literature review. *Archives of Internal Medicine*, 163(20), 2433–2445.

Bajaj, S., Ong, S. T., & Chandy, K. G. (2020). Contributions of natural products to ion channel pharmacology. *Natural Product Reports*, *37*(5), 703–716.

Behonick, G. S., Vu, C., Czarnecki, L., El-Ters, M., & Shanks, K. G. (2022). Two single-drug fatal intoxications by mitragynine. *Journal of Analytical Toxicology*, *46*(5), e110–e114.

Both, F. L., Kerber, V. A., Henriques, A. T., & Elisabetsky, E. (2002). Analgesic properties of umbellatine from Psychotria umbellata. *Pharmaceutical Biology*, *40*(5), 336–341.

Cavalcante-Silva, L. H. A., Falcão, M. A. P., Vieira, A. C. S., Viana, M. D. M., de Araújo-Júnior, J. X., Sousa, J. C. F., … & Alexandre-Moreira, M. S. (2014). Assessment of mechanisms involved in antinociception produced by the alkaloid caulerpine. *Molecules*, *19*(9), 14699–14709.

Chakrabarti, A., Singh, J., Saha, L., & Bhatia, A. (2018). Evaluation of berberin-a Nrf2 activator, on paclitaxel-induced neuropathic pain model in rat (P2. 104). *Neurology*, *90*(15_supplement), P2–104.

Chaubey, S., & Singh, L. (2025). Harmaline attenuates pain and inflammation: role of IL-1β, oxidative stress, nitric oxide and cyclo-oxygenase. *Naunyn-Schmiedeberg's Archives of Pharmacology*, 1–13.

Chen, C., Yu, Z., Li, Y., Fichna, J., & Storr, M. (2014). Effects of berberine in the gastrointestinal tract—A review of actions and therapeutic implications. *The American Journal of Chinese Medicine*, *42*(05), 1053–1070.

Chen, Q. Y., Zhang, Y., Ma, Y., & Zhuo, M. (2024). Inhibition of cortical synaptic transmission, behavioral nociceptive, and anxiodepressive-like responses by arecoline in adult mice. *Molecular Brain*, *17*(1), 39.

Chung, M., & Campbell, J. N. (2016). Use of capsaicin to treat pain: Mechanistic and therapeutic considerations. *Pharmaceuticals*, *9*(4), 66. https://doi.org/10.3390/ph9040066

da Silva Almeida, J.R.G., da Cruz Araújo, E.C., de Araújo Ribeiro, L.A., de Lima, J.T., Nunes, X.P., Lúcio, A.S.S.C., de Fátima, A.M., & Barbosa Filho, J.M. (2012). Antinociceptive activity of ethanol extract from *Duguetia chrysocarpa* Maas (Annonaceae). *Science World Journal 2012*, 859210

Danielewski, M., Zielińska, S., Słupski, W., Włodarczyk, M., Sozański, T., Ziółkowski, P., Szeląg, A., & Nowak, B. (2025). Sanguinarine–chelerythrine from coptis chinensis offers analgesic and anti-inflammatory effects without gastrotoxicity. *Pharmaceutics*, *17*(3), 323. https://doi.org/10.3390/pharmaceutics 17030323

Donald, G. R., de Carvalho, P. R., Fernandes, P. D., & Boylan, F. (2021). Antinociceptive activity of puberulin and choisyine from ethanol extract of *Choisya ternata Kunth* var. *Sundance*. *Biomedicine & Pharmacotherapy*, *141*, 111926.

Dong, R., Zhang, T., Lu, J., Wang, T., Yang, A., Pu, J., … & Luo, D. (2024). Secondary metabolites of an endophytic fungus from Ligularia fischeri. *Chemistry of Natural Compounds*, *60*(5), 986–987.

Dubey, T., Sahu, G., Kumari, S., Yadav, B. S., & Sahu, A. N. (2018). Role of herbal drugs on neurotransmitters for treating various CNS disorders: A review. *Indian Journal of Traditional Knowledge*, *17*(1), 113–121.

Edinoff, A. N., Patel, A. S., Baker, M. W., Lawson, J., Wolcott, C., Cornett, E. M., Sadegi, K., Kaye, A. M., & Kaye, A. D. (2021). Conolidine: A novel plant extract for chronic pain. *Anesthesiology and Pain Medicine*, *11*(6), e121438. https://doi.org/10.5812/aapm.121438

Ellis, C. R., Racz, R., Kruhlak, N. L., Kim, M. T., Zakharov, A. V., Southall, N., et al. (2020). Evaluating Kratom Alkaloids using PHASE. *PloS One*, 15(3), e0229646.

Fasakin, O. W., Oboh, G., Ademosun, A. O., & Lawal, A. O. (2022). The modulatory effects of alkaloid extracts of Cannabis sativa, Datura stramonium, Nicotiana tabacum and male Carica papaya on neurotransmitter, neurotrophic and neuroinflammatory systems linked to anxiety and depression. *Inflammopharmacology*, *30*(6), 2447–2476.

Feng, W., Kao, T. C., Jiang, J., Zeng, X., Chen, S., Zeng, J., … & Ma, X. (2024). The dynamic equilibrium between the protective and toxic effects of matrine in the development of liver injury: A systematic review and meta-analysis. *Frontiers in Pharmacology*, *15*, 1315584

Foss, J. D., Nayak, S. U., Tallarida, C. S., Farkas, D. J., Ward, S. J., & Rawls, S. M. (2020). Mitragynine, bioactive alkaloid of kratom, reduces chemotherapy-induced neuropathic pain in rats through α-adrenoceptor mechanism. *Drug and Alcohol Dependence*, *209*, 107946.

Friese, J., Gleitz, J., Gutser, U. T., Heubach, J. F., Matthiesen, T., Wilffert, B., & Selve, N. (1997). Aconitum sp. alkaloids: The modulation of voltage-dependent Na+ channels, toxicity and antinociceptive properties. *European Journal of Pharmacology*, *337*(2–3), 165–174.

Ghate, V., Alqahtani, M. S., Shah, H., & Veena, K. M. (2024). Oral mucosa and disorders: from drug discovery to development and translation. *Frontiers in Pharmacology*, *15*, 1445126.

Gopalakrishnan, L., Chatterjee, O., Ravishankar, N., Suresh, S., Raju, R., Mahadevan, A., & Prasad, T. K. (2022). Opioid receptors signaling network. *Journal of Cell Communication and Signaling*, *16*(3), 475–483.

Grama, P., Pepa, M., & Zah, C. (2020). Effect of caffeine on pain management. *Acta Medica Marisiensis*, *66*.

Gupta, A., Deka, R., Mehta, S. K., Kunal, K., & Singh, S. (2024). Modulatory effects of phytobioactives in bone metabolism and regeneration. In *Emerging Materials and Technologies for Bone Repair and Regeneration* (pp. 216–243). CRC Press.

Hafez, H. M., Abed El Baky, M. F., Mokhemer, S. A., Hassan, S. M., & Mohamed, M. Z. (2025). Vinpocetine alleviates valproic acid-induced hepatotoxicity and neurotoxicity through activation of cAMP and PI3K/AKT/CREB pathway in rats. *Journal of Biochemical and Molecular Toxicology*, *39*(6), e70316.

Huang, G., Zeng, D., Liang, T., Liu, Y., Cui, F., Zhao, H., & Lu, W. (2024). Recent advance on biological activity and toxicity of arecoline in edible areca (betel) nut: A review. *Foods*, *13*(23), 3825.

Huang, M., Luo, L., Zhang, Y., Wang, W., Dong, J., Du, W., … & Xu, T. (2019). Metabotropic glutamate receptor 5 signalling induced NMDA receptor subunits alterations during the development of morphine-induced antinociceptive tolerance in mouse cortex. *Biomedicine & Pharmacotherapy*, *110*, 717–726.

Hussain, G., Rasul, A., Anwar, H., Aziz, N., Razzaq, A., Wei, W., Ali, M., Li, J., & Li, X. (2018). Role of plant derived alkaloids and their mechanism in neurodegenerative disorders. *International Journal of Biological Sciences*, *14*(3), 341. https://doi.org/10.7150/ijbs.23247

Ilmie, M. U., Jaafar, H., Mansor, S. M., and Abdullah, J. M. (2015). Subchronic toxicity study of standardized methanolic extract of *Mitragyna Speciosa* Korth in Sprague-Dawley Rats. *Frontiers in Neuroscience*, *9*, 189.

Iwaoka, E., Wang, S., Matsuyoshi, N., Kogure, Y., Aoki, S., Yamamoto, S., … & Dai, Y. (2016). Evodiamine suppresses capsaicin-induced thermal hyperalgesia through activation and subsequent desensitization of the transient receptor potential V1 channels. *Journal of Natural Medicines*, *70*(1), 1–7.

Jiang, W., Tang, M., Yang, L., Zhao, X., Gao, J., Jiao, Y., … & Jiang, J. D. (2022). Analgesic alkaloids derived from traditional Chinese medicine in pain management. *Frontiers in Pharmacology*, *13*, 851508.

Johnson Jr, W. (2007). Final report on the safety assessment of capsicum annuum extract, capsicum annuum fruit extract, capsicum annuum resin, capsicum annuum fruit powder, capsicum frutescens fruit, capsicum frutescens fruit extract, capsicum frutescens resin, and capsaicin. *International Journal of Toxicology*, *26*(Suppl 1), 3–106.

Kaserer, T., Steinacher, T., Kainhofer, R., Erli, F., Sturm, S., Waltenberger, B., … & Spetea, M. (2020). Identification and characterization of plant-derived alkaloids, corydine and corydaline, as novel mu opioid receptor agonists. *Scientific Reports*, *10*(1), 13804.

Kataria, S., Patel, U., Yabut, K. et al. (2024). Recent advances in management of neuropathic, nociceptive, and chronic pain: A narrative review with focus on nanomedicine, gene therapy, stem cell therapy, and newer therapeutic options. *Current Pain and Headache Reports*, *28*, 321–333. https://doi.org/10.1007/s11916-024-01227-5

Kim, J., Bogdan, D. M., Elmes, M. W., Awwa, M., Yan, S., Che, J., Lee, G., Deutsch, D. G., Rizzo, R. C., Kaczocha, M., & Ojima, I. (2019). Incarvillateine produces antinociceptive and motor suppressive effects via adenosine receptor activation. *PLoS One*, *14*(6), e0218619. https://doi.org/10.1371/journal.pone.0218619

Klivinyi, C., Rumpold-Seitlinger, G., Dorn, C., Sampl, L., Sivro, N., Lang-Illievich, K., … & Bornemann-Cimenti, H. (2021). Perioperative use of physostigmine to reduce opioid consumption and peri-incisional hyperalgesia: a randomised controlled trial. *British Journal of Anaesthesia*, *126*(3), 700–705.

Kochegarov, A. A. (2003). Pharmacological modulators of voltage-gated calcium channels and their therapeutical application. *Cell Calcium*, *33*(3), 145–162.

Koutová, D., Maafi, N., Havelek, R., Opletal, L., Blunden, G., Řezáčová, M., & Cahlíková, L. (2019). Chemical and biological aspects of montanine-type alkaloids isolated from plants of the Amaryllidaceae family. *Molecules*, *25*(10), 2337. https://doi.org/10.3390/molecules25102337

Kozlakidis, Z., Cheong, I. H., & Wang, H. (2022). Betel nut and arecoline: Past, present, and future trends. *Innovations in Digital Health, Diagnostics, and Biomarkers*, *2*(64), 72. https://doi.org/10.36401/IDDB-22-05

Kumar, P., Kumar, V., Sharma, S., Sharma, R., & Warghat, A. R. (2025). Fritillaria steroidal alkaloids and their multi-target therapeutic mechanisms: insights from network pharmacology. *Naunyn-Schmiedeberg's Archives of Pharmacology*, *398*(3), 2209–2228.

Lai, C. K., and Chan, A. Y. W. (1999). Tetrahydropalmatine poisoning: Diagnoses of nine adult overdoses based on toxicology screens by HPLC with diode-array detection and gas chromatography–mass spectrometry. *Clinical Chemistry*, 45(2), 229–236.

León, F., Obeng, S., Mottinelli, M., Chen, Y., King, T. I., Berthold, E. C., et al. (2021). Activity of *Mitragyna Speciosa* ("Kratom") Alkaloids at Serotonin Receptors. *Journal of Medicinal Chemistry*, *64*(18), 13510–13523.

Li, J., Cao, Y., Li, L. N., Chu, X., Wang, Y. S., Cai, J. J., … & Fan, Z. K. (2023). Neuroprotective effects of oxymatrine via triggering autophagy and inhibiting apoptosis following spinal cord injury in rats. *Molecular Neurobiology*, *60*(8), 4450–4471.

Lima, J. A., Costa, T. W., Silva, L. L., Miranda, A. L. P., & Pinto, A. C. (2016). Antinociceptive and anti-inflammatory effects of a *Geissospermum vellosii* stem bark fraction. *Anais da Academia Brasileira de Ciências*, *88*(1), 237–248

Liu, J., Yao, Y., Ding, H., & Chen, R. (2014). Oxymatrine triggers apoptosis by regulating Bcl-2 family proteins and activating caspase-3/caspase-9 pathway in human leukemia HL-60 cells. *Tumor Biology*, *35*, 5409–5415.

Liu, Y. J., Peng, W., Hu, M. B., Xu, M., & Wu, C. J. (2016). The pharmacology, toxicology and potential applications of arecoline: A review. *Pharmaceutical Biology*, *54*(11), 2753–2760.

Luo, Y., Yang, Y., Schneider, C., & Balle, T. (2023). The anti-nociceptive effects of nicotine in humans: A systematic review and meta-analysis. *Pharmaceuticals*, *16*(12), 1665. https://doi.org/10.3390/ph16121665

Mahboubi, M. (2021). Pepper as analgesic and anti-inflammatory alternative and bio-enhancer agent for treatment of pain. *Proceedings of the National Academy of Sciences, India Section B: Biological Sciences*, *91*, 487–493. https://doi.org/10.1007/s40011-021-01243-0

Marchette, R. C., Vendruscolo, L. F., & Koob, G. F. (2025). The dynorphin/-opioid receptor system at the interface of hyperalgesia/hyperkatifeia and addiction. *Current Addiction Reports*, *12*(1), 11.

McCabe, D. J. (2022). Clinical effects from ingestion of lappaconitine, an Aconitum alkaloid with sodium channel blocking effects. *Journal of Medical Toxicology*, *18*(3), 243–247.

Meireles, V., Rosado, T., Barroso, M., Soares, S., Gonçalves, J., Luís, Â., … & Gallardo, E. (2019). *Mitragyna speciosa*: clinical, toxicological aspects and analysis in biological and non-biological samples. *Medicines*, *6*(1), 35.

Meng, J., Qiu, S., Zhang, L., You, M., Xing, H., & Zhu, J. (2022). Berberine alleviate cisplatin-induced peripheral neuropathy by modulating inflammation signal via TRPV1. *Frontiers in Pharmacology*, *12*, 774795.

Meredith, S. E., Juliano, L. M., Hughes, J. R., & Griffiths, R. R. (2013). Caffeine use disorder: A comprehensive review and research agenda. *Journal of Caffeine Research*, *3*(3), 114–130.

Mikkilineni, S., Cantuti-Castelvetri, I., Cahill, C. M., Balliedier, A., Greig, N. H., & Rogers, J. T. (2012). The anticholinesterase phenserine and its enantiomer posiphen as 5′ untranslated-region-directed translation blockers of the parkinson's alpha synuclein expression. *Parkinson's Disease*, *2012*(1), 142372.

Moore, R. A., Derry, S., Aldington, D., Cole, P., & Wiffen, P. J. (2012). Amitriptyline for neuropathic pain and fibromyalgia in adults. *Cochrane Database of Systematic Reviews*, *2012*(12), CD008242

Nakamura, M., Kido, K., Kinjo, J., & Nohara, T. (2000). Antinociceptive substances from Incarvillea delavayi. *Phytochemistry*, *53*(2), 253–256.

Nakatani, Y., Negoro, K., Yamauchi, M., Katasho, M., Ishikura, K. I., Iwaki, A., … & Amano, T. (2020). Neoline, an active ingredient of the processed aconite root in Goshajinkigan formulation, targets Nav1. 7 to ameliorate mechanical hyperalgesia in diabetic mice. *Journal of Ethnopharmacology*, *259*, 112963.

Obata, H. (2017). Analgesic mechanisms of antidepressants for neuropathic pain. *International Journal of Molecular Sciences*, 18(11), 2483.

Obeng, S., Kamble, S. H., Reeves, M. E., Restrepo, L. F., Patel, A., Behnke, M., … & McCurdy, C. R. (2019). Investigation of the adrenergic and opioid binding affinities, metabolic stability, plasma protein binding properties, and functional effects of selected indole-based kratom alkaloids. *Journal of Medicinal Chemistry*, *63*(1), 433–439.

Oh, Y. C., Choi, J. G., Lee, Y. S., Brice, O. O., Lee, S. C., Kwak, H. S., … & Kwon, D. Y. (2010). Tetrahydropalmatine inhibits pro-inflammatory mediators in lipopolysaccharide-stimulated THP-1 cells. *Journal of Medicinal Food*, *13*(5), 1125–1132.

Olofinsan, K., Abrahamse, H., & George, B. P. (2023). Therapeutic role of alkaloids and alkaloid derivatives in cancer management. *Molecules (Basel, Switzerland)*, *28*(14), 5578

Otimenyin, S. O., & Ior, L. D. (2022). Management of psychosis. *Complementary Therapies*, 85.

Ou, S., Zhao, Y. D., Xiao, Z., Wen, H. Z., Cui, J., & Ruan, H. Z. (2011). Effect of lappaconitine on neuropathic pain mediated by P2X3 receptor in rat dorsal root ganglion. *Neurochemistry International*, 58(5), 564–573.

Papsun, D. M., Chan-Hosokawa, A., Friederich, L., Brower, J., Graf, K., & Logan, B. (2019). The trouble with kratom: Analytical and interpretative issues involving mitragynine. *Journal of Analytical Toxicology*, 43(8), 615–629. https://doi.org/10.1093/jat/bkz064

Paul, B., Sribhashyam, S., & Majumdar, S. (2023). Opioid signaling and design of analgesics. *Progress in Molecular Biology and Translational Science*, *195*, 153–176.

Qin, G., Gui, B., Xie, J., Chen, L., Chen, L., Cui, Z., … & Tan, G. (2018). Tetrandrine alleviates nociception in a rat model of migraine via suppressing S100B and p-ERK activation in satellite glial cells of the trigeminal ganglia. *Journal of Molecular Neuroscience*, *64*, 29–38.

Radulovic, N., Blagojevic, P., Randjelovic, P., & Stojanovic, N. (2013). The last decade of antinociceptive alkaloids: Structure, synthesis, mechanism of action and prospect. *Current Topics in Medicinal Chemistry*, *13*(17), 2134–2170.

Rao, S., Liu, S., Zou, L., Jia, T., Zhao, S., Wu, B., … & Liang, S. (2017). The effect of sinomenine in diabetic neuropathic pain mediated by the P2X3 receptor in dorsal root ganglia. *Purinergic Signalling*, *13*(2), 227–235.

Rose, P., Nasim, J., & Zhu, Y. Z. (2024). Novel compounds from chemistry to druggable candidates. *Frontiers in Chemistry*, *12*, 1494407.

Ruiz-Miyazawa, K. W., Zarpelon, A. C., Pinho-Ribeiro, F. A., Pavao-de-Souza, G. F., Casagrande, R., & Verri Jr, W. A. (2015). Vinpocetine reduces carrageenan-induced inflammatory hyperalgesia in mice by inhibiting oxidative stress, cytokine production and NF-κB activation in the paw and spinal cord. *PloS One*, *10*(3), e0118942.

Schep, L. J., Slaughter, R. J., & Beasley, D. M. G. (2009). Nicotinic plant poisoning. *Clinical Toxicology*, *47*(8), 771–781.

Shi, Y., & Wu, W. (2023). Multimodal non-invasive non-pharmacological therapies for chronic pain: Mechanisms and progress. *BMC Medicine*, *21*, 372. https://doi.org/10.1186/s12916-023-03076-2

Singh, A. K., Kumar, S., & Vinayak, M. (2018). Recent development in antihyperalgesic effect of phytochemicals: Anti-inflammatory and neuro-modulatory actions. *Inflammation Research*, *67*, 633–654.

Sun, Q., Zhao, N., Zhang, T. J., Wang, S. Y., Wang, L., & Meng, F. H. (2024). Development of a novel UHPLC-MS/MS method for the quantification of corynoxeine: Application to pharmacokinetics and tissue distribution studies in normal and chronic unpredictable mild stress-induced depression rats. *Journal of Pharmaceutical and Biomedical Analysis*, *238*, 115850.

Tarselli, Michael, Raehal, Kirsten, Brasher, Alex, Streicher, John, Groer, Chad, Cameron, Michael, Bohn, Laura, & Micalizio, Glenn. (2011). Synthesis of conolidine, a potent non-opioid analgesic for tonic and persistent pain. *Nature Chemistry*, *3*, 449–453. https://doi.org/10.1038/nchem.1050

Till, C., & Jones, N. S. (2019). Tricyclic antidepressants for chronic pain: Limited efficacy and problematic adverse effects. *Pain Management*, 9(6), 557–560.

Wang, R., Han, L., Gao, Q., Chen, D., Wang, Y., Zhang, X., … & Bai, C. (2018). Progress on active analgesic components and mechanisms of commonly used traditional Chinese medicines: A comprehensive review. *Journal of Pharmacy & Pharmaceutical Sciences*, *21*, 437–480.

Wang, Y., Zhao, X., Wang, J., & Zhu, X. (2024). Norisoboldine reduces arthritis severity by attenuating inflammation, oxidative stress, and extracellular matrix degradation in a rat model of rheumatoid arthritis. *Journal of Inflammation Research*, *17*, 8839–8852.

Xie, M. X., Zhu, H. Q., Pang, R. P., Wen, B. T., & Liu, X. G. (2018). Mechanisms for therapeutic effect of bulleyaconitine A on chronic pain. *Molecular Pain*, *14*, 1744806918797243.

Yang, S., Yu, Z., Sun, W., Jiang, C., Ba, X., Zhou, Q., … and Hao, Y. (2020). The antiviral alkaloid berberine ameliorates neuropathic pain in rats with peripheral nerve injury. *Acta Neurologica Belgica*, 120(3), 557–564.

Yoo, J. H., Yang, E. M., Cho, J. H., Lee, J. H., Jeong, S. M., Nah, S. Y., … & Jang, C. G. (2006). Inhibitory effects of berberine against morphine-induced locomotor sensitization and analgesic tolerance in mice. *Neuroscience*, *142*(4), 953–961

You, L., Yang, C., Du, Y., Wang, W., Sun, M., Liu, J., … & Ni, J. (2020). A systematic review of the pharmacology, toxicology and pharmacokinetics of matrine. *Frontiers in Pharmacology*, *11*, 01067.

Yu, D., Thakor, D. K., Han, I., Ropper, A. E., Haragopal, H., Sidman, R. L., Zafonte, R., Schachter, S. C., & Teng, Y. D. (2013). Alleviation of chronic pain following rat spinal cord compression injury with multimodal actions of huperzine A. *Proceedings of the National Academy of Sciences*, *110*(8), E746–E755. https://doi.org/10.1073/pnas.1300083110

Yu, G., Qian, L., Yu, J., Tang, M., Wang, C., Zhou, Y., … & Tang, Z. (2019). Brucine alleviates neuropathic pain in mice via reducing the current of the sodium channel. *Journal of Ethnopharmacology*, *233*, 56–63.

Yun, D., Yoon, S. Y., Park, S. J., & Park, Y. J. (2021). The anticancer effect of natural plant alkaloid isoquinolines. *International Journal of Molecular Sciences*, *22*(4), 1653

11 Alkaloids as Antioxidants
Implications for Health and Disease Prevention

Shivam Verma, Shweta Singh Verma, Nisha Sharma, Jay Kumar, Amit Verma, and Prakash Chandra Gupta

11.1 INTRODUCTION

Oxidative stress has emerged as a pivotal contributor to the pathogenesis of numerous chronic and degenerative disorders, making it a major field of biomedical research over the past two decades. It is characterized by an imbalance between the production of reactive oxygen species (ROS) and the antioxidant defense mechanisms that preserve cellular and tissue integrity. ROS, such as superoxide anions, hydrogen peroxide, and hydroxyl radicals, are naturally generated during aerobic metabolism but can become pathologically elevated under stress conditions, leading to damage of lipids, proteins, and nucleic acids (Afzal et al., 2023).

Although the human body possesses endogenous antioxidant enzymes, including superoxide dismutase (SOD), catalase (CAT), and glutathione peroxidase (GPx), these defenses often prove insufficient when ROS generation surpasses physiological thresholds. The sustained oxidative burden not only disrupts redox homeostasis but also impairs intracellular signaling, contributing to disease progression. ROS are now recognized not only for their cytotoxic effects but also for their role in modulating various signaling pathways, which makes their regulation crucial to disease prevention and therapy (Jomova et al., 2024).

11.2 ALKALOIDS AS NATURAL ANTIOXIDANTS

In response to concerns regarding the toxicity and limited efficacy of synthetic antioxidants such as butylated hydroxytoluene (BHT) and butylated hydroxyanisole (BHA), increasing attention has turned to naturally occurring phytochemicals, particularly alkaloids. These nitrogen-containing compounds are widely distributed in the plant kingdom and exhibit diverse pharmacological properties, including potent antioxidant activity (Gulcin, 2025).

Structurally, alkaloids encompass various chemical families, including isoquinoline, indole, quinoline, pyrrolidine, piperidine, steroidal, and tropane alkaloids. This structural diversity contributes to a range of biological activities and physicochemical properties. Their widespread occurrence and historical use in traditional medicinal systems, such as Ayurveda, Traditional Chinese Medicine, and Unani, highlight their long-standing therapeutic relevance (Jang et al., 2009). Some alkaloid-rich plants with their therapeutic activity are summarized in Table 11.1.

11.3 MECHANISMS OF ANTIOXIDANT ACTION

The antioxidant activity of alkaloids involves both direct and indirect mechanisms.

 DOI: 10.1201/9781003539179-11

TABLE 11.1
List of Alkaloid-Rich Plants, Their Major Chemical Constituents, and Biological Activities

S. No.	Plants	Chemical Constituent	Activity
1	*Catharanthus roseus*	Vincristine and Vinblastine	Anti-Cancerous
2	*Corydalis sp.*	Tetrahydropalmatine	Neuroprotective
3	*Aconitum sp.*	Pseudaconitine	Anti-Bacterial
4	*Rauwolfia serpentina*	Reserpine	Anti-Hypertensive
5	*Coptis chinensis*	Berberine	Anti-Inflammatory
6	*Papaver somniferum*	Morphine	Analgesic
7	*Coffea arabica*	Caffeine	CNS stimulant
8	*Cinchona officinalis*	Quinine	Antimalarial
9	*Atropa belladonna*	Atropine	Anticholinergic
10	*Ephedra sinica*	Ephedrine	Bronchodilation, decongestant

11.3.1 Direct Mechanisms

Include free radical scavenging via electron or hydrogen atom donation, metal ion chelation (thereby preventing Fenton-type reactions), and inhibition of lipid peroxidation (Bogolitsyn et al., 2019).

11.3.2 Indirect Mechanisms

Involve the upregulation of antioxidant enzymes, modulation of redox-sensitive transcription factors like nuclear factor erythroid 2-related factor 2 (Nrf2), and attenuation of inflammation-related oxidative stress (Bogolitsyn et al., 2019). Recent research has underscored these mechanisms through various examples.

11.3.2.1 Oxidative Stress Reduction

Berberine enhances cellular defenses by activating the AMPK pathway, leading to reduced ROS levels and elevated antioxidant enzyme activity (García-Muñoz et al., 2024).

11.3.2.2 Neuroprotection

Certain alkaloids, such as harmine and harmaline, exhibit neuroprotective properties by targeting oxidative stress and mitochondrial dysfunction, critical features of neurodegenerative diseases (Niğdelioğlu Dolanbay et al., 2024).

11.4 HEALTH BENEFITS AND BIOLOGICAL ACTIVITIES

The antioxidant properties of alkaloids contribute significantly to their preventive roles in chronic diseases, including the following.

11.4.1 Cardiovascular Diseases and Type 2 Diabetes

These conditions are closely linked to oxidative stress and inflammation. Alkaloids such as berberine, piperine, and harmine have been demonstrated to mitigate lipid peroxidation and improve antioxidant status (Kamiloglu et al., 2022; Chen et al., 2016).

11.4.2 Multifunctional Roles

Beyond the antioxidant effect, various alkaloids have also shown antimicrobial and anti-inflammatory as well as cardioprotective activity (Atpadkar et al., 2023).

11.5 ALKALOIDS AND MITOCHONDRIAL PROTECTION

Mitochondrial impairment is a great contributor to the intracellular ROS production. Some of the alkaloids, including sinomenine and tetrandrine, help maintain the mitochondrial membrane potential and also decrease ROS formation by suppressing NADPH activity. Various models of liver damage, heart attack, and neurotoxicity confirm all the protective effects (Bribi, 2018).

11.6 STRUCTURE–ACTIVITY RELATIONSHIP (SAR)

Growing areas of research explore the SAR of alkaloids in relation to their antioxidant properties. Specific structural properties, including electron-donating substitutes, aromatic conjugation, and heteroaromatic rings, escalate their ROS scavenging and electron-donating abilities. Computational simulation and molecular docking studies play a crucial role in elucidating SAR characteristics (Gutiérrez-Grijalva et al., 2020).

11.7 DUAL ANTIOXIDANT AND ANTI-INFLAMMATORY EFFECTS

Alkaloids consistently demonstrate antioxidant and anti-inflammatory properties, making them valuable in conditions characterized by oxidative stress and inflammation. For instance,

- Tetrandrine can decrease oxidative stress indicators and also suppress inflammatory cytokines, such as TNF-α and IL-6.
- In experimental studies on neuroinflammation and rheumatoid arthritis, sinomenine demonstrated protective effects (Laghezza Masci et al., 2019).

11.8 ETHNOMEDICAL AND NUTRACEUTICAL APPLICATIONS

For a long time, traditional medicines have comprised alkaloid-rich plants, such as *Catharanthus roseus*, for managing diabetes, *Uncaria tomentosa* for treating inflammation, and *Peganum harmala* for treating neurological disorders. Modern pharmacological research has approved these applications (Adamski et al., 2020). Moreover, functional foods and drinks rich in alkaloids, such as yerba mate, green tea, and guarana, are generally consumed for their health-promoting properties. Compounds such as caffeine and theobromine exhibit mild antioxidant activity, which is related to both cardiovascular and neuroprotective actions (Dey & Mukherjee, 2018).

11.9 TECHNOLOGICAL ADVANCEMENTS AND CHALLENGES

Progression in extraction and delivery methods, such as supercritical fluid extraction and nanoencapsulation, as well as lipid-based systems, is overcoming the bioavailability challenges of alkaloids. These systems enhance pharmacokinetic profiles, reduce adverse effects, and also improve the therapeutic potential of alkaloid-based treatments (Mukherjee et al., 2017). Although challenges persist, including variability in natural sources, narrow therapeutic windows, regulatory hurdles, insufficient clinical trials, and potential interactions with cytochrome P450 enzymes that affect drug metabolism.

11.10 RATIONALE FOR INVESTIGATING ANTIOXIDANT PROPERTIES OF ALKALOIDS

Alkaloids exhibit antioxidant activity in managing oxidative stress-related conditions, such as cancer and neurodegenerative diseases, by counteracting oxidative damage through the scavenging of free radicals and the regulation of cellular stress pathways (Bagga et al., 2024). The following sections outline their applications in various contexts.

11.10.1 Antioxidant Mechanisms

Alkaloidal compounds from *Huperzia selago* and *Diphasiastrum complanatum* exhibited potent antioxidant effects, which protect brain macromolecules against oxidative injury (Czapski et al., 2014). *U. tomentosa* extract has been shown to lower ROS levels, indicating its potential in enhancing cellular defense (Azevedo et al., 2019).

11.10.2 Neuroprotection

Alkaloids are notable for their neuroprotective properties, acting on key pathways in neurodegenerative disorders such as Alzheimer's and Parkinson's disease. Their effects include regulating oxidative stress, protecting mitochondrial function, and reducing neuroinflammation (Niğdelioğlu Dolanbay et al., 2024). Certain alkaloids have demonstrated their potential in mitigating oxidative damage associated with amyloid proteins, which play a crucial role in neurodegenerative processes (Czapski et al., 2014).

11.10.3 Cancer Therapeutics

Certain alkaloidal compounds act as Nrf2 pathway inhibitors, which are important regulators of cellular defense against oxidative stress. By hindering this pathway, they can enhance the sensitivity of cancer cells to chemotherapy, thereby improving therapeutic effectiveness (Gjorgieva Ackova et al., 2023). Alkaloids such as berberine and evodiamine have been shown to have potential in cancer treatment by inducing oxidative stress within cancer cells, which subsequently leads to apoptosis (Gjorgieva Ackova et al., 2023). Although alkaloids exhibit considerable therapeutic responses, further research is needed to clarify their mechanism.

11.11 CHEMISTRY OF ALKALOIDS RELATED TO ANTIOXIDANT ACTIVITY

Alkaloids are a chemically diverse class of nitrogen-containing natural compounds, primarily produced by plants as secondary metabolites, but some are also derived from bacteria, fungi, and other sources. Their unique chemistry is both intriguing and biologically relevant, especially their antioxidant function, which is influenced by their structural properties. Many alkaloids possess phenolic or indole moieties, which are known for their ability to donate electrons or hydrogen atoms, thus neutralizing free radicals and ROS. The nitrogen atom within the alkaloid core can act as an electron donor or participate in redox reactions, making alkaloids excellent candidates for antioxidant action. Furthermore, alkaloids may participate in metal ion chelation (Desmet et al., 2021).

11.11.1 Mechanisms of Antioxidant Action

Alkaloids may act via several mechanisms.

11.11.1.1 Direct Radical Scavenging

Hydrogen atom donation or single-electron transfer to neutralize radicals like superoxide, hydroxyl, and peroxyl radicals (Lu et al., 2010).

11.11.1.2 Metal Chelation

Some alkaloids can chelate redox-active transition metals (Fe^{2+}, Cu^{2+}), preventing the Fenton reaction and subsequent hydroxyl radical formation (Parvin et al., 2022).

11.11.1.3 Modulation of Antioxidant Enzymes

Through transcriptional activation of Nrf2 or inhibition of pro-oxidant pathways such as NF-κB and MAPK (Xu et al., 2022).

11.11.1.4 Inhibition of Lipid Peroxidation

Preventing oxidative degradation of polyunsaturated fatty acids, which maintains membrane integrity (Catalá, 2013).

11.12 EXAMPLES OF ALKALOIDS AND THEIR ANTIOXIDANT CHEMISTRY

11.12.1 Berberine

It is a protoberberine alkaloid that contains a planar isoquinoline nucleus with methylenedioxy groups. It is capable of generating ROS in *Rhizophora mucronata* and suppressing oxidative markers through mitochondrial pathways. It can intercalate DNA and inhibit oxidative damage to nucleic acids (Gurudeeban et al., 2015).

11.12.2 Harmine and Harmaline

These are β-carboline alkaloids with a tricyclic indole system that confers excellent radical-stabilizing properties. Their antioxidant potential is often attributed to hydroxyl substitutions on the aromatic ring.

11.12.3 Yohimbine

It exhibits antioxidant action in addition to being an α-2 adrenergic receptor antagonist. Its fused indole–quinolizidine framework contributes to both receptor binding and free radical neutralization.

11.12.4 Sanguinarine

It is a benzophenanthridine alkaloid with a quaternary nitrogen structure that promotes redox cycling and ROS interaction.

11.12.5 Reserpine

It contains a yohimban skeleton and a methoxy-substituted indole ring, acting on vesicular monoamine transporters and modulating oxidative neurotransmitter metabolism.

11.12.6 Cytisine and Lupinine

Though less potent antioxidants, their nitrogen-containing ring systems allow moderate free radical interaction and possible synergistic effects when combined with flavonoids. Chemical Modifications and Structure–Activity Relationship (SAR) studies have shown that electron-donating groups at specific positions enhance activity. The increased planarity of the molecule correlates with better interaction with ROS. Methylation of phenolic groups reduces antioxidant potential. Quaternization of nitrogen enhances water solubility but may reduce membrane penetration. These findings offer a blueprint for the synthetic modification of natural alkaloids or the de novo synthesis of analogs to enhance antioxidant efficacy.

11.13 FUTURE DIRECTIONS IN ALKALOID CHEMISTRY

Modern tools, including molecular docking, quantum chemical calculations, and machine learning, are being applied to predict and design potent antioxidant alkaloids. Biotechnological synthesis through engineered microbial pathways offers a scalable and environmentally friendly route to rare and complex alkaloids. Combinatorial chemistry can further refine SAR models to develop high-potency antioxidant candidates with better bioavailability and safety profiles (Agrawal et al., 2024). The structure of some alkaloid-derived compounds is shown in Figure 11.1.

In conclusion, the chemistry of alkaloids is at the core of their antioxidant behavior. By understanding the structural features, reactive sites, and contributions of functional groups, researchers can develop innovative therapeutic agents for diseases related to oxidative stress. Integrating traditional knowledge with modern synthetic and computational chemistry will be essential to fully harness the antioxidant power of alkaloids.

11.14 STRUCTURAL FEATURES CONTRIBUTING TO ANTIOXIDANT POTENTIAL

The antioxidant potential of phenolic and anilinic compounds is significantly influenced by their structural features, particularly the presence and position of specific functional groups. The study highlights several key aspects that contribute to their radical scavenging and antioxidant activities.

11.14.1 The Presence of a Second Hydroxyl or Amino Group Is Important for Enhancing Antioxidant Activities

11.14.1.1 Ortho or Para Positioning

The most crucial structural feature for practical antiradical and antioxidant activities in phenolic compounds is the presence of a second hydroxyl group in the *ortho* or *para* position. Similarly, for aniline compounds, an amino group in these positions is highly effective.

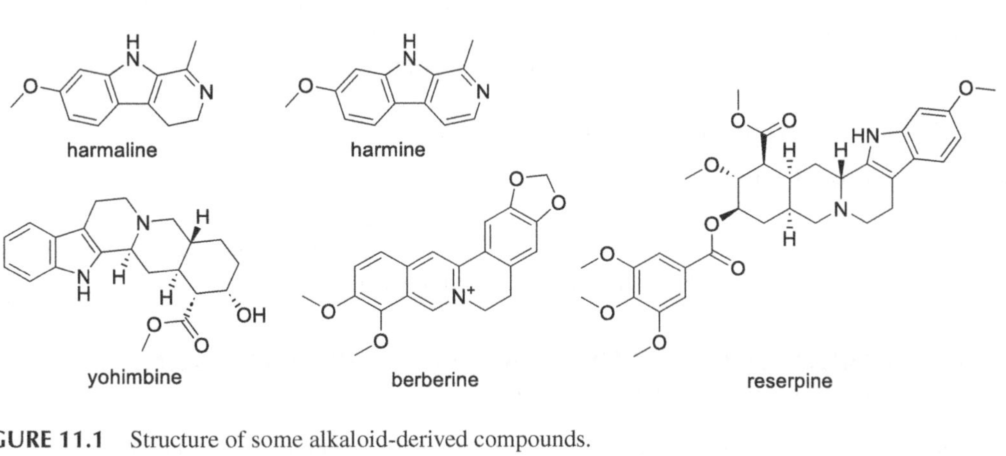

FIGURE 11.1 Structure of some alkaloid-derived compounds.

11.14.1.2 Electron-Donating Effect

This positioning is effective due to the strong electron-donating effect of these groups. For instance, the hydroxyl group in *the ortho* and *para* positions has a strong electron-donating ability ($\sigma+ = 0.92$, $\sigma+ = 0.61$). The amino group also exhibits a strong electron-donating effect ($\sigma+ = -1.30$, $\sigma+ = 0.86$).

11.14.1.3 Stable Quinone-Like Products

The presence of these groups in *ortho* or *para* positions facilitates a two-hydrogen-atom transfer process, leading to the formation of stable quinone-like products (for phenols) or iminoketones (for anilines), which are critical for their antioxidant action. Catechol and hydroquinone, for example, undergo this process to form stable *o* and *p*-quinones.

11.14.1.4 Contrast with Meta-Position

In contrast, a hydroxyl group in the *meta* position (e.g., in resorcinol) shows very low activity (2.5%) because it acts as an electron-withdrawing group ($\sigma m = 0.12$). Similarly, *m*-aminophenol showed only moderate activity (20.2%) (Platzer et al., 2022).

11.14.2 Electron-Donating Groups (EDG)

11.14.2.1 Alkyl and Methoxy Groups

Electron-donating groups, especially alkyl and methoxy groups, enhance the electron density of phenoxyl radicals, thereby improving radical scavenging and antioxidant activity.

11.14.2.2 Stabilization of Phenoxyl Radicals

These groups stabilize the resulting phenoxyl radicals through inductive effects (alkyl substituents) or resonance effects (methoxy or amino substituents), which lowers the O-H bond energy and boosts radical scavenging activity.

11.14.2.3 Positional Importance

For methoxy groups, their presence in *ortho* or *para*-positions is necessary for them to act as beneficial electron-donating groups ($\sigma+ = -0.78$, $\sigma+ = -0.51$). While *m*-methoxy groups typically act as electron-withdrawing groups ($\sigma m = 0.12$), the strong electron-withdrawing activity of an oxygen phenoxyl radical can cause them to become weak electron-donating groups ($\sigma m = -0.14$).

11.14.2.4 Additive Effect

The study suggests that alkyl and alkoxy groups, when in the correct positions, can enhance activity in an additive manner. For instance, BHT with three alkyl groups showed high activity (96.0%). Syringic acid, with two methoxy groups, also exhibited high activity (90.4%) (Platzer et al., 2022).

11.14.3 Electron Withdrawing Groups (EWG)

11.14.3.1 Destabilization

Conversely, electron-withdrawing groups tend to stabilize the phenols but destabilize the resulting radicals, thus reducing antioxidant potential. Monophenols with substituents such as carboxyl, ester, CHO, CH_3CO, SO_3H, NO_2, and Cl generally exhibited low activity (<3.6%).

11.15 EXAMPLES OF KEY STRUCTURAL CLASSES (ISOQUINOLINE, INDOLE, QUINOLIZIDINE, ETC.)

The structural classes of alkaloids, such as isoquinoline, indole, and quinolizidine, represent a broad spectrum of naturally occurring nitrogen-containing compounds with remarkable structural diversity and potent biological activities. These compounds are foundational in natural product chemistry due to their complex biosynthetic origins and structural motifs. In pharmacology, they serve as prototypes or direct agents in the development of therapeutics for diseases ranging from cancer to neurological and infectious disorders. Their structural uniqueness, biological specificity, and medicinal relevance continue to make them central to the fields of drug discovery and chemical biology. Furthermore, there is a comprehensive overview of each of these classes.

11.15.1 Isoquinoline Alkaloids

11.15.1.1 Diversity

Isoquinoline alkaloids are one of the widest groups of plant-based alkaloids, ranking just after indole alkaloids in structural diversity and prevalence. Among them, aporphinoids form an important subgroup, characterized by a tetracyclic framework biosynthesized from the amino acid tyrosine. Their structural diversity, shaped by various ring fusion and oxidative modifications, underlies their broad spectrum of biological activities (dos Santos & Vaz, 2019).

11.15.1.2 Biological Activities

Isoquinoline alkaloids represent a wide biological properties. Berberine is known for its antimicrobial and antidiabetic effects, but morphine and codeine are used as pain relievers. Certain compounds in this group interact with dopaminergic receptors, which is significant in neuropharmacology (Wang et al., 2024).

11.15.1.3 Chemotaxonomy

Since isoquinoline alkaloids are limited to certain plant families, consisting of Berberidaceae, Papaveraceae, and Menispermaceae, they are employed as chemotaxonomic markers. Their occurrence and absence offer helpful information for plant classification and evolutionary studies (dos Santos & Vaz, 2019).

11.15.2 Indole Alkaloids

11.15.2.1 Structural Motifs

Indole alkaloids originate from tryptophan and contain an indole nucleus, which is a bicyclic structure made up of six-membered benzene rings fused to five-membered pyrrole rings with nitrogen. This core structure enables biological interactions, primarily with neurotransmitter receptors, enzymes, and nucleic acids. Structural diversity is further enhanced through substitution at several sites on the indole ring (Michael, 2007).

11.15.2.2 Natural Sources

Indole alkaloids are distributed across plants, marine species, and amphibians. Members of the Apocynaceae family, such as *C. roseus*, yield anticancer compounds, including vincristine and vinblastine (Michael, 2007).

11.15.3 Quinolizidine Alkaloids

11.15.3.1 Chemical Diversity

Quinolizidines are marked by a bicyclic nitrogen-containing core procured from lysine and are grouped into 20 subclasses, containing matrine, lupanine, and sparteine. They are mainly prevalent in leguminous plants, such as *Lupinus, Sophora*, and *Cytisus* species. Their structural diversity arises from processes such as oxidation, ring fusion, and various functional group substitutions, which influence both solubility and biological activity (Cely-Veloza et al., 2023).

11.15.3.2 Bioactivity

Quinolizidine alkaloids have wide pharmacological activities, comprising cytotoxicity and anti-inflammatory as well as antiviral effects. Matrine is investigated for its anticancer and immunomodulatory potential, but cytisine functions as a partial agonist of the nicotinic receptor and is also applied in smoking cessation treatment. Their bioactivity exhibits remarkable structural features, making them promising candidates for managing both chronic and infectious diseases (Cely-Veloza et al., 2023).

11.16 MECHANISMS OF ANTIOXIDANT ACTION OF ALKALOIDS

11.16.1 The Antioxidant Action of Alkaloids: Mechanistic Insights and Therapeutic Implications

Alkaloids exhibit antioxidant properties through multiple biochemical and molecular pathways, which enable them to scavenge reactive species, protect biomolecules from oxidative injury, and maintain intracellular redox balance. These are evaluated based on their chemical structure and functional groups, as well as their capacity to interact with various cellular targets. This part focuses on the major pathways by which alkaloids reduce oxidative stress and strengthen the defense process. Structurally, alkaloids are a group of natural products that comprise pharmacological properties. Among all these, their antioxidant properties have garnered growing interest due to their capacity to counteract oxidative stress, a major factor in the development of various diseases, including cancer, neurodegenerative, and cardiovascular conditions. Examples, including berberine, quinolone, isoquinolone, and piperine, have demonstrated antioxidant activity through mechanisms such as enzyme regulation, free radical neutralization, and modulation of transcription factors (Sivashanmugam & Velmathi, 2021).

11.16.2 Enzyme Activation

One of the primary mechanisms by which alkaloids exert antioxidant activity is by enhancing the expression of endogenous antioxidant enzymes. Like berberine acquired from *Berberis* species, they have proven to boost the activity of SOD, CAT, and GPx; these play a crucial role in detoxifying ROS. These enzymes act as the first line of defense against oxidative stress by converting superoxide radicals into hydrogen peroxide, and hence preserving the redox balance within the cell. By enhancing enzymatic activity, berberine helps reduce oxidative injury and inflammation (García-Muñoz et al., 2024). Similarly, piperine from *Piper nigrum nourishes* antioxidant defense by structural analógs. The SAR activities of piperine indicate that alterations in the aromatic ring and methylenedioxy groups affect its redox properties, highlighting the potential for rational structural modifications to enhance its antioxidant effectiveness (Joshi et al., 2021).

11.16.3 Free Radical Scavenging via SPLET Mechanism

Another crucial antioxidant pathway is direct free radical scavenging, sometimes mediated through the sequential proton loss electron transfer (SPLET) mechanism. Viridicatol and 5-hydroxy-8-methoxy-4-phenylisoquinolin-1(2H)-one are good radical scavengers. In this SPLET mechanism, the phenolic group undergoes deprotonation first, and further electron transfer helps stabilize the radical. It is favorable for the polar solvents where there is a need for proton affinity (Dung et al., 2020).

Additionally, alkaloids exhibit reliable antioxidant activity in various solvents, including water, methanol, and DMSO, indicating solvent-independent antioxidant activity (Dung et al., 2020).

11.16.4 Modulation of Transcription Factors: The Nrf2 Pathway

By directly neutralizing radicals, alkaloids provide indirect antioxidant benefits by managing gene expression via transcription factors. From all studies, Nrf2 is the most extensively studied, serving as a key regulator of the antioxidant response element in DNA. Compounds, such as berberine, have been shown to activate Nrf2 signaling, as well as detoxification and antioxidant defense (Gjorgieva Ackova et al., 2023).

The activation of Nrf2 can benefit healthy cells by strengthening their defense against oxidative stress; sustained Nrf2 activation in cancer cells may encourage chemoresistance and support survival.

11.17 Evidence from *In Vitro* and *In Vivo* Studies

11.17.1 Comparative Analysis of *In Vitro* and *In Vivo* Antioxidant Activity of Alkaloids from *Uncaria Tomentosa*

The antioxidant activity of *U. tomentosa*, a plant with medicinal properties, has been confirmed by the *in vitro* and *in vivo* models. Even notable differences often arise between antioxidant properties observed in living models. This emphasizes the challenges of effectiveness and reveals the effects of metabolic and physiological, as well as molecular, processes in biological systems.

11.17.2 *In Vitro* Antioxidant Activity

In vitro assays serve as an essential starting point for evaluating the antioxidant activity of natural compounds. This evaluation is conducted in a precise and controlled manner to minimize biological variability. The commonly used methods are DPPH (2,2-diphenyl-1-picrylhydrazyl), ABTS (2,2-azino-bis(3-ethylbenzothiazoline-6-sulfonic acid)), and FRAP (ferric reducing antioxidant power) assays, which assess the compound's ability to scavenge free radicals. Using this method, *U. tomentosa* has demonstrated radical scavenging activity, indicating the presence of bioactive antioxidant components; *in vitro* results have shown a decrease in ROS (Azevedo et al., 2019).

In-depth studies on alkaloids, such as mitraphylline, a key bioactive agent in *U. tomentosa*, have not consistently demonstrated an antioxidant effect. This suggests that the *in vitro* antioxidant effect may result from a synergistic interaction between the components. This emphasizes a more targeted approach when evaluating plant-based antioxidants, particularly in the context of therapeutic development (Azevedo et al., 2019).

11.17.3 *In Vivo* Antioxidant Activity

In vivo studies also provided important insights into the activity of *U. tomentosa* within the complex biological environment of a living organism. The activity performed using rats that are exposed to stress and then *U. tomentosa* extract significantly improves the oxidative damage confirmed from the lipid peroxidation and protein carbonyl levels, and it was also seen that the extract also increases the activity of antioxidant enzymes such as SOD, CAT, and GPx, which reflects its potential to strengthen the defense mechanism (Oubaid et al., 2023).

These results are mediated by both direct radical scavenging and the regulation of redox-sensitive genes and signaling. Studies have shown that *U. tomentosa* stimulates the Nrf2 pathway, leading to increased expression of genes associated with detoxification and antioxidant defense (Onakpa et al., 2016). This gene-level modulation proves an extensive biological impact, highlighting how complex interactions *in vivo* may contribute to therapeutic outcomes that are not evident from *in vitro* assays alone.

Additionally, the physiological environment is crucial, as these alkaloids play a significant role in determining their effectiveness. Within the *in vivo* system, proteins, enzymes, and cellular structures can interact with these compounds, potentially enhancing or reducing their antioxidant activity. This was observed during the *in vivo* studies of the antioxidant effects of *U. tomentosa*, where the results varied according to the animal's stress condition, metabolic state, and organ-specific response (Pasqualetti et al., 2021). The results highlight the fact that *in vivo* findings cannot be directly translated into clinical outcomes.

11.17.4 Challenges in Translating *in vitro* Antioxidant Activity to Human Applications

Antioxidant exploration typically begins with *in vitro* experiments, providing crucial insights into the redox modulation of both natural and synthetic compounds. However, applying these results to living human contexts is a complex procedure that also needs to consider physiological, biological, and methodological factors. *In vitro* results do not directly translate to living systems, mainly due to hindrances related to bioavailability, metabolic processes, and variations in measurement methods (Kasote et al., 2015).

11.17.5 Correlation between *in vitro* and *in vivo* Studies

In vitro assays, such as DPPH, FRAP, and ABTS, are used to assess the radical scavenging properties of antioxidants. The results from the studies reflected that active compounds from various sources demonstrated excellent antioxidant potential under controlled experimental conditions (Oubaid et al., 2023; Power et al., 2013). While such results are encouraging, they often fail to replicate with the same magnitude in *in vivo* systems.

This disparity arises due to several factors. Notably, clinical studies indicate that the correlation between *in vitro* antioxidant scores and *in vivo* outcomes can be inconsistent or weak, largely due to the complexity of human physiology. As emphasized by Beretta and Facino, the choice of biomarkers, individual metabolism, and the form of compound administration greatly influence the observed effects in clinical trials. Consequently, a compound that appears highly active *in vitro* may exhibit minimal or no beneficial effect when administered orally, depending on its absorption, metabolism, and distribution throughout the body (Beretta & Facino, 2010).

11.17.6 Mechanisms of Action and Bioavailability

A major determinant of antioxidant efficacy *in vivo* is bioavailability, which refers to the proportion of a compound that enters systemic circulation and reaches target tissues in an active form. Phenolic compounds, for instance, are among the most potent antioxidants *in vitro*; however, their absorption in the gastrointestinal tract, chemical modifications during first-pass metabolism, and subsequent elimination can significantly alter their biological activity (Martins et al., 2016). Despite these challenges, some studies have successfully demonstrated the functional effects of antioxidants in human subjects; for example, they reported that consumption of antioxidant-rich extracts resulted in measurable reductions in circulating ROS and improvements in markers of oxidative stress, confirming that under certain conditions, dietary antioxidants can indeed confer health benefits. However, such findings are not universal, and the variability in study design, subject population, and compound formulation complicates the interpretation and generalization of results (Nemzer et al., 2014).

11.17.7 Need for Standardization and Further Research

One of the persistent limitations in antioxidant research is the lack of standardized methods for measuring antioxidant activity, both *in vitro* and *in vivo*. While *in vitro* assays are relatively simple and cost-effective, they often rely on chemical reactions that may not mimic physiological conditions. This makes it difficult to compare results across studies or to extrapolate them to clinical contexts (Power et al., 2013). Additionally, *in vivo* activity may differ in certain aspects, such as study duration, dosing, and biomarker consideration, which can further lead to inconsistent results. So, there is a need for well-designed studies to minimize these gaps. Martin et al. emphasized that confirmation of the antioxidant effect in human subjects is crucial. However, there is a need for optimal therapeutic dosing, specific delivery methods, and long-term safety. Further research should focus on the biochemical interactions of antioxidants, their synergistic actions, and their contributions to disease prevention and management (Martins et al., 2016).

11.18 IMPORTANT ALKALOIDS STUDIED FOR ANTIOXIDANT EFFECTS (BERBERINE, CAFFEINE, HARMINE, ETC.)

11.18.1 Antioxidant Mechanisms of Selected Alkaloids: Berberine, Caffeine, and Harmine

Alkaloids, a diverse class of naturally occurring compounds with various pharmacological properties, are particularly valued for their strong antioxidant potential. Bioactive compounds, such as berberine and caffeine, play a crucial role in mitigating oxidative stress, which is important for the prevention of neurodegenerative diseases, cancer, and other conditions. The antioxidant mechanism appears through several mechanisms, such as mitigating the ROS, regulating antioxidant gene expression, and suppressing the inflammatory pathways (Hussain et al., 2018).

11.18.1.1 Berberine's Mechanisms of Antioxidant Action

Berberine, an isoquinoline alkaloid derived from various medicinal plants like *Berberis vulgaris*, has been examined for its antioxidant properties. One of the primary mechanisms of berberine's action is its direct scavenging of ROS, thereby protecting cells from oxidative stress. Studies have indicated that, both *in vitro* and *in vivo*, berberine can scavenge free radicals and contribute to lipid peroxidation, DNA damage, and protein denaturation (Imenshahidi & Hosseinzadeh, 2020). Additionally, Berberine has the potential to strengthen the body's intrinsic antioxidant system by stimulating the Nrf2 signaling pathway, which leads to Nrf2 moving into the nucleus and promoting the transcription of antioxidant enzymes. Activation leads to Nrf2 moving into the nucleus and promotes the transcription of antioxidant enzymes, including SOD, CAT, and GPx. This will lead to an enhancement of cellular mechanisms and ensure a prolonged antioxidant response against ongoing oxidative stress (Ashrafizadeh et al., 2020).

Additionally, berberine exhibited a beneficial anti-inflammatory effect, which helped reduce oxidative stress. This leads to the suppression of pro-inflammatory cytokines, such as TNF-α, IL-1β, and IL-6; furthermore, it inhibits the NF-κB signaling pathway and reduces the inflammatory conditions that give rise to ROS production (Shakeri et al., 2024).

11.18.1.2 Caffeine's Antioxidant Properties

Caffeine is a methylxanthine alkaloid that is present in various beverages like tea, coffee, etc. It is recognized for its stimulant action on the central nervous system and reflects potent antioxidant properties. A vital mechanism is the antagonism of the adenosine receptor, which leads to decreased neuronal toxicity and enhances mitochondrial function, thereby lowering oxidative stress (Tian et al., 2023). Caffeine maintains the oxidative balance by inhibiting phosphodiesterase enzymes; furthermore, it leads to an increased level of adenosine monophosphate (cAMP), which modulates the expression of antioxidant enzymes and elevates the levels of glutathione, SOD, and CAT as well. This increased antioxidant activity further contributes to better redox homeostasis, demonstrating caffeine's therapeutic potential in managing oxidative stress associated with disorders such as neurodegenerative or cardiovascular diseases (Tian et al., 2023).

11.18.1.3 Harmine's Neuroprotective and Antioxidant Effects

Harmine, the β-carboline alkaloid found in plants such as *Peganum harmala* and *Banisteriopsis caapi,* is recognized for its neuroprotective and antioxidant activity as well. It protects the neuronal cells from oxidative stress, specifically in experimental models of neurodegenerative disorder. Currently, studies suggest that harmine influences pathways associated with berberine, including Nrf2 activation, ROS suppression, and improvement of mitochondrial function (Tian et al., 2023). Harmine regulates multiple neurotransmitter systems and supports neurogenesis; it also serves as a potential candidate for treating neurodegenerative diseases. Antioxidant property is the main contributor to the neuroprotective effect, specifically against the neuronal injury caused by oxidative stress.

11.18.2 Considerations for Therapeutic Use

Although alkaloids like berberine, caffeine, and harmine have shown considerable pharmacological effects, they also contribute to the multiple therapeutic effects. Factors such as dosage, absorption, and metabolic differences vary from person to person, which can affect the effectiveness of bioactive compounds. For example, berberine has low bioavailability, which can hinder its antioxidant effects unless it is properly formulated. Similarly, high caffeine consumption may lead to side effects such as insomnia, tachycardia, etc., overshadowing its antioxidant effect. Therefore, systematic clinical trials and in-depth pharmacokinetic studies are necessary to establish optimal therapeutic regimens and gain a deeper understanding of how these compounds behave in the human system (Och et al., 2022).

11.19 COMPARATIVE ANTIOXIDANT CAPACITY WITH STANDARD ANTIOXIDANTS

11.19.1 Comparative Insights into Natural and Synthetic Antioxidants

Antioxidants are crucial for preventing oxidative stress, a primary factor in the development of chronic conditions such as cancer, neurodegenerative disorders, and cardiovascular disorders. These are classified as natural and synthetic, also based on their sources, chemical structure, and mechanism of action, among other factors. There are differences in antioxidant capacity, which offer both strengths and drawbacks. Antioxidants from natural origin are the first choice, as they are safe and offer additional benefits. However, synthetic antioxidants are used in industries due to their strong potency and chemical stability (Tan et al., 2018).

11.19.2 Origin and Composition

Natural antioxidants are produced from sources such as plants, animals, microorganisms, etc.; they also serve as the organism's defense mechanism. This includes compounds such as flavonoids, polyphenols, vitamins, and others. Due to their structural diversity, natural antioxidants exhibit multifunctional activities, including anti-inflammatory, anticancer, and antimicrobial properties. Additionally, their natural origin confirms their biocompatibility and reduced toxicity; this will lead to enhanced utilization in pharmaceuticals and nutraceuticals (Koltover & Skipa, 2023).

Synthetic antioxidants are chemically developed compounds primarily used for industrial purposes, and they also find applications in cosmetics, pharmaceutical formulations, and food preservation. Some familiar examples include BHA and BHT, which primarily inhibit lipid peroxidation in processed foods and edible oils. Far from their natural origin, these antioxidants are mainly monofunctional and especially modified so that they meet stability demands, specifically under high temperatures or oxidative conditions (Koltover & Skipa, 2023).

11.19.3 Efficacy and Mechanism of Action

The basic mechanism of natural and synthetic antioxidants differs significantly. Natural antioxidants usually act through indirect pathways by enhancing the body's defense system. For example, they can activate signaling cascades, such as the Nrf2 pathway, which further enhances the expression of endogenous antioxidant enzymes, including SOD and CAT. Many natural compounds affect the composition of gut microbiota, which in turn influences overall metabolic health (Koltover & Skipa, 2023).

Antioxidants from synthetic origin act by direct mechanisms, such as neutralizing free radicals or binding to metal ions, which further block the oxidative chain reaction. This targeted approach makes them effective at preventing auto-oxidation. Their structure makes them precise and predictable in activity; this also limits their vast biological influence. Synthetic antioxidants function effectively in their specific application, but their limited integration with the biological processes leads to a reduction in therapeutic effectiveness.

11.19.4 Safety and Health Implications

The natural antioxidant safety profile is a point of discussion, as it is typically considered safe and effective due to its evolutionary alignment with human physiology. Some of them have been included in traditional diets and herbal medicines for a long time, with minimal side effects. Additionally, the studies have shown that these natural antioxidants may contribute to disease prevention through their anti-inflammatory and immunomodulatory effects (Fadhil et al., 2023).

In the case of synthetic antioxidants, they have some serious health-causing issues when used for a long time with high dosing frequency. Research studies suggest that compounds such as BHA and BHT may cause endocrine disruption and be carcinogenic under such conditions. Thus, regulatory agencies and industries reassess their extensive use, as natural ones in food, cosmetics, and healthcare products replace synthetic preservatives.

11.20 FUTURE RESEARCH DIRECTIONS

11.20.1 Formulation Strategies

To manage challenges regarding the poor bioavailability, novel drug delivery methods should be considered. Nanocarriers, such as liposomes, nanoparticles, and microemulsions, have demonstrated potential in enhancing the solubility, stability, and permeability of berberine. Formulating berberine into nanoformulations may result in increased absorption and bioavailability.

11.20.2 Biosynthetic Pathway Elucidation

Biosynthetic pathways of alkaloids involve genomic, transcriptomic, and metabolomic approaches, facilitating the optimized production as well as structural modification. Additionally, regulatory points should also be recognized within these pathways in medicinal plants, focusing on how biosynthesis shapes both structural diversity and biological function. These points should be included to advance biotechnological and synthetic approaches, enabling the production of structurally active alkaloids.

11.20.3 Structural Modifications

There is a need for structural derivatives of natural alkaloids to provide an approach to ameliorate their pharmacokinetic and pharmacodynamic properties. For example, modifications of aporphine and isoquinoline lead to increased solubility, improved stability, high target selectivity, and enhanced antioxidant properties. SAR is the principal approach for drug design.

11.21 CONCLUSION

There is evidence that plant-based alkaloids can contribute to antioxidant defense through ROS, modulation of redox-sensitive signaling, and upregulation of endogenous antioxidant systems, making them a promising agent for strategies in chronic disease prevention. Systematically, alkaloids aim to function at different levels of redox regulation; some of them directly scavenge free radicals, while others inhibit ROS production enzymes and also downregulate pro-inflammatory pathways, such as ERK and iNOS. Overall, alkaloids comprise a valuable source of bioactive compounds that can manage oxidative stress and inflammatory responses. Preclinical and clinical studies show their potential effects in preventing specific diseases and disorders.

REFERENCES

Adamski, Z., L. L. Blythe, L. Milella and S. A. Bufo. (2020). Biological activities of alkaloids: From toxicology to pharmacology. *Toxins* 12(4):210.

Afzal, S., A. S. Abdul Manap, A. Attiq, I. Albokhadaim, M. Kandeel and S. M. Alhojaily. 2023. From imbalance to impairment: The central role of reactive oxygen species in oxidative stress-induced disorders and therapeutic exploration. *Frontiers in Pharmacology* 14:1269581.

Agrawal, K., T. M. Patel, S. Thakur, K. Patel and S. Mittal. 2024. Quantum chemical modelling, molecular docking, synthesis and experimental antimicrobial activity of 1, 4-diazepan linked piperidine derivative. *Future Journal of Pharmaceutical Sciences* 10(1):80.

Ashrafizadeh, M., H. S. Fekri, Z. Ahmadi, T. Farkhondeh and S. Samarghandian. 2020. Therapeutic and biological activities of berberine: The involvement of Nrf2 signaling pathway. *Journal of Cellular Biochemistry* 121(2):1575–1585.

Atpadkar, P. P., S. Gopavaram and S. Chaudhary. 2023. Natural–product–inspired bioactive alkaloids agglomerated with potential antioxidant activity: Recent advancements on structure-activity relationship studies and future perspectives. *In Vitamins and Hormones* 121:355–393.

Azevedo, C. B., M. Roxo, C. M. Borges, H. Peixoto, E. J. Crevelin, B. W. Bertoni, S. Contini, A. A. Lopes, S. C. França, A. M. S. Pereira and M. Wink. 2019. Antioxidant activity of an aqueous leaf extract from Uncaria tomentosa and its major alkaloids mitraphylline and isomitraphylline in Caenorhabditis elegans. *Molecules* 24(18):3299.

Bagga, S., P. Singh, P. Kumar, S. Kumar, A. Singh, S. Pathak and A. Khatri. 2024. Integrating traditional and modern medicine: A holistic approach to patient care. *Asian Journal of Pharmaceutical Analysis* 14(3):185–190.

Beretta, G. and R. M. Facino 2010. Recent advances in the assessment of the antioxidant capacity of pharmaceutical drugs: From in vitro to in vivo evidence. *Analytical and Bioanalytical Chemistry* 398(1):67–75.

Bogolitsyn, K., A. Druzhinina, P. Kaplitsin, D. Ovchinnikov, A. Parshina and M. Kuznetsova. 2019. Relationship between radical scavenging activity and polymolecular properties of brown algae polyphenols. *Chemical Papers* 73(10):2377–2385.

Bribi, N. 2018. Pharmacological activity of alkaloids: A review. *Asian Journal of Botany* 1(1):1–6.

Catalá, A. (2013). Five decades with polyunsaturated fatty acids: Chemical synthesis, enzymatic formation, lipid peroxidation and its biological effects. *Journal of Lipids 2013*(1):710290.

Cely-Veloza, W., M. J. Kato and E. Coy-Barrera. 2023. Quinolizidine-type alkaloids: Chemodiversity, occurrence, and bioactivity. *ACS Omega* 8(31):27862–27893.

Chen, W., Z Jia, M. H. Pan and P. V. A. Babu. 2016. Natural products for the prevention of oxidative stress-related diseases: Mechanisms and strategies. *Oxidative Medicine and Cellular Longevity* 2016:4628502.

Czapski, G. A., W. Szypuła, M. Kudlik, B. Wileńska, M. Kania, W. Danikiewicz and A. Adamczyk. 2014. Assessment of antioxidative activity of alkaloids from Huperzia selago and Diphasiastrum complanatum using in vitro systems. *Folia Neuropathologica* 52(4):394–406.

Desmet, S., K. Morreel and R. Dauwe. 2021. Origin and function of structural diversity in the plant specialized metabolome. *Plants* 10(11):2393.

Dey, A. and A. Mukherjee. 2018. Plant-derived alkaloids: A promising window for neuroprotective drug discovery. In *Discovery and development of neuroprotective agents from natural products*, 237–320.

dos Santos, A. R. and N. P. Vaz. 2019. Isoquinoline alkaloids and chemotaxonomy. In *Biodiversity and Chemotaxonomy*, 167–193.

Dung, N. T., D. M. Thanh, N. T. Huong, P. T. Thuy, N. T. Hoan, D. T. M. Thanh, N. Van Trang and N. T. Son. 2020. Quinolone and isoquinolone alkaloids: The structural-electronic effects and the antioxidant mechanisms. *Structural Chemistry* 31(6):2435–2450.

Fadhil, K. A., T. Suryati and A. Jayanegara. 2023. Comparison between natural and synthetic antioxidants in beef products: A metaanalysis. *Jurnal Ilmu Produksi dan Teknologi Hasil Peternakan* 11(1):19–26.

García-Muñoz, A. M., D. Victoria-Montesinos, P. Ballester, B. Cerdá, P. Zafrilla. 2024. A descriptive review of the antioxidant effects and mechanisms of action of berberine and silymarin. *Molecules 29*(19): 4576.

Gjorgieva Ackova, D., V. Maksimova, K. Smilkov, B. Buttari, M. Arese and L. Saso. 2023. Alkaloids as natural NRF2 inhibitors: Chemoprevention and cytotoxic action in cancer. *Pharmaceuticals* 16(6):850.

Gulcin, İ. 2025. Antioxidants: A comprehensive review. *Archives of Toxicology* 99(5):1–105.

Gurudeeban, S., T. Ramanathan and K. Satyavani. 2015. Antimicrobial and radical scavenging effects of alkaloid extracts from Rhizophora mucronata. *Pharmaceutical Chemistry Journal* 49(1):34–37.

Gutiérrez-Grijalva, E. P., L. X. López-Martínez, L. A. Contreras-Angulo, C. A. Elizalde-Romero and J. B. Heredia, 2020. Plant alkaloids: Structures and bioactive properties. In *Plant-Derived Bioactives: Chemistry and Mode of Action*, 85–117

Hussain, G., A. Rasul, H. Anwar, N. Aziz, A. Razzaq, W. Wei, M. Ali, J. Li and X. Li. 2018. Role of plant derived alkaloids and their mechanism in neurodegenerative disorders. *International Journal of Biological Sciences* 14(3):341.

Imenshahidi, M. and H. Hosseinzadeh. 2020. Berberine neuroprotection and antioxidant activity. In *Oxidative stress and dietary antioxidants in neurological diseases*, 199–216.

Jang, M. H., H. Y. Kim, K. S. Kang, T. Yokozawa and J. H. Park. 2009. Hydroxyl radical scavenging activities of isoquinoline alkaloids isolated from Coptis chinensis. *Archives of Pharmacal Research* 32(3):341–345.

Jomova, K., S. Y. Alomar, S. H. Alwasel, E. Nepovimova, K. Kuca and M. Valko. 2024. Several lines of antioxidant defense against oxidative stress: Antioxidant enzymes, nanomaterials with multiple enzyme-mimicking activities, and low-molecular-weight antioxidants. *Archives of Toxicology* 98(5):323–1367.

Joshi, S., R. Kumar, and D. Sharma. 2021. An overview of bioactive alkaloid, semi-synthetic analogs, mechanism of action and structure-activity relationship of piperine as antioxidant. *Asian Journal of Chemistry* 33(9):2244–2250.

Kamiloglu, S., E. Capanoglu and S. M. Jafari. 2022. An overview of food bioactive compounds and their health-promoting features. In *Retention of Bioactives in Food Processing* (pp. 3–36). Springer.

Kasote, D. M., S. S. Katyare, M. V. Hegde and H. Bae. 2015. Significance of antioxidant potential of plants and its relevance to therapeutic applications. *International Journal of Biological Sciences* 11(8):982.

Koltover, V. K. and T. A. Skipa. 2023. Antioxidant pharmacology. In *Anti-Aging Pharmacology*, 341–365.

Laghezza Masci, V., S. Bernardini, L. Modesti, E. Ovidi and A. Tiezzi. 2019. Medicinal plants as a source of alkaloids. In *Medically Important Plant Biomes: Source of Secondary Metabolites*, 85–113.

Lü, J. M., Lin, P. H., Yao, Q. and Chen, C. 2010. Chemical and molecular mechanisms of antioxidants: Experimental approaches and model systems. *Journal of Cellular and Molecular Medicine* 14(4):840–860.

Martins, N., L. Barros and I. C. Ferreira. 2016. In vivo antioxidant activity of phenolic compounds: Facts and gaps. *Trends in Food Science & Technology* 48:1–12.

Michael, J. P. 2007. Indolizidines and quinolizidines: Natural products and beyond. *Beilstein Journal of Organic Chemistry* 3(1):27.

Mukherjee, P. K., S. Bahadur, R. K. Harwansh, S. Biswas and S. Banerjee. 2017. Paradigm shift in natural product research: Traditional medicine inspired approaches. *Phytochemistry Reviews* 16(5):803–826.

Nemzer, B., T. Chang, Z. Xie, Z. Pietrzkowski, T. Reyes and B. Ou. 2014. Decrease of free radical concentrations in humans following consumption of a high antioxidant capacity natural product. *Food Science & Nutrition* 2(6):647–654.

Niğdelioğlu Dolanbay, S. E. R. A. P., S. Şirin and B. E. L. M. A. Aslım. 2024. Neuroprotection Induced by Plant Alkaloids. In *Natural Molecules in Neuroprotection and Neurotoxicity*, 1159–1195.

Och, A., M. Och, R. Nowak, D. Podgórska and R. Podgórski. 2022. Berberine, a herbal metabolite in the metabolic syndrome: The risk factors, course, and consequences of the disease. *Molecules* 27(4):1351.

Onakpa, M. M., J. O. Ode, O. P. Ajagbonna and I. U. Asuzu 2016. In vivo and in vitro antioxidant effects of Icacina trichantha tuber extract. *Nigerian Veterinary Journal* 37(3):148–154.

Oubaid, E. N., A. R. Abu-Raghif and I. M. Al-Sudani. 2023. Phytochemical screening and antioxidant activity of Uncaria tomentosa extract: In vitro: and: In vivo: Studies. *Medical Journal of Babylon* 20(1):136–142.

Parvin, M. S., Chlebek, J., Hošťálková, A., Catapano, M. C., Lomozová, Z., Macáková, K. and Mladěnka, P. 2022. Interactions of isoquinoline alkaloids with transition metals iron and copper. *Molecules 27*(19):6429.

Pasqualetti, V., V. Locato, C. Fanali, N. Mulinacci, S. Cimini, A. M. Morgia, G. Pasqua and L. De Gara. 2021. Comparison between in vitro chemical and ex vivo biological assays to evaluate antioxidant capacity of botanical extracts. *Antioxidants* 10(7):1136.

Platzer, M., S. Kiese, T. Tybussek, T. Herfellner, F. Schneider, U. Schweiggert-Weisz and P. Eisner. 2022. Radical scavenging mechanisms of phenolic compounds: A quantitative structure-property relationship (QSPR) study. *Frontiers in Nutrition* 9:882458.

Power, O., P. Jakeman and R. J. FitzGerald 2013. Antioxidative peptides: Enzymatic production, in vitro and in vivo antioxidant activity and potential applications of milk-derived antioxidative peptides. *Amino Acids* 44(3):797–820.

Shakeri, F., S. Kiani, G. Rahimi and M. H. Boskabady. 2024. Anti-inflammatory, antioxidant, and immunomodulatory effects of Berberis vulgaris and its constituent berberine, experimental and clinical: A review. *Phytotherapy Research* 38(4):1882–1902.

Sivashanmugam, A. and S. Velmathi. 2021. Bioactive component of black pepper-piperine: Structure-activity relationship and its broad-spectrum activity—An overview. In *Chemistry of Biologically Potent Natural Products and Synthetic Compounds* (pp. 43–92). Wiley.

Tan, B. L., M. E. Norhaizan, W. P. P. Liew and H. Sulaiman Rahman. 2018. Antioxidant and oxidative stress: A mutual interplay in age-related diseases. *Frontiers in Pharmacology* 9:1162.

Tian, E., G. Sharma and C. Dai. 2023. Neuroprotective properties of berberine: Molecular mechanisms and clinical implications. *Antioxidants* 12(10):1883.

Wang, D., L. Qin, C. Jing, G. Wang, H. Zhou, P. Deng, S. Zhang, Y. Wang, Y. Ding, Z. Zhang and Z. Wu. 2024. Biologically active isoquinoline alkaloids covering 2019–2022. *Bioorganic Chemistry* 145:107252.

Xu, W., H. Lu, Y. Yuan, Z. Deng, L. Zheng and H. Li. 2022. The antioxidant and anti-inflammatory effects of flavonoids from propolis via Nrf2 and NF-κB pathways. *Foods* 11(16):2439.

12 Alkaloids in Traditional Medicine Systems
Insights from Ethnobotanical Studies

Monica Butnariu

12.1 INTRODUCTION

Plants represent those used traditionally or in alternative medicine for their therapeutic and beneficial properties on human health. These plants contain active chemical substances, for example, phytochemical compounds, essential oils, tannins, alkaloids (ALKs), and vitamins, which can have beneficial effects on the human body. The use of plants to treat various ailments or to maintain health is an old and widespread practice in many cultures around the world. Throughout history, people have begun to notice and learn that they can use various plants to relieve pain, treat disease, and support the healing process (Dini et al., 2022). Medicinal chains can have strong effects on the body and can interact with drugs or other health conditions, so it is important to provide complete information to health professionals. The ALKs are nitrogen-based molecules, mostly of plant origin. Their discovery, at the end of the 19th century, made it possible to develop many medicines. The ALKs represent a group of natural substances and a separate chapter, of particular importance in pharmacognosy and, at the same time, in therapeutics; this is because, on the one hand, these compounds are endowed with a strong pharmacological action, and on the other hand, some of them have narcotic properties, developing pharmacodependence (cocaine, morphine). They are complex nitrogenous organic substances, produced only by the plant organism, which present a more or less important alkaline reaction (Brazier-Hicks et al., 2007).

They are end products of nitrogen metabolism. They are nitrogenous, basic heterocyclic organic substances, with a complex and very varied chemical structure, endowed with pharmacological properties, being toxic in large doses. They are found in plant vacuoles, in the form of salts with different acids (benzoic, citric, tartaric acid, etc.) or in tannic combinations. However, they can also be found in the form of quaternary or tertiary bases. The ALKs have at least one heterocyclic nitrogen atom, this being mostly tertiary, rarely quaternary. Heterocycles can condense with each other or with other cycles so that ALK molecules can become polycyclic or monocyclic. Due to the nucleus grafting of numerous functional groups, some ALKs can have a phenolic character (morphine), others, such as atropine and reserpine, can form esters, ethers (codeine), or glycosidated ALKs (potato solanine) (Jensen et al., 2009). They usually have optical activity, being levogyric (the most common and with greater pharmacological activity) or dextrogyric, activity imprinted by the carbon atoms of the acids with which they are esterified: tropanoic acid, in the case of hyoscyamine and scopolamine. The orientation of the substances in S or R positions confers the modification of the pharmacological action. With few exceptions, the most important ALKs are used therapeutically in the form of pure ALKs (extracted from plant products or obtained by synthesis), which are considered independent medicines. At the same time, ALKs also constituted structural models for the

DOI: 10.1201/9781003539179-12

synthesis of major therapeutic substances. Thus, for example, cocaine constituted the structural model for the synthesis of modern local anesthetics, quinine for antimalarials, and morphine for opioid analgesics.

The history of ALKs is intertwined with that of human civilization. Plant products with ALKs have been used for thousands of years, with the first material evidence indicating their use first as arrow poisons, then for curative purposes and as sacred plants in various rituals. Other examples of ALKs present in the plant material are colchicine from the root of autumn sedge (*Colchicum autumnale*); capsaicin from hot pepper (*Capsicum anuum*); morphine, codeine, papaverine from poppy seeds (*Papaver somniferum*); the complex of ALKs from the sicaria horn (*Claviceps purpurea*): ergotoxin, ergometrine, with a beneficial effect in regulating the heart rate and in relieving pain (Doesburg-van Kleffens et al., 2023). Other ALKs, such as trigonelline from fenugreek seeds (*Trigonella foenum-graecum*), are used as a spice and can have an aphrodisiac effect. Atropine from nightshade (*Atropa belladonna*) is used for ophthalmological investigations. After a long period in which therapeutics relied almost exclusively on synthetic drugs, in recent years, plant products and the substances isolated from them have again captured the interest of the pharmaceutical industry. A significant contribution in this direction was the introduction of standardization procedures for extracts and medicines obtained from them, but above all the evidence of therapeutic efficiency brought through clinical studies, the identification of mechanisms of action, which allowed a more accurate delimitation of the fields of indication and through this, they ensured a rational use of them (Marín-Sáez et al., 2018). Through this additional knowledge, not only the quality of many drugs, but also their acceptance in the medical world has improved. On the other hand, the synthetic medicine began to show its reverse more and more often—the adverse effects—and what the mass media called the "green wave" invaded our lives; at the same time, we find that the patient, who until yesterday used almost exclusively "pills", sometimes unjustifiably, has returned to "natural", so that the doctor and the pharmacist are increasingly faced with the question of whether there is no and a herbal remedy.

As we know, the plants used in therapeutics contain complex mixtures of secondary metabolites that can interact with various relevant cellular components involved in the generation or progression of disorders or diseases, including cancer. Active plant principles that can selectively interact with certain receptors, enzymes, or ion channels have long been known and used in therapeutics, especially in the form of pure substances, being strongly active (Peron et al., 2024). This category includes ALKs; most plant products containing them are disappearing from pharmacies, being taken over by the extractive industry that processes them. In fact, this is not the case, as pure, new substances were separated from them, including reserpine, atropine, vincamine, vincristine, vinblastine, taxol, and camptothecin, which were discovered through screening programs and subsequently assimilated into chemotherapeutics, securing their well-deserved place in therapeutics. However, of the approximately 320 medicinal plants considered indispensable in medicine and phytotherapy, 7% are products that contain ALKs. Inside the cells, ALKs are frequently found in vacuoles; there are complex mechanisms for their uptake at this level and their subsequent sequestration/storage (Nowak & Selmar, 2016). These mechanisms allow potentially toxic levels of ALKs to be stored (in base form) safely away from the cytoplasm, where they are transported as salts.

12.2 PLANT ALKALOIDS

The incidence of ALKs in the vegetable kingdom is very high, being identified in over 1200 plant species, belonging to 519 genera from 91 families. Currently, over 200 ALKs are known. In general, ALKs are characteristic of a certain species or related species from the same family; very rarely are the same ALKs found in plants from different families. There are cases where several structurally unrelated ALKs are found in the same species (Misiurek et al., 2023). Some examples of plants containing ALKs are as follows.

Species of sage (*Salvia divinorum Epling* and *Játiva*)—popularly known as Sage is a psychoactive plant that can induce dissociative effects. The name of the species (divinorum) was given because it has been used in divination since ancient times. This form of sage has become increasingly popular in recent times, with users smoking or mixing the leaves or using the tinctures from the plant. The active substance from this species of sage is Salvinorin A and is present in the dry plant in a percentage of 0.18%. Among the effects induced by Salvinorin A, we mention the following:

- uncontrolled shaving
- memory loss
- the feeling that you are pushed and twisted by forces
- overlapping realities
- synesthetic experiences
- alteration of consciousness

These psychedelic experiences are subjective and differ from subject to subject. The following symptoms may also appear:

- bad condition
- break from reality
- sweating
- cold or warm skin
- loss of self-confidence
- concentration difficulties (Babu et al., 2008)

Mitragyna species *(Mitragyna speciosa Korth (Kratom))* is a plant with cocaine-like effects. The active substances are found in the leaves of the tree that grows in Southeast Asia. Kratom contains many ALKs, including mitragynine, mitraphylline, 7-hydroxymitraynine, raubasine (commonly known as *Rauwolfia serpentine*), and corynantheidine (yohimbe ALKs), substances that produce the following:

- feeling good;
- analgesic effect;
- antidepressant;

But also feeling of nausea, vomiting, constipation, and addiction. Among the manifestations that appear in withdrawal, we mention the following:

- depression;
- fatigue;
- restlessness;
- lacrimation and insomnia (León et al., 2021).

Mushrooms: *Amanita muscaria*, popularly known as the snake's hat, is a poisonous mushroom whose psychoactive substance is muscimol. The effects of consuming Amanita muscaria are unpredictable; they differ from case to case, depending on the body weight and the amount ingested. Among them, we mention the following:

- nausea;
- drowsiness;
- lowering of blood pressure;

- agitation;
- confusion;
- hallucinations;
- sweating;
- auditory and visual distortions;
- change of mood;
- euphoria;
- relaxation;
- ataxia;
- loss of balance.

In some cases of overdose, it may occur: delirium; agitation; confusion; depression; headache; retrograde amnesia, and even death (Tsujikawa et al., 2006).

Amanita pantherina var. pantherina, also known as the panther sponge, is a mushroom related to the snake's hat *(Macrolepiota procera)*, which is edible, and fly agaric *(Amanita muscaria)*, having similar but stronger effects, thanks to the increased content of ibotenic acid and muscimol (Okhovat et al., 2023).

All species of the genus Psilocybe contain psilocybin and psilocin. The effects are subjective, varying from person to person, and depending on the amount consumed, the following physical effects may occur:

- loss of appetite;
- the feeling of cooling the extremities;
- increase in heart rate;
- numbness of the mouth;
- nausea;
- increase in blood pressure;
- difficulties in moving the limbs;
- muscle relaxation;
- pupil dilation.

The psychic effects include the following:

- synesthesia;
- alteration of perception.

The toxicity of plants is also influenced by factors dependent on the body. Apart from the difference in receptivity, some animals do not have gustatory and olfactory sensitivity that would allow them to discern toxic plants (Gyekis et al., 2012). Gluttonous animals are those who suffer from "taste aberration" and, as such, present a much higher risk of intoxication. Toxic plants are also consumed if they are mixed with other plants whose taste and smell mask their organoleptic properties. The most frequent cases of plant poisoning are found in young animals, especially at the beginning of the grazing period. In order to reduce the incidence of plant poisoning, a series of measures are required: avoiding pastures where poisonings occurred in previous years; botanical analysis of the pastures and the removal of toxic plants, taking the animals out to graze in the case of using new pastures, for a few days; knowledge of the phytosanitary actions that were carried out on the pastures and the removal to pasture or the harvesting of fodder after the remanence period has passed; lands will not be allocated for the cultivation of fodder plants near enterprises that pollute the atmosphere (Lamontagne et al., 2022).

The laurel (*Datura stramonium*) is also popularly called ciumafaie, pig's apple, prickly apple, bolindarita, etc. The herbaceous, perennial plant grows in uncultivated areas, characterized by its unpleasant taste and smell. It features white flowers shaped like trumpets and fruits represented by tough capsules that open into four valves when dried. All parts of the plant are toxic, but especially the seeds, toxicity that is not lost under any form of preservation or drying.

Young plants are twice as toxic as adults. The most important ALKs contained are hyoscyamine, hyoscine, and scopolamine. *Hyoscyamus niger* (*H. niger, H. albus*) is also popularly called crazy, porcarita, screeching, ringing, etc. It is a herbaceous plant that has no preference for soil; it is toxic both when green and dry. In young plants, the main ALK is hyoscine, and in adult plants it is hyoscyamine. It has an unpleasant smell and taste and a hairy, sticky appearance. Besides the fact that it produces excitation of the central nervous system, it also produces mydriasis and diarrhea (Velázquez-Márquez et al., 2021).

Stingray (*Veratrum album, V. Nigrum* or *Stingray*) is a perennial, herbaceous plant with a robust stem, alternating leaves that have characteristic longitudinal veins, and bristles on the lower side. It grows on wet lands, meadows, and pastures, especially in mountain regions, and due to its roots, which are a rhizome, it tends to invade. Toxic principles are spread in all plant organs, but the highest concentration is found in the rhizome and young shoots. The toxic ALKs are of the jerveratrum type (jervina, rubujervina, isorubijervina, veratrobasine) and of the cerveratrum type (ALKs esters of germine: germine, germidine, germitethrine, germithrine, isogermidine, neogermbudine, proteverathridine; esters of protoverine: germbutrine, desacetylprotoverathrine, protoveratrine A, protoveratrin B; esters of veracevin: veratridine).

The toxicity of the plant is preserved by drying. It has a strong irritating effect on the skin and mucous membranes, even being able to penetrate the intact skin. It is an initial central nervous system stimulant, after which it has a strong anesthetic action (Seale et al., 2022).

Hemlock (*Conium maculatum*) is also popularly called bricinis, dudaie. It is a herbaceous plant that grows all over the country on lowlands, hills, mountains, in uncultivated places, next to roads, fences, and ditches.

Toxic principles (coniine, conhydrin, gamma conicein, pseudoconhydrin, N-methylconiine) are distributed in all the plant's organelles, but unripe seeds contain the largest amount of ALKs. By drying, the plant loses an important part of its toxicity, but it still remains dangerous. Toxic principles have irritating effects on the digestive mucosa and depressing effects on the peripheral nervous system (blockers of the motor nerve endings in the muscles), and in larger quantities, they cause paralysis of the nerve centers. Death occurs by asphyxiation due to the paralysis of the respiratory muscles. It has a characteristic smell, similar to mouse urine, which is also perceived in the milk of animals that have consumed hemlock (Green et al., 2024).

Wolfsbane (*Aconitum paniculata; A. tauricum; A. vulparia*) is also known as the aconite, monkshood, wolfsbane, leopard's bane, devil's helmet, or blue rocket, etc. It is found on pastures, meadows, and grows on moist soils, often on hills and mountains. The main toxic principle is aconitine. It also contains napelin, napellenin, aconin, neolin, benzoilacholine, neopelin, aconitic acid, and an enzyme—aconitase. They are present in all parts of the plant, but the roots and seeds contain the largest amounts. Through the action of these toxic principles, local irritations, lightning paralysis of the bulbar respiratory and cardio-circulatory nerve centers are produced (Kurtoğlu et al., 2012).

Horsetail *(Equisetrum palustre; E. Arvense)* is also popularly called colt's tail, bear's beard, bradisor, etc. It is a herbaceous plant that grows in moist, shady places. It prefers sandy soils, and it is found around swamps, on plains, hills, and mountains. The most toxic is *E. palustre*. Toxic principles: a thiaminase; ALKs—equisetin, palustrine, and nicotine; nicotinic acid; aconitic acid; equisetone saponin (with hemolyzing properties); heterosis: galuteolin, equisetrin, silicon dioxide. The toxic principles are much more active if molds like *Fussarium* parasitize the plant. The toxic principles in this plant produce nervous disorders, initially, excitement, depression, then paresis, paralysis, and a serious sensory imbalance (Müller et al., 2020).

Lupine (*Lupinus luteus—yellow lupine; L. albus—white lupine; L. angustipholius*—blue lupine, etc.) is also popularly called cafeluta or niprala. It is a herbaceous, perennial plant, growing spontaneously as well as cultivated, as a fodder, textile, honey plant, ornamental, or used as a green fertilizer for acidic or sandy soils. Toxic principles: lupinin (lupinotoxin), lupinidin, lupamine, octalupin, and spartein are found in all the organs of the plant, especially in the seeds.

The toxicity of plants in the genus *Lupinus* varies. The green plant is the most toxic. Exposure to drying conditions, such as sun or hot air above 90°C, reduces toxicity. Additionally, treatment with a weak solution of hydrochloric acid inactivates toxic principles.

The ALKs are much more active when the plant is parasitized by a fungus, *Phomosis leptostronsmiformis*. It has a toxic action on the liver, causing liver degeneration. It also causes heart failure, has a toxic effect on the nervous system, being cortical excitatory and producing depression of the bulbar nerve centers of breathing and circulation, it acts as an irritant on the kidneys, and gives a very bitter taste to milk (Valente et al., 2024).

Yew (*Taxus baccata*) is a shrub that grows in the mountainous region or is cultivated as an ornamental plant. It is one of the most toxic plants in our country. It contains the ALK taxine and the glycoside taxicatin. It acts similarly to morphine, an anesthetic and strong narcotic, and taxicatin has a cardiotoxic action. Both toxic principles act rapidly, blocking the bulbar, respiratory, and circulatory nerve centers (Brooks-Lim et al., 2022).

Senecio (*Senecio jacobea, S. crucifolium, S. vernalis*) is popularly called cruciulita, slaviosa, dead horse grass, etc. The main ALKs are Jacobina, Jacodina, and Jaconina. It is a herbaceous plant found everywhere, up to the sea, in sub-mountainous regions, on uncultivated land, in orchards, and on the sides of roads. The ALKs contained (in a number of 60) have hepatotoxic properties, causing serious irreversible damage, both in the liver and in other organs. In humans, the consumption of bread made from flour containing *Senecio* seeds in the mixture leads to the appearance of cirrhosis and even liver cancer (Christov et al., 2002).

Tobacco (*Nicotiana tabacum; N. rustica*) is also popularly called tobacco, mahorca, gypsy tobacco, etc. Tobacco is a cultivated plant, spread especially in the Southeast. Toxic principles (nicotine, nicotimine, nicoteine, nicotellina, anabazin) are found in all parts of the plant, but the highest concentrations are in the leaves. By drying, the percentage content increases due to the evaporation of water, then by keeping, the actual amount of nicotine decreases due to its volatilization. Nicotine acts on the nervous system as an exciter and secondarily acts as a vasoconstrictor, irritating the digestive and urinary mucosa, and reducing the amount of active hemoglobin (Yang et al., 2022).

12.3 THE PHARMACOLOGICAL ACTION OF ALKALOIDS

The pharmacological action of ALKs varies.

Sometimes there may be a relationship between structure and action. However, many times structurally related ALKs have different pharmacological actions, and at other times, ALKs with different structures possess similar pharmacological qualities. The ALKs act both on the central nervous system and on the autonomic nervous system (Shi et al., 2003).

1. *Action on the central nervous system*

   Some of these compounds act on the central nervous system as depressants. Among these, the following can be mentioned.

   - *Morphine*, which has analgesic, hypnotic, and narcotic action, causes a euphoric state, accompanied by difficulty in concentration, drowsiness, or sleep. Sometimes, especially at the beginning of the treatment, nausea or even vomiting, itching, dizziness, and a feeling of heat set in. Emotional reactions, fear, and affective disorders are diminished. Morphine has an antidiuretic action and causes a reduction in pain perception. The main disadvantage of morphine is that, upon repeated consumption, it causes pharmacological dependence, the body gets used to the drug, and to achieve the desired effect, it is necessary to increase the administered dose. Other undesirable actions include the depression of the respiratory center and the cough center, endocrine metabolism disorders, and disturbances in the regulation of body temperature, *among others*.

- *Codeine* does not have the narcotic properties of morphine, and the analgesic properties are greatly diminished. Serves as cough medicine (Lötsch et al., 2006).
- *Reserpine* has been used for centuries in India as a medicine for states of nervous overexcitation. This ALK acts on the central nervous system, having a hypotensive and tranquilizing role. It is used as an antidote in snake and insect bites, as an antiepileptic, febrifuge, antidysenteric, and anthelmintic. Secondary actions refer to the appearance of a chronic rhinitis, and sometimes, a chronic diarrhea. When using large doses, serious mental disorders can develop, which can degenerate into actual melancholy (Ingersoll et al., 1999).
- *Scopolamine* produces sedation of motor centers, depressing vegetative centers. It is used in Parkinsonism, delirium tremens, and in cases of excitement in the insane; it has an anti-vomiting action and is used to combat altitude sickness, sea sickness, and car sickness (Liem-Moolenaar et al., 2011).

Other ALKs act on the central nervous system as **stimulants**. Among them, the most important are the following:

- *Caffeine* is a cardiac tonic, diuretic, antidiarrheal ALK. Being a cerebral stimulant, it prevents sleep and comforts. It has found some applications in the food industry for the production of cocoa powder and chocolate (Gonzalez de Mejia & Ramirez-Mares, 2014).
- *Lobelina* is a stimulant of the central nervous system. Acting on the respiratory centers, it increases the frequency and amplitude of respiratory movements. It is administered as an antiasthmatic and to combat smoking by carrying out actions that cause vomiting (Martin et al., 2018).
- *Strychnine* acts as a general tonic for the body, raising muscle tone, exciting the vasomotor center, increasing the intensity of reflexes, regulating circulatory and respiratory functions, and serving as a good additive in detoxification cures. It is recommended for peripheral paralysis of toxic, rheumatic, or infectious origin. Renal elimination is slow and may accumulate. The ALK is contraindicated in hypertension, liver and kidney diseases, spastic paralysis, and states of excitement (Seeman & Tantillo, 2020).

2. *Action on the autonomic nervous system*
   Many ALKs act on the autonomic nervous system by inhibiting or stimulating the parasympathetic and sympathetic systems. They can be mentioned as parasympathomimetic or cholinergic:

   - *Pilocarpine* excites the secretions of some glands, especially sweating and salivation. It is used in ophthalmology as a miotic and in glaucoma to reduce intraocular pressure. Pilocarpine finds its widest use in multiple and different lotions for the treatment of capillary diseases, through the effect of activating blood circulation in the capillaries of the scalp. It causes salivary and lacrimal secretion, activates smooth muscles, increases body temperature, and produces accommodation paralysis of the eye and miosis. In cases of intoxication, abundant salivation, sweating, miosis, intestinal colic, bronchospasm, hypotension, bradycardia, and collapse occur. The lethal dose for an adult is about 6 mg, taken orally (Zhang et al., 2024).
   - *Arecoline* is used in China and India as an anthelmintic, increases intestinal peristalsis and exocrine gland secretions, and creates a state of euphoria, accompanied by the annihilation of hunger and thirst (Liu et al., 2016).
   - *Physostigmine* is a parasympathomimetic with indirect action. It causes miosis, increases exocrine gland secretions, activates intestinal peristalsis, and produces bronchial vasoconstriction. Physostigmine inhibits acetylcholinesterase in body fluids and tissues, thus preventing the degradation of acetylcholine. The main therapeutic actions of this ALK concern the pupil and the intestine. In ophthalmology, physostigmine is used as a miotic. Additionally, due to its effect on reducing intraocular pressure, it is administered in the treatment of glaucoma and corneal ulcers (Chiew et al., 2024).

Among the ALKs with parasympatholytic (anticholinergic) action, it is the most representative:

- *Atropine* relaxes the smooth muscles and decreases the secretion of exocrine glands. It is used to combat pain, nausea, and vomiting in cases of gastritis, ulcer, enterocolitis, spastic colitis, nephritis, biliary intestinal colic, and cardiospasm. In high doses, atropine produces hypoacidity. It is administered in bronchial asthma, in the treatment of the iris and cornea, to change the heart rhythm, in the initial treatment of myocardial infarction, in the prevention of seasickness, motion sickness, and altitude sickness, to stimulate the respiratory center, and in the treatment of Parkinsonism. Atropine is contraindicated in glaucoma, prostate adenoma, and mechanical stenosis of the gastrointestinal tract (Upadhyay & Beuerman, 2020).

It is an ALK with sympathomimetic action:

- *Ephedrine* has a vasoconstrictive effect and is administered to maintain blood pressure. It is also used in whooping cough, asthma, hay fever, and other allergic conditions. It is a central nervous system stimulant by activating adrenergic receptors (Kobayashi et al., 2003).

The most important ALK with sympatholytic action is:

- *Ergotamine* has a constricting effect on the muscles of the uterus and is used to ease childbirth. In addition, ergotamine produces a constriction of peripheral blood vessels, which can lead to gangrene of the extremities (Bigal & Tepper, 2003).

3. *Curative, antispastic, antineoplastic, and hypotensive local anesthetic action on the heart and uterus*
   Cocaine has been used since ancient times by the inhabitants of South America as a physical stimulant. Due to the stimulating effect on the central nervous system, this ALK alleviates fatigue. The toxic dose is very close to the therapeutic dose. Cocaine has a narcotic character; its repeated use leads to addiction (cocaine addiction). For this reason, today cocaine is used less and less as a local anesthetic (Chauvet et al., 2009).
   - *Tubocurarine* is a powerful poison that produces motor nerve paralysis. The indigenous people of South America used it to poison arrows (Yan & White, 2002).
   - *Papaverine* has an antispasmodic action that it exerts non-specifically, directly on smooth muscles. It is prescribed for gastrointestinal, gall bladder, and urinary disorders. It is a coronary dilator, increasing coronary perfusion. It is used in the treatment of vascular spasms accompanying pulmonary embolism and cerebrovascular thrombosis (Egbewande et al., 2019).
   - *Vinblastine* and *vincristine* were used for a long time in the traditional medicine of the countries of origin as purgatives, vermifuges, and febrifuges (roots) and antidiarrheals (aerial parts). These ALKs are used therapeutically in the treatment of many forms of human cancer. Vincristine is used in the treatment of acute leukemia, Hodgkin's disease, lymphoma (lymphosarcoma), reticular cell sarcoma, breast, cervical, and prostate carcinoma, and karyocarcinoma. Vinblastine has approximately the same indications, but is also prescribed in the treatment of testicular tumors. Both ALKs belong to the category of immunosuppressants (Wu et al., 2021).

The ALKs are used therapeutically as narcotics and sedatives. Cannabinoids are the main active compounds responsible for the medicinal and recreational properties of the cannabis plant (*Cannabis sativa*). They function as an immune system for the plant, fighting off predators, insects, and parasites. In the human body, cannabinoids interact with the endocannabinoid system, influencing both mental and psychological processes as well as pain modulation, inflammation, and the immune system (Cristino et al., 2020).

12.4 PHYSIOLOGICAL ROLE AND PHARMACOLOGICAL ACTION OF ALKALOIDS

Opinions regarding the physiological role of ALKs in plants are still poorly defined. The ALKs could be the plant's means of defense against insects or higher animals. Other theories consider ALKs as protein breakdown products, which the plant, due to the lack of excretion possibilities comparable to those of animals, deposits in its organs. The ALKs would therefore play the same role as urea and uric acid in the animal body. Serious objections can be raised against this opinion. First of all, it seems unlikely that the plant would build molecules, sometimes huge and complicated, to fix a single nitrogen atom. On the other hand, it cannot be explained why a physiological function of the importance of the excretory process (if it were really necessary for the plant) would be carried out in such diverse ways in related botanical species, when other processes of vitally important (carbon dioxide assimilation, glycolysis, etc.) are done in all plants, through an identical mechanism (Bharathi Priya et al., 2021).

The pyrozylidine ALKs (P-ALK), which are incriminated in a series of serious poisonings, are produced especially in Australia, affecting both humans and animals. They mainly come from weeds that grow alongside cereal crops, namely: *Heliotropum europaeum, Echium plantagineum, Symphytum* spp., *and Crotalaria retouch*ae. Practically, 13 plant families produce P-ALK in larger or smaller quantities, but in three of them, the amounts of ALKs are notable (*Compositae, Boraginaceae, Leguminaceae*). Toxicity to humans presents 1-hydroxymethyl-1,2-dehydropyrrolizidine, an active hepatotoxic, which transforms into pyrrolic compounds. These electrophilic compounds actually cause the most severe metabolic disturbances, affecting the hepatocyte, the circulatory system of the liver, producing liver necrosis and cirrhosis (Yang et al., 2024). The formed pyrroles attack the cellular macromolecules, including the sulfur bridges of DNA protein adducts. In this context, the mutagenic and carcinogenic effects of pyrrolizidine ALK derivatives, known as pyrones, are incriminated. The latter has a strong antimitotic effect. In humans, it was found that the major conditions related to chronic PA ingestion are circulatory (venous occlusions). The dose of 1-hydroxymethyl-1,2-dehydropyrrolizidine that causes chronic intoxication is 1mg/kg/day. The weeds from which P-ALK is derived are sometimes used as dietary supplements. The foods in which the highest amounts of PA were found are cereals (50–6000 mg/kg), honey, milk (more than 0.1% of the total amount of PA ingested), eggs (5–168 mg/kg), and meat patties.

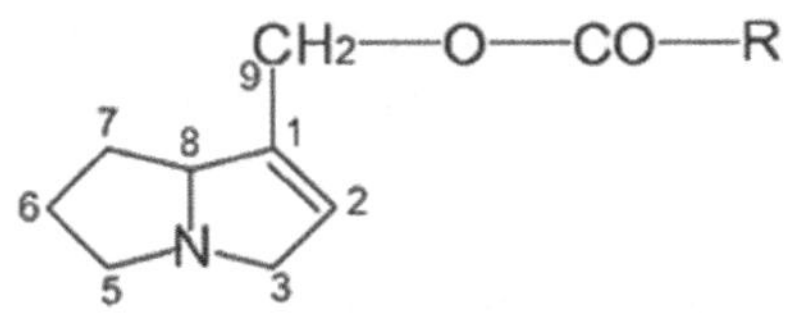

(General structure of pyrozylidine ALKs)

For derivatives, ester bonds are made at C7 and C9 (macrocyclic or noncyclic diesters are formed). Along with these forms, the N-oxides of these ALKs (less toxic) coexist in nature, which bioactivate in organisms, transform into ALKs (much more toxic), and cause severe poisoning (Ziegler & Facchini, 2008). Among the most active PAs, we mention: supinin, indicin, echimidine, and retrorsin. Metabolism of P-ALK is initially done by hydrolysis, catalyzed by esterases (necine and non-toxic necine acids are obtained). The attack of esterases is made difficult by the stereochemistry of P-ALK molecules, so that a large part reaches the liver, where it is hydroxylated at C3 and C8. Hydroxides are unstable; they dehydrate, the molecules rearrange, and become pyrroles (didehydro P-ALK). The allylic structure of pyrrolic ALKs gives them a special reactivity, being extremely toxic primary metabolites. However, the liver manages to hydrolyze them to some extent and transform them into pyrrolic alcohols, which are secondary metabolites, less toxic, but very persistent, in the sense that they cannot be eliminated. The greatest danger of these primary toxic metabolites is represented by their ability to bind the sulfhydryl groups of amino acids, proteins, as well as the nitrogenous bases of nucleic acids. Complexes are formed (at C7 or C9) with the exocyclic nitrogen atom of guanosine. Unfortunately, given the nature of PA contamination in the chain—encompassing grains, animal bodies, food products, and human bodies—it is currently impossible to provide an exact estimate of the toxicological risks, thereby hindering the application of 100% effective measures (Ke et al., 2021). *Belladonna* is toxic, and the ALK responsible for this toxicity is called hyoscyamine. It is a parasympathetic nervous system depressant that causes, among other things, a dilation of the pupil and an acceleration of the heart rate.

Belladonna also contains two other ALKs—atropine and some scopolamine—as well as coumarins and many mineral salts. All parts of the belladonna plant are toxic, but it is the fruit in particular that can cause accidents, especially to children who think they are edible. A few berries alone can be fatal if swallowed. The presence of various ALKs with attributes of biologically active substances in vegetable products (and sometimes in animal products) is of interest for the effects induced on the body when they enter food, but especially when they are used in pharmacology (Parks et al., 2023). Although they are found in small amounts in natural products, ALKs are important for their physiological/pharmacological role and, in limited situations, for their toxicological effects. These considerations were the basis for including ALKs in the group of biologically active substances, being considered physiologically active when they accompany nutrients from food, or pharmacologically active, when they are used therapeutically, but also toxicologically active in accidental circumstances. In general, it is known that in the initially isolated forms, ALK molecules show high toxicity. In low doses, however, it exhibits physiological, pharmacological (acute or chronic) activity with some delayed effects, or even toxicological activity (with some delayed effects).

Thus, the use of small amounts of caffeine from coffee, cocaine from coca leaves, and nicotine from tobacco is mentioned; these substances were accepted in various human cultures, often very old in time. It was difficult to assess the reason for the various effects, for example, stimulants, tonics, psychotropics, psychoactives, emetics, sedatives, analgesics, etc. (Domfeh et al., 2022).

It is now known that ALK molecules in their pure state are often very toxic, for example, strychnine, atropine, aconitine, cocaine, etc. In certain controlled doses, they lend themselves to therapeutic use; for example, morphine and codeine are used in some therapeutic protocols with an anesthetic purpose; quinine is considered an agent with antimalarial effects; vinblastine and vincristine are used as anticancer agents, etc. The investigations undertaken on the chemical compounds present in plants, in general, and in medicinal plants in particular, led to the isolation of a significant number of nitrogenous compounds, among them proteins, amines, amides, some vitamins, and antibiotics, as well as numerous ALKs.

From a biochemical point of view, ALKs, as they have been presented, are organic substances with nitrogen content in the molecule, mainly of vegetable origin but also of animal origin (Pingali et al., 2015). Their structure being extremely varied, a classification based on the structure has a conventional character, distinguished, according to Figure 12.1, into three categories.

In general, it is estimated that ALKs, as biologically active substances, are products of the secondary metabolism of higher plants. Their amount in the plant kingdom can go up to 8%. The distribution in the vegetable kingdom is very varied. For example, it is found in monocotyledonous plants belonging to the Liliaceae and Amarylidaceae families. More frequently, however, it is found in dicotyledonous plants belonging to the families *Solanaceae, Papaveraceae, Rutaceae, Rubiaceae, Leguminosae, Ranunculaceae, Compositae*, etc. The distribution of ALKs in plant organs differs greatly in terms of quantity. For example, it predominates in the roots or rhizomes of *A. belladonna, Rauwolfia* sp., *V. album*; in leaves of *Erythroxylon coca, D. stramonium, H. niger, N. tabacum,*

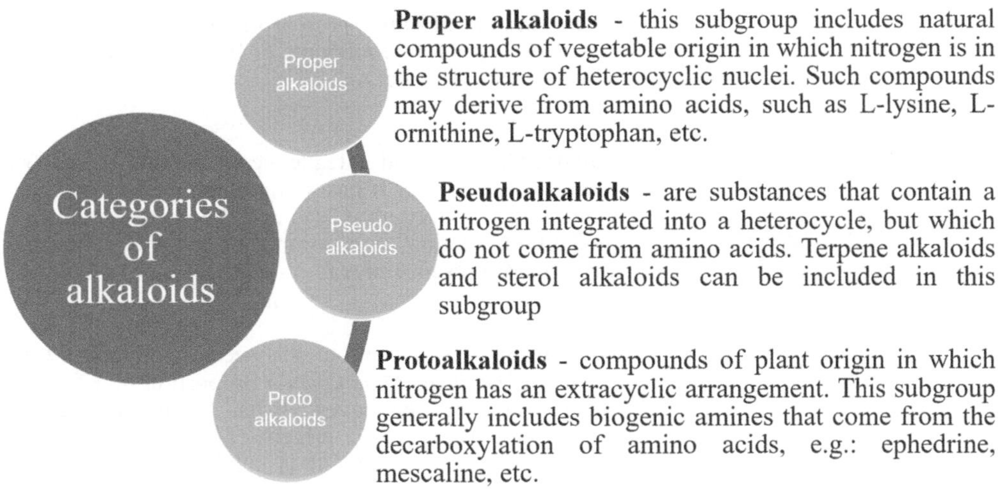

FIGURE 12.1 Categories of alkaloids.

etc.; in the bark of *Berberis vulgaris, Cinchon*a sp., etc.; in fruits of *P. somniferum*; in seeds of *C. autumnale, Coffea arabica, Theobroma cacao*, etc.

It should be noted that the number of ALKs differs greatly from one plant to another. Sometimes their number exceeds 40 molecular species (Deng et al., 2021). There is usually a main ALK and a number of secondary ALKs that are structurally related. For example, Opium contains morphine, thebaine, and codeine, as related substances, originating from the morphinan nucleus.

A similar example is presented by the ALKs of *Secale cornutum*, which have similar structures originating from the lysergic nucleus. In other situations, there is a circumscription of the distribution of ALKs; they are found preferentially (almost exclusively) in certain plants, for example, morphine in *P. somniferum*, vincristine in *Catharanthus roseus*, quinine in *Cinchona* sp. Also, there is a distribution of certain ALKs in plant species belonging to the same family; for example, hyoscyanine and scopolamine are found in plants of the *Solanaceae* family (Zhang et al., 2023).

Often, however, a certain ALK is found in many plant species; for example, nicotine is found in various plants that taxonomically belong to 13 botanical families.

12.5 THE ROLE OF ALKALOIDS IN TRADITIONAL MEDICINE SYSTEMS

Regarding the ALKs more frequently encountered in products used for human consumption, they are mentioned according to Figure 12.2.

The most important effects that characterize all these compounds are central and respiratory stimulant, bronchodilator, diuretic, cardiac stimulant, and striated muscle contraction stimulant. The intensity of action differs.

Thus, in terms of central and respiratory stimulant action, caffeine is more active than theophylline and theobromine (caffeine > theophylline > theobromine), and as diuretics and bronchodilators, theophylline is more active than caffeine and theobromine (theophylline > caffeine > theobromine) (Moreira et al., 2018).

Caffeine is well absorbed from the intestine. It is distributed throughout the body and achieves high concentrations in the brain. It is cleared almost entirely by hepatic metabolism. In adults, the half-life is about 4 hours. In smokers, it decreases to 3 hours or even less, and in the case of administration of oral contraceptives or deficient liver function, the half-life increases, reaching 13 hours. The main products of caffeine metabolism are 1-methyluric acid and 1-methylxanthine, both of which are eliminated in the urine. It is a psychomotor stimulant with a moderate intensity action that

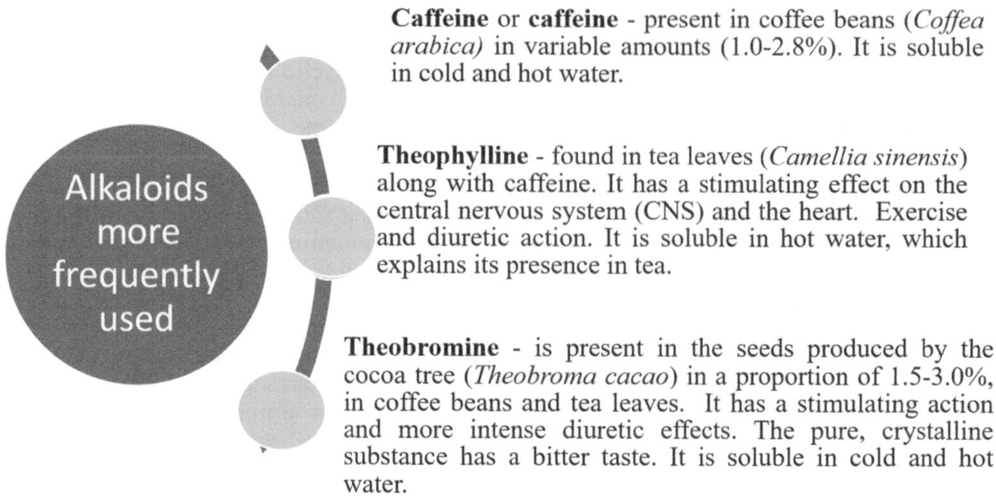

FIGURE 12.2 The alkaloids more frequently encountered in products used for human consumption.

lasts up to 8 hours. Usual doses of 100–200 mg cause a feeling of liveliness, alleviate fatigue, increase the capacity for physical and intellectual effort, and improve motor performance, except for those involving high-finesse activities.

A large single dose (250–300 mg) administered to those who do not normally consume coffee causes irritability, nervousness, tremors, insomnia, and epigastric burning. At usual doses, the respiratory stimulant effect (increasing the frequency and amplitude of respiratory movements), which is minimal, becomes relevant in conditions of depressed breathing; caffeine increases the reactivity of the bulbar centers to CO_2 (Jee et al., 2020).

It stimulates the contraction of the heart and causes a decrease in peripheral resistance by dilating the arterioles of the systemic circulation, while vasoconstriction occurs in the cerebral territory, with a decrease in blood flow and a decrease in the pressure of the cerebrospinal fluid. It causes vasodilatation by direct action on blood vessels, but the vascular effects are not evident at therapeutic doses. Blood pressure is influenced differently.

Thus, it increases as a result of the stimulation of the vasomotor center and the myocardium, but decreases due to direct vascular effects and by stimulating the vagal center. Caffeine stimulates chlorhydropeptic gastric secretion. It has a weak diuretic effect due to the increase in renal circulation and the inhibition of tubular reabsorption of salt.

It relaxes the bronchial smooth muscles. At high doses, it favors the contraction of striated muscles and causes a modest increase in basal metabolism.

Regarding the intimate mechanism of action of caffeine and other methylxanthines, direct effects have been described, as well as indirect effects, following the release of adrenaline and noradrenaline by the sympatho-medulla-adrenal system.

Psychomotor stimulation is explained by an increase in excitatory processes at the level of the cerebral cortex. High concentrations of methylxanthines cause an increase in the amount of cAMP in cells as a result of inhibiting phosphodiesterase (the enzyme responsible for inactivating cyclic adenylate). At concentrations close to the therapeutic ones, caffeine (like theophylline) competitively antagonizes the actions of adenosine (Reyes & Cornelis, 2018).

Caffeine releases calcium ions from the cytoplasm of the sarcoplasmic reticulum, which explains the favoring of striated muscle contraction. Caffeine is used as such (basic caffeine) or in the form of salts (citric caffeine and sodium benzoic caffeine) that contain the active ALK in a proportion of 50%. It is indicated in the treatment of acute intoxication with alcoholic beverages (e.g., 100–500 mg once, without exceeding 1.5 g/day), migraine, and other vascular headaches.

In migraine, it is advantageous to combine it with ergotamine, which favors intestinal absorption.

Small amounts of caffeine are included in various anti-neuralgic products due to a potentiating effect of analgesic-antipyretics by favoring their action of inhibiting PG synthesis. In doses of 3–6 mg/kg body, caffeine improves physical performance in sports that involve a high degree of resistance and a longer duration of exercise (athletics, cycling, swimming, tennis), being suggested three major hypotheses by which it is achieved this beneficial effect, namely: direct damage to skeletal muscle by altering key enzymes or systems that regulate carbohydrate catabolism; directly affecting the CNS and altering the individual's perception of the effort made and the degree of fatigue; saving glycogen reserves by using lipids as "fuel" for muscle exercise. The effect is sustained by increasing the plasma concentration of catecholamines and mobilizing free fatty acids from fat or muscle cells (Wikoff et al., 2017).

Currently, it is considered that the action of caffeine is more likely related to effects on the CNS and skeletal muscle. Until 2004, caffeine was on the list of banned substances by the International Olympic Committee. Currently, athletes are eliminated from the competition if their urinary levels exceed 12 µg caffeine/mL. This concentration is not reached by the quantities usually consumed, which are equivalent to about 8 cups of coffee. Very high doses of caffeine (300–600 mg) cause restlessness, anxiety, confusion, insomnia, palpitations, heartburn, headache, fine tremor of the extremities, and vision and hearing disorders. Severe acute caffeine intoxication is very rare, the lethal dose being around 150–200 mg/kg body (75 cups of coffee for a body weight of 50 kg). The sudden stop of caffeine consumption causes an abstinence syndrome manifested by irritability, nervousness, difficulty in working, headache, and can last for several days. The consumption of caffeine, as well as the drinks that contain it (coffee, tea), is contraindicated in coronary diseases, arrhythmias, ulcers, glaucoma, pregnancy, and breastfeeding (Schubert et al., 2017).

Orally administered **theophylline** is quickly and completely absorbed from the digestive tract under the conditions of the appropriate disintegration of the pharmaceutical product. The plasma concentration is maximum about 2 hours after administration on an empty stomach. The ALK binds to plasma proteins in a proportion of 56%. Purification is predominantly achieved through hepatic metabolism, which involves the intervention of cytochrome P450 (CYP450) isozymes. The metabolizing capacity is limited, which makes the clearance kinetics dose-dependent; 18% of theophylline is eliminated through urine, and the rest in the form of metabolites.

The average half-life is 8–9 hours for adults and 3–4 hours for children aged 5–7 years, with large individual variations, especially in children. Purification is much reduced in newborns. Severe cirrhosis considerably increases the T1/2 of theophylline (it can reach 25–28 hours), while heart failure and hypothyroidism increase it moderately (Camandola et al., 2019).

Metabolism of the ALK is intensified in smokers or in the case of association with phenobarbital, phenylbutazone, and alcohol.

Macrolide antibiotics (erythromycin) and cimetidine inhibit the metabolism of theophylline. It has a net bronchodilator action through a direct effect on the smooth muscles. It improves pulmonary function in bronchial asthma and can be effective in patients in whom sympathomimetics have become inactive. It also intervenes favorably by stimulating mucociliary clearance. In allergic asthma, theophylline diminishes the response to histamine, reduces airway obstruction, and inhibits eosinophil migration. Thus, in asthma, theophylline acts by increasing ventilation during hypoxia, removing diaphragm muscle fatigue, and preventing the release of pro-inflammatory autacoids from mast cells (Purkiewicz et al., 2022).

The central stimulating effect of theophylline may have favorable consequences in nocturnal asthma. It also has favorable hemodynamic effects, such as increasing the contraction force of the myocardium and dilating the pulmonary arteries, as well as anti-inflammatory and immunostimulating actions. Theophylline achieves all these actions through non-specific inhibition of phosphodiesterase isozymes and antagonism towards adenosine at the level of its membrane receptors. Lately, theophylline is no longer indicated as a curative medication in an asthma attack. However, it is recommended for patients who do not respond to systemic glucocorticoids or β-adrenergic stimulants, and also represents an alternative in situations where oral administration is preferred rather than inhalation. Available pharmaceutical preparations contain anhydrous theophylline in microcrystalline form that is almost completely absorbed and retard forms that ensure effective concentrations for up to 12 hours. Very widely used is a complex containing 86% theophylline and ethylenediamine, known as aminophylline. It is soluble in water and allows conditioning in the form of an injectable solution, administered i.v. (Salcedo & Kern, 2009).

The injection of aminophylline is done very slowly and with great care, being indicated in asthmatic illness that does not yield to sympathomimetics and glucocorticoids administered intravenously. With rapid intravenous injection of aminophylline, there is a risk of skin congestion, severe arrhythmia, hypotension, chest pain, convulsions, and even sudden death.

Aminophylline in injectable form is contraindicated in epilepsy, acute myocardial infarction, and in cases of allergy to theophylline and derivatives.

Theophylline frequently causes adverse reactions consisting of anorexia, nausea, palpitations, headache, and insomnia. At high doses, arrhythmias, tachycardia, and convulsions may occur. Administration of theophylline is contraindicated in patients with gastrointestinal ulcers, and its use in hyperthyroid, cardiac, hepatic, elderly, and newborns requires caution. At the same time, the possible drug interactions that theophylline/aminophylline can cause in various combinations must be taken into account.

Interplay

- increases the risk of toxic reactions;
- ventricular arrhythmias;
- increases the risk of convulsions;
- decreases the sedative effect of benzodiazepines;
- increases the elimination of lithium preparations;
- the action of propanol is antagonized;
- the medicine inhibits the metabolism of theophylline, so the doses of theophylline/aminophylline have to be increased;
- antagonizes the action of curarizing agents (Ram et al., 2005).

Theobromine, used years ago in the treatment of angina pectoris and cardiac edema (in doses of 300–500 mg), no longer has the same relevance in current therapeutics, disappearing from medical prescriptions.

In the case of using extracts from plant products containing methylxanthines, the effect (even at equivalent dosage) is different, because a multitude of other substances present in the composition of the extracted phytocomplex compete for its achievement.

Pharmacologically, the result will be the sum of the synergistic, complementary, or potentiating actions, as well as those of retardation or antagonism brought about by the other components in the extract.

Moreover, this is a problem that therapeutics constantly face, plant extracts frequently developing a much more complex (and often milder) action than the pure biogenic substance (in this case, the ALK). Medicines containing nitrogen in the molecule are not administered together with extracts/drinks based on *Colae semen, Coffeae semen, Cacao semen*, or *Theae folium*, because the tannins they contain precipitate the nitrogenous substances to a lesser or greater extent, diminishing their bioavailability (Martínez-Pinilla et al., 2015).

Tropanic alkaloids (> 200) represent the class of secondary metabolites that are naturally present in several plant species, mainly in plant species of the *Solanaceae* family (genus *Datura*), and are often used in medicine as various remedies.

One of the main tropane ALKs found in *Datura* plant species is scopolamine, also called hyoscine. Scopolamine (α-(hydroxymethyl)benzene acetic acid 9-methyl-3-oxa-9-azatricyclo non-7-yl ester) is a non-polar belladonna tropane ALK, and its structural formula is a tertiary amine L-(2)-scopolamine (Sharma et al., 2021).

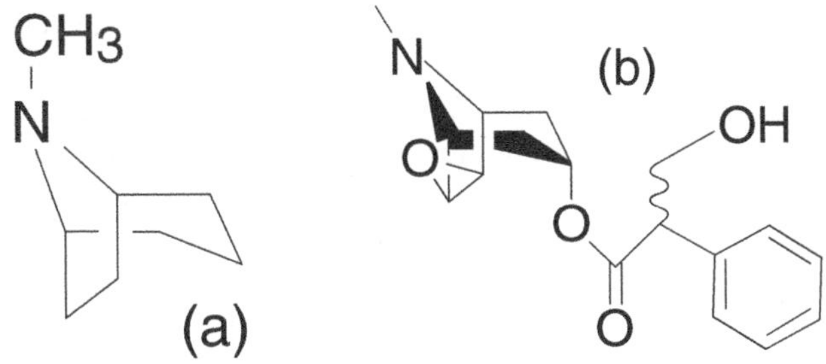

FIGURE 12.3 Basic chemical structures: (a) tropane alkaloids, (b) scopolamine.

This is widely found in plants of the *Solanaceae* family. The basic structure of tropane ALKs is shown in Figure 12.3a, while in Figure 12.3b, the chemical structure of scopolamine is presented.

This ALK is mainly used in the pharmaceutical industry in the treatment of asthma, sialorrhea, rashes, colds, nervous disorders, neuro-sedative Parkinson's disease, and motion sickness. This tropane ALK is an antimuscarinic, antispasmodic, anticholinergic, antispasmodic, antiasthmatic, antimicrobial agent, and used as an analgesic, narcotic, sedative, as well as for treatment following surgical procedures. Due to the very high demand in the pharmaceutical market, there is a need for inventions, new discoveries in terms of extraction methods, and quantitative and qualitative investigations to obtain scopolamine. According to preliminary statistics, the demand for scopolamine is 10 times higher than for atropine. Currently, scopolamine is generally extracted from plants that are part of the *Solanaceae* family. It is known that scopolamine is an ALK with toxic properties, which, by ingesting a high concentration, can lead to death (Kariñho-Betancourt et al., 2023).

In nature, scopolamine is found in the plant biomass of *Datura* plant species (*Datura metel* and *Datura innoxia*) belonging to the *Solanaceae* family. It is also known as the ALK plant—an alkaline organic compound that contains nitrogen in living organisms. Most are found in plants, and some are found in animals. This compound has a ring structure that is difficult to dissolve in water, and its acid can form a salt. It exhibits a certain rotation and optical absorption spectrum, and is the most bitter, colorless crystal, found in some liquids.

There are thousands of ALKs, which are synthesized from different amino acids or their direct derivatives. They are one of the secondary metabolites and have toxic or intense physiological effects on the body (Cárdenas et al., 2015).

The ALKs are found in leaves, bark, flowers, stems, seeds, and fruits of plants. A plant often contains several or even dozens of ALKs at the same time. For example, seven ALKs have been found in ephedra, and more than 60 ALKs have been isolated from the anticancer drug periwinnia. Many ALKs are good medicines for curing diseases. For example, coptidis rhizome root is the main component of berberine, which has antibacterial and anti-inflammatory effects. Reserpine in *Rauwolfia* can lower blood pressure. Garantamine from Lycoris sasiva has a curative effect for the sequelae of poliomyelitis. The morphine base contained in poppy husk is a famous analgesic. Quinine is a valuable antipyretic. Tricusperzine and vinblastine are effective against cancer. Colchicine (alkali) can produce polyploids by artificial mutation. Some ALKs can be used to make agricultural pesticides (Han et al., 2023).

ALKs have also been isolated in vertebrates and invertebrates, some of which are related to the plants they eat, and those found in flu, salamanders, and some fish are true animal metabolites.

12.6 CONCLUSION AND FUTURE REMARKS

The ALKs exhibit a wide range of biological effects and are therefore used as pharmaceuticals and therapeutics. A number of criteria were taken into account in defining the plants that produce poisoning. Thus, among the toxic plants were plants that cause diseases in the conditions of consumption by sensitive animals of fruits, seeds, aerial parts, roots; plants that contain factors that limit or inhibit the optimal morphoproductive performance of animals; plants that contain substances that interfere with the reproductive function and that, in certain circumstances, produce infertility. Many plants can produce pathological conditions, but the designation as a toxic plant is relative, as toxicity depends on various factors, including pedoclimatic conditions, season, meteorological conditions, soil fertilization practices, atmospheric pollution levels, and the plant's developmental stage. The toxicity of plants is also influenced by factors dependent on the body. Apart from the difference in receptivity, some animals do not have gustatory and olfactory sensitivity that would allow them to discern toxic plants. This category includes greedy animals, those who suffer from the perversion of taste. Toxic plants are also consumed if they are mixed with other plants whose taste and smell mask the organoleptic properties. The ALKs are used in medicine in their pure state, after industrial extraction. In therapy, they are used in extremely small doses; they have a beneficial effect, but when the dose is higher, they can become poisonous. The pharmaceutical forms in which they are found are tablets, dragees, suppositories, ampoules for injections (e.g., Scobutil, which is based on scopolamine extracted from *Solanaceae* plants). Other examples of ALKs present in the plant material are colchicine from the brandusa root or capsaicin from hot pepper, as well as morphine, codeine, papaverine, from poppy seeds, as well as the complex of ALKs from the rye horn: ergotoxin, ergotamine, ergometrine, with a beneficial effect in regulating the heart rate and in relieving pain. Other ALKs, such as trigonelline from fenugreek, are used as a spice and have an aphrodisiac effect. Atropine from *A. belladonna* is used for ophthalmological investigations. Herbal teas are a natural and effective way to relieve the symptoms of panic attacks. Medicinal herbs such as chamomile, lavender, passionflower, and nightshade have calming and anxiolytic properties that help reduce anxiety and stress. The beneficial effects of these plants are due to bioactive compounds such as flavonoids and essential oils that act on the central nervous system.

COMPETING INTERESTS

The authors declare that they have no competing interest.

DECLARATION OF GENERATIVE AI AND AI-ASSISTED TECHNOLOGIES IN THE WRITING PROCESS

During the preparation of this work, the book chapter author(s) have not taken any assistance from AI, and the book chapter author(s) reviewed and edited the content as needed and take(s) full responsibility for the content of the publication.

REFERENCES

Babu, K. M., McCurdy, C. R., & Boyer, E. W. (2008). Opioid receptors and legal highs: Salvia divinorum and Kratom. *Clinical Toxicology (Philadelphia, PA)*, *46*(2), 146–152. https://doi.org/10.1080/15563650701241795

Bharathi Priya, L., Huang, C. Y., Hu, R. M., Balasubramanian, B., & Baskaran, R. (2021). An updated review on pharmacological properties of neferine-A bisbenzylisoquinoline alkaloid from Nelumbo nucifera. *Journal of Food Biochemistry*, *45*(12), e13986. https://doi.org/10.1111/jfbc.13986

Bigal, M. E., & Tepper, S. J. (2003). Ergotamine and dihydroergotamine: a review. *Current Pain and Headache Reports*, *7*(1), 55–62. https://doi.org/10.1007/s11916-003-0011-7

Brazier-Hicks, M., Offen, W. A., Gershater, M. C., Revett, T. J., Lim, E. K., Bowles, D. J., Davies, G. J., & Edwards, R. (2007). Characterization and engineering of the bifunctional N- and O-glucosyltransferase involved in xenobiotic metabolism in plants. *Proceedings of the National Academy of Sciences of the United States of America*, *104*(51), 20238–20243. https://doi.org/10.1073/pnas.0706421104

Brooks-Lim, E. W. L., Mérette, S. A., Hawkins, B. J., Maxwell, C., Washbrook, A., & Shapiro, A. M. (2022). Fatal ingestion of Taxus baccata: English yew. *Journal of Forensic Sciences*, *67*(2), 820–826. https://doi.org/10.1111/1556-4029.14941

Camandola, S., Plick, N., & Mattson, M. P. (2019). Impact of coffee and cacao purine metabolites on neuroplasticity and neurodegenerative disease. *Neurochemical Research*, *44*(1), 214–227. https://doi.org/10.1007/s11064-018-2492-0

Cárdenas, P. D., Sonawane, P. D., Heinig, U., Bocobza, S. E., Burdman, S., & Aharoni, A. (2015). The bitter side of the nightshades: Genomics drives discovery in Solanaceae steroidal alkaloid metabolism. *Phytochemistry*, *113*, 24–32. https://doi.org/10.1016/j.phytochem.2014.12.010

Chauvet, C., Lardeux, V., Goldberg, S. R., Jaber, M., & Solinas, M. (2009). Environmental enrichment reduces cocaine seeking and reinstatement induced by cues and stress but not by cocaine. *Neuropsychopharmacology: Official Publication of the American College of Neuropsychopharmacology*, *34*(13), 2767–2778. https://doi.org/10.1038/npp.2009.127

Chiew, A. L., Holford, A. G., Chan, B. S. H., & Isoardi, K. Z. (2024). Rivastigmine for the management of anticholinergic delirium. *Clinical Toxicology (Philadelphia, PA)*, *62*(2), 82–87. https://doi.org/10.1080/15563650.2024.2319854

Christov, V. S., Mikhova, B. P., & Evstatieva, L. N. (2002). Alkaloids from Senecio aquaticus. *Fitoterapia*, *73*(2), 171–173. https://doi.org/10.1016/s0367-326x(02)00012-6

Cristino, L., Bisogno, T., & Di Marzo, V. (2020). Cannabinoids and the expanded endocannabinoid system in neurological disorders. Nature reviews. *Neurology*, *16*(1), 9–29. https://doi.org/10.1038/s41582-019-0284-z

Deng, A. P., Zhang, Y., Zhou, L., Kang, C. Z., Lv, C. G., Kang, L. P., Nan, T. G., Zhan, Z. L., Guo, L. P., & Huang, L. Q. (2021). Systematic review of the alkaloid constituents in several important medicinal plants of the Genus Corydalis. *Phytochemistry*, *183*, 112644. https://doi.org/10.1016/j.phytochem.2020.112644

Dini, S., Chen, Q., Fatemi, F., & Asri, Y. (2022). Phytochemical and biological activities of some Iranian medicinal plants. *Pharmaceutical Biology*, *60*(1), 664–689. https://doi.org/10.1080/13880209.2022.2046112

Doesburg-van Kleffens, M., Zimmermann-Klemd, A. M., & Gründemann, C. (2023). An overview on the hallucinogenic peyote and its alkaloid mescaline: The importance of context, ceremony and culture. *Molecules (Basel, Switzerland)*, *28*(24), 7942. https://doi.org/10.3390/molecules28247942

Domfeh, S. A., Narkwa, P. W., Quaye, O., Kusi, K. A., Addy, B. S., Lant, S., Sumner, R. P., Maluquer de Motes, C., Awandare, G. A., Ansah, C., & Mutocheluh, M. (2022). The pharmacologically active alkaloid cryptolepine activates a type 1 interferon response that is independent of MAVS and STING pathways. *Journal of Immunology Research*, *2022*, 8873536. https://doi.org/10.1155/2022/8873536

Egbewande, F. A., Coster, M. J., Jenkins, I. D., & Davis, R. A. (2019). Reaction of papaverine with baran Diversinates™. *Molecules (Basel, Switzerland)*, *24*(21), 3938. https://doi.org/10.3390/molecules24213938

Gonzalez de Mejia, E., & Ramirez-Mares, M. V. (2014). Impact of caffeine and coffee on our health. *Trends in Endocrinology and Metabolism: TEM*, *25*(10), 489–492. https://doi.org/10.1016/j.tem.2014.07.003

Green, B. T., Lee, S. T., Welch, K. D., Cook, D., & Stonecipher, C. A. (2024). The actions of varenicline on alkaloids from Conium maculatum (poison hemlock), Lupinus sulphureus (sulphur lupine) and Nicotiana glauca (tree tobacco). *Toxicon: Official Journal of the International Society on Toxinology*, *252*, 108184. https://doi.org/10.1016/j.toxicon.2024.108184

Gyekis, J. P., Dingman, M. A., Revitsky, A. R., Bryant, B. P., Vandenbergh, D. J., Frank, M. E., & Blizard, D. A. (2012). Gustatory, trigeminal, and olfactory aspects of nicotine intake in three mouse strains. *Behavior Genetics*, *42*(5), 820–829. https://doi.org/10.1007/s10519-012-9546-x

Han, J., Zhang, S., He, J., & Li, T. (2023). Piperine: Chemistry and biology. *Toxins*, *15*(12), 696. https://doi.org/10.3390/toxins15120696

Ingersoll, R. T., Montgomery, E. B., Jr, & Aposhian, H. V. (1999). Central nervous system toxicity of manganese. II: Cocaine or reserpine inhibit manganese concentration in the rat brain. *Neurotoxicology*, *20*(2–3), 467–476.

Jee, H. J., Lee, S. G., Bormate, K. J., & Jung, Y. S. (2020). Effect of caffeine consumption on the risk for neurological and psychiatric disorders: Sex differences in human. *Nutrients*, *12*(10), 3080. https://doi.org/10.3390/nu12103080

Jensen, P. H., Jacobsen, O. S., Henriksen, T., Strobel, B. W., & Hansen, H. C. (2009). Degradation of the potato glycoalkaloids--alpha-solanine and alpha-chaconine in groundwater. *Bulletin of Environmental Contamination and Toxicology*, *82*(6), 668–672. https://doi.org/10.1007/s00128-009-9698-4

Kariñho-Betancourt, E., Vázquez-Lobo, A., & Núñez-Farfán, J. (2023). Effect of plant defenses and plant nutrients on the performance of specialist and generalist herbivores of *Datura*: A macroevolutionary study. *Plants (Basel, Switzerland)*, *12*(14), 2611. https://doi.org/10.3390/plants12142611

Ke, S., Xu, T., Min, Y., Wan, Z., Yang, Z., & Wang, K. (2021). Marine alkaloid pityriacitrin and its analogues: Discovery, structures, synthetic methods and biological properties. *Mini Reviews in Medicinal Chemistry*, *21*(2), 233–244. https://doi.org/10.2174/1389557520666201116144156

Kobayashi, S., Endou, M., Sakuraya, F., Matsuda, N., Zhang, X. H., Azuma, M., Echigo, N., Kemmotsu, O., Hattori, Y., & Gando, S. (2003). The sympathomimetic actions of l-ephedrine and d-pseudoephedrine: direct receptor activation or norepinephrine release?. *Anesthesia and Analgesia*, *97*(5), 1239–1245. https://doi.org/10.1213/01.ANE.0000092917.96558.3C

Kurtoğlu, S., Sen, B., Melikoğlu, G., Becker, H., Zapp, J., Kiemer, A. K., Meriçli, F., & Meriçli, A. H. (2012). Diterpenoid alkaloids of Aconitum vulparia Rchb. *Zeitschrift fur Naturforschung. C, Journal of Biosciences*, *67*(3–4), 103–107.

Lamontagne, J., Sas, E., Vincent, G., Shang, K., Pitre, F. E., & Labrecque, M. (2022). Valuable alkaloids content is preserved in *Camptotheca acuminata* and *Morus alba* grown in trace elements contaminated soil. *International Journal of Phytoremediation*, *24*(14), 1533–1542. https://doi.org/10.1080/15226514.2022.2043238

León, F., Obeng, S., Mottinelli, M., Chen, Y., King, T. I., Berthold, E. C., Kamble, S. H., Restrepo, L. F., Patel, A., Gamez-Jimenez, L. R., Lopera-Londoño, C., Hiranita, T., Sharma, A., Hampson, A. J., Canal, C. E., McMahon, L. R., & McCurdy, C. R. (2021). Activity of *Mitragyna speciosa* ("Kratom") alkaloids at serotonin receptors. *Journal of Medicinal Chemistry*, *64*(18), 13510–13523. https://doi.org/10.1021/acs.jmedchem.1c00726

Liem-Moolenaar, M., de Boer, P., Timmers, M., Schoemaker, R. C., van Hasselt, J. G., Schmidt, S., & van Gerven, J. M. (2011). Pharmacokinetic-pharmacodynamic relationships of central nervous system effects of scopolamine in healthy subjects. *British Journal of Clinical Pharmacology*, *71*(6), 886–898. https://doi.org/10.1111/j.1365-2125.2011.03936.x

Liu, Y. J., Peng, W., Hu, M. B., Xu, M., & Wu, C. J. (2016). The pharmacology, toxicology and potential applications of arecoline: A review. *Pharmaceutical Biology*, *54*(11), 2753–2760. https://doi.org/10.3109/13880209.2016.1160251

Lötsch, J., Skarke, C., Schmidt, H., Rohrbacher, M., Hofmann, U., Schwab, M., & Geisslinger, G. (2006). Evidence for morphine-independent central nervous opioid effects after administration of codeine: Contribution of other codeine metabolites. *Clinical Pharmacology and Therapeutics*, *79*(1), 35–48.

Marín-Sáez, J., Romero-González, R., Garrido Frenich, A., & Egea-González, F. J. (2018). Screening of drugs and homeopathic products from Atropa belladonna seed extracts: Tropane alkaloids determination and untargeted analysis. *Drug Testing and Analysis*, *10*(10), 1579–1589. https://doi.org/10.1002/dta.2416

Martin, C. A., Nuzzo, P. A., Ranseen, J. D., Kleven, M. S., Guenthner, G., Williams, Y., Walsh, S. L., & Dwoskin, L. P. (2018). Lobeline effects on cognitive performance in adult ADHD. *Journal of Attention Disorders*, *22*(14), 1361–1366. https://doi.org/10.1177/1087054713497791

Martínez-Pinilla, E., Oñatibia-Astibia, A., & Franco, R. (2015). The relevance of theobromine for the beneficial effects of cocoa consumption. *Frontiers in Pharmacology*, *6*, 30. https://doi.org/10.3389/fphar.2015.00030

Misiurek, J., Plech, T., Kaproń, B., Makuch-Kocka, A., Szultka-Młyńska, M., Buszewski, B., & Petruczynik, A. (2023). Determination of some isoquinoline alkaloids in extracts obtained from selected plants of the ranunculaceae, papaveraceae and fumarioideae families by liquid chromatography and in vitro and in vivo investigations of their cytotoxic activity. *Molecules (Basel, Switzerland)*, *28*(8), 3503. https://doi.org/10.3390/molecules28083503

Moreira, R., Pereira, D. M., Valentão, P., & Andrade, P. B. (2018). Pyrrolizidine alkaloids: Chemistry, pharmacology, toxicology and food safety. *International Journal of Molecular Sciences*, *19*(6), 1668. https://doi.org/10.3390/ijms19061668

Müller, J., Puttich, P. M., & Beuerle, T. (2020). Variation of the main alkaloid content in *Equisetum palustre* L. in the light of its ontogeny. *Toxins*, *12*(11), 710. https://doi.org/10.3390/toxins12110710

Nowak, M., & Selmar, D. (2016). Cellular distribution of alkaloids and their translocation via phloem and xylem: the importance of compartment pH. *Plant Biology (Stuttgart, Germany)*, *18*(6), 879–882. https://doi.org/10.1111/plb.12504

Okhovat, A., Cruces, W., Docampo-Palacios, M. L., Ray, K. P., & Ramirez, G. A. (2023). Psychoactive isoxazoles, muscimol, and isoxazole derivatives from the amanita (agaricomycetes) species: Review of new trends in synthesis, dosage, and biological properties. *International Journal of Medicinal Mushrooms*, *25*(9), 1–10. https://doi.org/10.1615/IntJMedMushrooms.2023049458

Parks, H. M., Cinelli, M. A., Bedewitz, M. A., Grabar, J. M., Hurney, S. M., Walker, K. D., Jones, A. D., & Barry, C. S. (2023). Redirecting tropane alkaloid metabolism reveals pyrrolidine alkaloid diversity in Atropa belladonna. *The New Phytologist*, *237*(5), 1810–1825. https://doi.org/10.1111/nph.18651

Peron, G., Prasad Phuyal, G., Hošek, J., Adhikari, R., & Dall'Acqua, S. (2024). Identification of hydroxyquinazoline alkaloids from Justicia adhatoda L. leaves, a traditional natural remedy with NF-κB and AP-1-mediated anti-inflammatory properties and antioxidant activity. *Journal of Ethnopharmacology*, *331*, 118345. https://doi.org/10.1016/j.jep.2024.118345

Pingali, S., Donahue, J. P., & Payton-Stewart, F. (2015). Tetrahydroberberine, a pharmacologically active naturally occurring alkaloid. *Acta Crystallographica. Section C, Structural Chemistry*, *71*(Pt 4), 262–265. https://doi.org/10.1107/S2053229615004076

Purkiewicz, A., Pietrzak-Fiećko, R., Sörgel, F., & Kinzig, M. (2022). Caffeine, paraxanthine, theophylline, and theobromine content in human milk. *Nutrients*, *14*(11), 2196. https://doi.org/10.3390/nu14112196

Ram, F. S., Jardin, J. R., Atallah, A., Castro, A. A., Mazzini, R., Goldstein, R., Lacasse, Y., & Cendon, S. (2005). Efficacy of theophylline in people with stable chronic obstructive pulmonary disease: A systematic review and meta-analysis. *Respiratory Medicine*, *99*(2), 135–144. https://doi.org/10.1016/j.rmed.2004.10.009

Reyes, C. M., & Cornelis, M. C. (2018). Caffeine in the diet: Country-level consumption and guidelines. *Nutrients*, *10*(11), 1772. https://doi.org/10.3390/nu10111772

Salcedo, J., & Kern, M. J. (2009). Effects of caffeine and theophylline on coronary hyperemia induced by adenosine or dipyridamole. *Catheterization and Cardiovascular Interventions: Official Journal of the Society for Cardiac Angiography & Interventions*, *74*(4), 598–605. https://doi.org/10.1002/ccd.22030

Schubert, M. M., Irwin, C., Seay, R. F., Clarke, H. E., Allegro, D., & Desbrow, B. (2017). Caffeine, coffee, and appetite control: A review. *International Journal of Food Sciences and Nutrition*, *68*(8), 901–912. https://doi.org/10.1080/09637486.2017.1320537

Seale, J. T., Carpenter, J. E., Eisenstat, M. D., Kiernan, E. A., Morgan, B. W., Nogee, D. P., Pu, X., Therriault, C. A., Yeh, M., & McDougal, O. M. (2022). *Veratrum parviflorum* poisoning: identification of steroidal alkaloids in patient blood and breast milk. *Clinical Toxicology (Philadelphia, PA)*, *60*(12), 1309–1317. https://doi.org/10.1080/15563650.2022.2132166

Seeman, J. I., & Tantillo, D. J. (2020). From decades to minutes: Steps toward the structure of strychnine 1910–1948 and the application of today's technology. *Angewandte Chemie (International ed. in English)*, *59*(27), 10702–10721. https://doi.org/10.1002/anie.201916566

Sharma, M., Dhaliwal, I., Rana, K., Delta, A. K., & Kaushik, P. (2021). Phytochemistry, pharmacology, and toxicology of *Datura* species: A review. *Antioxidants (Basel, Switzerland)*, *10*(8), 1291. https://doi.org/10.3390/antiox10081291

Shi, J. S., Yu, J. X., Chen, X. P., & Xu, R. X. (2003). Pharmacological actions of Uncaria alkaloids, rhynchophylline and isorhynchophylline. *Acta Pharmacologica Sinica*, *24*(2), 97–101.

Tsujikawa, K., Mohri, H., Kuwayama, K., Miyaguchi, H., Iwata, Y., Gohda, A., Fukushima, S., Inoue, H., & Kishi, T. (2006). Analysis of hallucinogenic constituents in Amanita mushrooms circulated in Japan. *Forensic Science International*, *164*(2–3), 172–178. https://doi.org/10.1016/j.forsciint.2006.01.004

Upadhyay, A., & Beuerman, R. W. (2020). Biological mechanisms of atropine control of Myopia. *Eye & Contact Lens*, *46*(3), 129–135. https://doi.org/10.1097/ICL.0000000000000677

Valente, I. M., Monteiro, A., Sousa, C., Miranda, C., Maia, M. R. G., Castro, C., Cabrita, A. R. J., Trindade, H., & Fonseca, A. J. M. (2024). Agronomic, nutritional traits, and alkaloids of *Lupinus albus, Lupinus angustifolius* and *Lupinus luteus* Genotypes: Effect of sowing dates and locations. *ACS Agricultural Science & Technology*, *4*(4), 450–462. https://doi.org/10.1021/acsagscitech.3c00581

Velázquez-Márquez, S., De-la-Cruz, I. M., Tapia-López, R., & Núñez-Farfán, J. (2021). Tropane alkaloids and terpenes synthase genes of *Datura stramonium* (Solanaceae). *PeerJ*, *9*, e11466. https://doi.org/10.7717/peerj.11466

Wikoff, D., Welsh, B. T., Henderson, R., Brorby, G. P., Britt, J., Myers, E., Goldberger, J., Lieberman, H. R., O'Brien, C., Peck, J., Tenenbein, M., Weaver, C., Harvey, S., Urban, J., & Doepker, C. (2017). Systematic review of the potential adverse effects of caffeine consumption in healthy adults, pregnant women, adolescents, and children. *Food and Chemical Toxicology: An International Journal Published for the British Industrial Biological Research Association*, *109*(Pt 1), 585–648. https://doi.org/10.1016/j.fct.2017.04.002

Wu, T., Wu, Y., Chen, S., Wu, J., Zhu, W., Liu, H., Chen, M., & Xu, B. (2021). Curative effect and survival assessment comparing gemcitabine and cisplatin versus methotrexate, vinblastine, doxorubicin and cisplatin as neoadjuvant therapy for bladder cancer: A systematic review and meta-analysis. *Frontiers in Oncology*, *11*, 678896. https://doi.org/10.3389/fonc.2021.678896

Yan, D., & White, M. M. (2002). Interaction of d-tubocurarine analogs with mutant 5-HT(3) receptors. *Neuropharmacology*, *43*(3), 367–373. https://doi.org/10.1016/s0028-3908(02)00125-9

Yang, G. Y., Dai, J. M., Mi, Q. L., Li, Z. J., Li, X. M., Zhang, J. D., Wang, J., Li, Y. K., Wang, W. G., Zhou, M., & Hu, Q. F. (2022). Cyclopiazonic acid type indole alkaloids from Nicotiana tabacum-derived fungus Aspergillus versicolor and their anti-tobacco mosaic virus activities. *Phytochemistry*, *198*, 113137. https://doi.org/10.1016/j.phytochem.2022.113137

Yang, N., Guo, J., Zhang, J., Gao, S., Xiang, Q., Wen, J., Huang, Y., Rao, C., & Chen, Y. (2024). A toxicological review of alkaloids. *Drug and Chemical Toxicology*, *47*(6), 1267–1281. https://doi.org/10.1080/01480545.2024.2326051

Zhang, F., Qiu, F., Zeng, J., Xu, Z., Tang, Y., Zhao, T., Gou, Y., Su, F., Wang, S., Sun, X., Xue, Z., Wang, W., Yang, C., Zeng, L., Lan, X., Chen, M., Zhou, J., & Liao, Z. (2023). Revealing evolution of tropane alkaloid biosynthesis by analyzing two genomes in the Solanaceae family. *Nature Communications*, *14*(1), 1446. https://doi.org/10.1038/s41467-023-37133-4

Zhang, X., Xiong, X., Zhang, H., Huang, T., & Zhou, X. (2024). Pilocarpine in the treatment of presbyopia: Progress, issues, and future prospects. *Drugs & Aging*, *41*(11), 897–906. https://doi.org/10.1007/s40266-024-01155-4

Ziegler, J., & Facchini, P. J. (2008). Alkaloid biosynthesis: metabolism and trafficking. *Annual Review of Plant Biology*, *59*, 735–769. https://doi.org/10.1146/annurev.arplant.59.032607.092730

13 Alkaloids and Drug Discovery
Approaches and Success Stories

Preeti Saxena, Sheetal Tyagi, and Anshika Sharma

13.1 INTRODUCTION

Since ancient times, humans have utilized plants that contain alkaloids for both medicinal and recreational purposes. For instance, Mesopotamia has been aware of medicinal plants since around 2000 BC (Tadeusz, 2007). The Odyssey of Homer referred to a gift given to Helen by the Egyptian queen, a drug that brings oblivion. It is believed that the gift was an opium-containing drug (Hesse, 2002). Ephedra and opium poppies were used medicinally, according to a Chinese book on houseplants from the 1st to 3rd century BC. Additionally, Indigenous South Americans have been using coca leaves since ancient times.

13.1.1 Historical Significance

In 1805, morphine was the first reported alkaloid isolated in its crude form from opium and, to date, is of medicinal importance (Devereaux et al., 2018). After a few years, Quinine, a significant antimalarial drug, was discovered, and the world now recognizes the significance of alkaloids (Kaufman & Rúveda, 2005). Alkaloids are found throughout the biome, and a small number of terrestrial isolations have also been documented from marine sources such as sponges, tunicates, corals, and mammals, as well as from plants and a variety of microorganisms (Cushnie et al., 2014). The French scientists Pierre Joseph Pelletier and Joseph Bienaimé Caventou made significant improvements to the chemical reactions of alkaloids in the early years of their development. They made a significant contribution to the chemistry of alkaloids and discovered quinine (1820) and strychnine (1818), respectively, in the early years of development. Xanthine (1817), atropine (1819), caffeine (1820), coniine (1827), nicotine (1828), colchicine (1833), sparteine (1851), and cocaine (1860) are just a few more to the list discovered around that time. Advances in spectroscopic and chromatographic techniques have accelerated the study of alkaloids. By 2008, more than 12,000 alkaloids had been reported and identified in the 20th century. A large class of organic compounds that occur naturally and have a nitrogen atom of some sort is known as alkaloids. Alkaloids are another name for certain artificial substances with a similar structure (Lewis, 1998).

Plant alkaloids exhibit mild to extreme physiological effects on humans when they are consumed. These physiological effects range from light stimulants such as caffeine to poisons such as coniine (the poison in hemlock). Different organisms like bacteria, fungi, plants, and animals produce a variety of alkaloids (Kinghorn et al., 2011). The discovery of penicillin in 1929 by Alexander Fleming, a Scottish bacteriologist, from the fungus (microorganism) Penicillium notatum, is a well-known natural discovery. The structure of Penicillin, Norcardicin, Imipenem, and Aztreonam is shown in Figure 13.1.

A few alkaloids called pinnatoxins, pinnamine, and halochlorine are present in marine organisms like sponges and shellfish. They are found to be highly effective in treating inflammatory conditions of the heart and non-heart (Kuramoto et al., 2004). Many ladybird beetles use alkaloids as defensive chemicals, and when they are molested, they release bitter alkaloids that contain hemolymph. Due to

DOI: 10.1201/9781003539179-13

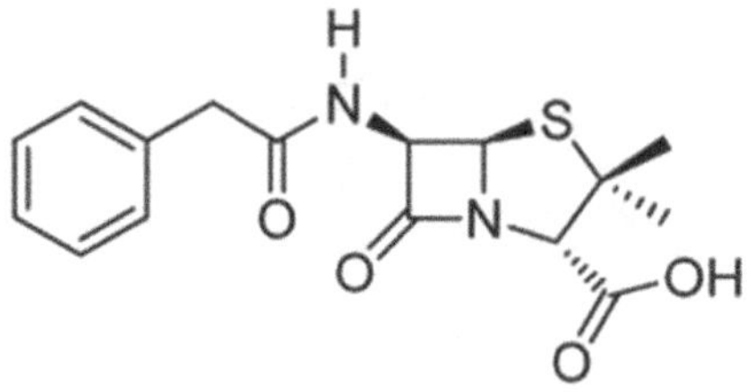

Penicillin

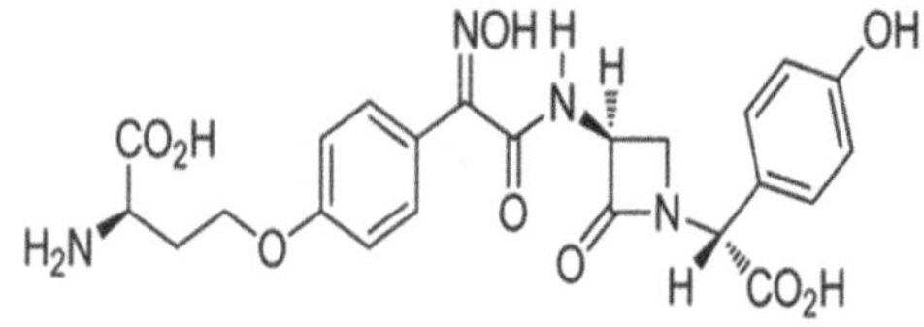

Norcardicin

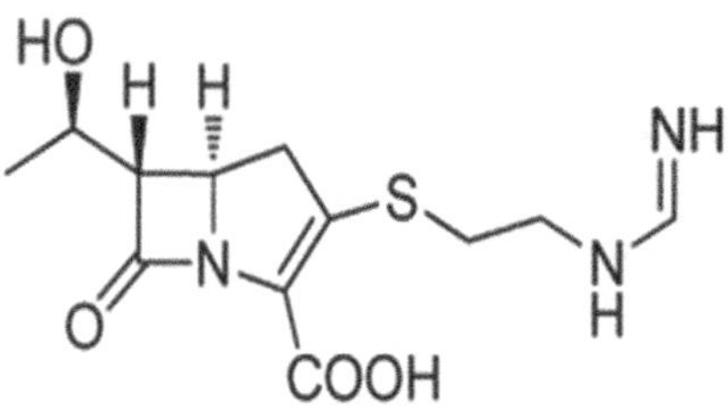

Imipenem

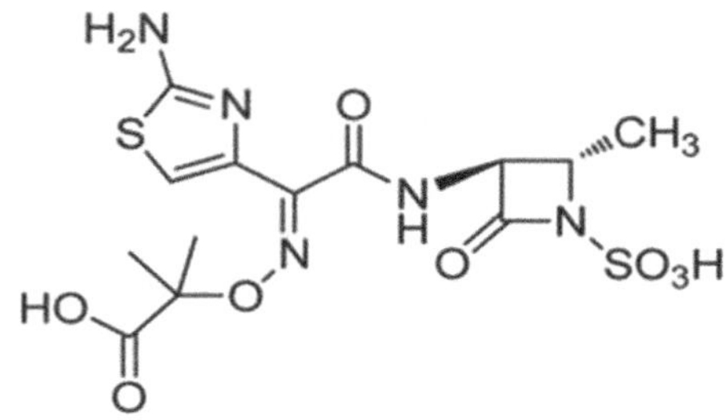

Aztreonam

FIGURE 13.1 Structure of penicillin, norcardicin, imipenem, and aztreonam.

the presence of proteinaceous compounds and alkaloids such as cis- and trans-2-methyl-6-alkylpiperidines, which are employed both offensively and defensively, few ant species have toxic venom (Braekman et al., 1998).

13.1.2 Properties

The presence of nitrogen in alkaloids sets them apart from any other molecules. This nitrogen is often basic in nature and comes from amino acids, incorporated into a heterocyclic ring. Alkaloids almost always have physiological activity in animals, although some have limited effects (Seigler, 2012). Most of the alkaloids are crystalline in nature and unite with acids to form salts. In the plant, they may exist in the free state, as salts or as N-oxides.

Colored alkaloids, although relatively rare, are not unknown; for example, berberine and the salts of sanguinarine, which are yellow and copper-red in color, respectively (Abbott, 1970). Due to the basic character of alkaloids, they readily react with mineral acids (hydrochlorides, sulfates, nitrates) or organic acids (tartrates, sulfamates, and maleates) to form salts.

Under normal circumstances, oxygen molecules are typically colorless crystals. Alkaloids that do not contain oxygen, like nicotine or coniine, are usually flammable, colorless, and greasy liquids.

The presence of tertiary nitrogen in the structure of a few alkaloids generally makes them optically active. This results in various isomeric forms having different physical, chemical, and pharmacological properties. For example, (+)-tubocurarine isolated from Chondrodendron tomentosum has muscle relaxant activity. In contrast, its levo isomer is less active (Cushnie et al., 2014).

Recently, the official definition of alkaloids has been revised to include pure structures that are cyclic compounds containing nitrogen, which are found in living organisms. Typically, alkaloids are basic and can accept hydrogen because the nitrogen is in the neutral state. Alkaloids are synthesized in plants starting with amino acids, which provide the needed nitrogen for the alkaloid structure. It is challenging to distinguish alkaloids from other naturally occurring substances that contain nitrogen. While some alkaloids are neutral or even slightly acidic, the majority are basic. Alkaloids may contain sulfur or oxygen in addition to carbon, hydrogen, and nitrogen (Manske & Holmes, 2014). According to reports, higher plant species from the families of Berberidaceae, Amaryllidaceae, Liliaceae, Leguminaceae, Papaveraceae, Ranunculaceae, and Solanaceae are specifically rich in alkaloids (Matsuura & Fett-Neto, 2015). Alkaloids play an essential role in both human medicine and in an organism's natural defense; 20% of the known secondary metabolites found in plants are made up of alkaloids. These alkaloids protect plants from predators and regulate their growth (Chik et al., 2013). Therapeutically, alkaloids are particularly well known as anaesthetics, cardioprotective, and anti-inflammatory agents. Morphine, strychnine, quinine, ephedrine, and nicotine are a few famous alkaloids used in clinical settings (Kurek, 2019).

13.1.2.1 Physical Properties

Alkaloids tend to dissolve well in water, ethanol, and a variety of organic solvents. Most of them have relatively high melting points, usually above 200°C, and their boiling points are often even higher, typically exceeding 300°C. When isolated from solutions, these compounds often solidify into crystals, commonly appearing as thin needles or flat plates.

13.1.2.2 Chemical Properties

Alkaloids exhibit several key chemical characteristics. They possess basicity due to their ability to accept a proton (H^+), forming a conjugate acid, which makes them inherently basic in nature. Their reactive behavior allows them to interact with acids to yield salts readily and also form various derivatives when combined with other substances. However, these compounds are often unstable when exposed to heat, oxygen, or light, as such conditions may lead to their degradation or decomposition. Furthermore, many alkaloids demonstrate optical activity, meaning they can rotate plane-polarized light—a property that signifies the presence of chiral centers in their structure.

13.1.3 Relevance of Alkaloids in Modern Drug Discovery Pipelines

Alkaloids are highly relevant in modern drug discovery pipelines due to their diverse chemical structures, wide range of biological activities, and potential as lead compounds for new drugs. Their unique properties, like the ability to act as hydrogen bond acceptors or donors, make them ideal for interacting with biological targets. This has led to their use in developing drugs for various conditions, including cancer, tuberculosis, and smoking cessation. Moreover, Alkaloid drug discovery is hampered by several significant experimental limitations, including the need to incorporate a selectivity index. Other limitations include the depletion of biological resources, testing feasibility, dereplication for known compounds and those that disrupt bioassays, and continuous complexity.

In medical science, alkaloids are considered very important compounds due to their physiological effects. They are utilized as treatments for pain relief, stimulants, muscle relaxants, sedatives, anesthetics, antimalarial agents, antimicrobial substances, antidiabetic medications, anticancer treatments, anti-HIV drugs, and antioxidants, among others. Proaporphines and crotsparine extracted from Cocculus sparciflorus demonstrated notable hypotensive and anticancer properties (Bhakuni & Dhar, 1968). A molluscicidal alkaloid derived from the root of Galipea racemosa is a derivative of homoerythrine. Nowadays, people all over the world consume the stimulants nicotine in cigarettes and caffeine in coffee, tea, and cacao. Alkaloids with narcotic, hallucinogenic, or analgesic effects—such as quinine, atropine, and morphine—have been used in medicine. While some alkaloids, like cocaine, are abused as illegal drugs, others have been used as model compounds for contemporary synthetic drugs. Some alkaloids, like strychnine and coniine, are too toxic to be used therapeutically. Plant components are also still being studied in an effort to find new biologically active compounds. As a cytostatic drug, Taxol is used to treat ovarian, breast, neck, and lung cancers (Fielding et al., 2020).

In 1969, Hartwell and Abbot reported that coniine, tropine, vendoline, morphine, and tyrosine showed significant anticancer activity. Nicotiana tabacum belongs to the Solanaceae family and gives nicotine, which is a highly addictive stimulant of the nervous system in small doses. In contrast, its higher doses can be highly toxic.

13.1.3.1 Alkaloids in Contemporary Drug Discovery

In addition to many antibiotics and certain substances from marine sources, nearly 60 alkaloids derived from plants are currently approved as medications in several nations. They originate from amino acids known as L-amino acids (tryptophan, ornithine, lysine, phenylalanine, proline, glutamine, histidine, etc.) and have attachments from the shikimate pathway, terpenoid biosynthesis, and polyketide biosynthesis that produce a variety of heterocyclic nuclei. Alkaloids are physiologically important metabolites that are frequently very strong and, in certain cases, unique in their mechanism of action and use (Funayama & Cordell, 2014). New alkaloids exhibiting a wide variety of modern biological activities have been studied and isolated by scientists. In 2020 alone, the Journal of Natural Products published findings on 316 newly discovered alkaloids with 124 metabolites exhibiting 25 different biological activities (Cordell, 2017).

New classes of antimicrobial agents are constantly needed to treat bacterial and fungal infections that have become resistant to the available medications (Newman & Cragg, 2020). Infectious disease-related deaths were the second most common cause of death worldwide and the third most common in developed countries, with bacterial infections killing 17 million patients annually (Kraus, 2008). Enterococcus fecium, Staphylococcus aureus, Klebsiella pneumoniae, Acinetobacter baumannii, Pseudomonas aeruginosa, and Enterobacter spp. are among the ESKAPE pathogens that the WHO has identified as being of primary concern and are responsible for more than 700,000 deaths annually (Mulani et al., 2019).

The development of virulent genes containing protective mechanisms, such as beta-lactamases and biofilm formation, has led to the persistent rise of antibiotic-resistant strains of microorganisms, which is the main obstacle in the search for new antibiotic drugs. The development of antibiotic resistance necessitates more controlled use and administration of antibiotics, while preserving their availability in middle- and low-income nations. To ensure that everyone has access to appropriate and efficient antibiotic treatment therapy, this should be combined with better local diagnostic testing (Nathan & Cars, 2014), along with strictly limiting the use of 30 different antibiotics, as growth promoters for plants and animals (Casewell et al., 2003). Among the chemically based methods for overcoming drug resistance are:

1. using bioinformatics or in-field techniques to find new antibiotic scaffolds,
2. combining antimicrobial agents in therapy,
3. using adjuvants in either combination, as well as alternately for treatment, and
4. altering the structure of already-existing antibiotics.

13.1.4 Pharmacological Actions

The effect of alkaloids on herbivores, toxicity for vertebrates, cytotoxic, mutagenic, or carcinogenic activity, antibacterial, antifungal, and antiviral properties, and their possible roles as phytoalexins are well known. Many alkaloids possess a level of toxicity that can be lethal to animals if eaten. Certain alkaloids, such as nicotine and anabasine, are used as insecticides. Many alkaloids act on the nervous system, one of the two important information systems in animals. Plants containing protoberberine alkaloids are reported to be used in Chinese folk medicine as analgesics, antiseptics, sedatives, and stimulants. In Indian and Islamic folk medicine, such plants are utilized for bleeding disorders, eye diseases, as well as for their antiseptics, sedatives, stomachic, and uterine muscle-relaxing properties. Quaternary alkaloids and their tetrahydro derivatives show significant biological and therapeutic effects. For example, palmatine, jatrorrhizine, and tetrahydropalmatine have been noted to exhibit in vitro antimalarial activity. In China, tetrahydropalmatine is used as a pain reliever and has been reported to display bradycardial, hypotensive, and sedative effects (Da-cunha et al., 2005).

13.1.5 Natural Product-Based Drug Discovery Resurgence

In an ideal scenario, the discovery of drugs should assess medicinal agents and align them with the present and anticipated health challenges faced by humanity. When it comes to targeted drug development, the vast majority of individuals on the globe have no say in how sickness priorities are determined. Since 1992, the World Health Organization (WHO), World Bank, and Harvard School of Public Health have been conducting the Global Burden of Disease (GBD) analysis of data project, which encompasses all aspects of human diseases, both communicable and non-communicable, their changing demographics, the causes of death rates, and the forecasting of needs (Murray, 1994). "It is time for the global health community to change direction," read the editorial in The Lancet's special issue on the 2019 GBD in October 2020. The majority population's health care needs are not being met by the current drug discovery system (Karimkhani et al., 2016). To meet the demands for these medicinal agents, both synthetic and natural compounds are desperately needed for biological evaluation. One of the main natural resources that can be utilized for the development of new drugs is alkaloids.

The almost 60 plant-derived alkaloids that are approved as medicines in different countries include a few antibiotics and a few compounds from marine sources. Their biosynthesis involves the use of tryptophan, ornithine, lysine, phenylalanine, proline, glutamine, histidine, and other L-amino acids. Attachments also produce a range of heterocyclic nuclei through the shikimate pathway, polyketide biosynthesis, and terpenoid biosynthesis. In some cases, a pre-existing nucleus (e.g., terpenoid alkaloids) is supplemented with the nitrogen atom. Scientists are actively examining these pathway relationships in addition to the metabolic pool's size and complexity (Khamkar et al., 2015). Alkaloids are physiologically important metabolites that are frequently very strong and, in certain cases, unique in their mode of action and use.

13.2 SOURCES OF ALKALOIDS

Some authors argue that alkaloids are mainly related to the vegetable kingdom and are found as salts or may be formed biosynthetically from an amino acid. These components define what may be referred to as true alkaloids. Many authors distinguish, in addition, proto-alkaloids and pseudo-alkaloids. It is very rare to get a single alkaloid from plants; generally, they produce a complex mixture with one major component. Hyoscyamine is an example of a single alkaloid obtained from the leaves of belladonna. It is not uncommon to find several dozen alkaloids in one drug. All alkaloids found in a particular plant have a common biogenetic source, even if their structures initially appear quite different. The concentration of alkaloids can be the same or differ significantly among various parts of the same plant (Zhou et al., 2012).

13.2.1 Plant-Derived Alkaloids

Plant-derived alkaloids are typically found in all plant parts, but occasionally they are concentrated in just one organ, excluding other organs from their presence. For instance, the green portions of the potato plant contain the toxic alkaloid solanine, while the edible tubers are free of alkaloids. Not all alkaloids are produced in the same organ; for example, nicotine in tobacco is made in the roots and then moved to the leaves for storage (Rice, 2008).

13.2.2 Animal Derived Alkaloids

Along with other toxins, the skin of amphibians contains about 300 alkaloids, which belong to over 24 classes. The skin of frogs in the genus Phyllobates was found to contain a highly toxic and poisonous alkaloid (Evans, 2009). Reptile skin contains a variety of antimicrobial alkaloids. Mammal morphine is one of the few indole and isoquinoline alkaloids that have been isolated from mammals.

13.2.3 Microbial and Marine Sources

Indole alkaloids derived from the marine fungus Aspergillus versicolor decreased NO levels by inhibiting iNOS activity in lipopolysaccharide (LPS)-stimulated RAW 264.7 macrophages (Li et al., 2018). The NO levels generated by LPS-stimulated macrophages were also reduced by two N,N′-ketal quinazoline alkaloids, the enantiomers (±)-penicamide A [5: (+)-penicamide A; 6: (-)-penicamide A], which were extracted from the fungus Penicillium canescens that was isolated from the ascidian Styela plicata (Chen et al., 2019). LPS-stimulated RAW 264.7 and BV2 cells produced less NO and PGE2 as a result of the diketopiperazine alkaloids 5-prenyl-dihydrovariecolorin F (7) and 5-prenyl-dihydrorubrumazine A (8), which were isolated from Aspergillus species present in seaweed and showed an inhibitory effect on the activity of iNOS and COX-2. In LPS-stimulated RAW 264.7 cells, neoechinulin A (9), an indolic alkaloid originating from marine fungi Eurotium spp., inhibited the expression of iNOS and COX-2, which in turn decreased the production of NO and PGE2. It also decreased the production of IL-1β and TNF-α (Kim et al., 2013). In addition to preventing IκB-α phosphorylation and subsequent degradation, the anti-inflammatory properties were linked to blocking the attachment of the NF-κB p65 subunit to nuclear DNA, which suppressed the pro-inflammatory reaction mediated by the NF-κB pathway. Neoechinulin A additionally showed the capacity to prevent MAPK p38 phosphorylation, which is another element that contributes to the induction of a pro-inflammatory response.

13.3 APPROACHES IN ALKALOID-BASED DRUG DISCOVERY

13.3.1 Ethnopharmacology and Traditional Medicine Knowledge

Ethnopharmacology plays a major role in the development of new therapeutic medications. For generations, traditional healers have utilized the medicinal properties of local plants, many of which are now recognized to have pharmacological potential. For example, Catharanthus roseus, commonly referred to as Madagascar periwinkle, helped discover vincristine and vinblastine, two drugs that fundamentally altered the treatment of cancer. The study of indigenous cultures' use of natural substances for therapeutic purposes and traditional medical practices is known as ethnopharmacology. According to the theory, traditional knowledge and customs can offer important insights into possible novel medications and therapeutic agents (Pirintsos et al., 2022). To find, separate, and evaluate bioactive substances from natural sources, this multidisciplinary method integrates anthropology, chemistry, pharmacology, botany, and other disciplines. Included are a few crucial actions. Knowing how various cultures use plants and other natural substances for medicinal purposes is known as ethnobotanical knowledge. Identification of Bioactive Compounds: Separating and describing the active ingredients that give traditional medicines their therapeutic effects. Many diseases, especially infectious and chronic ones, have benefited from ethnopharmacology's contributions to treatment. Bioactive compounds with therapeutic potential against diseases like cancer, diabetes, and infectious diseases have been discovered in many traditional remedies. Several compounds derived from plants have been developed into chemotherapeutic agents for the treatment of cancer. For instance, paclitaxel, a substance derived from Taxus brevifolia, is now a crucial medication for the treatment of breast and ovarian cancer (Wani et al., 1971). Ethnopharmacological drugs and traditional–modern applications are shown in Table 13.1.

13.3.2 In Silico Drug Design and Molecular Docking

Utilizing in silico techniques is very beneficial because it eliminates the need for intensive in vitro screening of sizable compound libraries and can identify significant compounds, especially alkaloids. They allow only those candidates that have a high probability of being active to be tested in vitro, use minimal resources, and save time and effort that would otherwise be wasted isolating pure metabolites (Olğaç et al., 2017). If a protein structure for an active site is available, probing with the

TABLE 13.1
Ethnopharmacological Drugs and Traditional and Modern Applications

Traditional Medicinal Plant	Derived Drug	Structure	Traditional Use	Modern Application	References
Cinchona officinalis	Quinine	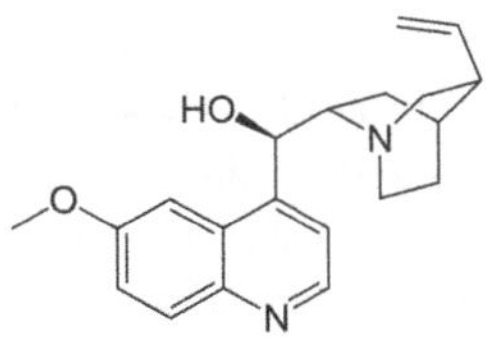	Fever, Malaria	Antimalarial	Jose (2020)
Taxus brevifolia	Paclitaxel	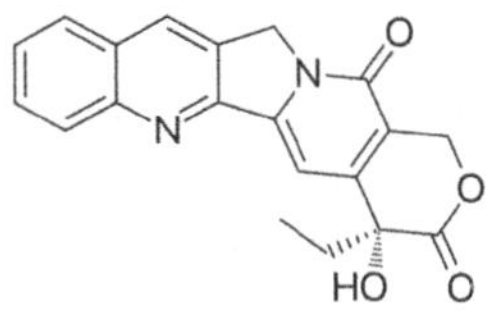	Tumor treatment	Anticancer (breast, ovarian cancer)	Barti et al. (2018)
Papaver somniferum	Morphine	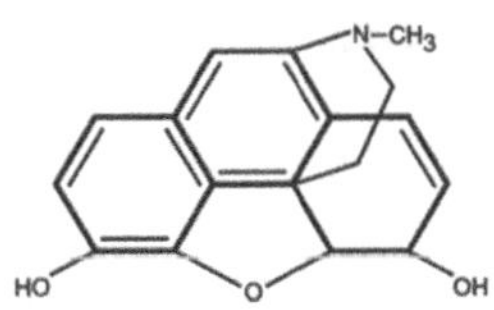	Pain relief	Analgesic in severe pain management	Butnariu et al. (2022)
Salix alba	Aspirin	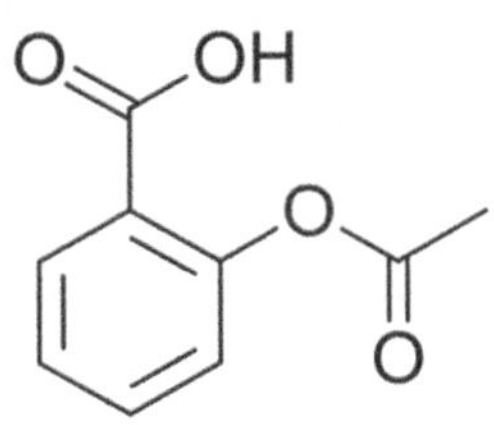	Pain, fever	Anti-inflammatory, analgesic	Montinari et al. (2019)
Withania somnifera	Withaferin	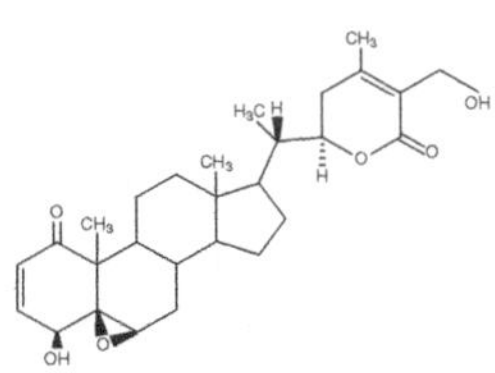	A General tonic	Adaptogen, anti-stress	Atteeq (2022)
Rauvolfia serpentina	Reserpine	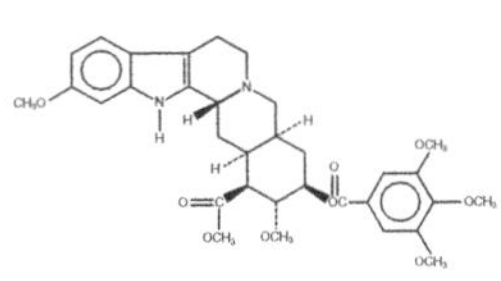	Hypertension	Antihypertensive, antipsychotic	Lobay (2015)
Erythroxylum coca	Cocaine	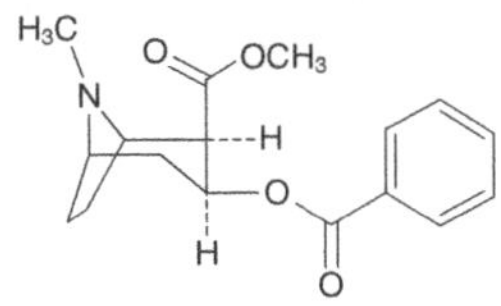	Anesthetic	Local anesthetic in surgeries	Biondich and Joslin (2016)
Camptotheca acuminata	Camptothecin	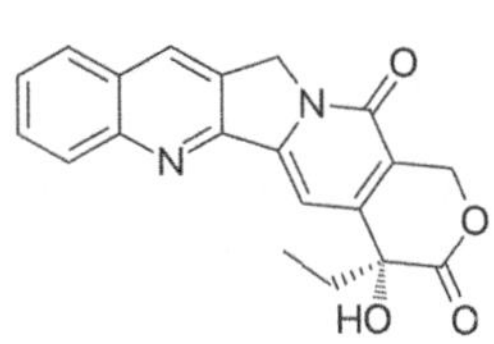	Cancer	Anticancer (colon, lung cancer)	Ghanbari-Movahed et al. (2021)

range of known alkaloid structures is a useful way to generate potential metabolites for in vitro biological tests and provides the chance to design inhibitors.

13.3.3 Combinatorial Biosynthesis and Semi-Synthetic Modification

The highly concentrated process of biosynthesis, which includes several enzyme enhancements that produce molecules with complex structures, produces alkaloids. The first step in the production of amines is the decarboxylation of starting amino acids. These amines subsequently participate in cyclization, methylation, and oxidation processes to produce the characteristic heterocyclic cores of alkaloids. In order to create indole alkaloids, like those made by alkaloid biosynthesis, it is a highly specialized process that involves a series of enzyme reactions that result in structurally complicated compounds. The precursors of amino acids are the first in the route; they are decarboxylated to produce amines. Alkaloids' characteristic heterocyclic cores are the result of these amines' involvement in cyclization, methylation, and oxidation activities. Attempts to produce indole alkaloids, such as *Catharanthus roseus*, which also produces the anti-cancer drugs vincristine and vinblastine, first decarboxylate tryptophan to produce tryptamine (Daley & Cordell, 2021).

13.4 ALKALOIDS AS LEAD COMPOUNDS IN THERAPEUTICS

Alkaloids derived from plants have been used as medicine since ancient times and are widely used today. The ethnobotanical use of plants high in alkaloids cleared the path for the pharmacological properties of these compounds to be characterized, isolated, and evaluated, ultimately resulting in the development of several currently prescribed drugs. In addition to their many uses, alkaloids are currently being studied for their potential as chemotherapy agents, neuroprotective agents, and antidiabetic agents.

13.4.1 Anti-Inflammatory and Neuroprotective Applications

Alkaloids exhibit significant anti-inflammatory and neuroprotective properties. Berberine is a promising option for treating prolonged inflammation because it has been demonstrated to inhibit COX-2 expression and nuclear factor-κB (NF-κB) signaling. In neurodegenerative diseases, alkaloids such as galantamine, derived from Galanthus species, act as acetylcholinesterase inhibitors, providing symptomatic relief in Alzheimer's disease (Zhang et al., 2018). Tetrahydropalmatine, a bis-benzylisoquinoline alkaloid, has neuroprotective effects through the modulation of dopamine and glutamate signaling, suggesting its potential in treating Parkinson's disease and related disorders.

13.4.2 Antiviral Properties

Natural alkaloids have been explored for their antiviral activities against various RNA and DNA viruses. Berberine, harmine, and lycorine have been shown to inhibit viral replication by targeting essential enzymes or host factors critical for the viral life cycle. In the context of the SARS-CoV-2 pandemic, alkaloids like quinine as well as berberine have been highlighted for their potential to block viral entry and replication (Dey et al., 2017).

13.4.3 Anti-Diabetic Effects

Several alkaloids have shown promise in managing diabetes by modulating insulin signaling and glucose metabolism. Berberine enhances insulin sensitivity and reduces blood glucose levels by activating the AMP-activated protein kinase (AMPK) pathway. Trigonelline and piperine have also been shown to improve beta-cell function and reduce oxidative stress, which helps manage diabetes-related problems (Pervaiz et al., 2019).

13.4.4 Cancer Therapy

Alkaloids have been thoroughly researched as anticancer agents because they can interfere with several processes that lead to tumorigenesis and tumor progression. Vinblastine and Vincristine, which are derived from Catharanthus roseus, change the dynamics of microtubules, resulting in apoptosis and cell cycle arrest in cancer cells. Berberine, an isoquinoline alkaloid, exerts anti-proliferative effects by inhibiting cyclooxygenase-2 (COX-2) and telomerase activity, targeting cancer stem cells and reducing angiogenesis. Additionally, camptothecin and its derivatives, including topotecan and irinotecan, are potent inhibitors of topoisomerase I, a critical enzyme for DNA replication in cancer cells (Butaye et al., 2003). Recent advancements have also explored the potential of metalloalkaloid complexes, such as those derived from berberine and luotonin A, to enhance therapeutic outcomes in chemotherapy while mitigating drug resistance. Furthermore, steroidal alkaloids from the Solanaceae family plants have shown selective anticancer activity, with in vitro and in vivo studies supporting their potential to inhibit various cancer cell lines.

13.5 SUCCESS STORIES OF ALKALOID-DERIVED DRUGS

Pharmacogenetic research has also significantly contributed to overcoming challenges related to the pharmacokinetics and bioavailability of alkaloids. Many alkaloids suffer from poor solubility and rapid metabolism, which limits their therapeutic potential. Nanotechnology-based approaches, such as liposomal and nanoparticle formulations, have been developed to increase the delivery and bioavailability of alkaloids. For example, berberine-loaded nanoparticles have shown improved solubility and sustained release, enhancing their efficacy in treating metabolic and inflammatory disorders (Martens & Demain, 2017).

The integration of advanced delivery systems, such as transdermal patches and controlled release formulations, into alkaloid therapeutics exemplifies the synergy between Pharmacognostic research and pharmaceutical technology. These innovations not only improve patient compliance but also optimize the therapeutic efficacy of alkaloid-based drugs (Wohlleben et al., 2016). Therapeutic applications of alkaloids are shown in Table 13.2.

TABLE 13.2
Therapeutic Applications of Alkaloids: Diseases, Mechanisms, and Uses

Disease	Key Alkaloid	Mechanism of Action	Current Application	References
Cancer	Vinblastine	Microtubule disruption	Chemotherapy	Mondal et al. (2019)
Asthma	Ephedrine	Beta-adrenergic receptor stimulation	Bronchodilator	Pache et al. (2025)
Bacterial Infections	Sanguinarine	Inhibition of bacterial replication	Antibacterial agent	Pervaiz et al. (2019)
Tuberculosis	Piperine	Disruption of bacterial membrane function	Antitubercular agent	Kittakoop et al. (2014)
Inflammatory Diseases	Berberine	NF-kB and COX-2 inhibition	Anti-inflammatory agent	Aryal et al. (2022)
Malaria	Quinine	Heme polymerization inhibition	Antimalarial drug	Aniszewski (2015)
Neurodegenerative Diseases	Huperzine A	Inhibition of acetylcholinesterase	Neuroprotection	Kong et al. (2021)
Alzheimer's Disease	Galantamine	Acetylcholinesterase inhibition	Symptomatic treatment	Kong et al. (2021)

13.5.1 Historical Milestones in Alkaloid-Based Drug Development Studies

13.5.1.1 Quinine

As of 2006, the WHO discontinued recommending quinine as the initial therapy for malaria due to the availability of other drugs that are equally effective and have fewer side effects. It should only be used when artemisinins are not accessible, they say. Quinine is also used to treat lupus and arthritis (Devereaux et al., 2018).

Since 2010, quinine has been less commonly prescribed as a contraindication for nighttime leg cramps because the Food and Drug Administration of the United States (FDA) has warned that this practice is linked to potentially fatal side effects (Malaria, 2005). Additionally, monoamine oxidase (MAO), a digestive enzyme that eliminates neurotransmitters from the brain, can be competitively inhibited by quinine (Reyburn et al., 2009). Like antidepressants that inhibit MAO, it may be used as a treatment for people with psychological disorders because it is an MAO inhibitor (Mitsui et al., 1989).

13.5.1.2 Opium Poppy (*Papaver sominiferum*)

It is the most widely used medicinal herb and a member of the Papaveraceae family. The opium poppy plant contains three alkaloids: morphine, codeine, and thebaine. This plant's common medicinal property is pain relief, and early civilizations have used it for this purpose up to this point (Mishra et al., 2013). For thousands of years, traditional medicine has made use of it. Opium poppies were a common component of many medicines in the ancient Greek Hippocratic tradition (Mavrogenis et al., 2019). The poppy flower is commonly used in herbal medicine to treat chronic pain, anxiety, and insomnia. The petals can be used to make medicinal tinctures, oils, and teas (Bedikian et al., 2006). Some cultures consume the seeds raw or incorporate them into salads, muffins, and wholesome breakfast recipes.

13.5.1.3 Vincristine

Catharanthus roseus, a periwinkle plant, is its original source. The anti-tumor properties of this compound have been discovered to exhibit strong anticancer activity against various cell types in vitro, with high activity against tumor types that are resistant to multiple drugs. In contrast, Ruskin and Aruna demonstrated that the ethanolic extract of vincristine exhibited significant antitumor activity in vitro and in vivo in the Ehrlich carcinoma tumor model (Ruskin & Aruna, 2014).

13.5.2 Clinical Development and Commercial Success

Traditional and natural remedies are the source of many contemporary pharmaceutical treatments. For thousands of years, people have used plants and herbs to treat illnesses. Over 53,000 plant species from around the world have been utilized in conventional medicine (Pan et al., 2014). Berberis vulgaris, for example, is a herb that has long been used to prevent coronary artery disease and has anti-inflammatory and immunomodulatory properties. It is rich in berberine alkaloids. Diabetes and disorders of the central nervous system (CNS) can be effectively treated with the trigonelline alkaloid, which was extracted from Trigonella foenum-graecum in China (Oguntibeju, 2018). In Tanzania, oxindole and L-stachydrine from *Capparis tomentosa* have long been used to treat inflammation because of the synergistic effects of secondary ingredients (Vos et al., 2020).

13.6 CHALLENGES IN ALKALOID-BASED DRUG DEVELOPMENT

Despite their therapeutic promise, alkaloids face several challenges that limit their widespread application.

13.6.1 Challenges in Biodiversity Conservation

Preserving India's remarkable biodiversity is crucial, especially its endangered and rare medicinal plant species. These invaluable plant resources have been healing sources for centuries, but they now confront significant threats to their survival.

13.6.2 Pharmacokinetics and Bioavailability

Many alkaloids exhibit poor water solubility, limited bioavailability, and rapid metabolism, reducing their therapeutic efficacy in vivo. Addressing these limitations requires advanced formulation strategies, such as nanoencapsulation or prodrug development (Aryal et al., 2022).

13.6.3 Toxicity and Side Effects

Some alkaloids, such as aconitine and strychnine, are associated with significant toxicity at therapeutic doses, necessitating careful dose optimization and risk assessment during drug development (Dey et al., 2017).

13.6.4 Over-Exploitation

A significant challenge is the excessive harvesting of medicinal plants driven by global demand for herbal remedies. Traditional medicines like Ayurveda and Unani have deep-rooted traditions in India, creating immense pressure on medicinal plant populations. Unregulated harvesting and extraction can lead to population depletion and even the extinction of rare species (Kala et al., 2006).

13.6.5 Unsustainable Extraction

Unsustainable extraction methods exacerbate the issue. Medicinal plant species are often entirely uprooted rather than selectively harvested for leaves, bark, or fruits. This unscientific approach disrupts natural reproduction, hindering the plants' ability to rebound (Zhang et al., 2018).

13.6.6 Biodiversity Loss

The consequences of these challenges extend to the broader issue of biodiversity loss. As medicinal plants face habitat degradation and overharvesting, the delicate balance of ecosystems is further destabilized, impacting various species within these ecosystems (Rajeswara et al., 2012).

13.6.7 Certification and Standardization

Certification and Standardization, particularly in the context of organic and fair trade practices, hold significant importance in the medicinal plant industry. Organic certifications ensure that medicinal plants are grown using environmentally sustainable methods, without synthetic pesticides or fertilizers, promoting the production of clean and natural herbal products. These certifications enhance the quality and sustainability of medicinal plant cultivation and meet the increasing global demand for ethical and environmentally responsible herbal products (Raju & Das, 2024).

13.6.8 The Absence of Standardized Plant Identification and Processing

The quality and effectiveness of herbal medicines depend heavily on robust quality control and standardization procedures. Nevertheless, the herbal medicine sector faces a pressing challenge due to the need for standardized plant identification and processing protocols. This gap undermines the uniformity and trustworthiness of herbal products, with significant implications for public health.

13.6.9 Challenges in Plant Identification

Accurate identification of medicinal plants is a fundamental step in herbal medicine production and a central challenge. The diversity within plant species and their components necessitates precise identification since different species or plant parts can have distinct chemical profiles and therapeutic attributes. Unfortunately, the absence of universally accepted standards for plant identification leads to discrepancies and inconsistencies (Sahoo et al., 2010).

13.6.10 Taxonomic Confusion

Adding complexity is the prevalent taxonomic confusion within medicinal plants. These plants often have multiple vernacular names that vary across regions and languages, contributing to the need for standardized nomenclature. This complication further impedes the identification process, increasing the risk of misidentifications and affecting the quality and safety of herbal products (Sahil et al., 2011).

13.6.11 Climate Change

Climate change compounds these challenges. Shifts in temperature and precipitation patterns alter the distribution and growth of medicinal plant species (Shah et al., 2010). Species already endangered may struggle to adapt to changing conditions, increasing the risk of extinction (Sen et al., 2011).

13.6.12 Adulteration and Contamination in Herbal Products

As revealed by systematic reviews, the herbal medicine industry faces significant challenges related to adulteration and contamination (Posadzki et al., 2013).

Adulteration can manifest in various forms, from unintentional incidents to deliberate actions for economic gain. It often involves the introduction of substances for fraudulent purposes, further complicated by instances of mislabeling that misrepresent product composition. On the other hand, contamination arises from factors that include pesticide residues, heavy metals, and microbial contaminants (Zhang et al., 2012). These contaminants jeopardize product safety and compromise the therapeutic effectiveness of herbal remedies. Consumer health risks include allergies, poisonings, and infections, making it an urgent public health concern. Addressing these concerns requires stringent quality control measures, adherence to good agricultural, collection, processing, and manufacturing practices, and continuous monitoring, as illustrated in recent research (Crighton et al., 2019). Recent advancements in analytical techniques have revealed the extent of adulteration and contamination in herbal products. Commercial herbal products' DNA-based authentication has exposed extensive global adulteration. While traditional analytical chemistry methods have effectively detected contaminants and toxins in herbal medicines, they often struggle to identify the source of contamination (Ichim, 2019).

13.7 CONCLUSION

Alkaloids have long held a central place in drug discovery, offering a rich reservoir of structurally diverse and biologically potent compounds. Their natural origin, complex chemical architecture, and pronounced pharmacological activities make them invaluable templates for therapeutic development. Over time, a wide array of approaches—including traditional extraction, synthetic modifications, and modern computational methods—have significantly enhanced the discovery and optimization of alkaloid-based drugs.

Various success stories, such as quinine, Opium poppy, vincristine, and more recent semi-synthetic derivatives, underscore the continuing relevance of alkaloids in treating a broad spectrum of diseases, including cancer, infectious diseases, and neurological disorders. Advances in biotechnology, metabolic engineering, and machine learning are further expanding the potential of alkaloids in the 21st century.

COMPETING INTERESTS

The authors declare that they have no competing interest.

DECLARATION OF GENERATIVE AI AND AI-ASSISTED TECHNOLOGIES IN THE WRITING PROCESS

During the preparation of this work, the book chapter author(s) have not taken any assistance from AI, and the book chapter author(s) reviewed and edited the content as needed and take(s) full responsibility for the content of the publication.

REFERENCES

Abbott, H. (1970). Antineoplastic principles in plants: Recent developments in the field. *Advances in Pharmacology*, *7*, 117–209.

Aniszewski. (2015). Applied potential and current applications of alkaloids. *Alkaloids*, *2*, 345–420.

Aryal, B., Raut, B. K., Bhattarai, S., Bhandari, S., Tandan, P., Gyawali, K., Sharma, K., Ranabhat, D., Thapa, R., Aryal, D., & Ojha, A., 2022. Potential therapeutic applications of plant-derived alkaloids against inflammatory and neurodegenerative diseases. *Evidence-Based Complementary and Alternative Medicine*, *2022*(1), 7299778.

Atteeq, M. (2022). Evaluating anticancer properties of Withaferin A—A potent phytochemical. *Frontiers in Pharmacology*, *13*, 975320.

Barti, H., Pant, B., Joshi, S., & Kumar, S. (2018). Ethnopharmacology: Harnessing traditional knowledge for advancements in modern drug discovery. *Nanotechnology Perceptions*, *20*(7), 90–105.

Bedikian, A. Y., Vardeleon, A., Smith, T., Campbell, S., & Namdari, R. (2006). Pharmacokinetics and urinary excretion of vincristine sulfate liposomes injection in metastatic melanoma patients. *The Journal of Clinical Pharmacology*, *46*(7), 727–737.

Bhakuni, D. S., & Dhar, M. M. (1968). Crotsparine, a new proaporphine alkaloid from Croton sparsiflorus Morong. *Experientia*, *24*(1), 10–11.

Braekman, J. C., Daloze, D., & Pasteels, J. M. (1998). Alkaloids in animals. In *Alkaloids: Biochemistry, Ecology, and Medicinal Applications* (pp. 349–378). Springer US.

Butaye, P., Devriese, L. A., & Haesebrouck, F. (2003). Antimicrobial growth promoters used in animal feed: Effects of less well known antibiotics on gram-positive bacteria. *Clinical Microbiology Reviews*, *16*(2), 175–188.

Butnariu et al. (2022). Papaver plants: Current insights on phytochemical and nutritional composition along with biotechnological applications. *Oxidative Medicine and Cellular Longevity*, *2022*(1), 2041769.

Biondich, A. S., & Joslin, J. D. (2016). Coca: The history and medical significance of an ancient Andean tradition. *Emergency Medicine International*, *2016*(1), 4048764.

Casewell, M., Friis, C., Marco, E., McMullin, P., & Phillips, I. (2003). The European ban on growth-promoting antibiotics and emerging consequences for human and animal health. *Journal of Antimicrobial Chemotherapy*, *52*(2), 159–161.

Chen, S., Jiang, M., Chen, B., Salaenoi, J., Niaz, S. I., He, J., & Liu, L. (2019). Penicamide A, a unique N, N′-ketal quinazolinone alkaloid from ascidian-derived fungus Penicillium sp. 4829. *Marine Drugs*, *17*(9), 522.

Chik, S. C., Or, T. C., Luo, D., Yang, C. L., & Lau, A. S. (2013). Pharmacological effects of active compounds on neurodegenerative disease with Gastrodia and Uncaria decoction, a commonly used poststroke decoction. *The Scientific World Journal*, *2013*(1), 896873.

Cordell, G. A. (2017). Cognate and cognitive ecopharmacognosy—In an anthropogenic era. *Phytochemistry Letters*, *20*, 540–549.

Crighton, E., Coghlan, M. L., Farrington, R., Hoban, C. L., Power, M. W., Nash, C., Mullaney, I., Byard, R. W., Trengove, R., Musgrave, I. F., & Bunce, M. (2019). Toxicological screening and DNA sequencing detects contamination and adulteration in regulated herbal medicines and supplements for diet, weight loss and cardiovascular health. *Journal of Pharmaceutical and Biomedical Analysis*, *176*, 112834.

Cushnie, T. T., Cushnie, B., & Lamb, A. J. (2014). Alkaloids: An overview of their antibacterial, antibiotic-enhancing and antivirulence activities. *International Journal of Antimicrobial Agents*, *44*(5), 377–386.

Da-Cunha, E. V. L., Fechine, I. M., Guedes, D. N., Barbosa-Filho, J. M., & Da Silva, M. S. (2005). Protoberberine alkaloids. *The Alkaloids: Chemistry and Biology*, *62*, 1–75.

Daley, S. K., & Cordell, G. A. (2021). Alkaloids in contemporary drug discovery to meet global disease needs. *Molecules*, *26*(13), 3800.

Devereaux, A. L. et al. (2018). DARK classics in chemical neuroscience: Morphine. *ACS Chemical Neuroscience*, *9*(10), 2395–2407.

Dey, A., Mukherjee, A., & Chaudhury, M. (2017). Alkaloids from Apocynaceae: Origin, pharmacotherapeutic properties, and structure-activity studies. *Studies in Natural Products Chemistry*, *52*, 373–488.

Evans, W. C. (2009). *Trease and Evans' Pharmacognosy*. Elsevier Health Sciences.

Fielding, B. C., da Silva Maia Bezerra Filho, C., Ismail, N. S., & Sousa, D. P. D. (2020). Alkaloids: Therapeutic potential against human coronaviruses. *Molecules*, *25*(23), 5496.

Funayama, S., & Cordell, G. A. (2014). *Alkaloids: A Treasury of Poisons and Medicines*. Elsevier.

Ghanbari-Movahed, M., Kaceli, T., Mondal, A., Farzaei, M. H., & Bishayee, A. (2021). Recent advances in improved anticancer efficacies of camptothecin nano-formulations: A systematic review. *Biomedicines*, *9*(5), 480.

Hesse, M. (2002). *Alkaloids: Nature's Curse or Blessing?*. John Wiley & Sons.

Ichim, M. C. (2019). The DNA-based authentication of commercial herbal products reveals their globally widespread adulteration. *Frontiers in Pharmacology*, *10*, 1227.

Jose, W. M. (2020). Taxanes–the backbone of medical oncology. *Indian Journal of Medical and Paediatric Oncology*, *41*(02), 221–234.

Kala, C. P., Dhyani, P. P., & Sajwan, B. S. (2006). Developing the medicinal plants sector in northern India: Challenges and opportunities. *Journal of Ethnobiology and Ethnomedicine*, *2*(1), 32.

Karimkhani, C., Wanga, V., Coffeng, L. E., Naghavi, P., Dellavalle, R. P., & Naghavi, M. (2016). Global burden of cutaneous leishmaniasis: A cross-sectional analysis from the Global Burden of Disease Study 2013. *The Lancet Infectious Diseases*, *16*(5), 584–591.

Kaufman, T. S., & Rúveda, E. A. (2005). The quest for quinine: Those who won the battles and those who won the war. *Angewandte Chemie International Edition*, *44*(6), 854–885.

Khamkar, A. D., Motghare, V. M., & Deshpande, R. (2015). Ethnopharmacology—A novel approach for drug discovery. *Indian Journal of Pharmacy and Pharmacology*, *2*(4), 222–225.

Kim, K. S., Cui, X., Lee, D. S., Sohn, J. H., Yim, J. H., Kim, Y. C., & Oh, H. (2013). Anti-inflammatory effect of neoechinulin a from the marine fungus Eurotium sp. SF-5989 through the suppression of NF-κB and p 38 MAPK pathways in lipopolysaccharide-stimulated RAW264. 7 macrophages. *Molecules*, *18*(11), 13245–13259.

Kinghorn et al. (2011). The relevance of higher plants in lead compound discovery programs. *Journal of Natural Products*, *74*(6), 1539–1555.

Kittakoop, P., Mahidol, C., & Ruchirawat, S. (2014). Alkaloids as important scaffolds in therapeutic drugs for the treatments of cancer, tuberculosis, and smoking cessation. *Current Topics in Medicinal Chemistry*, *14*(2), 239–252.

Kong, Y. R., Tay, K. C., Su, Y. X., Wong, C. K., Tan, W. N., & Khaw, K. Y. (2021). Potential of naturally derived alkaloids as multi-targeted therapeutic agents for neurodegenerative diseases. *Molecules*, *26*(3), 728.

Kraus, C. N. (2008). Low hanging fruit in infectious disease drug development. *Current Opinion in Microbiology*, *11*(5), 434–438.

Kuramoto, M., Arimoto, H., & Uemura, D. (2004). Bioactive alkaloids from the sea: A review. *Marine Drugs*, *2*(1), 39–54.

Kurek, J. (Ed.). (2019). *Alkaloids: Their Importance in Nature and Human Life*. BoD–Books on Demand.

Lewis, R. A. (1998). *Lewis' Dictionary of Toxicology*. CRC Press.

Li, H., Sun, W., Deng, M., Zhou, Q., Wang, J., Liu, J., Chen, C., Qi, C., Luo, Z., Xue, Y., & Zhu, H. (2018). Asperversiamides, linearly fused prenylated indole alkaloids from the marine-derived fungus Aspergillus versicolor. *The Journal of Organic Chemistry*, *83*(15), 8483–8492.

Lobay, D. (2015). Rauwolfia in the treatment of hypertension. *Integrative Medicine: A Clinician's Journal*, *14*(3), 40.

Malaria, S. E. A. Q. A. (2005). Artesunate versus quinine for treatment of severe falciparum malaria: A randomised trial. *The Lancet*, *366*(9487), 717–725.

Manske, R. H. F., & Holmes, H. L. (Eds.). (2014). *The Alkaloids: Chemistry and Physiology*. Elsevier.

Martens, E., & Demain, A. L. (2017). The antibiotic resistance crisis, with a focus on the United States. *The Journal of Antibiotics*, *70*(5), 520–526.

Matsuura, H. N., & Fett-Neto, A. G. (2015). Plant alkaloids: Main features, toxicity, and mechanisms of action. In *Plant Toxins* (pp. 1–15). Springer.

Mavrogenis, A. F., Saranteas, T., Markatos, K., Kotsiou, A., & Tesseromatis, C. (2019). Pharmacies for pain and trauma in ancient Greece. *International Orthopaedics*, *43*(6), 1529–1536.

Mishra, B. K., Rastogi, A., Siddiqui, A., Srivastava, M., Verma, N., Pandey, R., Sharma, N. C., & Shukla, S. (2013). Opium poppy: Genetic upgradation through intervention of plant breeding techniques. In *Plant Breeding from Laboratories to Fields*. IntechOpen.

Mitsui, N., Noro, T., Kuroyanagi, M., Miyase, T., Umehara, K., & Ueno, A. (1989). Monoamine oxidase inhibitors from Cinchonae Cortex. *Chemical and Pharmaceutical Bulletin*, *37*(2), 363–366.

Mondal, A., Gandhi, A., Fimognari, C., Atanasov, A. G., & Bishayee, A. (2019). Alkaloids for cancer prevention and therapy: Current progress and future perspectives. *European Journal of Pharmacology*, *858*, 172472.

Montinari, M. R., Minelli, S., & De Caterina, R. (2019). The first 3500 years of aspirin history from its roots–A concise summary. *Vascular Pharmacology*, *113*, 1–8.

Mulani, M. S., Kamble, E. E., Kumkar, S. N., Tawre, M. S., & Pardesi, K. R. (2019). Emerging strategies to combat ESKAPE pathogens in the era of antimicrobial resistance: A review. *Frontiers in Microbiology*, *10*, 539.

Murray, C. J. L. (1994). A comprehensive assessment of mortality and disability from disease, injuries and risk factors in 1990 and projected to 2020. *The Global Burden of Disease*, *1*.

Nathan, C., & Cars, O. (2014). Antibiotic resistance—problems, progress, and prospects. *New England Journal of Medicine*, *371*(19), 1761–1763.

Newman, D. J., & Cragg, G. M. (2020). Natural products as sources of new drugs over the nearly four decades from 01/1981 to 09/2019. *Journal of Natural Products*, *83*(3), 770–803.

Oguntibeju, O. O. (2018). Medicinal plants with anti-inflammatory activities from selected countries and regions of Africa. *Journal of Inflammation Research 11*, 307–317.

Olğaç, A., Orhan, I. E., & Banoglu, E. (2017). The potential role of in silico approaches to identify novel bioactive molecules from natural resources. *Future Medicinal Chemistry*, *9*(14), 1665–1686.

Pache, M. M., Thorat, A. S., Mule, R. R., Pradhan, K. H., Watane, R. G., & Jagtap, M. N. (2025). Recent advances in the use of alkaloids as therapeutic agents: A pharmacognostic perspective. *World Journal of Pharmaceutical Research*, *14*(2), 538–552.

Pan, S. Y., Litscher, G., Gao, S. H., Zhou, S. F., Yu, Z. L., Chen, H. Q., Zhang, S. F., Tang, M. K., Sun, J. N., & Ko, K. M. (2014). Historical perspective of traditional indigenous medical practices: The current renaissance and conservation of herbal resources. *Evidence-Based Complementary and Alternative Medicine*, *2014*(1), 525340.

Pervaiz, A., Khan, H., & Amin, S. (2019). Therapeutic potential of alkaloids as antibacterial agents: Drugs of future. *Current Bioactive Compounds*, *15*(1), 31–40.

Pirintsos, S., Panagiotopoulos, A., Bariotakis, M., Daskalakis, V., Lionis, C., Sourvinos, G., Karakasiliotis, I., Kampa, M., & Castanas, E. (2022). From traditional ethnopharmacology to modern natural drug discovery: A methodology discussion and specific examples. *Molecules*, *27*(13), 4060.

Posadzki, P., Watson, L., & Ernst, E. (2013). Contamination and adulteration of herbal medicinal products (HMPs): An overview of systematic reviews. *European Journal of Clinical Pharmacology*, *69*(3), 295–307.

Rajeswara, R., Syamasundar, K. V., Rajput, D. K., Nagaraju, G., & Adinarayana, G. (2012). Biodiversity, conservation and cultivation of medicinal plants. *World*, *422000*(77000), 18–22.

Raju, S., & Das, M. (2024). Medicinal plants industry in India: Challenges, opportunities and sustainability. *Medicinal Plants - International Journal of Phytomedicines and Related Industries*, *16*(1), 1–14.

Reyburn, H., Mtove, G., Hendriksen, I., & von Seidlein, L. (2009). Oral quinine for the treatment of uncomplicated malaria. *BMJ*, *339*, b2066.

Rice, L. B. (2008). Federal funding for the study of antimicrobial resistance in nosocomial pathogens: No ESKAPE. *The Journal of Infectious Diseases*, *197*(8), 1079–1081.

Ruskin, R. S., & Aruna, S. R. (2014). In-vitro and in-vivo Antitumor Activity of Catharanthus Roseus. *International Research Journal of Pharmaceutical and Applied Sciences*, 1–4.

Sahil, K., Sudeep, B., & Akanksha, M. (2011). Standardization of medicinal plant materials. *International Journal of Research in Ayurveda & Pharmacy*, *2*(4), 1100–1109.

Sahoo, N., Manchikanti, P., & Dey, S. (2010). Herbal drugs: Standards and regulation. *Fitoterapia*, *81*(6), 462–471.

Seigler, D. S. (2012). *Plant Secondary Metabolism*. Springer Science & Business Media.

Sen, S., Chakraborty, R., & De, B. (2011). Challenges and opportunities in the advancement of herbal medicine: India's position and role in a global context. *Journal of Herbal Medicine*, *1*(3–4), 67–75.

Shah, S., Saravanan, R., & Gajbhiye, N. A. (2010). Phytochemical and physiological changes in Ashwagandha (Withania somnifera Dunal) under soil moisture stress. *Brazilian Journal of Plant Physiology*, *22*, 255–261.

Tadeusz, A. (2007). Alkaloids-secrets of life. In *Alkaloid Chemistry, Biological Significance, Applications and Ecological Role*. Elsevier.

Vos, T., Lim, S. S., Abbafati, C., Abbas, K. M., Abbasi, M., Abbasifard, M., Abbasi-Kangevari, M., Abbastabar, H., Abd-Allah, F., Abdelalim, A., & Abdollahi, M. (2020). Global burden of 369 diseases and injuries in 204 countries and territories, 1990–2019: A systematic analysis for the global burden of disease study 2019. *The Lancet*, *396*(10258), 1204–1222.

Wani, M. C., Taylor, H. L., Wall, M. E., Coggon, P., & McPhail, A. T. (1971). Plant antitumor agents. VI. Isolation and structure of taxol, a novel antileukemic and antitumor agent from Taxus brevifolia. *Journal of the American Chemical Society*, *93*(9), 2325–2327.

Wohlleben, W., Mast, Y., Stegmann, E., & Ziemert, N. (2016). Antibiotic drug discovery. *Microbial Biotechnology*, *9*(5), 541–548.

Zhang, J., Wider, B., Shang, H., Li, X., & Ernst, E. (2012). Quality of herbal medicines: Challenges and solutions. *Complementary Therapies in Medicine*, *20*(1–2), 100–106.

Zhang, Y. B., Luo, D., Yang, L., Cheng, W., He, L. J., Kuang, G. K., Li, M. M., Li, Y. L., & Wang, G. C. (2018). Matrine-type alkaloids from the roots of Sophora flavescens and their antiviral activities against the hepatitis B virus. *Journal of Natural Products*, *81*(10), 2259–2265.

Zhou, J., Chan, L., & Zhou, S. (2012). Trigonelline: A plant alkaloid with therapeutic potential for diabetes and central nervous system disease. *Current Medical Chemistry*, *19*(21), 3523–3531.

14 Synergistic Effects and Combinations with Alkaloids
Enhanced Therapeutic Potency

Shivam Verma, Zeeshan Ansari, Swetza Singh, Jay Kumar, Nisha Sharma, and Prakash Chandra Gupta

14.1 INTRODUCTION

14.1.1 Overview of Alkaloids

Plant-derived alkaloids represent one of the largest groups of secondary metabolites, known for their therapeutic properties in both traditional and modern medicine. Alkaloids are naturally occurring compounds, primarily derived from plants, composed of organic molecules containing nitrogen, and known for their diverse pharmacological properties (Matsuura & Fett-Neto, 2015). The classification of alkaloids is based on the presence of a basic nitrogen atom located anywhere in the molecule, excluding nitrogen involved in peptide bonds. It is estimated that around 12,000 natural products are examples of plant alkaloids (Gutiérrez-Grijalva et al., 2020). True alkaloids are defined by their structure, which includes a carbon framework with a nitrogen atom derived from an amino acid and incorporated into a heterocyclic ring. They are categorized based on their biosynthetic origin and the type of heterocyclic ring they contain (O'Connor, 2010). The medicinal potential of alkaloids enhances their industrial relevance. Many studies have explored the therapeutic properties of various plant-derived alkaloids. Table 14.1 provides an overview of alkaloid-rich plants, highlighting their major chemical constituents and their associated biological activities. These naturally occurring chemical compounds typically contain basic nitrogen atoms but may also include neutral or mildly acidic components (Manske, 2014; McNaught & Wilkinson, 1997).

Alkaloids are vital for both human health and an organism's defense mechanisms. About 20% of the secondary metabolites that are known to exist in plants are alkaloids. Alkaloids in plants control their growth and shield them from predators. Alkaloids are especially well-known for their analgesic, anticancer, CNS stimulant, antihypertensive, and anti-inflammatory properties in medicine. Widely used in clinical settings, alkaloids such as nicotine, ephedrine, quinine, strychnine, and morphine are well-known for their therapeutic applications. Recently, bioactive natural compounds have attracted increasing interest, both for their potential in drug development and their relevance in the expanding field of ethnopharmacology, which explores traditional medicine systems. Recently discovered or re-examined alkaloids from nature between 2014 and 2020 made up the 27,683 alkaloids that were included in the Dictionary of Natural Products (DNP) as of October 25, 2020 (Heinrich et al., 2021).

Throughout history, alkaloids have played a crucial role in traditional medicine, having been frequently utilized as herbal remedies in their natural state for thousands of years (Bernardini et al., 2018). In contemporary pharmaceutical science, alkaloids are often chemically modified to enhance their bioactivity and, in some instances, completely synthesized to improve their therapeutic effectiveness (Cushnie et al., 2014). This chapter focuses on exploring how alkaloids can be effectively

DOI: 10.1201/9781003539179-14

TABLE 14.1
List of Alkaloid-Rich Plants, Their Major Chemical Constituents, and Biological Activities

S. No.	Plants	Chemical Constituents	Activities	References
1	*Aconitum sp.*	Pseudaconitine	Anti-bacterial	Kumar et al. (2021)
2	*Alstonia yunnanensis*	Acetoxytabernosine	Anti-cancerous	Lai et al. (2023)
3	*Atropa belladonna*	Atropine	Anticholinergic	Schmoll et al. (2022)
4	*Capsicum annuum*	Capsaicin	Analgesic	Catalfamo et al. (2022)
5	*Catharanthus roseus*	Vincristine and Vinblastine	Anti-cancerous	Paul et al. (2023)
6	*Cinchona officinalis*	Quinine	Antimalarial	Aslam et al. (2023)
7	*Claviceps purpurea*	Ergotamine	Vasoconstrictor	Ashfaq et al. (2024)
8	*Coffea arabica*	Caffeine	CNS stimulant	Diaz-Gómez et al. (2025)
9	*Colchicum autumnale*	Colchicine	Anti-inflammatory, Anti-arthritic	Sharma and Singh (2023)
10	*Coptis chinensis*	Berberine	Anti-inflammatory	Jeong et al. (2009)
11	*Corydalis sp.*	Tetrahydropalmatine	Neuroprotective	Lyu et al. (2023)
12	*Datura stramonium*	Scopolamine	Antiemetic, sedative	Shim et al. (2022)
13	*Ephedra sinica*	Ephedrine	Bronchodilation, decongestant	Snell (2023)
14	*Nicotiana tabacum*	Nicotine	CNS stimulant	Sanjeev and Kapgate (2021)
15	*Papaver somniferum*	Morphine	Analgesic	Butnariu et al. (2022)
16	*Rauwolfia serpentina*	Reserpine	Anti-hypertensive	Paul et al. (2022)
17	*Stephania japonica*	Tetradrine	Anti-inflammatory	Xue et al. (2008)
18	*Sophora alopecuroides*	Aloprine	Anti-inflammatory	Ye et al. (2020)
19	*Sophora flavescentis*	Oxymatrine	Anti-inflammatory	Ma et al. (2017)

combined with other pharmacological agents to enhance therapeutic benefits. By analyzing recent studies and real-world clinical applications, we aim to uncover the potential of such combinations in advancing treatment strategies and achieving better patient outcomes.

14.2 THERAPEUTIC EFFECTS

Alkaloids are naturally occurring organic compounds, primarily derived from plants, that exhibit significant therapeutic effects and have been extensively utilized in medicine. Their pharmacological activities include analgesic, antimalarial, anticancer, antimicrobial, and cardiovascular benefits. For example, morphine, which is extracted from the opium poppy (*Papaver somniferum*), is a powerful analgesic that works by binding to opioid receptors in the central nervous system, helping to alleviate severe pain. Similarly, quinine, extracted from the bark of the Cinchona tree, is an effective antimalarial agent that disrupts the replication of *Plasmodium* parasites. In oncology, vincristine and vinblastine, obtained from the Madagascar periwinkle (*Catharanthus roseus*), inhibit cell division by targeting microtubules, making them invaluable in treating leukemia and lymphomas. Berberine, an antimicrobial alkaloid extracted from *Berberis* species, interferes with microbial DNA replication, effectively targeting bacterial and protozoal infections (Sanati et al., 2018). Additionally, reserpine, an alkaloid extracted from *Rauwolfia serpentina*, reduces blood pressure by depleting catecholamines at nerve terminals.

In contrast, caffeine, a stimulant derived from coffee and tea plants, promotes alertness by blocking adenosine receptors in the central nervous system. Piperine, a pyridine alkaloid from *Piper longum* L., exhibits diverse anticonvulsant properties. Beyond its ability to reduce seizures, it is also clinically utilized for managing lipids, relieving pain, and alleviating anxiety (Ren et al., 2019). These examples underscore that alkaloids play a crucial role in modern medicine, providing treatments for a range of health conditions (Roy, 2017).

14.3 SYNERGISTIC EFFECT

Synergy originates from the Greek word "synergos" which translates to "working together". When the effect of one drug is enhanced or amplified by the presence of another, the drugs are considered synergistic. In a synergistic combination, both drugs may act in the same way, or one might be inactive on its own but still boosts the effect of the other when administered together. Synergism may be of two types:

1. Additive
   When the two drugs produce the same effect.
   Effect of drug A + B = Effect of drug A + Effect of drug B
2. Supraadditive
   The combined effect is greater than the sum of the individual effects of the components.
   Effect of drug A + B > Effect of drug A + Effect of drug B

This always occurs when one component, by itself, has no effect but amplifies the effect of the other (potentiation).

14.3.1 Combinations Involving Alkaloids

The combination of different phytoconstituents can work through various mechanisms, such as targeting multiple biological pathways, enhancing drug absorption, or mitigating adverse reactions (Ayaz et al., 2022). These combinations can be particularly useful in the treatment of complex and chronic diseases, such as cancer, rheumatoid arthritis, inflammation, atherosclerosis, diabetes, cardiovascular disorders, and infectious diseases, where mono-drug therapy often falls short due to issues like resistance, ineffectiveness, and side effects (Yang et al., 2014).

This section focuses on alkaloid–alkaloid combinations, in which two or more alkaloids are combined to realize magnified effects, and alkaloid–non-alkaloid combinations, in which alkaloids are combined with other classes of drugs to produce synergistic therapeutic strategies.

14.3.1.1 Alkaloid–Alkaloid Combinations

The combination of two or more alkaloids has attracted attention in recent years due to their synergistic effects and improved therapeutic benefits (Pezzani et al., 2019). Here are a few notable examples.

14.3.1.1.1 Camptothecin and Etoposide

Mechanism: Camptothecin specifically inhibits the activity of topoisomerase I, while etoposide acts by targeting topoisomerase II (Andrade-Pavón & Gómez-García, 2021). Topoisomerases are a vital group of enzymes that are integral to several DNA functions, including the processes of DNA replication and transcription (Yakkala et al., 2023). Their inhibition disrupts cancer cell growth by interfering with different phases of the cell cycle (Kashifa Fathima et al., 2022).

Clinical implications: This combined therapy is utilized in the treatment of cancers such as lung, ovarian, and testicular cancer (Yakkala et al., 2023), particularly in situations where tumors have become resistant to single-agent therapies (Kamle et al., 2024).

14.3.1.1.2 Ephedrine and Caffeine

Mechanism: Ephedrine, a sympathomimetic alkaloid, influences adrenergic receptors, notably α and β receptors. This results in an elevated release of norepinephrine and a stronger activation of the sympathetic nervous system, which drives thermogenesis and affects cardiovascular function

(Stohs et al., 2020), while caffeine is classified as a CNS stimulant (Fiani et al., 2021). The combination of these two alkaloids is well-established for increasing energy expenditure and enhancing fat oxidation, making them popular ingredients in weight-loss and performance-boosting supplements (Magkos & Kavouras, 2004). Caffeine enhances the thermogenic and lipolytic effects of ephedrine (Dulloo et al., 1992).

Clinical implications: This combination is commonly used for weight management and enhancing athletic performance; however, it requires caution due to potential cardiovascular risks (Gutiérrez-Hellín & Varillas-Delgado, 2021).

14.3.1.1.3 Morphine and Codeine

Mechanism: Morphine is a full agonist at μ-opioid receptors, activating them with high affinity (Listos et al., 2019). On the other hand, codeine is a prodrug and a weaker agonist at these receptors, with its analgesic effects primarily arising from its conversion to morphine in the liver via CYP2D6 (Ćelić et al., 2009). When used together, their synergistic interaction enhances the overall pain-relief efficacy (Miranda et al., 2013).

Clinical implications: The morphine-codeine combination is widely employed in managing moderate to severe pain. Studies suggest that their synergistic effect allows for lower doses of both, which may reduce the potential risks (Miranda et al., 2013).

14.3.1.2 Alkaloid Non-Alkaloid Combinations

The therapeutic potential of alkaloids can be further enhanced when combined with non-alkaloid compounds, as they target multiple biological pathways. Below, we explore various alkaloid–non-alkaloid combinations. Additionally, Table 14.2 outlines selected combinations, highlighting the active components and the mechanisms responsible for their synergistic effects.

14.3.1.2.1 Berberine and Metformin

Mechanism: Berberine, a plant-based alkaloid, promotes glucose uptake and insulin sensitivity by activating AMP-activated protein kinase (AMPK) (Christodoulou et al., 2019). Metformin, a commonly prescribed biguanide, works by decreasing hepatic glucose production and also activates AMPK (Mobasher, 2021). Studies indicate that these two compounds together exhibit a synergistic effect in enhancing glycemic control in type 2 diabetes patients.

Clinical implications: The berberine-metformin combination offers a promising treatment for type 2 diabetes, improving blood sugar control and reducing the gastrointestinal side effects typically associated with higher metformin doses (Lyu et al., 2022). Additionally, this combination may help lower lipid levels, providing potential cardiovascular benefits (Wang et al., 2018).

14.3.1.2.2 Quinine and Doxycycline

Mechanism: Quinine, an alkaloid derived from the cinchona tree's bark, serves as a key antimalarial agent by disrupting the Plasmodium parasite's hemoglobin digestion process (Rawe & McDonnell, 2020). Doxycycline, a tetracycline antibiotic, targets the parasite by inhibiting its protein synthesis (Anandabaskar, 2021). Together, they provide a powerful dual mechanism, attacking the parasite at different stages of its life cycle (Lopes et al., 2021).

Clinical implications: This combination has proven effective in treating multidrug-resistant strains of malaria. The use of combination therapy enhances treatment effectiveness and delays the development of resistance to the individual drugs involved (Trasia, 2021). The synergy between quinine and doxycycline not only increases therapeutic efficacy but also helps reduce the duration of therapy, thereby minimizing the risk of adverse effects associated with long-term quinine use (Ejaz et al., 2007).

TABLE 14.2
Selected Alkaloidal and Non-Alkaloidal Compound Combinations Exhibiting Synergistic Effects

S. No.	Active Constituent (Alkaloidal)	Active Constituent (Non-Alkaloidal)	Mechanism of Synergy	References
1	Palmatine	Stigmasterol	The combination of drugs decreased both the quantity and size of tumors. It also led to a notable reduction in serum levels of enzymes like glutamate pyruvate transaminase, alkaline phosphatase, glutamate oxalate transaminase, and bilirubin. Additionally, it boosted the activity of oxidative enzymes, including glutathione, superoxide dismutase, and catalase, while inhibiting lipid peroxidation.	Aljarba et al. (2021)
2	Berberine	Fluconazole	The combination of berberine and fluconazole resulted in an increased mitochondrial membrane potential, a reduction in intracellular ATP levels, inhibition of ATP synthase activity, and an elevation in the production of endogenous reactive oxygen species in fluconazole-resistant strains.	Li et al. (2013)
3	Berberine	ER antagonists	Berberine decreased the expression of EGFR, HER2, Bcl-2, and COX-2, while increasing the levels of IFN-β and p21.	Liu et al. (2009)
4	Piperine	Paclitaxel	ROS-mediated activation of apoptosis.	Rawat et al. (2020)
5	Piperlongumine	Docetaxel	Decrease in docetaxel efflux *in vitro*, increase in docetaxel bioavailability *in vivo*, and alteration of the expression of proliferative markers.	Patel et al. (2015)
6	Dendrobine	Cisplatin	Induction of proapoptotic Bax and Bim protein activation and JNK/p38 stress signaling pathways.	Song et al. (2019)
7	Piperine	Curcumin	Piperine inhibits drug metabolism, enhancing curcumin absorption	Ahmadi et al. (2023)
8	Ergotamine	Caffeine	Alpha-adrenergic receptor action and CNS stimulation	Arnold et al. (2014)
9	Vincristine	Prednisone	Microtubule interference and anti-inflammatory action	Škubník et al. (2021)
10	Codeine	Paracetamol	Combined action on opioid receptors and prostaglandin pathways	Mattia and Coluzzi (2015)

14.3.1.2.3 Morphine and Ibuprofen

Mechanism: Morphine, an opioid alkaloid, exerts its analgesic effects through activation of mu-opioid receptors in the central nervous system (Badshah et al., 2024). Ibuprofen inhibits cyclo-oxygenase (COX) enzymes, reducing the production of inflammatory prostaglandins (Halim et al., 2022). The combination of central analgesia from morphine and peripheral anti-inflammatory effects from ibuprofen results in enhanced pain relief (Safaie et al., 2022).

Clinical implications: This combination is widely used in the management of moderate to severe pain, particularly in post-surgical or cancer-related pain (Mestdagh et al., 2023). The synergistic effect allows for reduced morphine dosage, which helps minimize the risk of opioid-related side effects, including respiratory depression and tolerance (Ramirez et al., 2020).

Zhang et al. used three alkaloid combinations (Berberine, Hypaconitine, Skimmianine) for the treatment of 2,4,6-trinitrobenzenesulfonic acid-induced Ulcerative Colitis. They found that hypaconitine and skimminanine showed analgesic effects and inhibited gastrointestinal transit. All three alkaloids, in combination, target inflammation, diarrhea, and bowel pain. It also enhanced the therapeutic effect while minimizing the side effects. The effect of any one alkaloid is not as effective as when in combination. All three alkaloids, in combination, reflected a strong synergistic effect on colitis treatment (Zhang et al., 2011).

Feng et al. evaluated the effect of the combination of alkaloids and triterpenes extract of *Alstonia scholaris* for non-small cell lung cancer (NSCLC). Together, they demonstrated immunomodulatory effects and reduced apoptosis in tumor-bearing mice and A549 cells. The combined action of alkaloids and triterpenes demonstrated a synergistic effect in inhibiting tumor growth, suggesting potential benefits for both the prevention and treatment of NSCLC (Feng et al., 2013).

Alshabi et al. evaluated the antidiabetic effect of pioglitazone given with caffeine. They found that co-administration of pioglitazone with caffeine enhances both the antidiabetic effect and bioavailability. The clearance of pioglitazone slows down when given along with caffeine, leading to a prolonged and potentiated antidiabetic effect of pioglitazone (Alshabi et al., 2021).

Research led by Wan et al. extracted alkaloids from *Piper longum* and explored their potential to enhance the therapeutic effects of chemotherapy drugs in treating cervical cancer. They identified five unique dimeric amide alkaloids that exhibited strong synergistic effects when combined with chemotherapeutic agents such as paclitaxel, adriamycin, and vincristine. Interestingly, these alkaloids also improved the performance of paclitaxel in cases where cancer cells had developed resistance to the drug. One specific dimeric amide alkaloid, when combined with paclitaxel, significantly enhanced cancer cell death by modulating the Src/ERK/STAT3 signaling pathway (Wan et al., 2023).

Ortiz et al. tested Candimine from *Hippeastrum recompense* for its synergistic effect with benznidazole for anti-parasitic activity. The results show that, in combination at a concentration of 4 × IC50, this may be a successful novel treatment for chronic CD (Ortiz et al., 2023).

14.4 THE OPTIMAL PROCEDURE FOR OPTIMIZING MEDICATION COMBINATIONS

The initial stage in designing an experimental platform and choosing medications for drug combination trials is literature mining. Significant work will be saved in subsequent optimization procedures if we can exclude any drug or combination composition from the drug candidate library before tests begin. Table 14.3 summarizes the key methods, their principles, and applications in optimizing drug combinations, highlighting approaches that enhance the overall process efficiency. The two-step approach increases the likelihood that the optimization result will be used in a clinical setting. Though pharmacodynamic and pharmacokinetic differences may alter the ideal dose ratios, synergisms often found *in vitro* are equally likely to act *in vivo*. Consequently, preclinical methods are employed for pre-experimental data mining and synergy discovery, whereas preclinical or clinical methods may be utilized for dose optimization, depending on the study's focus (Wang et al., 2021).

TABLE 14.3
Strategies for Optimizing Synergistic Combinations

S. No.	Methods	Application	Principles	References
1	Database modeling	Reduce drug library	Using databases that are accessible online to forecast synergism	Tang et al. (2013)
2	Metadata analysis	Reduce drug library	Compiling medical records and using statistical methods for analysis	Kerrien et al. (2012)
3	Experiment-based bioinformatics	Refining regimen	Using genomic, transcriptomic, and proteomic data to analyze experimental results	Kim et al. (2016)
4	Conventional biological screening	Refining regimen/ Dose optimization	Examining every potential combination	Wang et al. (2021)
5	Model-based prediction	Refining regimen/ Dose optimization	Predicting the effectiveness of combinations by evaluating pairwise interactions	Zimmer et al. (2016)

14.4.1 Step 0: Identifying the First Drug Library

The first step in studying drug combination optimization is identifying suitable drug candidates, as well as determining the target disease and experimental platform. Since no actual experiments are needed for this step, it is designated as Step 0. The objective of this step is to eliminate combinations that are clinically restricted due to adverse effects, like antagonism or combined toxicity, while selecting the most promising drug candidates for further experimental refinement. In addition to ruling out candidate medications with possible DDIs, selecting medications that may work in concert can occasionally be more effective in creating a smaller pharmacological library for the experimental phase. By using statistical analysis, specifically meta-analyses, to compare published articles and clinical data, potential synergisms can be found. The advancement of information technologies makes it easier to gather and examine published research, offering a methodical way to use meta-analyses (Mathieson et al., 2019).

14.4.2 Step 1: Identifying Synergistic Practices

Step 1 marks the beginning of real optimization following the pre-experiment studies. Two to six medications are frequently included in clinical combinational regimens. Thus, a brute-force screening approach will be effective and practical, particularly when the initial drug library is carefully selected and of manageable size, especially when paired with high-throughput platforms. This tactic also works when only one or two medications remain variable in the regimens, with the majority of the medications being fixed. An essential step in the optimization process is figuring out the dose range or levels for each medicine in the combination. It influences the effectiveness of identifying clinically appropriate regimens. Each drug's dose levels should ideally be within its therapeutic window, as established by evaluating its single-drug response curve. However, not all medications exhibit a typical response curve that can be used to identify the therapeutic window (Held et al., 2013; Laird et al., 2015).

14.4.3 Step 2: Adjusting the Dosages of the Drugs

One medicine may have distinct effects at different dosages in both clinical settings and studies. It should come as no surprise that drug dosages can also influence interactions within a combinatorial regimen, which in turn can impact treatment efficacy and outcome. For this reason, dose modifications are just as crucial as regimen optimization. The number of dose levels that each medication in the regimen can be taken determines the number of possible dose combinations. Since the medicine dosage is a continuous variable, the number is theoretically unlimited. However, in reality, dosages are typically restricted to a few distinct numbers for every medication to streamline computation, support clinical prescriptions, and ease testing. Dose optimization can be conducted in preclinical settings using cell lines and animal models, patient-derived samples, or, in some cases, directly with patients, depending on whether the findings are intended for direct clinical application (Wang et al., 2021).

14.5 CLINICAL BENEFITS

In many therapeutic areas, the clinical benefits associated with synergistic combinations of alkaloids are now receiving increased attention. These combinations offer a viable strategy to enhance patient outcomes by improving potency, reducing adverse effects, and combating drug resistance. The key clinical advantages of synergistic alkaloid therapies are given below.

1. *Combination of morphine with cannabinoids enhanced efficacy in pain management*
   Notably, morphine and other opioid analgesics are good and effective means of assuring symptomatic relief. Morphine, along with cannabis, augment their analgesic efficacy by allowing lower doses of morphine to provide the same pain relief, thus decreasing the risk of dependency and other side effects. This particular way of treatment using painful combinations is even more helpful in chronic pain, particularly in cancer and neuropathy (Babalonis & Walsh, 2020; Chen et al., 2019).
2. *Combination of piperine and curcumin improved bioavailability and absorption*
   Although curcumin has low bioavailability, its anti-inflammatory and antioxidant qualities make it an intriguing option for treating inflammatory illnesses. Piperine, an alkaloid of black pepper, greatly enhances the absorption of curcumin by inhibiting metabolic enzymes (Haq et al., 2021). Patients with inflammatory bowel disease and arthritis may benefit most from this combination, as a long-lasting anti-inflammatory effect is necessary in these situations (Pratti et al., 2024).
3. *Combination of quinine and artemisinin addresses antimicrobial resistance*
   The synergistic effects of combining quinine with artemisinin have been observed in treating malaria, particularly forms resistant to conventional treatment (Ain et al., 2024). Artemisinin acts quickly against the malaria parasites, while quinine provides sustained activity against malaria; hence, this combination is quite effective against drug-resistant strains. This combination helps manage severe cases, thereby reducing mortality and improving recovery outcomes in endemic areas (Achan et al., 2011).
4. *Combination of berberine with probiotics supports gastrointestinal health*
   According to all reports, berberine by itself has the potential to disturb the gut microbiota. Nonetheless, it promotes overall gut health by specifically eliminating harmful bacteria without upsetting the beneficial species when combined with probiotics (Wang et al., 2022). By altering the balance of the microbiome and reducing gastrointestinal tract symptoms, this combination can significantly help patients with stomach infections, irritable bowel syndrome (IBS), and various other gastrointestinal disorders (Huang et al., 2022).

14.6 FUTURE DIRECTIONS

The increasing evidence supporting the synergistic effects of alkaloids combined with other agents opens several promising pathways for both research and clinical use. Future studies should focus on developing advanced drug delivery systems, such as hydrogels, liposomes, or nanoparticles, to enhance the bioavailability and targeted delivery of alkaloid combinations. This would improve their therapeutic efficacy and address challenges related to their pharmacokinetic limitations. Additionally, personalized medicine strategies, which tailor combinations based on individual genetic profiles and disease characteristics, could greatly enhance treatment outcomes.

Incorporating alkaloids with modern synthetic drugs or biologics, like immune checkpoint inhibitors or monoclonal antibodies, might lead to innovative and more effective treatment approaches. To confirm preclinical findings and ensure the safety and effectiveness of alkaloid combinations, more clinical trials are necessary. These trials should involve diverse populations to evaluate the effects of these combinations on varying genetic profiles, comorbidities, and potential drug interactions. Furthermore, mechanistic studies are crucial for fully understanding how alkaloids synergize with other substances, potentially minimizing adverse effects and optimizing combination therapies.

Although alkaloids have been widely studied, many other plant-derived compounds are still underexplored. Expanding the search for synergistic combinations with lesser-known alkaloids could lead to the discovery of new, highly potent therapeutic agents.

14.7 CONCLUSION

In this chapter, we explored how alkaloids, when combined with both alkaloidal and non-alkaloidal agents, can unlock remarkable therapeutic potential. These natural compounds have demonstrated an impressive ability to enhance bioavailability and modify pharmacokinetics, providing tangible benefits to society. Notably, our studies have identified specific alkaloid combinations that demonstrate exceptional efficacy in clinical settings, particularly in the treatment of oncology and infectious diseases. For example, these synergistic interactions not only amplify the therapeutic impact of individual drugs but also present a promising approach to addressing drug resistance, a persistent and critical challenge in modern medicine.

Looking ahead, we emphasized the importance of further research in this area, particularly in the development of advanced drug delivery systems and personalized treatment strategies. By integrating alkaloids with synthetic drugs and biologics, we can create innovative therapies tailored to the unique needs of individual patients.

Reflecting on these findings, it is evident that studying synergistic combinations of alkaloids is essential for driving progress in pharmacological research and improving patient outcomes. This journey is far from over, and as we continue to investigate these powerful interactions, we contribute to both the scientific understanding of alkaloids and the resolution of pressing clinical challenges. The possibilities are vast, and the potential to transform treatment paradigms and enhance patient care is truly inspiring.

COMPETING INTERESTS

The authors declare that they have no competing interest.

DECLARATION OF GENERATIVE AI AND AI-ASSISTED TECHNOLOGIES IN THE WRITING PROCESS

During the preparation of this work, the book chapter author(s) have not taken any assistance from AI, and the book chapter author(s) reviewed and edited the content as needed and take(s) full responsibility for the content of the publication.

REFERENCES

Achan, J., Talisuna, A. O., Erhart, A., Yeka, A., Tibenderana, J. K., Baliraine, F. N., Rosenthal, P. J., & D'Alessandro, U. (2011). Quinine, an old anti-malarial drug in a modern world: Role in the treatment of malaria. *Malaria Journal*, 10, 1–12.

Ahmadi, F., Akbari, J., Saeedi, M., Seyedabadi, M., Ebrahimnejad, P., Ghasemi, S., & Nokhodchi, A. (2023). Efficient synergistic combination effect of curcumin with piperine by polymeric magnetic nanoparticles for breast cancer treatment. *Journal of Drug Delivery Science and Technology*, 86, 104624.

Ain, Q. T., Saleem, N., Munawar, N., Nawaz, R., Naseer, F., & Ahmed, S. (2024). Quest for malaria management using natural remedies. *Frontiers in Pharmacology*, 15, 1359890.

Aljarba, N. H., Ali, H., & Alkahtani, S. (2021). Synergistic dose permutation of isolated alkaloid and sterol for anticancer effect on young Swiss albino mice. *Drug Design, Development and Therapy*, 15, 4043–4052.

Alshabi, A. M., Alkahtani, S. A., Shaikh, I. A., & Habeeb, M. S. (2021). Caffeine modulates pharmacokinetic and pharmacodynamic profiles of pioglitazone in diabetic rats: Impact on therapeutics. *Saudi Medical Journal*, 42(2), 151.

Anandabaskar, N. (2021). Protein synthesis inhibitors. In *Introduction to Basics of Pharmacology and Toxicology: Volume 2: Essentials of Systemic Pharmacology: From Principles to Practice* (pp. 835–868). Springer.

Andrade-Pavón, D., & Gómez-García, O. (2021). Etoposide and camptothecin reduce growth, viability, the generation of petite mutants, and recognize the active Site of DNA topoisomerase I and II enzymes in Candida Glabrata. *Indian Journal of Microbiology*, 61(3), 306–314.

Arnold, A. C., Ramirez, C. E., Choi, L., Okamoto, L. E., Gamboa, A., Diedrich, A., … & Shibao, C. A. (2014). Combination ergotamine and caffeine improves seated blood pressure and presyncopal symptoms in autonomic failure. *Frontiers in Physiology*, 5, 270.

Ashfaq, M., Mushtaq, I., Mehmood, M. A., Ahmad, F., & Rehman, S. (2024). Ergot alkaloids from Claviceps: Production and pharmacological properties. In *Fungal Secondary Metabolites* (pp. 241–257). Elsevier.

Aslam, S., Jabeen, T., Ahmad, M., & AL-Huqail, A. A. (2023). Cinchona. In *Essentials of Medicinal and Aromatic Crops* (pp. 221–248). Springer International Publishing.

Ayaz, M., et al. (2022). Underlying anticancer mechanisms and synergistic combinations of phytochemicals with cancer chemotherapeutics: Potential benefits and risks. *Journal of Food Quality*, 1, 1189034.

Babalonis, S., & Walsh, S. L. (2020). Therapeutic potential of opioid/cannabinoid combinations in humans: Review of the evidence. *European Neuropsychopharmacology*, 36, 206–216.

Badshah, I., et al. (2024). Molecular mechanisms of morphine tolerance and dependence; novel insights and future perspectives. *Molecular and Cellular Biochemistry*, 479(6), 1457–1485.

Bernardini, S., Tiezzi, A., Laghezza Masci, V., & Ovidi, E. (2018). Natural products for human health: An historical overview of the drug discovery approaches. *Natural Product Research*, 32(16), 1926–1950.

Butnariu, M., Quispe, C., Herrera-Bravo, J., Pentea, M., Sarac, I., Küşümler, A. S., … & Cho, W. C. (2022). Papaver plants: Current insights on phytochemical and nutritional composition along with biotechnological applications. *Oxidative Medicine and Cellular Longevity*, 2022(1), 2041769.

Catalfamo, L. M., Marrone, G., Basilicata, M., Vivarini, I., Paolino, V., Della-Morte, D., … & Noce, A. (2022). The utility of Capsicum annuum L. in internal medicine and in dentistry: A comprehensive review. *International Journal of Environmental Research and Public Health*, 19(18), 11187.

Ćelić, I., et al. (2009). Resolving issues about efficacy and safety of low-dose codeine in combination analgesic drugs: A systematic review. *Pain and Therapy*, 9, 171–194.

Chen, X., Cowan, A., Inan, S., Geller, E. B., Meissler, J. J., Rawls, S. M., Tallarida, R. J., Tallarida, C. S., Watson, M. N., Adler, M. W., & Eisenstein, T. K. (2019). Opioid-sparing effects of cannabinoids on morphine analgesia: Participation of CB1 and CB2 receptors. *British Journal of Pharmacology*, 176(17), 3378–3389.

Christodoulou, M.-I., et al. (2019). Natural alkaloids intervening the insulin pathway: New hopes for antidiabetic agents? *Current Medicinal Chemistry*, 26(32), 5982–6015.

Cushnie, T. T., Cushnie, B., & Lamb, A. J. (2014). Alkaloids: An overview of their antibacterial, antibiotic-enhancing and antivirulence activities. *International Journal of Antimicrobial Agents*, 44(5), 377–386.

Diaz-Gómez, R., Sandoval-Cortes, J., & Aguilar, C. N. (2025). Coffea arabica. In *Coffee in Health and Disease Prevention* (pp. 11–17). Academic Press.

Dulloo, A. G., Seydoux, J., & Girardier, L. (1992). Potentiation of the thermogenic antiobesity effects of ephedrine by dietary methylxanthines: Adenosine antagonism or phosphodiesterase inhibition? *Metabolism*, 41(11), 1233–1241.

Ejaz, A., et al. (2007). Treatment of uncomplicated Plasmodium falciparum malaria with quinine-doxycycline combination therapy. JPMA. *The Journal of the Pakistan Medical Association*, 57(10), 502.

Feng, L., Chen, Y., Yuan, L., Liu, X., Gu, J. F., Zhang, M. H., & Wang, Y. (2013). A combination of alkaloids and triterpenes of Alstonia scholaris (Linn.) R. Br. leaves enhances immunomodulatory activity in C57BL/6 mice and induces apoptosis in the A549 cell line. *Molecules*, 18(11), 13920–13939.

Fiani, B., et al. (2021). The neurophysiology of caffeine as a central nervous system stimulant and the resultant effects on cognitive function. *Cureus*, 13(5), 1–8.

Gutiérrez-Grijalva, E. P., López-Martínez, L. X., Contreras-Angulo, L. A., Elizalde-Romero, C. A., & Heredia, J. B. (2020). Plant alkaloids: Structures and bioactive properties. In *Plant-Derived Bioactives* (pp. 85–117). Springer.

Gutiérrez-Hellín, J., & Varillas-Delgado, D. (2021). Energy drinks and sports performance, cardiovascular risk, and genetic associations; future prospects. *Nutrients*, 13(3), 715.

Halim, P. A., El-Nassan, H. B., & El-Dash, Y. S. (2022). Design and synthesis of novel ibuprofen derivatives as selective COX-2 inhibitors and potential anti-inflammatory agents: Evaluation of PGE2, TNF-α, IL-6 and histopathological study. *Medicinal Chemistry*, 18(4), 427–443.

Haq, I. U., Imran, M., Nadeem, M., Tufail, T., Gondal, T. A., & Mubarak, M. S. (2021). Piperine: A review of its biological effects. *Phytotherapy Research*, 35(2), 680–700.

Heinrich, M., Mah, J., & Amirkia, V. (2021). Alkaloids used as medicines: Structural phytochemistry meets biodiversity—An update and forward look. *Molecules*, 26(7), 1836.

Held, M. A., Langdon, C. G., Platt, J. T., Graham-Steed, T., Liu, Z., Chakraborty, A., & Stern, D. F. (2013). Genotype-selective combination therapies for melanoma identified by high-throughput drug screening. *Cancer Discovery*, 3(1), 52–67.

Huang, C., Sun, Y., Liao, S. R., Chen, Z. X., Lin, H. F., & Shen, W. Z. (2022). Suppression of berberine and probiotics (in vitro and in vivo) on the growth of colon cancer with modulation of gut microbiota and butyrate production. *Frontiers in Microbiology*, 13, 869931.

Jeong, H. W., Hsu, K. C., Lee, J.-W. et al. (2009). Berberine suppresses proinflammatory responses through AMPK activation in macrophages. *American Journal of Physiology-Endocrinology and Metabolism*, 296(4), E955–E964.

Kamle, M., et al. (2024). Camptothecin and its derivatives: Advancements, mechanisms and clinical potential in cancer therapy. *Medical Oncology*, 41(11), 263.

Kashifa Fathima, J., et al. (2022). The effectiveness of various chemotherapeutic agents in cancer treatment. *Current Pharmacology Reports*, 8(4), 236–252.

Kerrien, S., Aranda, B., Breuza, L., Bridge, A. Broackes-Carter, F. Chen, C., Duesbury, M., Dumousseau, M., Feuermann, M., Hinz, U., & Jandrasits, C. (2012). Hermjakob, The IntAct molecular interaction database in 2012, *Nucleic Acids Research*, 40(D1), D841–D846.

Kim, K. T., Lee, H. W., Lee, H. O., Song, H. J., Jeong, D. E., Shin, S., ... & Park, W. Y. (2016). Application of single-cell RNA sequencing in optimizing a combinatorial therapeutic strategy in metastatic renal cell carcinoma. *Genome Biology*, 17, 1–17.

Kumar, S., Javed, M. S., Kumar, P., Gupta, S., Kumar, R., & Singh, P. K. (2021). In-vitro antifungal and anti-bacterial activity of chloroform extract from tubers of Aconitum laeve Royle: Endangered species, India. *Materials Today: Proceedings*, 34, 563–568.

Lai, Q., Yang, C. J., Zhang, Q., Zhuang, M., Ma, Y. H., Lin, C. Y., ... & Yin, J. L. (2023). Alkaloid from Alstonia yunnanensis diels root against gastrointestinal cancer: Acetoxytabernosine inhibits apoptosis in hepato-cellular carcinoma cells. *Frontiers in Pharmacology*, 13, 1085309.

Laird, G. M., Bullen, C. K., Rosenbloom, D. I., Martin, A. R., Hill, A. L., Durand, C. M., & Siliciano, R. F. (2015). Ex vivo analysis identifies effective HIV-1 latency–reversing drug combinations. *The Journal of Clinical Investigation*, 125(5), 1901–1912.

Li, D. D., Xu, Y., Zhang, D. Z., Quan, H., Mylonakis, E., Hu, D. D., ... & Jiang, Y. Y. (2013). Fluconazole assists berberine to kill fluconazole-resistant Candida albicans. *Antimicrobial Agents and Chemotherapy*, 57(12), 6016–6027.

Listos, J., et al. (2019). The mechanisms involved in morphine addiction: An overview. *International Journal of Molecular Sciences*, 20(17), 4302.

Liu, J., He, C., Zhou, K., Wang, J., & Kang, J. X. (2009). Coptis extracts enhance the anticancer effect of estrogen receptor antagonists on human breast cancer cells. *Biochemical and Biophysical Research Communications*, 378(2), 174–178.

Lopes, E. A., Santos, M. M., & Mori, M. (2021). Pharmacological treatment of malaria. In *Antiprotozoal Drug Development and Delivery* (pp. 219–240). Springer.

Lyu, Y., et al. (2022). Effects of combination treatment with metformin and berberine on hypoglycemic activity and gut microbiota modulation in db/db mice. *Phytomedicine*, 101, 154099.

Lyu, Y., Wang, Y., Guo, J., Wang, Y., Lu, Y., Hao, Z., ... & Shi, J. (2023). Integrating serum pharmacochemistry and network pharmacology to reveal the active constituents and mechanism of Corydalis Rhizoma in treating Alzheimer's disease. *Frontiers in Aging Neuroscience*, *15*, 1285549.

Ma, A., Yang, Y., Wang, Q., Wang, Y., Wen, J., & Zhang, Y. (2017). Anti inflammatory effects of oxymatrine on rheumatoid arthritis in rats via regulating the imbalance between Treg and Th17 cells. *Molecular Medicine Reports*, 15(6), 3615–3622.

Magkos, F., & Kavouras, S. A. (2004). Caffeine and ephedrine: Physiological, metabolic and performance-enhancing effects. *Sports Medicine*, 34, 871–889.

Manske R. H. F., & Holmes, H. L., editors. (2014). *The Alkaloids: Chemistry and Physiology*. Elsevier.

Mathieson, S., Kasch, R., Maher, C. G., Pinto, R. Z., McLachlan, A. J., Koes, B. W., & Lin, C. W. C. (2019). Combination drug therapy for the management of low back pain and sciatica: Systematic review and meta-analysis. *The Journal of Pain*, 20(1), 1–15.

Matsuura, H. N., & Fett-Neto, A. G. (2015). Plant alkaloids: Main features, toxicity, and mechanisms of action. In Carlini, C., & Ligabue-Braun, R. (eds.), *Gopalakrishnakone P Plant toxins toxinology* (pp. 1–5). Springer.

Mattia, C., & Coluzzi, F. (2015). A look inside the association codeine-paracetamol: Clinical pharmacology supports analgesic efficacy. *European Review for Medical and Pharmacological Sciences*, 19(3), 507–516.

McNaught, A. D., & Wilkinson, A. (1997). *Compendium of Chemical Terminology*. Blackwell Science, Oxford.

Mestdagh, F., Steyaert, A., & Lavand'homme, P. (2023). Cancer pain management: A narrative review of current concepts, strategies, and techniques. *Current Oncology*, 30(7), 6838–6858.

Miranda, H. F., et al. (2013). Systemic synergism between codeine and morphine in three pain models in mice. *Pharmacological Reports*, 65(1), 80–88.

Mobasher, M. A. (2021). Metformin: An AMPK-dependent antidiabetic drug with novel medical applications. *Egyptian Journal of Cancer and Biomedical Research*, 5(2), 1–12.

O'Connor, S. E. (2010). Alkaloids. In *Comprehensive Natural Products II* (pp. 977–1007). Elsevier.

Ortiz, J. E., Piñeiro, M., Martinez-Peinado, N., Barrera, P., Sosa, M., Bastida, J., Alonso-Padilla, J., & Feresin, G. E. (2023). Candimine from Hippeastrum escoipense (Amaryllidaceae): Anti-Trypanosoma cruzi activity and synergistic effect with benznidazole. *Phytomedicine*, 114, 154788.

Patel, K., Chowdhury, N., Doddapaneni, R., Boakye, C. H., Godugu, C., & Singh, M. (2015). Piperlongumine for enhancing oral bioavailability and cytotoxicity of docetaxel in triple-negative breast cancer. *Journal of Pharmaceutical Sciences*, 104(12), 4417–4426.

Paul, A., Acharya, K., & Chakraborty, N. (2023). Biosynthesis, extraction, detection and pharmacological attributes of vinblastine and vincristine, two important chemotherapeutic alkaloids of *Catharanthus roseus* (L.) G. Don: A review. *South African Journal of Botany*, 161, 365–376.

Paul, S., Thilagar, S., Nambirajan, G., Elangovan, A., Lakshmanan, D. K., Ravichandran, G., ... & Murugesan, S. (2022). Rauwolfia serpentina: A potential plant to treat insomnia disorder. *Sleep and Vigilance*, *6*(1), 31–40

Pezzani, R., et al. (2019). Synergistic effects of plant derivatives and conventional chemotherapeutic agents: An update on the cancer perspective. *Medicina*, 55(4), 110.

Pratti, V. L., Thomas, M., Bhoite, R., & Satyavrat, V. (2024). Investigating bioavailability of curcumin and piperine combination in comparison to turmeric rhizomes: An in vitro Study. *Journal of Experimental Pharmacology*, 16, 37–47.

Ramirez, M. F., Kamdar, B. B., & Cata, J. P. (2020). Optimizing perioperative use of opioids: A multimodal approach. *Current Anesthesiology Reports*, 2020(10), 404–415.

Rawat, L., Hegde, H., Hoti, S. L., & Nayak, V. (2020). Piperlongumine induces ROS mediated cell death and synergizes paclitaxel in human intestinal cancer cells. *Biomedicine & Pharmacotherapy*, 128, 110243.

Rawe, S. L., & McDonnell, C. (2020). The cinchona alkaloids and the aminoquinolines. In *Antimalarial Agents* (pp. 65–98). Elsevier.

Ren, T. J., Hu, M. Y., Cheng, Y., Shek, T. L., Xiao, M., Ho, N. J., et al. (2019). Piperine-loaded nanoparticles with enhanced dissolution and oral bioavailability for epilepsy control. *European Journal of Pharmaceutical Sciences*, 137, 104988.

Roy, A. (2017). A review on the alkaloid an important therapeutic compound from plant. *International Journal of Plant Biotechnology*, 3(2), 1–9.

Safaie, A., et al. (2022). Intravenous morphine plus ibuprofen or ketorolac versus intravenous morphine alone in reducing renal colic pain intensity in emergency department: A randomized, double-blind clinical trial. *Turkish Journal of Emergency Medicine*, 22(1), 8–14.

Sanati, S., Razavi, B. M., & Hosseinzadeh, H. (2018). A review of the effects of Capsicum annuum L. and its constituent, capsaicin, in metabolic syndrome. *Iranian Journal of Basic Medical Sciences*, 21(5), 439.

Sanjeev, A., & Kapgate, S. M. (2021). Tobacco (Nicotiana tabacum) - A systemic review. *Population*, 51, 7.

Schmoll, S., Romanek, K., Zorn, G., Eiglmeier, H., & Eyer, F. (2022). Anticholinergic syndrome after atropine overdose in a supposedly homeopathic solution: A case report. *Clinical Toxicology*, 60(2), 252–254.

Sharma, N., & Singh, R. (2023). Colchicum autumnale (Naked Ladies). In *Exploring Poisonous Plants* (pp. 195–202). CRC Press.

Shim, K. H., Kang, M. J., Sharma, N., & An, S. S. A. (2022). Beauty of the beast: Anticholinergic tropane alkaloids in therapeutics. *Natural Products and Bioprospecting*, 12(1), 33.

Škubník, J., Pavlíčková, V. S., Ruml, T., & Rimpelová, S. (2021). Vincristine in combination therapy of cancer: Emerging trends in clinics. *Biology*, 10(9), 849.

Snell, N. (2023). Ephedra sinica: The source of ephedrine, pseudoephedrine and amphetamines. In *Modern Medicines from Plants* (pp. 137–143). CRC Press.

Song, T. H., Chen, X. X., Lee, C. K. F., Sze, S. C. W., Feng, Y. B., Yang, Z. J., … & Zhang, K. Y. (2019). Dendrobine targeting JNK stress signaling to sensitize chemotoxicity of cisplatin against non-small cell lung cancer cells in vitro and in vivo. *Phytomedicine*, 53, 18–27.

Stohs, S. J., Shara, M., & Ray, S. D. (2020). p-Synephrine, ephedrine, p-octopamine and m-synephrine: Comparative mechanistic, physiological and pharmacological properties. *Phytotherapy Research*, 34(8), 1838–1846.

Tang, J., Karhinen, L., Xu, T., Szwajda, A., Yadav, B., & Aittokallio, T. (2013). Target inhibition networks: Predicting selective combinations of druggable targets to block cancer survival pathways. *PLoS Computational Biology*, 9(9), e1003226.

Trasia, R. F. (2021). Use of combination therapy in malaria treatment and prevention in Indonesia. *Journal of Pharmaceutical and Sciences*, 4(1), 29–33.

Wan, Q., Xu, J., Zhu, C., Liu, X., Tu, Y., Lei, J., & Yu, J. (2023). Alkaloids from Piper longum exhibit anti-inflammatory activity and synergistic effects with chemotherapeutic agents against cervical cancer cells. *Journal of Agricultural and Food Chemistry*, 71(27), 10349–10360.

Wang, B., Warden, A. R., & Ding, X. (2021). The optimization of combinatorial drug therapies: Strategies and laboratorial platforms. *Drug Discovery Today*, 26(11), 2646–2659.

Wang, H., et al. (2018). Metformin and berberine, two versatile drugs in treatment of common metabolic diseases. *Oncotarget*, 9(11), 10135.

Wang, S., Ren, H., Zhong, H., Zhao, X., Li, C., Ma, J., … & Wang, W. (2022). Combined berberine and probiotic treatment as an effective regimen for improving postprandial hyperlipidemia in type 2 diabetes patients: A double blinded placebo controlled randomized study. *Gut Microbes*, 14(1), 2003176.

Xue, Y., Wang, Y., Feng, D., Xiao, B., & Xu, L. (2008). Tetrandrine suppresses lipopolysaccharide-induced microglial activation by inhibiting NF-κB pathway. *Acta Pharmacologica Sinica*, 29(2), 245–251.

Yakkala, P. A., et al. (2023). Prospects of topoisomerase inhibitors as promising anti-cancer agents. *Pharmaceuticals*, 16(10), 1456.

Yang, Y., et al. (2014). Synergy effects of herb extracts: Pharmacokinetics and pharmacodynamic basis. *Fitoterapia*, 92, 133–147.

Ye, Y., Wang, Y., Yang, Y., & Tao, L. (2020). Aloperine suppresses LPS-induced macrophage activation through inhibiting the TLR4/NF-κB pathway, *Inflammation Research*, 69(4), 375–383.

Zhang, M., Long, Y., Sun, Y., Wang, Y., Li, Q., Wu, H., Guo, Z., Li, Y., Niu, Y., Li, C., & Liu, L. (2011). Evidence for the complementary and synergistic effects of the three-alkaloid combination regimen containing berberine, hypaconitine and skimmianine on the ulcerative colitis rats induced by trinitrobenzene-sulfonic acid. *European Journal of Pharmacology*, 651(1–3), 187–196.

Zimmer, A., Katzir, I., Dekel, E., Mayo, A. E., & Alon, U. (2016). Prediction of multidimensional drug dose responses based on measurements of drug pairs. *Proceedings of the National Academy of Sciences*, 113(37), 10442–10447.

15 Future Directions and Challenges in Alkaloid Research

Opportunities for Innovation and Sustainability

Monica Butnariu

15.1 INTRODUCTION

Most alkaloids (ALKs) are solid (morphine, codeine) or can be found in liquid form (nicotine from tobacco leaves). All ALKs have a bitter taste, and the non-specific antidote for ALK poisoning is tannic acid (Chida, 2011). The name of ALKs ends with the suffix "-ine" (nicotine, morphine, atropine, cocaine). Each plant, in principle, shows interest in a certain active substance, which can be isolated by industrial processes (Fournier, 2001). Although the physiological action of many active plant principles has been known since ancient times – for example, it is known that the Chinese have long used opium as a sleeping pill – the chemical research of ALKs is still a concern of many organic chemists (Baeradeh et al., 2022). The ALKs are alkaline organic compounds that contain nitrogen in living organisms. This compound features a ring structure that is difficult to dissolve in water, and its acid can form a salt. It exhibits a certain rotation and optical absorption spectrum, and is the most bitter, colorless crystal, with some liquids (Ain et al., 2016). Examples of active principles can be listed: ALKs, glycosides, saponins, bitter principles, tannins, aromatic substances, volatile or essential oils, fatty oils, vegetable mucilages, and antiseptics (Chrzanowska et al., 2024). The complex structure of ALKs has made their research very difficult; there are still ALKs of imperfectly known, or even unknown, structure. For this reason, it is customary to classify ALKs according to their basic character, origin, and toxicity in a single class. However, from a structural point of view, ALKs are part of different classes of substances (Choi & Curhan, 2007).

Sometimes, certain types of ALKs present both optical isomerism and stereoisomerism, both of which are important, especially from the perspective of biological activities; thus, with tropane ALKs, the intensity of action and toxicity is conferred by optical isomerism (Abels et al., 2014). The fresh product contains L (–) hyoscyamine, which, when administered to humans, inhibits gastric, salivary, and sweating secretions, reducing at the same time the tone and motility of the digestive tract, biliary and urinary tracts, and bronchial muscles, being a spasmolytic (Lynch et al. 2007). The drying, storage, preservation, and defective processing of the raw material (*Belladonnae radix* et *folium*) determines the racemization of the ALK, the resulting racemic being known and used in therapeutics under the name of atropine (Abdel-Salam & Nammas, 2010). In general, optically active ALKs are found in plants in the form of levorotatory isomers, which present the most intense pharmacological action. In contrast, dextrorotatory isomers, with few exceptions (D-tubocurarine), are either slightly active or inactive. The World Health Organization (WHO) defines an ALK as any substance that, when absorbed by a living organism, modifies one or more of its functions. From a pharmacological point

DOI: 10.1201/9781003539179-15

of view, ALKs are substances used in medicine and whose abusive administration (consumption) can create physical and/or mental dependence or serious disorders of mental activity, perception, behavior, or consciousness. In therapeutics, ALKs are used in the form of salts, choosing those salts that are well crystallized, more stable, and that can be obtained on an industrial scale with maximum yields. For example, morphine is used in the form of hydrochloride, codeine–phosphate, caffeine–citrate, atropine–sulfate, and pilocarpine–nitrate. They have molecular masses between 100 and 900 daltons, they show characteristic spectra in UV, IR, and NMR, and some ALKs are characterized by a specific fluorescence in UV (Bartok et al., 2000). The ALKs are found in the form of primary (RNH_2), secondary (R_2NH), or tertiary (R_3N) amines, respectively, quaternary ammonium bases.

15.2 THE BASIC CHARACTER OF ALKALOIDS

The most important chemical property of ALKs is their basic character, as it appears in the definition of this class of substances. By basicity, in the case of ALKs, it is understood as their ability to acquire a proton in aqueous solution (Tarzi & Erra-Balsells, 2006). It is imprinted by the pair of non-participating electrons from the nitrogen atom. It varies depending on the nitrogen binding mode in the molecule, the type of heterocycle, and the nature of the functional groups adjacent to the nitrogen atom (Wallace et al., 2023). The alkalinity of the ALK is indicated by the alkalinity constant, respectively, dissociation: Kb or pKb. To illustrate how the basicity of ALKs should be understood, ALKs with pKb values lower than 3 (coniine) are considered strong bases, medium bases are those whose pKb values are placed between 3 and 7 (ALKs from opium and *Solanaceae*); weakly basic ALKs have a pKb between 7 and 10 (ALKs from *Chinae cortex*), and very weakly basic ones (purine ALKs) between 10 and 12 (Avdeef et al., 1999). If the substituents adjacent to the nitrogen are electron-withdrawing groups (alkyl groups), the basicity of the compound will increase, because it increases the availability of electrons from the nitrogen atom to fix protons. If, on the other hand, electron-attracting functional groups (carbonyl, carboxyl groups) are found adjacent to the nitrogen, the availability of electrons from the nitrogen will be reduced, so that the basicity will decrease, sometimes to zero. This is the case of amide or purine ALKs (Rajkamal & Kim, 2023). For example, caffeine, although it has four nitrogen atoms (four non-participating electron doublets), has a neutral reaction, a fact that can be explained by the general effect achieved by the rejection of the electron doublets by the methyl groups, on the one hand, and the attraction exerted by the groups carbonyl, on the other hand (Chapman & Miller, 1974).

The ALKs with non-aromatized nuclei are stronger bases than those with aromatic nuclei. Thus, for example, piperidine (saturated) is more basic than its aromatic derivative, pyridine, because the pair of non-participating electrons from nitrogen is not involved in aromatization, being available for fixing the proton. Similarly, pyrrolidine is a stronger base than pyrrole. In the case of indole, the influence of aromatization is even more obvious, having a weak acid character due to the tendency of nitrogen to lose a proton. The group of strong bases also includes non-oxygenated ALKs (nicotine), ALKs derived from substituted amines, and those in which the heterocyclic nitrogen is found in the form of an ammonium salt (serpentine) (Bailey et al., 2018). Basicity is also influenced by steric hindrances, especially in the case of complex polycyclic molecules. The basic character of ALKs allows the formation of salts with inorganic (sulfuric, hydrochloric, nitric) or organic (tartaric, malic) acids (Ulgen & Gorrod, 2000). The salts of the ALKs are strongly ionized in water, from which the ALKs are displaced with the help of alkaline hydroxides or ammonia. The extraction and purification of ALKs is based on this property.

15.3 PREPARATION OF ALKALOIDS

The extraction of ALKs is basically based on the differential solubility of ALKs' free bases and salts in water and organic solvents (Al-Maaieh & Flanagan, 2005). Since plant products often contain significant amounts of lipids (as in the case of seeds), but also waxes, pigments, terpenes or other lipophilic substances that can interfere with the extraction process, usually before proceeding to the actual extraction, the plant material is subjected to degreasing with petroleum ether, hexane, mineral spirits, solvents in which ALKs are soluble only exceptionally, under neutral environmental conditions (Abd El-Mawla, 2010). At the same time, prior to the extraction, the operation of spraying the plant products is carried out, in order to ensure the penetration of the solvent inside the cells and the "flow" of the active principles from the plant material into the extraction medium. The degree of pulverization is decisive for this process: thus, if the vegetable product is coarsely pulverized, a total extraction is not ensured. In addition, the plant material occupies a large volume, and the extractive solutions obtained are diluted (Akveran et al., 2018). If the vegetable product is pulverized very finely, it forms a shell with the solvent that can block the machinery, but at the same time prevents the circulation of the solvent through the vegetable material. There is no universally applicable preparation method. At the laboratory and industrial level, four processes are used, each with modifications depending on the characteristics of the product to be processed, as follows:

- The drug is extracted with methanol or ethanol, after which the solvent is removed; the concentrate is treated with an acid solution (HCl 1–3%, H_2SO_4 1–2%, organic acids 2–5%), when the ALKs become water-soluble salts and are delipidated with an immiscible organic solvent.
- The aqueous phase is separated, alkalized to release the base ALKs (in the form of salts), which are then extracted with a non-polar organic solvent (toluene, chloroform, ether). The total crude ALK can then be purified by the following methods: (1) fractional crystallization, (2) successive transitions from base ALKs to salts and vice versa, and (3) column chromatography (Macabeo et al. 2005).
- The vegetable powder is treated with a base (ammonia, potassium carbonate, lime water) and extracted with a non-polar organic solvent. Acidified water is added over the organic extract, when the ALKs are passed in the form of salt, becoming water-soluble, the ballast substances remaining in the non-polar phase. Finally, the salt ALKs are converted back into bases by alkalization and extraction with a non-polar organic solvent, the latter being removed, when the total crude ALK is obtained. Quaternary ALKs are not extracted by this process. In the case of ALKs with a labile chemical structure in an alkaline environment, such as ester ALKs (hyoscyamine), a weak base is used in a dilute solution (sodium carbonate). At the same time, in an alkaline environment, ALKs with phenolic hydroxyl groups (morphine) can form water-soluble phenolates, which allows their separation from the rest of the non-phenolic ALKs (Cao et al., 2016).
- The pulverized plant product is extracted with acidified water, the resulting extract is alkalized and exhausted with a non-polar organic solvent. For purification, the basic ALKs can be salted again with the help of acidic aqueous solutions, finally being brought back to the basic stage, obtaining the total gross ALK (Kempf et al., 2011).
- The plant material is easily acidified by treatment with ammonium sulfate, when the ALKs are fixed in the form of salt in the plant product. Next, the vegetable product is extracted with an organic solvent, which removes the acidic and neutral ballast substances; the vegetable residue is treated with a basic solution, and the basic ALKs are extracted with a non-polar organic solvent. Extraction with non-polar organic solvents is carried out cold, by percolation, stirring, or in a Soxhlet apparatus, and that with polar organic solvents can be carried out hot (reflux) or cold (maceration, percolation, stirring). In the case of quaternary ALKs that are soluble in water and insoluble in non-polar solvents, regardless of the form of salt or base, the separation is done by precipitating them with Mayer's reagent, ammonium reineckate, and picric acid. After filtration, the precipitates thus obtained are dissolved in an alcohol-water-acetone or acetone-water mixture and decomposed by passing through a column of ion-exchange resins. The extraction of volatile ALKs (nicotine, sparteine) is carried out by entrainment with water vapor of the corresponding vegetable products treated with alkaline hydroxides or sodium carbonate (Al-Kamil et al., 2024).

- The identification of ALKs is based on a series of reactions that they give with some substances called general reagents of ALKs. Thus, ALKs can give color reactions with concentrated strong acids, especially concentrated sulfuric acid, and precipitation reactions with salts of heavy metals, tannins, saturated solutions of inorganic acids with high molecular weight, double iodides, or tetraphenylborate or reineckate of ammonium (Abdul et al., 2024).

The specificity of the reagents is not absolute, because other nitrogenous or non-nitrogenous substances (some coumarins, furanocoumarins, hydroxyflavones, lignans) can react with them, giving false positive results. At the same time, some ALKs, such as the purine ALKs, do not precipitate with these general reagents. Of the 16,000 known ALKs, few are formalized in European Pharmacopoeia (Abdelrheem et al., 2021); all of them are characterized by the presence of an intense pharmacological action, often doubled by toxicity, which is based on the transformation of alkaloidsal salts administered p.o. in a lipophilic base, under the action of the alkalinity existing in the small intestine, which results in a rapid and almost total resorption of the ALK (ALKs with quaternary nitrogen or those possessing free phenolic groups are an exception).

15.4 THE RANGE OF PHARMACOLOGICAL ACTIVITIES OF ALKALOIDS

Each ALK acts on a neurotransmitter or receptor at the level of synapses in neurons, which generally leads to long-term adverse effects. The action is very different, depending on the chemical structure of the ALK, and affects various devices, systems, and functions of the human body. An ALK can have multiple actions or ambivalent effects, depending on the dose.

On the other hand, almost all ALKs cross the blood–brain barrier, which explains the existence in many cases of an action at the CNS level. The ALKs react directly with a multitude of receptors, enzymes, and ion channels, functioning either as agonists (+) or as antagonists, respectively inhibitors (−) (Abacioğlu et al., 2001).

Examples in this sense can be the following.

- *The ALK–receptor interactions*: Morphine with opioid receptors (+); rye horn ALKs with dopaminergic (+,−) and αadrenergic (+) receptors; tubocurarine with nicotinergic/cholinergic receptors (−); strychnine with glycocol receptors (−); Pilocarpine with muscarinic receptors (+); caffeine, theophylline with purinergic/adenosine receptors (−); ephedrine/cocaine/capsaicin with presynaptic receptors (Abdelhamid et al., 1991).
- *The ALK–enzyme interactions*: Papaverine with phosphodiesterase (−); harminic ALKs with MAO (−); galantamine with acetylcholinesterase (−).
- *The ALK–ion channel interactions*: Aconitine and *Veratrum* ALKs inhibit the closing of ion channels; tetrodotoxin blocks sodium (Na^+) channels (Duan et al., 2005).

Due to this very varied way of reaction, at an appropriate dosage, many ALKs develop a therapeutic activity. The palette of fields of use is extremely varied, and ALKs can act from analeptics to antihypertensives or from cytostatics to sedatives, as can be seen from the following examples.

- *Action at SNV level – sympatholytic (adrenolytic)*: ALKs from the ergotoxine and ergotamine group, yohimbine, reserpine;
  - indirect sympathomimetics: ephedrine, sea buckthorn;
  - direct parasympathomimetics: nicotine, pilocarpine, arecoline, and indirect: physostigmine;
  - parasympatholytics: atropine;
  - ganglioplegic (ganglionic blocker): nicotine, lobeline, coniine;
  - myotonolytics (curarizing): Curara, D-tubocurarin, C-toxiferin;
  - antivertigo: atropine, scopolamine (Angrist et al., 1977).

- *Neurological – sedatives/tranquilizers*: reserpine, *Corydalidis radix, Eschscholtziae* extractum.
- *Cardiovascular – antiarrhythmics*: quinidine, sparteine, ajmaline; antihypertensives: reserpine, serpentine, rescinamine, raubazine, ester ALKs from *Veratrum*;
- antihypotensives: ephedrine, caffeine, sparteine, *Guarana*, extracts from *Ephedra, Coffea, Cola, Sarothamnus*;
  - coronary vasodilators: theophylline, theobromine, quinidine, raubazine (ajmalicine);
  - peripheral vasodilators: hydrogenated ALKs from *Secale cornutum*;
  - cerebral vasodilators: vincamine, dihydroergotoxine;
  - antimigraines: ergotamine, dihydroergotamine (Bagri & Deshmukh, 2022).
- *Analeptics – by direct central stimulation*: caffeine, theophylline, theobromine, strychnine, *Cola* extracts, lobeline.
- *Antitussives – expectorants*: *Ipecacuanhae radix*, emetine;
  - central antitussives: codeine, noscapine;
  - broncholytics: theophylline, atropine, ephedrine, papaverine, caffeine;
  - rhinological: ephedrine, pseudoephedrine.
- *Gastroenterological – antigastric/antiulcer*: atropine, *Belladonna* extract;
  - antidiarrheal: Opium;
  - choleretic-cholagogue: chelidonin;
  - antiemetics: scopolamine.
- *Spasmolytic – musculotropic*: papaverine
  - neurotropics: atropine.
- *Antiphlogistics (anti-gout)*: colchicine.
- *Cytostatics*: vinblastine, vincristine, colchicine, demecolcine, camptothecin, Taxol.
- *Ophthalmic – miotic*: pilocarpine, physostigmine;
  - mydriatics: atropine;
  - antimicrobials: berberine (Abd El-Rehim et al., 2013).
- *Analgesics*: Opium, morphine, codeine, scopolamine, chelidonine, *Aconitum* extracts.
- *Antipyretics*: quinine.
- *Andrological*: yohimbine, *Yohimbe cortex*.
- *Antiparasitic – amoebicides*: emetine, dehydroemetine;
  - antimalarials: quinine.

ALKs are biogenic substances of therapeutic interest; they are very widespread in the plant kingdom, but unevenly. Rich in alkaloids are some plants from the Papaveraceae (poppy), Rubiaceae (quinine tree), and Solanaceae (tobacco, potatoes, etc.) families.

Depending on the nature of the groups existing in their molecule, ALKs can be esters (hyoscyamine, atropine, reserpine), alcohols (aconine), amides (ergotamine), phenols (morphine, cepheline), or ethers (codeine). Some of them are glycosidated, as is the case with sterol glycoalkaloids from *Liliaceae* and *Solanaceae* species (Cárdenas et al., 2015).

In these plants, there is not a single species of ALK, but several, even up to 20, usually chemically related.

15.5 PHYTOTOXICOLOGY OF ALKALOIDS

Phytotoxicology is particularly important, especially for herbivores, and, in part, for carnivores. The frequency of plant poisoning is still relatively high. There are many plants that can produce pathological states. However, the nomination as a toxic plant is relative, because the toxicity is dependent on the pedoclimatic conditions, the season, the meteorological conditions, the actions of soil fertilization, the degree of atmospheric pollution, and the stage of development (Flade et al., 2019). A number of criteria were taken into account in defining the plants that produce poisoning. Thus, among the toxic plants were included plants that cause diseases in the conditions of consumption by sensitive animals of fruits, seeds, aerial parts, roots; plants that contain factors that limit or inhibit the optimal morphoproductive performance of animals; plants that contain substances that interfere with the reproductive function and that, in certain circumstances, produce infertility. The ALKs are found throughout the plant, where they are formed in tissues and then located in various parts, including the root, fruit, and stem (Acito et al., 2022). In plants, ALKs are not found in the free state, but in the form of salts. Plant acids, which neutralize ALKs, are acids with a more complex structure, such as fumaric acid, chelidonic acid, quinic acid, and others. The concentration of ALKs in plants varies depending on several factors, including geographical region and climate. Finely chopped plants are treated with alkali to liberate the ALKs from their salts. They can be separated, either by extraction with ether or chloroform or by distillation with water vapor. Sometimes, the separation is done by passing the ALK into a sparingly soluble salt. Most ALKs are crystalline substances, hardly soluble in water. With acids, they form soluble salts from which they can be set free again by alkalis. Some give color reactions, but not always characteristic. Titanium, phosphotungstic acid, phosphomolybdenic acid, iodine–potassium iodide solution, etc., precipitate the ALKs from their solutions. According to the nature of the heterocycle contained in the ALK molecule, several subclasses are distinguished: ALKs of the pyridine and piperidine type, those of the quinoline, isoquinoline type, etc., those of the pyrimidine type, etc., then ALKs with condensed cyclic systems, such as those with pyrrolidine nuclei or condensed piperidinics, etc. (Kast et al., 2018).

Coniine, $C_8H_{17}N$, is one of the ALKs found in hemlock. Structurally, coniine is a derivative of piperidine and is α-propyl-piperidine. Coniin is a colorless liquid, hardly soluble in water, but soluble in alcohol. It is characterized by a pronounced toxicity on the body, which is manifested by the paralysis of the motor nerves and muscles. Large amounts cause death by paralyzing the respiratory centers (Barrera-Adame et al., 2024).

Nicotine, $C_{10}H_{14}N_2$, is found in tobacco leaves in the form of salts of citric or malic acids. Being volatile, its separation, after alkalinization of the aqueous solution, is done by steam distillation and extraction with a solvent. As a structure, nicotine is 1-methyl-2-pyridine-pyrrolidine, that is, it contains a hydrogenated pyrrolic nucleus, methylated at the nitrogen atom, and a pyridinic nucleus linked to C_3 with the hydrogenated pyrrolic (pyrrolidine) nucleus.

Natural nicotine is levorotatory; its salts are dextrorotatory. Nicotine is a colorless oil with an unpleasant smell, very soluble in water and organic solvents. In the air, it darkens in color.

In small amounts, nicotine is exciting; in large quantities, it is toxic, especially the levorotatory isomer (it has a paralyzing effect on the nervous system). Nicotine extracts are used as insecticides (Abdelkader et al., 2010).

Atropine, $C_{17}H_{23}O_2N$, is found in horseradish (*Atropa belladonna*), laurel, etc. In its molecule, there is a piperidine nucleus condensed with a pyrrolidine nucleus. By treating atropine with hydrochloric acid or sodium hydroxide, a hydrolysis occurs, with the formation of an amino-alcohol, tropine, and (±) tropic acid. The reaction is reversible. Therefore, atropine is an ester of (±) tropic acid with tropine. Atropine has a characteristic physiological action: in small quantities, it causes dilation of the pupil of the eye; in large doses, it is very toxic because it can act on the central nervous system, paralyzing it (Zeng et al., 2021).

Cocaine, $C_{17}H_{21}O_4N$, is the ALK extracted from a *Peruvian shrub*. It contains a molecule with a piperidine nucleus fused with a pyrrolidine nucleus. It is a double ester. Cocaine is a crystalline substance, soluble in alcohol, and is left-handed. In small amounts, cocaine is a stimulant. In large quantities, it is toxic (produces paralysis of the central nervous system). In medicine, it is widely used as a local anesthetic (it has a paralyzing action on the peripheral nerves) (Abdul-Hussein et al., 2013).

Quinine, $C_{20}H_{24}O_2N_2$, together with cinchonine, is widely distributed in the bark of the quinine tree, from which they are extracted. Quinine is a crystalline substance with a bitter taste, slightly soluble in water but soluble in alcohol. Crystallizes with three water molecules. Alcoholic solution of quinine is levogyric. Quinine contains a quinoline ring and a non-aromatic part in the molecule.

Being a weak base, it can form salts with mineral acids. Solutions of quinine salts give blue fluorescence, after which they are recognized (Boratyński, 2015).

If a solution of quinine is treated with chlorine water and then with excess ammonia, a green coloration is obtained, a behavior that distinguishes quinine from cinchonine. An important priority of quinine is its action to combat malaria; it is also used as an antipyretic (it produces a decrease in the febrile temperature of the human body). Small amounts of quinine have an exciting action on the muscles (Cinconine has the same behavior, but less pronounced) (Tasler & Bringmann, 2002).

Cinchonine, $C_{19}H_{22}ON_2$, is a crystalline substance with a m.p. = 264° C. It is insoluble in water and alkalis, soluble in acids, alcohols, and chloroform. It has a structure similar to quinine, with the difference that it lacks the methoxy group (at position 6 of the quinoline ring) (Parveen et al., 2024).

Papaverine, $C_{20}H_{21}O_4N$, is an ALK found in opium. Large quantities are obtained by synthesis. Papaverine is a benzylisoquinoline derivative. It is a crystallized substance, with m.p.=147° C, insoluble in water, soluble in chloroform. It is optically inactive.

It has a narcotic action, similar to morphine, and a paralyzing action, similar to codeine, but to a lesser extent (Abbas & Mostafa, 2007).

Morphine, $C_{17}H_{19}O_2N$, like codeine, are ALKs also extracted from opium, forming the so-called "morphine alkaloids" (Aabom et al., 2020).

Their structure is similar, namely, morphine, is a tertiary amine, methylated at the nitrogen atom, which also contains two hydroxyl groups, and **codeine, $C_{18}H_{21}O_3N$,** is a methyl ether of morphine at the phenolic –OH group. Morphine acts on the central nervous system, in small quantities, as a sedative, in large quantities, as a strong narcotic. Codeine has no narcotic action (Abdel Shaheed et al., 2019).

Codeine is a natural ALK from the morphine subgroup, being the most commonly used opioid in the world. It is found in opium in a concentration of 0.3%, but most of the marketed codeine is obtained by synthesis from morphine, through a methylation process. Codeine is well absorbed after oral administration, with a bioavailability of 60–70%. Codeine is considered a prodrug, being metabolized in vivo into its active compounds – morphine and codeine-6-glucuronide. The percentage of non-biotransformed codeine excreted in the urine within 24 h is approximately 6–8% of the total amount and can increase to 10% in conditions of low urinary pH. The conversion of codeine into morphine takes place in the liver and is catalyzed by the cytochrome P450 enzyme CYP2D6 (Ashraf et al., 2024).

Some drugs can inhibit or even block this enzyme, thus blocking the conversion of codeine into morphine. Both ALKs have applications in medicine (Abdolmohammad-Zadeh et al., 2019).

Strychnine, $C_{21}H_{22}O_2N_2$, is an ALK with a very complex structure, consisting of seven cycles. Strychnine is extracted from the leaves and fruits of the *Strychnos nox-vomica* plant. It was also obtained through synthesis. It is a very strong poison, which acts on the central nervous system, producing convulsions. In small quantities, it is used in medicine as a body tonic. Known as the "devil's weed" because of its ability to induce hallucinations, *Datura* – or *Datura stramonium* – grows all over the world and almost everywhere around us. But nowadays, *Datura* is no longer reserved only for witches; it is also a medicinal plant. The toxicity of the date is due to the three tropane-derived ALKs: hyoscyamine, atropine, and scopolamine (Abdul-Ghani et al., 2004). The ALK medicines derived from plants and other natural products are illustrated in Figure 15.1.

The latter represents more than a third of the ALKs contained in the fresh aerial parts of the plant. Its effect is parasympatholytic, that is, it paralyzes the parasympathetic nervous system. In case of intoxication following the consumption of *Datura*, an abnormal dilation of the pupils is observed, accompanied by hallucinations and delirium. However, the presence of a significant amount of scopolamine also results in sedative properties of the central nervous system. *Datura* is cultivated for pharmaceutical purposes, preferably on light, calcareous soil. This cultivation made it possible to improve the species by favoring the richest in ALKs. *Datura* leaves are preferably harvested before

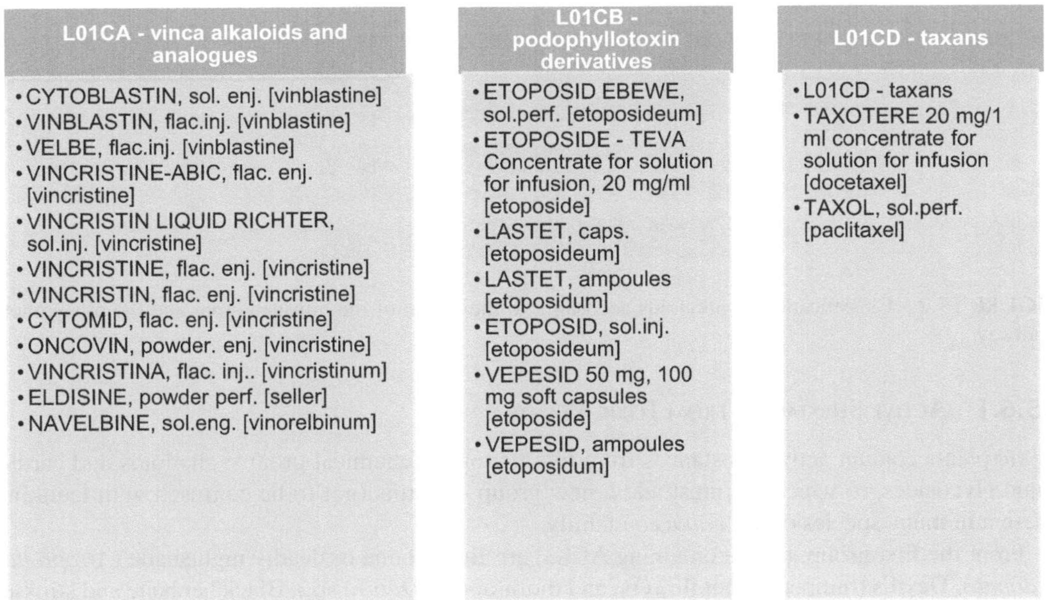

FIGURE 15.1 Alkaloid medicines from plants and other natural products.

flowering. First, the ones at the base are removed, and later, after a few weeks, the ones at the top (Adegoke & Alo, 2013). They are then quickly dried to preserve the qualities of the active principles, in the shade or in warm air with good ventilation. The ALK is the organic substance produced by plants, often used in therapy, due to the beneficial effects it produces on the body (Abadie et al., 2021). In modern pharmaceutical research, new drugs are tested on humans only after numerous stages of pharmacological and especially toxicological validation results from clinical experiments (An et al., 2010).

In the case of medicinal plants used in the treatment of benign (curable) conditions, these stages can be shortened, according to the European legislation on phytotherapeutic medicines.

15.6 PLANTS CONTAINING ALKALOIDS

The ALKs (from Alkali; Greek eidos – aspect) are heterocyclic organic substances of vegetable origin, characterized by their basic nature, resulting from the secondary metabolism of plants. They exhibit characteristic reactions and have an action on animal organisms, often toxic in nature. The ALKs may have therapeutic effects in stimulating appetite and in alleviating symptoms of mental disorders, epilepsy, asthma, cancer, and Alzheimer's disease. It can also ease menstrual pain and reduce the risk of obesity and diabetes. However, further research is needed to adequately demonstrate its therapeutic potential and long-term risks. Depending on the origin of the nitrogen atom and the biosynthetic pathway, they can be classified according to Figure 15.2.

The notion of ALK was established at the beginning of the 19th century. In the past, ALKs referred to all nitrogenous substances with a basic character that had a physiological action on animal organisms. In the group of ALKs, there were thus some amines, amides, amino alcohols, amino acids, purines, etc. In most cases, ALKs were isolated from Angiospermae, 10–15% of these plants being able to synthesize ALKs, some families even having a pronounced biosynthesis tendency: *Annonaceae, Apocynaceae, Asteraceae* (subfamily *Senecioneae*), *Berberidaceae, Boraginaceae, Convolvulaceae, Erytroxylaceae, Loganiaceae, Magnoliaceae, Papaveraceae, Solanaceae* (*Dicotyledons*), *Amarylidaceae*, and *Liliaceae* (*Monocotyledons*) (Anke et al., 2008).

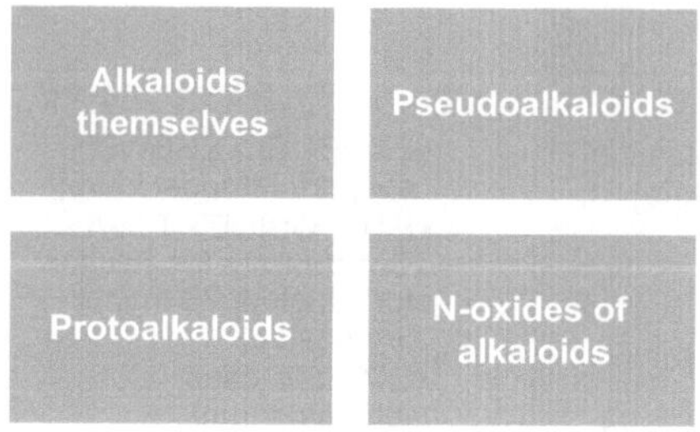

FIGURE 15.2 Classification of alkaloids according to the origin of the nitrogen atom and the biosynthetic pathway.

15.6.1 Active Substances from Toxic Plants

Toxic plants contain active substances from two important chemical groups: alkaloids and cardiotonic glycosides, to which we must add a new group – lectins (not to be confused with lecithins) present in many species of the *Fabaceae* family.

From the first group, those containing ALKs, are Belladonna or deadly nightshade (*Atropa belladonna*), Devil's trumpets, moonflowers, and thorn-apple (*Datura* sp.), Black henbane and stinking nightshade (*Hyoscyamus* sp.), Aconite, monkshood, wolfsbane, leopard's bane, devil's helmet, or blue rocket (*Aconitum* sp.), poison hemlock (*Conium maculatum*), and the exotic species of *Strychnos*. sp., which is part of curare, one of the most powerful poisons with which pygmies poison their arrowheads (Anwar et al., 2024). To all these, we must add the species *Erythroxylon coca, Tobacco, Cannabis indica, Garden Poppy*, and many others, which are considered drugs and which produce slow death or "white death" especially among teenagers and young people.

The plant products that contain ALKs specific to the flora of Israel are from the tropaneic groups: *Daturae folia* (02–0.6% atropine); nicotinic: *Nicotianae folia* (2–8% nicotine); quinolizidine and isoquinolinic: *Lupini semina* (2–3% lupanin), *Papaveris capita* (10–12% opium), *Withaniae somniferae radix* (10–13% withanolide), *Glauci herba* (2.6% glaucine); Indolics: *Claviceps africanae* (dihydroergoline), *Catharanthi herba* (8% vincristine, vinblastine), *Passiflorea herba* (0.05% harman, harmin, harmol); purines: *Theae folia* and *Coffea semina* (3.5–4% caffeine). Some of them are very toxic (spp. *Datura, Mandragora autumnalis*) (Bekkouche et al., 2001). The richest families in ALKs are *Solanaceae* (7 sp.), *Papaveraceae* (1 sp.), *Ranunculaceae* (5 sp.). The 41 alkaloidogenic species synthesize 136 ALKs, grouped into 17 groups. It is found that 24 ALKs are synthesized by 2–5 species, included in one or more families, close or distant from Yedere's systematic point of view. The remaining 112 ALKs are synthesized by only one species (Ablajan et al., 2021). Also, there are some plant species that can synthesize ALKs from different groups, such as

- *Galanthus nivcilis* synthesizes ALKs with an enclolic nucleus, namely, Maonine, Crielin, Tezetine, Narvedine, Nartazine, Lucorine (Arumugam, 2024);
- *Corydalis cava* contains ALKs with a quinoliziclinic nucleus, such as Dehydrocorydaline, Canadine, Caseanine, Cotibulbine, (−) Coripalmine, (+) Coripalmine, Isocoribulbine, (+) Isocoripalmine, (+) Tetrahydropalmitine, as well as ALKs with an isoquinolinic nucleus, such as (+) Coridine, Corytuberine, (+) Glaucine. They also synthesize derivatives from Protopine, such as Coricavamin and (−) Coricavidin (Chlebek et al., 2016).
- *Conium maculatum* synthesizes ALKs with pyriclinic nucleus as: (−) Conicicine, (+) Conhydrine, (+) N-methyl coniine, conhydrine, coniine, gamma-coniceine, N(−) N-methyl coniine, N-methyl coniine, Pseudoconhydrine, y-Coniceine (Vetter, 2004).
- *Chelidonium majus* contains ALKs with an isoquinolinic nucleus, such as (+) Chelidonine, Homochelidonine, Methoxy-Chelidonine, Oxychelidonine, oxysanguinine, as well as derivatives from Protopine, such as alpha-Alocriplopine and beta-Alocriplopine (Bozhadze et al., 2013).

- *Berberis vulgaris* synthesizes ALKs with a quinolizidine nucleus, such as Berberine and Colymbomine, as well as ALKs with an isoquinolinic nucleus, such as Berbamine and Oxycantine. Rincon minor synthesizes ALKs with an indolic nucleus, such as Vincamine, Yincanorin, and 14-Epivincamine. The alkaloidogenic plants presented in the work are ordered by families, genera, and species, all in alphabetical order (Arayne et al., 2007).

For each species presented, the ALKs they synthesize, as well as the phytocenotic value, are specified. Some ALKogenic species are very common, having a wide general distribution within the region, and participate as basic elements in the constitution of some phytocenoses. So, for example:

- Alkaloidogenic species frequently participate in the formation of the segetal associations from the *Tribulo-Eragrostion alliance* and *Caucalidion lappulae*: *Senecio jacobaea, Senecio vulgaris, Papaver dubiurn, Papaver rhoeas, Glaucium corniculatum, Delphinium consolida, Linaria vulgaris, Daucus carota* (Weber & Opatz, 2019).
- Other alkaloidogenic species participate in the formation of ruderal associations, from the *Chenopodietalia* order, such as *Chelidonium majus, Fumaria parviflora, Datura stramonium, Hyoscyamus niger, Conium maculatum, Urtica dioica*, and *Urtica urens* (Zhang et al., 2020).

A number of these species are edifying for the associations in which they live, for example, *Conium maculatum* for the *Conieturn maculati* association. From the perspective of alkaloidogenic species, it can be observed that most ruderal plants participate as component elements for the *Onopordetum acanthii* association, with different phytocenotic values, such as *Hyoscyamus niger*, an indicator species for the *Onopordion alliance, Datura stramonium* indicators for the *Chenopodietea class, Urtica dioica* for the orders: *Artemisietalia, Fagetalia*, and for the *Salicion alliance*, and the *Fumaria species*, frequently accompanying within the order *Polygono-Chenopodietalia*. Among the alkaloidogenic marsh species, we mention *Equisetum arvense*, a characteristic species for the *Secalinetea* class, and *Equisetum palustre*, characteristic of the *Magnocaricion alliance*. If *Zyis glomerata* is a common and frequent species in phytocoenoses from the *Arrhenatherea* class, and Lolium perenne is an indicator species for the *Lolietum perennae* association and for *Lolio-Potentilletum anserinae*, from the *Plantaginetea* class (Yamada & Sato, 2021).

Overview of alkaloidogenic species:

- *Artemisia scoparia* (Fam. *Compositae*), (ruderal associations), contains choline.
- *Arum 1naettlatum* L. (Fam. *Araceae*) (*Querco-Fagetea*) contains choline.
- *Atropa belladona* L. (Fam. *Solanaceae*) (*Fagion*) contains apoatropine, aposcopolamine, atropine, noratropine, norscopolamine, scopolamine, scopine, tropine.
- *Berberis vulgaris* L. (Fam. *Berberidaceae*) (*Berberidion*) contains berberine, berberubin, bervulcin, columbamin, isoteirandrin, jatrorrhizin, acanthine, bargustanine, berbamine, berlambine.
- *Centaurium rythraea* Rafn. (Fam. *Gentianaceae*) (*Festucion*) contains ALK E-2, gentianidin, gentianin, gentioflavin.
- *Chelidonium majus* L. (Fam. *Papaveraceae*) (Arction) contains alocryptopine, berberine, chelamidine, chelamine, chelidonine, chelitutin, chelirubin, choline, coptisine, chorisamine, dihydrosanguinarine, homochelidonine, hydroxychelidonine, hydroxysangunarine, protopine, sanguinarine, sparteine, stylopin.
- *Coniwn maculatum* L. (Fam. *Umbelliferae*) contains (+)-Conhydrin, gamma Conicein, (+)-coniine, (−)-coniine, (+)-N-methyl-coniine, (−)-N-methyl coniine, pseudoconhydrin.
- *Consolida regalis* S. F. Gray (Fam. *Ranunculaceae*) (*Caucalion*) contains ALKs A, C, E, delcosine, delsosine, licotonine.
- *Corydalis bulbosa* (L.) Pers. (Fam. *Papaveraceae*) (*Carpinion*) contains bulbocapnin, coridalin, stylopin.

- *Corydalis solida* (L.) Swartz. (Fam. *Papaveraceae*) (*Fagetalia, Carpinion*), contains: bulbocapnin, canadine, coridalin, stylopin.
- *Crataegus monogyna* Jacq. (Fam. *Rosaceae*) (*Prunetalia, Querco-Fagetea*), contains phenethylamine.
- *Cynoglossum officinale* L. (Fam. *Boraginaceae*) (*Onopordion*) contains cynoglossophine, echinatin, heliosupin, heliotridin, heliotrin, lasiocarpin, platyphyllin, viridiflorin.
- *Datura stramonium* L. (Fam. *Solanaceae*) (*Chenopodietea*) contains apoatropine, aposcopolamine, atropine, noratropine, norscopolamine.
- *Echium vulgare* L. (Fam. *Boraginaceae*) (*Sedo-Scleranthetalia*), *Festuco-Brometea*, contains heliosupine.
- *Frangula alnus* Mill. (Fam. *Rhamnaceae*) (*Salici-Franguletum*) contains franganin and frangufolin. lG. Lois's vaillant smoke. (Fam. *Papaveraceae*) (*Polygono-Chenopodietalia*) contains coptisine, fumaridine, fumvaline, protopine, sanguinarine.
- *Galanthus nivalis* L. (Fam. *Amaryllidaceae*) (*Fagetalia*) contains galantamine, ALK X, criveline, haemanthamine, hipeastrine, homolycorine, lycorenine, lycorine, magnarcine, nartasine, navedine, tazetine.
- *Galega officinalis* L. (Fam. *Leguminosae*) contains 2,3-(alpha-hydroxy trimethyl:1)-4-quinazolone, peganin.
- *Genista tinctoria* L. (Fam. *Leguminosae*) contains anagirine, cytisine, isopartein, lupanin, N-methylcytisine, thermopsin.
- *Hyoscyamus niger* L. (Fam. *Solanaceae*) (*Chenopodietea*) contains atropamine, tetramethyldiaminobutane, tropine.
- *Isopyrum thalictroides* L. (Fam. *Ranunculaceae*) (*Fagetalla*) contains magnoflorin.
- *Leonurus cardiaca* L. (Fam. *Labiatae*) (*Arction*) contains stachydrine.
- *Linaria vulgaris* Mill. (Fam. *Scrophulariaceae*) (*Onopordion, Secalietea*), contains: peganin.
- *Lolium perenne* L. (Fam. *Gramineae*) (*Plantaginetalia*) contains anloline, perloline.
- *Lunaria annua* L. (Fam. *Cruciferae*) (*Acerion*) contains lunarin.
- *Nigella arvensis* L. (Fam. *Ranunculaceae*) (*Caucalion*) contains magnoflorin.
- *Papaver dubium* L. (Fam. *Papaveraceae*) (*Caucalion*) contains ALK R-A, R-U, aporheidine, aporheine, aporphine, benzophenanthridine, benzylisoquinoline, berberine, coptisine, durhirheine, mecambrine, oemerine, papaverrubine A. D, E, protoberberine, protopine, rhoeadine, rhoeadine-type alkaloids, rhoeagenine, sanguinarine.
- *Papaver rhoeas* L. (Corn poppy) (Fam. *Papaveraceae*) (*Secalinetea*) contains ALKs: rhoeadic acid, papaveric acid, chelerythrine, coptisine, glaudine, isorhoeadine, isorhoeagenine, papaverrubin E, papaverubin A, E, porphyroxin, protopin-rhoeadine, rhoeadine, rhoeagenin, rhoeearnbin I and II, sanguinarine, synactin, thebaine.
- *Physalis alkekengi* L. (Fam. *Solanaceae*), companion in forest associations, contains three alpha(tigloyloxy)-tropanes (Tan & Takayama, 2019).
- *Plantago major* L. (Fam. *Plantaginaceae*) contains plantagonin.
- *Portulaca oleracea* L. (Fam. *Portulacaceae*) contains norepinephrine.
- *Scopolia carniolica* Jacq. (Fam. *Solanaceae*) (*Fagion, Arction*) contains atropine, cuscohygrin, hyoscyamine, pseudotropine, scopolamine, three alpha-tigloyloxytropane, tropine.
- *Sedum acre* L. (Fam. *Crassulacea*) (*Scleranthetea*) contains isopelletierin and sedamine.
- *Senecio jacobaea* L. (Fam. *Compositae*) (*Arrhenatherion*) contains senecionin and seneciphylline.
- *Senecio vulgaris* L. (Fam. *Compositae*) (*Chenopodietea*) contains retrorsine, senecionine, and seneciphylline.
- *Solanum dulcamara* L. (Fam. *Solanaceae*) contains 15 alpha-hydroxysoladulcidine, 15-alpha-hydroxysolasodine, 15-alpha-hydroxytomatidenol, 15-alpha-hydroxytomatidine, soldaulcamaridin, soladulcidin, solanidine, tomatid-5-en-3 beta-ol, 5-tomatidenol-(3-beta), 3,5-tomatidine, tomatidine.

- *Solanum nigrum* L. (Fam. *Solanaceae*) (*Chenopodietea*) contains solasodine.
- *Thalictrum aquilegifolium* L. (Fam. *Ranunculaceae*) (*Filipendulo-Petasition*), contains: berberine, magnoflorine.
- *Thalictrum minus* L. (Fam. *Ranunculaceae*) contains aconitine, berberine, canadine beta-methochloride, magnoflorine, O-methyl-thalmetine. talicarpin, talictrimin, talmelatin, talmidine.
- *Valeriana officinalis* L. (Fam. V*alerianaceae*), companion in *Festucetalia*, contains ALK, A, B, chatinine, choline, N-p-hydroxyphenythyl actinidin, valerine.
- *Vinca minor* L. (Fam. *Apocynaceae*) (Carpinion) contains 1,2-dehydroaspidospermidine, eburnamenine, eburnamine, eburnamonine, epivincamine, 14-epi vincamine, isoeburnamine, isovincamine, methoxy vincamine, N-methylaspidosperrnidine, minovincin, minovin, reserpine, strictamine, vincadiformin, vincadine, vincamine, vincamidine, vincaminine, vincaminoridine, vincaminorein, vincaminorine, vincarorma, vincine, vincicine, vincoridine, vincorine, vinoxin, vintsine (Cai et al., 2020).

The ALKs are characteristics for a certain species or for related species from the same family. The same ALKs are much less common in plants from different families. Species containing only one ALK are very rare. For example, only ricin was identified in *Ricinus communis*. Usually, a species contains several structurally related ALKs. There are also cases where several types of structurally unrelated ALKs coexist in the same species. The role of ALKs in plant life is still being discussed. Some authors consider ALKs to be waste products, resulting from protein metabolism. Others, considering the toxic effects they have on animal organisms, view them as protective substances against certain pests. It exerts considerable physiological effects on humans and animals. The ALKs are used therapeutically as narcotics and sedatives. Some ALKs are strong poisons, for example, curare, which can paralyze the nervous system. It is considered that ALKs are a means of defense of plants against insects. Some ALKs, for example, nicotine, play a role in redox enzymatic processes. The ALKs are a large group of natural compounds containing nitrogen, mainly of plant origin. Natural ALKs serve as models for the creation of new drugs, often more effective and at the same time simpler in structure. Also, there is a group of antimitotic drugs, derived from the sasquiu plant (*Vinca rosea*). These ALKs, which are frequently administered intravenously, are used especially in the treatment of leukemias and lymphomas (Kim et al., 2016). Examples: vinblastine, vincristine, and vindesine.

15.6.2 Intoxication with Plants Containing Heterocyclic Alkaloids with a Tropane Nucleus

15.6.2.1 *Solanaceae* Family: Belladonna or Deadly Nightshade (*Atropa belladonna* L.)

Belladonna or deadly nightshade has been known since antiquity and used therapeutically due to the presence of tropane ALKs in its leaves and roots. It is also known under the popular names of the wolf's cherry, the lady of the forest, and the empress. The plant is part of the medicines "Distonocalm" and "Foladon". The powder of the leaves is used in making anti-asthmatic cigarettes. It is recommended to place crops in submontane areas and intramontane depressions, in sheltered and sunny places. The species is picky about temperature, growing uniformly at temperatures higher than 5°C. It does not tolerate low and wet places. It prefers fertile, well-structured, and loose soils. In terms of light, it has low demands, succeeding in semi-shady or shady places. Marjoram is a perennial plant, growing in the form of a vigorous, erect, and highly branched bush. The stems can reach heights of 100–150 cm. The leaves at the base are alternate, and the upper ones are arranged in pairs. The flowers are solitary, attached with a peduncle of 1–2 cm. The corolla can be purple-brown to reddish-brown, with purple veins inside. It blooms from June to August. The fruit is a berry, similar to a cherry, initially green, and when ripe, it can turn black; it also has a persistent calyx. The seeds are reniform, with a mass of 1000 seeds of about one gram. Being a perennial plant, gourds are cultivated outside the summer season. It is not pretentious to the preceding plant, succeeding after any other culture. The best forerunners are legumes for grains, cereals, gorse, and other plants that leave the soil clean of weeds and without plant debris. He can return to the same field after 5–6 years. Horsetail responds very well to the application of organic fertilizers, such as manure. The plants react to the administration of fertilizers by increasing the production of leaves and also by increasing the content of ALKs. This will be applied in the fall before plowing. The recommended amount of manure is 20–30 t/ha. After harvesting the preceding plant, weeding is recommended, followed by deep plowing, carried out at 22–25 cm, after plowing. The land will be shredded with a disc harrow, after which the seedbed will be prepared with a combiner or with a disc harrow in aggregate with a harrow with adjustable tines, followed by rolling. It can be sown late in autumn in November or early spring in March. The seed used to establish the crop must be healthy, come from seed lots cultivated in an ecological system, be certified, and be produced in the previous year. The physical purity of the seeds must be at least 85%, and the germination at least 80%. The amount of seed used for one hectare is 4–6 kg; the seed of a row indicator plant (lettuce) can be added in the amount of 1–1.5 kg/ha. It will be sown with universal seeders (type SUP-21, SUP-29), at a distance between rows of 50 cm, for the execution of the grids and the bouquets after the emergence. The density must be 25–30 plants/m^2, and the sowing depth 1–2 cm. The species can also be propagated by seeding. For one hectare, approximately 200 square meters of cold layers (1–1.5 kg of seed) are required to obtain about 80,000–100,000 seedlings per hectare. To combat weeds, weeding between rows and weeding between plants in a row is recommended whenever necessary. After the plants have sprouted, thinning will be carried out in order to obtain a more harmonious development of the plants (Zhang et al., 2022).

In the second year, early spring, it is recommended to carry out a harrowing work, perpendicular to the direction of the rows, and in the containers, they make a grid to keep the crop free of weeds and to loosen the soil. Fighting diseases and pests is carried out through preventive methods, by maintaining cultural hygiene, balanced fertilization, and removing plants with symptoms of diseases or attacked by pests. To combat diseases such as stem and leaf rot or downy mildew (*Verticillium* spp., *Fusarium* spp., *Sclerotinia* spp., *Phytophthora* spp.), treatments can be applied to the vegetation with copper-based products (Zeama Bordelaise, Champion 70 PU) at a concentration of 1%. To combat pests such as fleas (Epithrix atropae), the vegetation can be sprayed with extracts based on black wormwood or mouse tail. It is recommended that the leaves be harvested at the beginning of flowering, and the herb be harvested at fruiting. In the first years of cultivation, one harvest of leaves and one of grass can be done, at fruiting, and when the crop is discontinued, it is recommended to harvest the roots. The plant material is harvested by hand with a sickle and placed in baskets without stuffing them. Harvesting must be done in good weather, after the dew has lifted. Immediately after harvesting, the product will be conditioned and dried. Drying in the shade is done in a thin layer, in well-ventilated, darkened rooms. Artificial drying will be performed at temperatures of 40–45°C. The drying yield is 1. The roots will be removed with the plow without a handle or with the beet displacer. They will be shaken off the ground, washed, threshed, and dried, naturally or artificially. During artificial drying, the drying temperature must be a maximum of 50–60°C, in order not to affect the ALK content (Fattahi et al., 2021).

Since the plant is toxic, starting from harvesting and until handing over the harvest to the beneficiaries, special measures will be taken to protect the personnel to avoid toxic effects, such as the following.

1. Avoiding direct skin contact with plant material.
2. During the execution of various works, smoking, drinking water, or eating food is prohibited.

3. Washing hands after work.
4. The use of a mask or a gauze applied to the mouth during harvesting in order to prevent accidental poisoning.

The reception conditions for plant material are as follows.

- *For dry leaves*: Max. 5% Browned leaves; max. 3% other parts of the plant; max. 0% organic and mineral foreign bodies; max. 13% moisture; minimum 0.3% hyoscyamine content;
- *For dry herb*: Max. 5% Browned leaves and branches; max. 1% organic and mineral foreign bodies; max. 13% moisture; minimum 0.33% hyoscyamine content;
- *For fresh herbs*: The maximum is 1% organic foreign bodies. 0.25% mineral foreign bodies; minimum 0.5 content in hyoscyamine;
- *For roots*: Max. 5% browned roots; max 1 organic and mineral foreign bodies; max. 13% moisture; max. 0.45% content in hyoscyamine;

The production of dried leaves can be about 600 kg/ha, and that of roots 1000–1500 kg/ha (Zhao et al., 2020).

15.6.2.2 *Atropa belladonna* Poisoning: Toxic Principles

Hyoscyamine (by drying, it turns into its dextrorotatory isomer, resulting in a raceme, atropine), scopolamine, belladonna, hyoscine, apoatropine, and nicotine are secondary metabolites. They are found in all parts of the plant, but the most toxic are the roots, then the leaves, flowers, stems, and, finally, the fruits. The toxic principles are not destroyed by preservation. The toxic dose is 120–180 g of dried leaves for horses, 60–90 g of roots, and 60 g of leaves for large ruminants. Sheep and goats are less sensitive, and rabbits are more resistant. In general, due to its repulsive smell, the plant is avoided by animals, but it is consumed in mowed or silage fodder (even 1% content in fodder is dangerous). Hyoscyamine and atropine are ALKs that act on the vegetative nervous system as parasympathetic by blocking cholinergic receptors and on the central nervous system, initially exciting, then depressing. In horses, symptoms appear 3–5 h after consuming the plant. The first sign is the refusal of food, then there is drying of the mucous membranes, mydriasis, the inability to adapt to light, a state of restlessness that is accentuated, symptoms of colic, bloating of the abdomen without noticing sensitivity to palpation, the absence of intestinal peristalsis, and constipation with the elimination of dry stools. After a period, the animal goes into a state of general depression with paresis, paralysis, and immobility. Death occurs through the paralysis of the bulbar respiratory centers. In large ruminants, the clinical symptoms evolve more rapidly, the tympanism is much more pronounced, and the signs of nervous excitement reach the intensity of rabies attacks. In small ruminants, the clinical picture is most often manifested only by mydriasis. In birds, it appears to affect the nervous system after consuming fruits or seeds. Anatomopathological changes – they are not characteristic.

The diagnosis is established based on the clinical signs, to some extent characteristic (mydriasis, drying of the mucous membranes), and the results of the botanical analysis of the pasture or the administered feed. The prognosis is reserved. The first objective of the treatment is the evacuation of the contents from the digestive tube through gastric lavage and purgatives. To precipitate the ALKs, tannin is administered. Pharmacodynamic antidotes are used: pilocarpine, arecoline, and eserine (physostigmine). Next, a symptomatic treatment is performed to support the cardiac and respiratory activities, to counteract the manifestations of excitement or nervous depression (Bedewitz et al., 2014).

15.6.3 Poisoning with Plants Containing Heterocyclic Alkaloids with a Tropolonic Nucleus

15.6.3.1 Family Liliaceae – Crocus, Meadow Saffron, Naked Boys, or Naked Ladies (*Colchicum Autumnale*)

This plant grows in meadows near beech trees and along roadsides. The plant blooms in autumn and bears fruit in spring.

It is one of the most toxic plants in Romania. The toxic principle, present in all the organs of the palm, is **Colchicine** – an ALK that has no alkaline reaction and does not form salts with acids. Through drying, colchicine is converted to demecolcine, which is equally toxic. Poisoning with autumn sawfly is found in the pasture, in spring and autumn, in the youth, and in the stable throughout the year, in all age groups. The most sensitive are cattle and pigs. Symptoms appear 2–48 h after ingestion. Colchicine has a cerebral, cerebellar, and bulbar paralyzing effect, abolishes voluntary movements, is analgesic on the peripheral nervous system, and is irritating to the mucous membranes of the digestive tube and kidneys. The clinical picture is present in all species: digestive signs (anorexia, ptyalism, hemorrhagic diarrhea with a fetid smell, tenesmus), nervous signs (tremors, depression, decreased myoclonic reflexes, wobbly gait). Hypothermia, anuria, and circulatory collapse gradually set in. Death occurs through paralysis of the bulbar respiratory centers. Mortality is high (Finkelstein et al., 2010).

The anatomopathological changes are uncharacteristic, like a hemorrhagic enteritis. The treatment consists of the administration of drugs that facilitate the evacuation of the toxic (emetics, purgatives), the precipitation of the toxic (tannin), and that combat the symptoms (diuretics, mucilaginous protectors for the lining of the digestive tube, rehydrating solutions, tonics, cardiac, general stimulants). The plant species of the *Aconitum* genus, popularly known as "omag", are deceptively beautiful and extremely toxic. Traditionally used in Oriental medicine, but also in Europe for its therapeutic properties, this herb requires careful handling and proper processing to reduce its toxicity. Plants belonging to the *Aconitum* genus are extremely toxic, with the roots and tubers being the parts with the highest concentration of the toxin. Severe aconitine poisoning can occur after accidental ingestion of the wild plant or after consuming a decoction based on *Aconitum* roots. In traditional Chinese medicine, the roots of this plant were used only after processing to reduce the content of toxic ALKs. But using a higher than recommended dose and inadequate processing increases the risk of poisoning (Nett & Sattely, 2021).

15.6.3.2 Aconitum Alkaloids

Aconitine and related ALKs, found in *Aconitum* species, are potent cardiotoxins and neurotoxins. These ALKs include aconitine, mesaconitine, hypaconitine, and other ALKs. The ALKs represent up to 1.5% of the dry weight of the plant. Aconitine is hydrolyzed to picraconitine, which hydrolyzes to aconine.

15.6.3.2.1 Aconitum napellus – *Mechanisms of Toxicity*

The cardiotoxicity and neurotoxicity of aconitine and related ALKs are due to their action on the sodium channels of the cell membranes of excitable tissues, including the myocardium, nerves, and muscles. Clinical symptoms in patients suffering from poisoning with Aconitum sp. predominantly present with a combination of neurological, cardiovascular, and gastrointestinal symptoms. Neurological features include paresthesia and numbness of the face, perioral area, and all four limbs, as well as muscle weakness in all four limbs (Li et al., 2022).

Cardiovascular features include hypotension, chest pain, palpitations, bradycardia, sinus tachycardia, ventricular tachycardia, and ventricular fibrillation. Gastrointestinal manifestations include nausea, vomiting, abdominal pain, and diarrhea. Aconitum has been used since ancient times as a poison used for hunting with spears and arrows. The plant was also believed to keep werewolves at bay and was also used against wolves. In ancient Rome, homage was used as a method of execution.

Aconitum's therapeutic uses included the treatment of joint and muscle pain. As a tincture applied to the skin, it is claimed to slow the heart rate in heart patients. Other uses included reducing fever and cold symptoms. Aconitum roots contain aconitine, mesaconitine, hypaconitine, and other related ALKs, which are known cardiotoxins and neurotoxins. Patients predominantly present with neurological, cardiovascular, and gastrointestinal features. It has no antidote, and the treatment is only supportive (Gao et al., 2022).

15.6.3.3 Poisoning with *Claviceps purpurea*

Ergotism is an intoxication resulting from the consumption of cereals (especially rye) contaminated with the fungus *C. purpurea*. This can also be caused by the ingestion of drugs derived from contaminated grains or interactions between different drugs that generate toxic ALKs (ergotine, ergotamine). Cases of ergotism have been known since the Middle Ages. At that time, the degree of grain contamination was 25%. The minimum ratio of contaminated grains to generate intoxication is 2%. The development cycle of *C. purpurea* fungi at the level of rye ears, instead of grains, is that the sclerotia of this fungus appear. They represent a form of resistance of the fungi, which, by germination, will give rise to a vegetative mycelium. Afterwards, they will fall to the ground, where they will remain until the following spring, when they will germinate. Upon germination, the sclerotium forms extensions of approximately 4 cm. They have club-shaped formations at the ends and are pink in color. They are responsible for the multiplication of this fungus. The elements that take part in this process are surrounded by a sweet liquid, which attracts insects. When the insects consume this liquid, they take over the surface of the body conidia (immobile spores that contribute to the asexual reproduction of this fungus). Conidia germinate and form filaments, which can contaminate neighboring ears. At the end of this process, sclerotia of this fungus are formed and thus, the cycle of multiplication resumes (Chen et al., 2019).

15.6.3.3.1 Alkaloids Produced by C. purpurea *Fungi*

Environmental factors influence the composition of the ALK content produced by this mushroom. In Europe, the predominant ALKs are of the ergotamine type, and in Ukraine, the predominant ones are of the ergotoxin type. Specifically, three chemical varieties of ALKs have been identified:

- Type G1 is present in field areas and contains ergotoxinic and ergotamine ALKs.
- Type G2 is present in areas with a higher degree of humidity and low light and contains ALKs such as ergocristine, ergosine, and ergocriptine.
- Type G3 is present in areas with salt marshes, being characteristic of the species *Spartina angelica*, and contains ALKs of the type ergocriptine and ergocristine (Dung et al., 2021).

15.6.3.4 The Effects of Alkaloids on the Human Body

The effects of ergot or rye ALKs are mainly on dopaminergic, adrenergic, and serotonergic receptors. They can have an agonistic or antagonistic role, depending on the type and dose of the ALK, as well as the organ on which it acts.

Natural ALKs have effects on the pregnant uterus. In small doses, they increase the frequency and force of contraction of the uterus and subsequently cause its relaxation. At higher doses, uterine contractions are stimulated, becoming more frequent and longer. In medical practice, ALKs are used to maintain contractions, but also to prevent postpartum bleeding. Other ALKs, such as ergotamine, have vasoconstrictive effects, acting on alpha 2 adrenergic receptors. They are used in medical practice in hypotensive states. In Parkinson's disease, bromocriptine is used, which is a dopamine D2 receptor antagonist. Bromocriptine is used together with levodopa to reduce the amount of levodopa given to patients with Parkinson's. In large doses or if administered over a longer period of time, these ALKs have toxic effects and produce ergotism.

15.6.3.4.1 Clinical Forms of Ergotism

Two types of ergotism are known:

- Ergotism with gangrene – also called "Saint Anthony's Fire";
- Ergotism with convulsions – also called "Saint Andrew's Fire" (Ayarragaray, 2014).

Ergotism with gangrene, the symptoms of this pathology are localized in certain areas of the body, unlike ergotism with convulsions. Thus, the affected areas become swollen, painful, and patients complain of burning sensations. At this level, the sensation of numbness and cyanosis appear, the area becomes dark, even black, and takes on a dry appearance. In advanced stages, the muscle tissue detaches without bleeding. The toxins responsible for poisoning affect the nervous, vascular, and endocrine systems. They generate vasoconstriction, which can lead to gangrene. Necrosis can also occur in the fingers.

Ergotism with convulsions is characterized by the onset of a heightened sensation of thirst. Other symptoms present are decreased appetite and growth retardation. Affected individuals experience convulsions, muscle numbness, and painful muscle contractions.

Medicinal ergotism, this form of ergotism occurs as a result of

1. Overdose of drugs containing ergotamine or dihydroergotamine. These drugs are mainly administered in the treatment of migraines. It is important that patients follow the doses recommended by the doctor. In addition to this use, this type of medication is also administered to patients with hypotension.
2. Drug combinations that intensify the effects of ergotamine or its natural or synthetic derivatives (Flamm, 2023).

The effects of ergotamine ALKs are increased when administered together with some medications. The processes underlying the potentiations of the effects of ALKs are represented by the inhibition of hepatic metabolism and the production of vasoconstriction in the blood vessels. Thus, hepatic metabolism is inhibited when taking macrolide antibiotics (erythromycin, telithromycin, clarithromycin, etc.), antifungals (voriconazole), nitrates (nitroglycerin), antiretrovirals such as protease inhibitors administered in HIV (ritonavir, nelfinavir), and antiepileptics (stiripentol). If the liver has low or no metabolic function, the ALKs remain in the body because they are no longer eliminated. As a result of this, they reach toxic levels in the body and cause the symptoms listed above. Another way in which the specific symptoms of ergotism appear is caused by the association of ALKs with drugs that act on blood vessels. Such drugs are beta-blockers and migraine drugs (sumatriptan). By the occurrence of vasoconstriction, stasis occurs, and thus, ischemia occurs, followed by gangrene.

In the case of the administration of drugs based on ergot ALKs in the treatment of migraines, some side effects may occur: nausea, vomiting, abdominal pain, muscle cramps, especially of the lower limbs, diarrhea, and vertigo. At high doses, the symptoms are increased heart rate, tingling, burning sensations, the skin becomes cold, hallucinations, confusion, and even coma may occur (Grzybowski et al., 2021).

Ergotic ALKs have a stimulating effect on the central nervous system, and later, their inhibition occurs. These ALKs have an inhibitory effect on the pituitary gland and implicitly on the release of prolactin. As a result, a delay in lactate secretion or even agallation occurs. The diagnosis is confirmed based on the symptoms, signs, and the history taken by the doctor. To confirm the diagnosis, tests can be done to detect toxic ALKs in grains or foods that are thought to be contaminated. Treatment currently used consists of intravenous or intra-arterial administration of a sodium nitroprusside solution. Nitroglycerin can also be given, but it is not as effective in treating the symptoms of ergotism. It is essential to stop eating contaminated grains or foods before starting treatment. Prevention can be achieved by testing for ergot ALKs in cereals, especially corned rye. Currently, the maximum allowed proportion of toxic ALKs is 0.1% for edible wheat and 0.05% for flour. It is recommended to consume only from safe and verified sources (Belser-Ehrlich et al., 2013).

The physiological influence of nicotine resides in increasing blood pressure and heart rate. In this situation, the release of adrenaline also increases. Numerous metabolic changes occur. There is also a transient effect of improving memory and concentration. Smoking (tobacco), however, also induces psychological dependence due to the presence of nicotine. Quitting smoking can cause undesirable phenomena, for example, headaches. In extreme situations, depression can occur that can last 3–4 days or even a few weeks after stopping smoking. A particular effect of nicotine resides in the abnormal production of dopamine, which creates a state of euphoria. When the dopamine level decreases, the smoker is in a position to resume smoking. Overdose and nicotine poisoning occur in the case of persistent active smoking. In such situations, characteristic symptoms appear: daytime hyperactivity, nausea, vomiting, pallor, palpitations/tachycardia, headache, insomnia, lipothymia, vertigo, abdominal pain, general weakness, etc. Nicotine addiction, although with fewer consequences, has some similarities with heroin and cocaine addiction. The chronic use of cigarettes has been proven to be harmful both because of the nicotine and especially the combustion products from the cigarette paper (generally polycyclic aromatic hydrocarbons) (Castro et al., 2023). The effects, in the long run, reside in accentuated respiratory disorders, accompanied by lung cancer, etc. Nicotyrin has a structure similar to nicotine, being a bisdehydronicotinic derivative. It is formed as an intermediate compound during nicotine biosynthesis. It appears as a colorless liquid (Stefaniak et al., 2022).

Morphine is a compound, used as medicine, that has an opioid agonist effect. It acts mainly on the CNS. Its effects are manifested by supraspinal analgesia, euphoria, respiratory depression, miosis (shrinkage of the eye pupil), and inhibition of digestive motility (followed by constipation. The effects are due to binding to the μ (mu), κ (kappa), and δ (delta) receptors. The action of morphine on cognitive ability is negative regarding anterograde and retrograde memory. Repeated use gives analgesic effects, but produces addiction (dependency). Biotransformation (in metabolic processes) is carried out by conjugation reactions, for example, conjugation with glucuronic acid (results in 2-,3-,6-glucuronidemorphine). Also, biotransformation can be done by N-demethylation (in reduced proportion). Excretion is carried out by kidney (90%) and the biliary system (10%). The therapeutic administration (in the form of a medicinal preparation) of morphine is done orally or by subcutaneous, intravenous, or even inhalational injection (using a nebulizer) (Kupnicka et al., 2020).

Codeine is also an ALK obtained from opium, in the extract of which it is present in a proportion of 0.7–2.5%. Codeine is also known as methylmorphine. It is also an analgesic, but weaker than morphine. It is used in antitussive and antidiarrheal medicinal preparations.

The biotransformation occurs at the level of the liver, being inactivated by conjugation with glucuronic acid, resulting in 6-glucuronide-codeine (inactivation in the proportion of approx. 80% by glucuron-conjugation). Codeine was also obtained through laboratory synthesis (later and industrially) through O-methylation of morphine (Eissing et al., 2012).

Thebaine is also known as codeine-methyl-enol-ester. The name Thebean comes from the ancient city of Thebes in Egypt. This ALK has structural similarities with morphine and codeine. It is obtained from the extract of *Papaver bracteatum* (*Iranian poppy*). Other compounds with ALK attributes can be obtained industrially from the original extract, such as oxycodone, oxymorphone, nalbuphine, naloxone, etorphine, etc. (Tarkesh Esfahani et al., 2021).

Papaverine, with the chemical name 1-(3,4-dimethoxybenzyl)-6,7-dimethylquinoline, is an isoquinoline ALK isolated from poppy latex (*Papaver somniferum*).

Structurally, papaverine has a 1-benzylisoquinolinic nucleus, on which methoxy groups (-O-CH_3) are grafted. Papaverine was discovered in 1848 by Georg Merck (1825–1873), who was a student of the German chemists Justus von Liebig and August von Hofmann. The discoverer of papaverine was the son of Emanuel Merck (1794–1855), who was the founder of the "Merck Corporation" – a German company for chemical and pharmaceutical products. It is mainly used in analgesic treatments for visceral spasms of the gastrointestinal tract, biliary, and ureteral crises. It is also used as a vasodilator in cerebral and coronary disorders, as well as in subarachnoid hemorrhages. It is also used in erectile dysfunctions (ensuring blood flow in the Corpora cavernosa). The mechanism of action in vivo is not fully known. However, a selective inhibition of the enzyme phosphodiesterase has been observed, which is the cause of the increase in the level of cyclic adenosine monophosphate (cyclic AMP). The less common side effects of papaverine include polymorphic ventricular tachycardia and an increase in certain enzyme markers (transaminases and alkaline phosphatases). Drowsiness and dizzinesshave also been observed. After chronic administration, in mice, motor and cognitive deficits and increased anxiety were observed. Narcotine – benzoquinolinic ALK obtained from poppy latex, is also present in other plants from the *Papaveraceae* family. It has an analgesic effect. It was initially used as an antitussive medicine. It also has properties as an antiproliferative agent (Ashrafi et al., 2023).

Laudanosine is a benzyltetrahydroisoquinolinic ALK, also known as N-methyl-1,2,3,4-tetrahydropapaverine. It is found in small amounts in poppy latex (0.1%) from where it was first isolated in 1871. As expected, knowing the structure, through partial dehydrogenation, laudanosine is transformed into papaverine. Laudanosine interacts with GABA receptors and glycine receptors, opioid and nicotinic receptors (Billiot et al., 2006).

Tomatidin was isolated from tomato leaves (*Solanum lycopersicum*) and other plants of the *Solanaceae* family. Tomatidine is the aglycone of tomatine, the heteroside that also contains tetraglucosides (Glc-Glc-Gal-Xyl). The effects of the glycoalkaloids solanidine and tomatidine were also studied in relation to embryology in female mice. It was found that it produces a decrease in the weight of conception products (more accentuated with solanidine – 36.1 compared to tomatidine – 11.9), a decrease in the number of conception products (with solanidine – 27.0, and with tomatidine – 15.5) (Fujimaki et al., 2022). Other ALKs with a steroid nucleus are veratridine – extracted from the dead plant (*Veratrum album*) from the family *Liliaceae* and *Schoenocaulon*; conesin – obtained from the bark and seeds of the Indian shrub *Holarrhena antidysenterica* from the *Apocynaceae* family; muldamine – isolated from *Veratrum californicum*, having teratogenic effects; zigacin – isolated in 1913 from plants belonging to the genus *Zigadenus*, etc. Other ALKs include *Piperine* – an ALK with a piperidine heterocycle present in peppercorns. It is insoluble in water. Piperine gives a specific taste to peppercorns. Hordein is present in small amounts in barley grains and in larger amounts in germinated barley. It stimulates the activity of the heart and produces an increase in blood pressure. It also activates diuresis (urine elimination). Without being ALKs (we note, by definition, the N-heterocyclic character of ALKs), compounds with a homocyclic structure, basic character, and biologically active properties are also known (Derosa et al., 2016). One such non-ALK compound is capsaicin, which is an aromatic amide. It was isolated from the fruits of pepper (*Capsicum annum*). The presence of capsaicin gives the spicy taste of these vegetables.

Solanidine was isolated from potato (*Solanum tuberosum*), where it is found in greater quantity in the skin of the tubers of younger plants. The properties of solanidine reveal similarities with steroids. Solanidine is the aglycone of solanine – the heteroside that also contains a triglyceride (Glc-Gal-Rha) (Attoumbré et al., 2013).

15.7 MEDICINAL PLANTS USED IN THE PREPARATION

The herbs used in panic attack teas contain bioactive compounds with scientifically proven anxiolytic and sedative properties. These herbs work by modulating neurotransmitters and reducing the stress response.

Chamomile tea: Chamomile contains apigenin, a flavonoid with strong anxiolytic effects that help calm the central nervous system. This tea reduces muscle tension and promotes relaxation, making it ideal for people who frequently experience panic attacks. Regular consumption of chamomile tea can improve sleep quality and reduce the frequency of anxiety episodes.

Lavender tea: Lavender is recognized for its calming and sedative properties. Lavender essential oil has anxiolytic and antidepressant effects, helping to reduce stress and nervous tension. Lavender tea can be consumed in the evening for a peaceful and restful sleep.

Passiflora tea: Passiflora or passion flower contains flavonoids and ALKs that act on GABA receptors in the brain, having effects similar to anxiolytic drugs. This tea reduces anxiety and promotes relaxation without significant side effects.

Wormwood tea: Wormwood or lemon balm has calming and anxiolytic properties due to active compounds such as rosmarinic acid. This tea reduces cortisol levels and relieves physical symptoms of anxiety, such as palpitations and sweating.

Peppermint tea: Peppermint has relaxing and antispasmodic properties that help calm the nervous and digestive systems. Peppermint essential oil reduces muscle tension and promotes general well-being, making it helpful in managing panic attacks (Reinhard & Zoller, 2021).

Valerian tea: Valerian is a medicinal plant with strong sedative properties, traditionally used to treat anxiety and insomnia. Valerian root contains compounds that increase GABA levels in the brain, thereby reducing agitation and restlessness.

Ashwaganda tea: Ashwaganda is an adaptogenic herb that helps the body cope better with stress. This tea reduces cortisol levels and relieves symptoms of anxiety, while improving focus and mental energy. Combining different herbs can amplify the therapeutic effects and provide additional benefits in managing panic attacks. Synergies between plants can increase the bioavailability of active compounds and improve treatment efficacy.

Chamomile and lavender: This combination harnesses the anxiolytic properties of chamomile and the calming effects of lavender. Apigenin in chamomile works synergistically with lavender essential oils to reduce anxiety and promote relaxation. The mixture has a pleasant taste and can be consumed several times a day, being particularly effective in the evening for improving the quality of sleep.

Evening primrose and passionflower: The combination of evening primrose and passionflower creates a powerful blend to calm the nervous system. The rosmarinic acids in the evening primrose complement the flavonoids in the passionflower, offering an enhanced anxiolytic effect. This mixture is especially recommended for moments of intense stress or before situations that can trigger panic attacks (Kaltner et al., 2018).

Peppermint and valerian: The antispasmodic properties of peppermint blend perfectly with the sedative effects of valerian, creating an effective blend for reducing anxiety and associated physical symptoms. This combination is ideal for people who experience both mental and physical symptoms during panic attacks.

Recommended custom blends: Creating a custom blend should take into account specific symptoms and individual preferences. The recommended proportion is two parts main plant and one part secondary plant (Bodi et al., 2014).

For optimal effect, the herbs must be fresh and of high quality, and the mixture can include a maximum of three different types of herbs to avoid unwanted interactions.

15.8 SAFETY CONSIDERATIONS

Although herbal teas are considered safe, they can have side effects or interact with certain medications. It is important that they are consumed responsibly and that specialists are consulted before starting a cure with medicinal teas.

Potential side effects: Excessive consumption of medicinal teas can cause side effects such as nausea, dizziness, or drowsiness. Some people may develop allergic reactions to certain plants, manifested by skin rashes or difficulty breathing. Symptoms usually disappear when consumption is discontinued, but caution and moderation in use are required (Osman et al., 2019).

Drug interactions: Herbs can interact with anxiolytic or antidepressant medications, altering their effectiveness. Valerian and passionflower can potentiate the effect of sedatives, while ashwaganda can interfere with thyroid medications. It is essential to inform the doctor about the consumption of medicinal teas when following a medical treatment.

Precautions in pregnancy and breastfeeding: Many herbs are contraindicated during pregnancy and breastfeeding due to potential effects on fetal development or transmission through breast milk. Chamomile and peppermint are considered safe in moderate amounts, but other herbs such as valerian or passionflower should be avoided during this time.

Recommended daily limits: For optimal safety and effectiveness, consumption of medicinal teas should be limited to 2–3 cups per day (Gou et al., 2021).

Each tea must be prepared correctly, respecting the recommended doses and infusion time. Long periods of consumption should be alternated with breaks of 1–2 weeks to avoid the development of tolerance and prevent adverse effects.

15.9 CONCLUSION AND FUTURE REMARKS

The ALKs are compounds that contain nitrogen, have an alkaline reaction, and are very strong poisons. For the time being, there is no herbal analgesic stronger than morphine and no cytostatic stronger than Taxol for treating ovarian or breast carcinoma. After, through the use of ALKs isolated from plant material, therapeutics accumulated in the last century, sufficient experience in use, pharmaceutical chemistry tried to optimize either the solubility or to modulate/enhance the pharmacological action of the ALKs through semisynthesis, in some cases designing new structures, starting from a natural substance. The main goal was to increase the activity of the substance by improving bioavailability and pharmacokinetics, as well as tolerability, by reducing side effects. Thus, starting from atropine, ipratropin-hydrobromide was obtained. By introducing the isopropyl lipophilic residue, the semi-synthetic substance became suitable for prophylactic and curative treatment by inhalation of non-obstructive bronchitis and bronchial asthma, as well as for intranasal application in the treatment of chronic rhinitis. Analogously, when the butyl residue is introduced into the scopolamine molecule, the resulting substance has a spasmolytic action used in the treatment of biliary colic and spasms occurring in the digestive tract or uterine ones, being devoid of the effects existing at the CNS level in the case of the precursor, scopolamine. Very successful were the semi-synthetic modifications brought to the quinoline core of quinine, which led to the obtaining of current, much less toxic, antimalarials. Derivatives of 4-aminoquinoline (chloroquine) and 8-aminoquinoline (primaquine) were obtained in this way, constituting second-generation antimalarials with a much reduced toxicity compared to that of quinine. The hydrogenation of the double bond from the C9–C10 carbon atoms of ergotamine led to the reduction of the agonistic action at the level of α-adrenergic receptors, so that the α-blocking action reaches its maximum expression, a fact that makes dihydroergotamine an ideal antihypotonic. Of course, successful structures were not always obtained through semisynthesis. In the attempts to acetylate morphine, which should have reduced the potential for the development of pharmacodependency, we reached diacetyl-morphine = heroin, with a narcotic potential several times greater than that of the precursor ALK. Analogously, trying to obtain a more effective cardiovascular drug, Hofmann synthesized lysergic acid diethylamide (LSD) starting from lysergic acid, the first being considered even after 70 years as the most powerful hallucinogen, without any therapeutic use.

COMPETING INTERESTS

The authors declare that they have no competing interest.

DECLARATION OF GENERATIVE AI AND AI-ASSISTED TECHNOLOGIES IN THE WRITING PROCESS

During the preparation of this work, the book chapter author(s) have not taken any assistance from AI, and the book chapter author(s) reviewed and edited the content as needed and take(s) full responsibility for the content of the publication.

REFERENCES

Akveran, G., Köse, K., & Köse, D. A. (2018). Solvent effect on endosulfan adsorption onto polymeric arginine-methacrylate cryogels. *Environmental Science and Pollution Research International*, *25*(25), 25458–25467. https://doi.org/10.1007/s11356-018-2531-z

Aabom, B., Laier, G., Christensen, P. L., Karlsson, T., Jensen, M. B., & Hedal, B. (2020). Oral morphine drops for prompt relief of breathlessness in patients with advanced cancer - a randomized, double blinded, crossover trial of morphine sulfate oral drops vs. morphine hydrochloride drops with ethanol (red morphine drops). *Supportive Care in Cancer: Official Journal of the Multinational Association of Supportive Care in Cancer*, *28*(7), 3421–3428. https://doi.org/10.1007/s00520-019-05116-1

Abacioğlu, N., Ozmen, R., Cakici, I., Tunçtan, B., & Kanzik, I. (2001). Role of L-arginine/nitric oxide pathway in the antinociceptive activities of morphine and mepyramine in mice. *Arzneimittel-Forschung*, *51*(12), 977–983. https://doi.org/10.1055/s-0031-1300148

Abadie, S., Leduc, C., Lintner, K., & Bedos, P. (2021). The alkaloid centcyamine increases expression of klotho and lamin B1, slowing the onset of skin ageing in vitro and in vivo. *International Journal of Cosmetic Science*, *43*(5), 561–572. https://doi.org/10.1111/ics.12731

Abbas, M. N., & Mostafa, G. A. (2007). Papaverine PVC membrane ion-selective electrodes based on its ion-exchangers with tetraphenylborate and tetrathiocyanate anions. *Annali di chimica*, *97*(8), 771–780. https://doi.org/10.1002/adic.200790060

Abd El-Mawla, A. M. (2010). Effect of certain elicitors on production of pyrrolizidine alkaloids in hairy root cultures of Echium rauwolfii. *Die Pharmazie*, *65*(3), 224–226.

Abd El-Rehim, H. A., Swilem, A. E., Klingner, A., Hegazy, S. A., & Hamed, A. A. (2013). Developing the potential ophthalmic applications of pilocarpine entrapped into polyvinylpyrrolidone-poly(acrylic acid) nanogel dispersions prepared by γ radiation. *Biomacromolecules*, *14*(3), 688–698. https://doi.org/10.1021/bm301742m

Abdel Shaheed, C., Maher, C. G., & McLachlan, A. J. (2019). Efficacy and safety of low-dose codeine-containing combination analgesics for pain: Systematic review and meta-analysis. *The Clinical Journal of Pain*, *35*(10), 836–843. https://doi.org/10.1097/AJP.0000000000000746

Abdelhamid, E. E., Sultana, M., Portoghese, P. S., & Takemori, A. E. (1991). Selective blockage of delta opioid receptors prevents the development of morphine tolerance and dependence in mice. *The Journal of Pharmacology and Experimental Therapeutics*, *258*(1), 299–303.

Abdelkader, M. S., Lockwood, B., & Sansongsak, P. (2010). Uptake of nicotine from suspension culture of Nicotiana tabacum by molecularly imprinted polymers. *The Journal of Pharmacy and Pharmacology*, *62*(5), 633–637. https://doi.org/10.1211/jpp.62.05.0011

Abdelrheem, D. A., Abd El-Mageed, H. R., Mohamed, H. S., Rahman, A. A., Elsayed, K. N. M., & Ahmed, S. A. (2021). Bis-indole alkaloid caulerpin from a new source *Sargassum platycarpum*: isolation, characterization, *in vitro* anticancer activity, binding with nucleobases by DFT calculations and MD simulation. *Journal of Biomolecular Structure & Dynamics*, *39*(14), 5137–5147. https://doi.org/10.1080/07391102.2020.1784285

Abdel-Salam, Z., & Nammas, W. (2010). Early atropine is safer than conventional atropine administration in the elderly undergoing dobutamine stress echocardiography. *Kardiologia polska*, *68*(4), 422–428.

Abdolmohammad-Zadeh, H., Zamani, A., & Shamsi, Z. (2019). Preconcentration of morphine and codeine using a magnetite/reduced graphene oxide/silver nano-composite and their determination by high-performance liquid chromatography. *Journal of Chromatography. A*, *1590*, 2–9. https://doi.org/10.1016/j.chroma.2018.12.064

Abdul, D. S., Pillai, P. G., & Sigamani, S. D. (2024). Ethanolic extracts of Sterculia guttata: Exploring the neuroprotective effects on memory and cognitive impairment in scopolamine-induced Alzheimer's disease rats. *Pakistan Journal of Pharmaceutical Sciences*, *37*(5), 927–938.

Abdul-Ghani, A. S., Abu-Hijleh, A. L., & Qazzaz, M. (2004). Effect of bis(acetato)tetrakis(imidazole) copper(II) in delaying the onset and reducing the mortality rate of strychnine- and thiosemicarbazide- induced convulsions. *Biological Trace Element Research*, *101*(1), 87–95. https://doi.org/10.1385/BTER:101:1:87

Abdul-Hussein, S., Andréll, J., & Tate, C. G. (2013). Thermostabilisation of the serotonin transporter in a cocaine-bound conformation. *Journal of Molecular Biology*, *425*(12), 2198–2207. https://doi.org/10.1016/j.jmb.2013.03.025

Abels, F., Lindemann, C., & Schneider, C. (2014). A general strategy for the catalytic, highly enantio- and diastereoselective synthesis of indolizidine-based alkaloids. *Chemistry (Weinheim an der Bergstrasse, Germany)*, *20*(7), 1964–1979. https://doi.org/10.1002/chem.201304086

Ablajan, N., Zhao, B., Zhao, J. Y., Wang, B. L., Sagdullaev, S. S., & Aisa, H. A. (2021). Diterpenoid alkaloids from Aconitum barbatum var. puberulum Ledeb. *Phytochemistry*, *181*, 112567. https://doi.org/10.1016/j.phytochem.2020.112567

Acito, M., Russo, C., Fatigoni, C., Mercanti, F., Moretti, M., & Villarini, M. (2022). Cytotoxicity and genotoxicity of *Senecio vulgaris* L. extracts: An in vitro assessment in HepG2 liver cells. *International Journal of Environmental Research and Public Health*, *19*(22), 14824. https://doi.org/10.3390/ijerph192214824

Adegoke, S. A., & Alo, L. A. (2013). Datura stramonium poisoning in children. *Nigerian Journal of Clinical Practice*, *16*(1), 116–118. https://doi.org/10.4103/1119-3077.106783

Ain, Q. U., Khan, H., Mubarak, M. S., & Pervaiz, A. (2016). Plant alkaloids as antiplatelet agent: Drugs of the future in the light of recent developments. *Frontiers in Pharmacology*, *7*, 292. https://doi.org/10.3389/fphar.2016.00292

Al-Kamil, R., Abed, M., Al-Awad, S., Al-Salman, H., Hussein, H., Hilyail, D., & Shari, F. (2024). Isolation, characterization, and antihypertensive activity alkaloids extracted from the leaves of the Alstonia Scholaris plant. *Georgian Medical News*, *352–353*, 213–217.

Al-Maaieh, A., & Flanagan, D. R. (2005). New drug salt formation in biodegradable microspheres. *International Journal of Pharmaceutics*, *303*(1-2), 153–159. https://doi.org/10.1016/j.ijpharm.2005.06.029

An, J. J., Zhou, J. L., Li, H. J., Jiang, Y., & Li, P. (2010). Puqienine E: An angiotensin converting enzyme inhibitory steroidal alkaloid from Fritillaria puqiensis. *Fitoterapia*, *81*(3), 149–152. https://doi.org/10.1016/j.fitote.2009.08.012

Angrist, B., Rotrosen, J., Kleinberg, D., Merriam, V., & Gershon, S. (1977). Dopaminergic agonist properties of ephedrine--theoretical implications. *Psychopharmacology*, *55*(2), 115–120. https://doi.org/10.1007/BF01457845

Anke, S., Gondé, D., Kaltenegger, E., Hänsch, R., Theuring, C., & Ober, D. (2008). Pyrrolizidine alkaloid biosynthesis in Phalaenopsis orchids: developmental expression of alkaloid-specific homospermidine synthase in root tips and young flower buds. *Plant Physiology*, *148*(2), 751–760. https://doi.org/10.1104/pp.108.124859

Anwar, S., Saleem, H., Azmat, T., Khurshid, U., Khan, K. M., Chohan, T. A., Khursheed, A., Alamri, A., & Awadh Ali, N. A. (2024). *Crotalaria burhia* Buch. -Ham.: A comprehensive review of its botany, traditional uses, phytochemistry, and pharmacology. *Natural Product Research*, 1–16. https://doi.org/10.1080/14786419.2024.2370045

Arayne, M. S., Sultana, N., & Bahadur, S. S. (2007). The berberis story: Berberis vulgaris in therapeutics. *Pakistan Journal of Pharmaceutical Sciences*, *20*(1), 83–92.

Arumugam P. (2024). Investigating the neuroprotective effects of alkaloids from Galanthus woronowii as a potential treatment for Alzheimer's disease. *Natural Product Research*, 1–2. https://doi.org/10.1080/14786419.2024.2431124

Ashraf, M. W., Poikola, S., Neuvonen, M., Kiiski, J. I., Kontinen, V. K., Olkkola, K. T., Backman, J. T., Niemi, M., & Saari, T. I. (2024). Population pharmacokinetic quantification of CYP2D6 activity in codeine metabolism in ambulatory surgical patients for model-informed precision dosing. *Clinical Pharmacokinetics*, *63*(11), 1547–1560. https://doi.org/10.1007/s40262-024-01433-9

Ashrafi, S., Alam, S., Sultana, A., Raj, A., Emon, N. U., Richi, F. T., Sharmin, T., Moon, M., Park, M. N., & Kim, B. (2023). Papaverine: A miraculous alkaloid from opium and its multimedicinal application. *Molecules (Basel, Switzerland)*, *28*(7), 3149. https://doi.org/10.3390/molecules28073149

Attoumbré, J., Giordanengo, P., & Baltora-Rosset, S. (2013). Solanidine isolation from Solanum tuberosum by centrifugal partition chromatography. *Journal of Separation Science*, *36*(14), 2379–2385. https://doi.org/10.1002/jssc.201300188

Avdeef, A., Box, K. J., Comer, J. E., Gilges, M., Hadley, M., Hibbert, C., Patterson, W., & Tam, K. Y. (1999). PH-metric log P 11. pKa determination of water-insoluble drugs in organic solvent-water mixtures. *Journal of Pharmaceutical and Biomedical Analysis*, *20*(4), 631–641. https://doi.org/10.1016/s0731-7085(98)00235-0

Ayarragaray J. E. (2014). Ergotism: A change of persepective. *Annals of Vascular Surgery*, *28*(1), 265–268. https://doi.org/10.1016/j.avsg.2013.02.005

Baeradeh, N., Ghoddusi Johari, M., Moftakhar, L., Rezaeianzadeh, R., Hosseini, S. V., & Rezaianzadeh, A. (2022). The prevalence and predictors of cardiovascular diseases in Kherameh cohort study: A population-based study on 10,663 people in southern Iran. *BMC Cardiovascular Disorders*, *22*(1), 244. https://doi.org/10.1186/s12872-022-02683-w

Bagri, K., & Deshmukh, R. (2022). Vinpocetine restores cognitive and motor functions in Traumatic brain injury challenged rats. *Inflammopharmacology*, *30*(6), 2243–2259. https://doi.org/10.1007/s10787-022-01059-y

Bailey, H. E., Wang, Y. L., Lynch, S. R., & Fayer, M. D. (2018). Dynamics and microstructures of nicotine/water binary Mixtures near the lower critical solution temperature. *The Journal of Physical Chemistry. B*, *122*(41), 9538–9548. https://doi.org/10.1021/acs.jpcb.8b06205

Barrera-Adame, D. A., Schuster, S., & Niedermeyer, T. H. J. (2024). Mass spectrometry imaging of coniine and other hemlock alkaloids after on-tissue derivatization reveals distinct alkaloid distributions in the plant. *Journal of Natural Products*, *87*(10), 2376–2383. https://doi.org/10.1021/acs.jnatprod.4c00445

Bartok, T., Szollosi, G., Felfoldi, K., Bartok, M., & Thiel, J. (2000). New results on the mass spectra of cinchona alkaloids. *Journal of Mass Spectrometry: JMS*, *35*(6), 711–717. https://doi.org/10.1002/1096-9888(200006)35:6<711::AID-JMS998>3.0.CO;2-B

Bedewitz, M. A., Góngora-Castillo, E., Uebler, J. B., Gonzales-Vigil, E., Wiegert-Rininger, K. E., Childs, K. L., Hamilton, J. P., Vaillancourt, B., Yeo, Y. S., Chappell, J., DellaPenna, D., Jones, A. D., Buell, C. R., & Barry, C. S. (2014). A root-expressed L-phenylalanine:4-hydroxyphenylpyruvate aminotransferase is required for tropane alkaloid biosynthesis in Atropa belladonna. *The Plant Cell*, *26*(9), 3745–3762. https://doi.org/10.1105/tpc.114.130534

Bekkouche, K., Daali, Y., Cherkaoui, S., Veuthey, J. L., & Christen, P. (2001). Calystegine distribution in some solanaceous species. *Phytochemistry*, *58*(3), 455–462. https://doi.org/10.1016/s0031-9422(01)00283-7

Belser-Ehrlich, S., Harper, A., Hussey, J., & Hallock, R. (2013). Human and cattle ergotism since 1900: Symptoms, outbreaks, and regulations. *Toxicology and Industrial Health*, *29*(4), 307–316. https://doi.org/10.1177/0748233711432570

Billiot, F. H., Billiot, E. J., Ng, Y. K., & Warner, I. M. (2006). Chiral separation of norlaudanosoline, laudanosoline, laudanosine, chlorthalidone, and three benzoin derivatives using amino acid based molecular micelles. *Journal of Chromatographic Science*, *44*(2), 64–69. https://doi.org/10.1093/chromsci/44.2.64

Bodi, D., Ronczka, S., Gottschalk, C., Behr, N., Skibba, A., Wagner, M., Lahrssen-Wiederholt, M., Preiss-Weigert, A., & These, A. (2014). Determination of pyrrolizidine alkaloids in tea, herbal drugs and honey. *Food Additives & Contaminants. Part A, Chemistry, Analysis, Control, Exposure & Risk Assessment*, *31*(11), 1886–1895. https://doi.org/10.1080/19440049.2014.964337

Boratyński P. J. (2015). Dimeric Cinchona alkaloids. *Molecular Diversity*, *19*(2), 385–422. https://doi.org/10.1007/s11030-014-9563-1

Bozhadze, A. D., Vachnadze, V. I. U., Dzhokhadze, M. S., Berashvili, D. T., & Bakuridze, A. D. (2013). Study on the separation process of pharmacological active total alkaloids from *Chelidonium majus L.* growing in Georgia. *Georgian Medical News*, 217, 61–65.

Cai, X. H., Zhang, H. Y., & Xie, B. (2020). Matrine-family alkaloids: Versatile precursors for bioactive modifications. *Medicinal Chemistry (Shariqah (United Arab Emirates))*, *16*(4), 431–453. https://doi.org/10.2174/1573406415666190507121744

Cao, J., Shen, H. M., Wang, Q., Qian, Y., Guo, H. C., Li, K., Qiao, X., Guo, D. A., Luo, X. D., & Ye, M. (2016). Characterization of chemical constituents and rats metabolites of an alkaloidal extract of Alstonia scholaris leaves by liquid chromatography coupled with mass spectrometry. *Journal of Chromatography. B, Analytical Technologies in the Biomedical and Life Sciences*, *1026*, 43–55. https://doi.org/10.1016/j.jchromb.2015.07.044

Cárdenas, P. D., Sonawane, P. D., Heinig, U., Bocobza, S. E., Burdman, S., & Aharoni, A. (2015). The bitter side of the nightshades: Genomics drives discovery in Solanaceae steroidal alkaloid metabolism. *Phytochemistry*, *113*, 24–32. https://doi.org/10.1016/j.phytochem.2014.12.010

Castro, E. M., Lotfipour, S., & Leslie, F. M. (2023). Nicotine on the developing brain. *Pharmacological Research*, *190*, 106716. https://doi.org/10.1016/j.phrs.2023.106716

Chapman, R. A., & Miller, D. J. (1974). Structure-activity relations for caffeine: A comparative study of the inotropic effects of the methylxanthines, imidazoles and related compounds on the frog's heart. *The Journal of Physiology*, *242*(3), 615–634. https://doi.org/10.1113/jphysiol.1974.sp010726

Chen, J. J., Han, M. Y., Gong, T., Qiao, Y. M., Yang, J. L., & Zhu, P. (2019). Epigenetic modification enhances ergot alkaloid production of Claviceps purpurea. *Biotechnology Letters*, *41*(12), 1439–1449. https://doi.org/10.1007/s10529-019-02750-x

Chida N. (2011). Recent advances in the synthesis of morphine and related alkaloids. *Topics in Current Chemistry*, *299*, 1–28. https://doi.org/10.1007/128_2010_73

Chlebek, J., De Simone, A., Hošťálková, A., Opletal, L., Pérez, C., Pérez, D. I., Havlíková, L., Cahlíková, L., & Andrisano, V. (2016). Application of BACE1 immobilized enzyme reactor for the characterization of multifunctional alkaloids from Corydalis cava (Fumariaceae) as Alzheimer's disease targets. *Fitoterapia*, *109*, 241–247. https://doi.org/10.1016/j.fitote.2016.01.008

Choi, H. K., & Curhan, G. (2007). Coffee, tea, and caffeine consumption and serum uric acid level: The third national health and nutrition examination survey. *Arthritis and Rheumatism*, *57*(5), 816–821. https://doi.org/10.1002/art.22762

Chrzanowska, E., Denisow, B., Ekiert, H., & Pietrzyk, Ł. (2024). Metabolites obtained from Boraginaceae plants as potential cosmetic ingredients-A review. *Molecules (Basel, Switzerland)*, *29*(21), 5088. https://doi.org/10.3390/molecules29215088

Derosa, G., Maffioli, P., & Sahebkar, A. (2016). Piperine and its role in chronic diseases. *Advances in Experimental Medicine and Biology*, *928*, 173–184. https://doi.org/10.1007/978-3-319-41334-1_8

Duan, J. H., Xing, J. L., Yang, J., & Hu, S. J. (2005). Different firing patterns induced by veratridine and aconitine in injured dorsal root ganglion neurons. *Sheng li xue bao : [Acta Physiologica Sinica]*, *57*(2), 169–174.

Dung, J. K. S., Duringer, J. M., Kaur, N., Scott, J. C., Frost, K. E., Walenta, D. L., Alderman, S. C., Craig, A. M., & Hamm, P. B. (2021). Molecular and alkaloid characterization of *Claviceps purpurea* Sensu Lato from grass seed production areas of the U.S. Pacific Northwest. *Phytopathology*, *111*(5), 831–841. https://doi.org/10.1094/PHYTO-07-20-0289-R

Eissing, T., Lippert, J., & Willmann, S. (2012). Pharmacogenomics of codeine, morphine, and morphine-6-glucuronide: model-based analysis of the influence of CYP2D6 activity, UGT2B7 activity, renal impairment, and CYP3A4 inhibition. *Molecular Diagnosis & Therapy*, *16*(1), 43–53.

Fattahi, F., Shojaeiyan, A., Palazon, J., Moyano, E., & Torras-Claveria, L. (2021). Methyl-β-cyclodextrin and coronatine as new elicitors of tropane alkaloid biosynthesis in Atropa acuminata and Atropa belladonna hairy root cultures. *Physiologia Plantarum*, *172*(4), 2098–2111. https://doi.org/10.1111/ppl.13444

Finkelstein, Y., Aks, S. E., Hutson, J. R., Juurlink, D. N., Nguyen, P., Dubnov-Raz, G., Pollak, U., Koren, G., & Bentur, Y. (2010). Colchicine poisoning: the dark side of an ancient drug. *Clinical Toxicology (Philadelphia, PA)*, *48*(5), 407–414. https://doi.org/10.3109/15563650.2010.495348

Flade, J., Beschow, H., Wensch-Dorendorf, M., Plescher, A., & Wätjen, W. (2019). Occurrence of nine pyrrolizidine alkaloids in *Senecio vulgaris* L. depending on developmental stage and season. *Plants (Basel, Switzerland)*, *8*(3), 54. https://doi.org/10.3390/plants8030054

Flamm H. (2023). Der Ergotismus – ein Ackerunkraut aus Mesopotamien wurde in Europa zum noch immer aktuellen Epidemie-Erreger [Ergotism-a weed from Mesopotamia became an epidemic pathogen in Europe]. *Wiener medizinische Wochenschrift (1946)*, *173*(15–16), 374–392. https://doi.org/10.1007/s10354-022-00960-z

Fournier J. (2001). Découverte des alcaloïdes des marqueurs pour l'histoire de la chimie organique [Alcaloids discovery, markers for the history of organic chemistry]. *Revue d'histoire de la pharmacie*, *49*(331), 315–332.

Fujimaki, J., Sayama, N., Shiotani, S., Suzuki, T., Nonaka, M., Uezono, Y., Oyabu, M., Kamei, Y., Nukaya, H., Wakabayashi, K., Morita, A., Sato, T., & Miura, S. (2022). The steroidal alkaloid tomatidine and tomatidine-rich tomato leaf extract suppress the human gastric cancer-derived 85As2 cells in vitro and in vivo via modulation of interferon-stimulated genes. *Nutrients*, *14*(5), 1023. https://doi.org/10.3390/nu14051023

Gao, Y., Fan, H., Nie, A., Yang, K., Xing, H., Gao, Z., Yang, L., Wang, Z., & Zhang, L. (2022). Aconitine: A review of its pharmacokinetics, pharmacology, toxicology and detoxification. *Journal of Ethnopharmacology*, *293*, 115270. https://doi.org/10.1016/j.jep.2022.115270

Gou, Z. P., Zhao, Y. L., Zou, L. L., Wang, Y., Shu, S. Q., Zhu, X. H., Zheng, L., Shen, Q., Luo, Z., Miao, J., Wang, Y. S., Luo, X. D., & Feng, P. (2021). The safety and tolerability of alkaloids from *Alstonia scholaris* leaves in healthy Chinese volunteers: A single-centre, randomized, double-blind, placebo-controlled phase I clinical trial. *Pharmaceutical Biology*, *59*(1), 484–493. https://doi.org/10.1080/13880209.2021.1893349

Grzybowski, A., Pawlikowska-Łagód, K., & Polak, A. (2021). Ergotism and Saint Anthony's fire. *Clinics in Dermatology*, *39*(6), 1088–1094. https://doi.org/10.1016/j.clindermatol.2021.07.009

Kaltner, F., Rychlik, M., Gareis, M., & Gottschalk, C. (2018). Influence of storage on the stability of toxic pyrrolizidine alkaloids and their N-oxides in peppermint tea, hay, and honey. *Journal of Agricultural and Food Chemistry*, *66*(20), 5221–5228. https://doi.org/10.1021/acs.jafc.7b06036

Kast, C., Kilchenmann, V., Reinhard, H., Droz, B., Lucchetti, M. A., Dübecke, A., Beckh, G., & Zoller, O. (2018). Chemical fingerprinting identifies Echium vulgare, Eupatorium cannabinum and Senecio spp. as plant species mainly responsible for pyrrolizidine alkaloids in bee-collected pollen. *Food Additives &*

Contaminants. Part A, Chemistry, Analysis, Control, Exposure & Risk Assessment, *35*(2), 316–327. https://doi.org/10.1080/19440049.2017.1378443

Kempf, M., Wittig, M., Reinhard, A., von der Ohe, K., Blacquière, T., Raezke, K. P., Michel, R., Schreier, P., & Beuerle, T. (2011). Pyrrolizidine alkaloids in honey: Comparison of analytical methods. *Food Additives & Contaminants. Part A, Chemistry, Analysis, Control, Exposure & Risk Assessment*, *28*(3), 332–347. https://doi.org/10.1080/19440049.2010.521772

Kim, N., Estrada, O., Chavez, B., Stewart, C., & D'Auria, J. C. (2016). Tropane and granatane alkaloid biosynthesis: A systematic analysis. *Molecules (Basel, Switzerland)*, *21*(11), 1510. https://doi.org/10.3390/molecules21111510

Kupnicka, P., Kojder, K., Metryka, E., Kapczuk, P., Jeżewski, D., Gutowska, I., Goschorska, M., Chlubek, D., & Baranowska-Bosiacka, I. (2020). Morphine-element interactions - The influence of selected chemical elements on neural pathways associated with addiction. *Journal of Trace Elements in Medicine and Biology: Organ of the Society for Minerals and Trace Elements (GMS)*, *60*, 126495. https://doi.org/10.1016/j.jtemb.2020.126495

Li, S., Yu, L., Shi, Q., Liu, Y., Zhang, Y., Wang, S., & Lai, X. (2022). An insight into current advances on pharmacology, pharmacokinetics, toxicity and detoxification of aconitine. *Biomedicine & Pharmacotherapy = Biomedecine & Pharmacotherapie*, *151*, 113115. https://doi.org/10.1016/j.biopha.2022.113115

Lynch, C. R., Khandekar, S., Lynch, S. M., & Disario, J. A. (2007). Sublingual L-hyoscyamine for duodenal antimotility during ERCP: A prospective randomized double-blinded study. *Gastrointestinal Endoscopy*, *66*(4), 748–752. https://doi.org/10.1016/j.gie.2007.02.052

Macabeo, A. P., Krohn, K., Gehle, D., Read, R. W., Brophy, J. J., Cordell, G. A., Franzblau, S. G., & Aguinaldo, A. M. (2005). Indole alkaloids from the leaves of Philippine Alstonia scholaris. *Phytochemistry*, *66*(10), 1158–1162. https://doi.org/10.1016/j.phytochem.2005.02.018

Nett, R. S., & Sattely, E. S. (2021). Total biosynthesis of the tubulin-binding alkaloid colchicine. *Journal of the American Chemical Society*, *143*(46), 19454–19465. https://doi.org/10.1021/jacs.1c08659

Osman, A. G., Haider, S., Chittiboyina, A. G., & Khan, I. A. (2019). Utility of alkaloids as chemical and biomarkers for quality, efficacy, and safety assessment of botanical ingredients. *Phytomedicine: International Journal of Phytotherapy and Phytopharmacology*, *54*, 347–356. https://doi.org/10.1016/j.phymed.2018.03.064

Parveen, S., Maurya, N., Meena, A., & Luqman, S. (2024). Cinchonine: A versatile pharmacological agent derived from natural cinchona alkaloids. *Current Topics in Medicinal Chemistry*, *24*(4), 343–363. https://doi.org/10.2174/0115680266270796231109171808

Rajkamal, A., & Kim, H. (2023). Theoretical verification on adsorptive removal of caffeine by carbon and nitrogen-based surfaces: Role of charge transfer, π electron occupancy, and temperature. *Chemosphere*, *339*, 139667. https://doi.org/10.1016/j.chemosphere.2023.139667

Reinhard, H., & Zoller, O. (2021). Pyrrolizidine alkaloids in tea, herbal tea and iced tea beverages- survey and transfer rates. *Food Additives & Contaminants. Part A, Chemistry, Analysis, Control, Exposure & Risk Assessment*, *38*(11), 1914–1933. https://doi.org/10.1080/19440049.2021.1941302

Stefaniak, A. B., Ranpara, A. C., Virji, M. A., & LeBouf, R. F. (2022). Influence of E-liquid humectants, nicotine, and flavorings on aerosol particle size distribution and implications for modeling respiratory deposition. *Frontiers in Public Health*, *10*, 782068. https://doi.org/10.3389/fpubh.2022.782068

Tan, M. A., & Takayama, H. (2019). Recent progress in the chemistry of pandanus alkaloids. *The Alkaloids. Chemistry and Biology*, *82*, 1–28. https://doi.org/10.1016/bs.alkal.2018.12.001

Tarkesh Esfahani, S., Karimzadeh, G., Naghavi, M. R., & Vrieling, K. (2021). Altered gene expression and root thebaine production in polyploidized and methyl jasmonate-elicited Papaver bracteatum Lindl. *Plant Physiology and Biochemistry: PPB*, *158*, 334–341. https://doi.org/10.1016/j.plaphy.2020.11.021

Tarzi, O. I., & Erra-Balsells, R. (2006). Effect of chlorine as substituent on the photochemistry and acid-base properties of beta-carboline alkaloids. *Journal of Photochemistry and Photobiology. B, Biology*, *82*(2), 79–93. https://doi.org/10.1016/j.jphotobiol.2005.09.001

Tasler, S., & Bringmann, G. (2002). Biarylic biscarbazole alkaloids: occurrence, stereochemistry, synthesis, and bioactivity. *Chemical Record (New York, NY)*, *2*(2), 113–126. https://doi.org/10.1002/tcr.10014

Ulgen, M., & Gorrod, J. W. (2000). Studies on the in vitro hepatic microsomal formation of amides during the metabolism of certain secondary and tertiary benzylic amines. *European Journal of Drug Metabolism and Pharmacokinetics*, *25*(2), 145–150. https://doi.org/10.1007/BF03190081

Vetter J. (2004). Poison hemlock (Conium maculatum L.). *Food and Chemical Toxicology: An International Journal Published for the British Industrial Biological Research Association*, *42*(9), 1373–1382. https://doi.org/10.1016/j.fct.2004.04.009

Wallace, M., Abiama, N., & Chipembere, M. (2023). Measurement of the pK_a values of organic molecules in aqueous-organic solvent mixtures by ^{1}H NMR without external calibrants. *Analytical Chemistry*, *95*(42), 15628–15635. https://doi.org/10.1021/acs.analchem.3c02771

Weber, C., & Opatz, T. (2019). Bisbenzylisoquinoline alkaloids. *The Alkaloids. Chemistry and Biology*, *81*, 1–114. https://doi.org/10.1016/bs.alkal.2018.07.001

Yamada, Y., & Sato, F. (2021). Transcription factors in alkaloid engineering. *Biomolecules*, *11*(11), 1719. https://doi.org/10.3390/biom11111719

Zeng, L., Zhang, Q., Jiang, C., Zheng, Y., Zuo, Y., Qin, J., Liao, Z., & Deng, H. (2021). Development of *Atropa belladonna* L. plants with high-yield hyoscyamine and without its derivatives using the CRISPR/Cas9 system. *International Journal of Molecular Sciences*, *22*(4), 1731. https://doi.org/10.3390/ijms22041731

Zhang, Q., Liang, M., Zeng, J., Yang, C., Qin, J., Qiang, W., Lan, X., Chen, M., Lin, M., & Liao, Z. (2022). Engineering tropane alkaloid production and glyphosate resistance by overexpressing AbCaM1 and G2-EPSPS in Atropa belladonna. *Metabolic Engineering*, *72*, 237–246. https://doi.org/10.1016/j.ymben.2022.03.014

Zhang, R., Guo, Q., Kennelly, E. J., Long, C., & Chai, X. (2020). Diverse alkaloids and biological activities of Fumaria (Papaveraceae): An ethnomedicinal group. *Fitoterapia*, *146*, 104697. https://doi.org/10.1016/j.fitote.2020.104697

Zhao, T., Li, S., Wang, J., Zhou, Q., Yang, C., Bai, F., Lan, X., Chen, M., & Liao, Z. (2020). Engineering tropane alkaloid production based on metabolic characterization of ornithine decarboxylase in *Atropa belladonna*. *ACS Synthetic Biology*, *9*(2), 437–448. https://doi.org/10.1021/acssynbio.9b00461

Index

Pages in *italics* refer to figures and pages in **bold** refer to tables.

B

C

D

E

F

G

N

O

Q

U

V

W

X

Y

Z